DISTILLATION DESIGN AND CONTROL USING ASPEN^{TM} SIMULATION

DISTILLATION DESIGN AND CONTROL USING ASPEN™ SIMULATION

WILLIAM L. LUYBEN
Lehigh University
Bethlehem, Pennsylvania

AIChE®

A joint publication of the American Institute of Chemical Engineers and John Wiley and Sons, Inc.

Published by John Wiley & Sons, Inc., Hoboken, New Jersey
Published simultaneously in Canada

Library of Congress Cataloging-in-Publication Data:

Luyben, William L.
Distillation design and control using Aspen simulation / William L. Luyben.
p. cm.
Includes index.
ISBN-13: 978-0-471-77888-2 (cloth)
ISBN-10: 0-471-77888-5 (cloth)
1. Distillation apparatus—Design and construction. 2. Chemical process control—Simulation methods. I. Title.

TP159.D5L89 2006
660′.28425- -dc22
2005023397

Printed in the United States of America

10 9 8 7 6 5 4 3 2 1

Dedicated to the memory of the pioneers of
Lehigh Chemical Engineering:
Alan Foust, Len Wenzel, and Curt Clump

CONTENTS

PREFACE

The rapid increase in the price of crude oil in recent years (as of 2005) and the resulting "sticker shock" at the gas pump have caused the scientific and engineering communities to finally understand that it is time for some reality checks on our priorities. Energy is the real problem that faces the world, and it will not be solved by the recent fads of biotechnology or nanotechnology. Energy consumption is the main producer of carbon dioxide, so it is directly linked with the problem of global warming.

A complete reassessment of our energy supply and consumption systems is required. Our terribly inefficient use of energy in all aspects of our modern society must be halted. We waste energy in our *transportation* system with poor-mileage SUVs and inadequate railroad systems. We waste energy in our *water* systems by using energy to produce potable water, and then flush most of it down the toilet. This loads up our *waste disposal* plants, which consume more energy. We waste energy in our *food* supply system by requiring large amounts of energy for fertilizer, herbicides, pesticides, tillage, and transporting and packaging our food for consumer convenience. The old farmers' markets provided better food at lower cost and required much less energy.

One of the most important technologies in our energy supply system is distillation. Essentially all our transportation fuel goes through at least one distillation column on its way from crude oil to the gasoline pump. Large distillation columns called "pipestills" separate the crude into various petroleum fractions based on boiling points. Intermediate fractions go directly to gasoline. Heavy fractions are catalytically or thermally "cracked" to form more gasoline. Light fractions are combined to form more gasoline. Distillation is used in all of these operations.

Even when we begin to switch to renewable sources of energy such as biomass, the most likely transportation fuel will be methanol. The most likely process is the partial oxidation of biomass to produce synthesis gas (a mixture of hydrogen, carbon monoxide, and carbon dioxide) and the subsequent reaction of these components to produce methanol and water. Distillation to separate methanol from water is an important part of this process. Distillation is also used to produce the oxygen used in the partial oxidation reactor.

Therefore distillation is, and will remain in the twenty-first century, the premier separation method in the chemical and petroleum industries. Its importance is unquestionable in helping to provide food, heat, shelter, clothing, and transportation in our modern society. It is involved in supplying much of our energy needs. The distillation columns in operation around the world number in the tens of thousands.

The analysis, design, operation, control, and optimization of distillation columns have been extensively studied for almost a century. Until the advent of computers, hand calculations and graphical methods were developed and widely applied in these studies. Since about 1950, analog and digital computer simulations have been used to solve many engineering problems. Distillation analysis involves iterative vapor–liquid phase equilibrium calculations and tray-to-tray component balances that are ideal for digital computation.

Initially most engineers wrote their own programs to solve both the nonlinear algebraic equations that describe the steady-state operation of a distillation column and the nonlinear ordinary differential equations that describe its dynamic behavior. Many chemical and petroleum companies developed their own in-house steady-state process simulation programs in which distillation was an important unit operation. Commercial *steady-state* simulators took over around the mid-1980s and now dominate the field.

Commercial *dynamic* simulators were developed quite a bit later. They had to wait for advancements in computer technology to provide the very fast computers required. The current state of the art is that both steady-state and dynamic simulations of distillation columns are widely used in industry and in universities.

My own technical experience has pretty much followed this history of distillation simulation. My practical experience started back in a high-school chemistry class in which we performed batch distillations. Next came an exposure to some distillation theory and running a pilot-scale batch distillation column as an undergraduate at Penn State, learning from Arthur Rose and "Black" Mike Cannon. Then there were 5 years of industrial experience in Exxon refineries as a technical service engineer on pipestills, vacuum columns, light-end units, and alkylation units, all of which used distillation extensively.

During this period the only use of computers that I was aware of was for solving linear programming problems associated with refinery planning and scheduling. It was not until returning to graduate school in 1960 that I personally started to use analog and digital computers. Bob Pigford taught us how to program a Bendix G12 digital computer, which used paper tape and had such limited memory that programs were severely restricted in length and memory requirements. Dave Lamb taught us analog simulation. Jack Gerster taught us distillation practice.

Next there were 4 years working in the Engineering Department of DuPont on process control problems, many of which involved distillation columns. Both analog and digital simulations were heavily used. A wealth of knowledge was available from a stable of outstanding engineers: Page Buckley, Joe Coughlin, J. B. Jones, Neal O'Brien, and Tom Keane, to mention only a few.

Finally, there have been over 35 years of teaching and research at Lehigh in which many undergraduate and graduate students have used simulations of distillation columns in isolation and in plantwide environments to learn basic distillation principles and to develop effective control structures for a variety of distillation column configurations. Both homegrown and commercial simulators have been used in graduate research and in the undergraduate senior design course.

The purpose of this book is to try to capture some of this extensive experience with distillation design and control so that it is available to students and young engineers when they face problems with distillation columns. This book covers much more than just the mechanics of using a simulator. It uses simulation to guide in developing the optimum economic steady-state design of distillation systems, using simple and practical approaches. Then it uses simulation to develop effective control structures for dynamic control. Questions are addressed as to whether to use single-end control or dual-composition control, where to locate temperature control trays, and how excess degrees of freedom should be fixed.

There is no claim that the material is all new. The steady-state methods are discussed in most design textbooks. Most of the dynamic material is scattered around in a number of papers and books. What is claimed is that this book pulls this material together in a coordinated, easily accessible way. Another unique feature is the combination of design and control of distillation columns in a single book.

There are three steps in developing a process design. The first is "conceptual design," in which simple approximate methods are used to develop a preliminary flowsheet. This step for distillation systems is covered very thoroughly in another textbook.[1] The next step is *preliminary design*, in which rigorous simulation methods are used to evaluate both steady-state and dynamic performance of the proposed flowsheet. The final step is "detailed design," in which the hardware is specified in great detail, with specifics such as types of trays, number of sieve tray holes, feed and reflux piping, pumps, heat exchanger areas, and valve sizes. This book deals with the second stage, preliminary design.

The subject of distillation simulation is a very broad one, which would require many volumes to cover comprehensively. The resulting encyclopedia-like books would be too formidable for a beginning engineer to try to tackle. Therefore this book is restricted in its scope to only those aspects that I have found to be the most fundamental and the most useful. Only continuous distillation columns are considered. The area of batch distillation is very extensive and should be dealt with in another book. Only staged columns are considered. They have been successfully applied for many years. Rate-based models are fundamentally more rigorous, but they require that more parameters be known or estimated.

Only rigorous simulations are used in this book. The book by Doherty and Malone is highly recommended for a detailed coverage of approximate methods for conceptual steady-state design of distillation systems.

I hope that the reader finds this book useful and readable. It is a labor of love that is aimed at taking some of the mystery and magic out of designing and operating a distillation column.

WILLIAM L. LUYBEN

[1]M. F. Doherty and M. F. Malone, *Conceptual Design of Distillation Systems*, McGraw-Hill, 2001.

CHAPTER 1

FUNDAMENTALS OF VAPOR–LIQUID PHASE EQUILIBRIUM (VLE)

Distillation occupies a very important position in chemical engineering. Distillation and chemical reactors represent the backbone of what distinguishes chemical engineering from other engineering disciplines. Operations involving heat transfer and fluid mechanics are common to several disciplines. But distillation is uniquely under the purview of chemical engineers.

The basis of distillation is phase equilibrium, specifically, vapor–liquid (phase) equilibrium (VLE) and in some cases vapor–liquid–liquid (phase) equilibrium (VLLE). Distillation can effect a separation among chemical components only if the compositions of the vapor and liquid phases that are in phase equilibrium with each other are different. A reasonable understanding of VLE is essential for the analysis, design, and control of distillation columns.

The fundamentals of VLE are briefly reviewed in this chapter.

1.1 VAPOR PRESSURE

Vapor pressure is a physical property of a pure chemical component. It is the pressure that a pure component exerts at a given temperature when both liquid and vapor phases are present. Laboratory vapor pressure data, usually generated by chemists, are available for most of the chemical components of importance in industry.

Vapor pressure depends **only** on temperature. It does not depend on composition because it is a pure component property. This dependence is normally a strong one with an exponential increase in vapor pressure with increasing temperature. Figure 1.1 gives two typical vapor pressure curves, one for benzene and one for toluene. The natural log of the vapor pressures of the two components are plotted against the reciprocal of the

Distillation Design and Control Using Aspen™ Simulation, By William L. Luyben

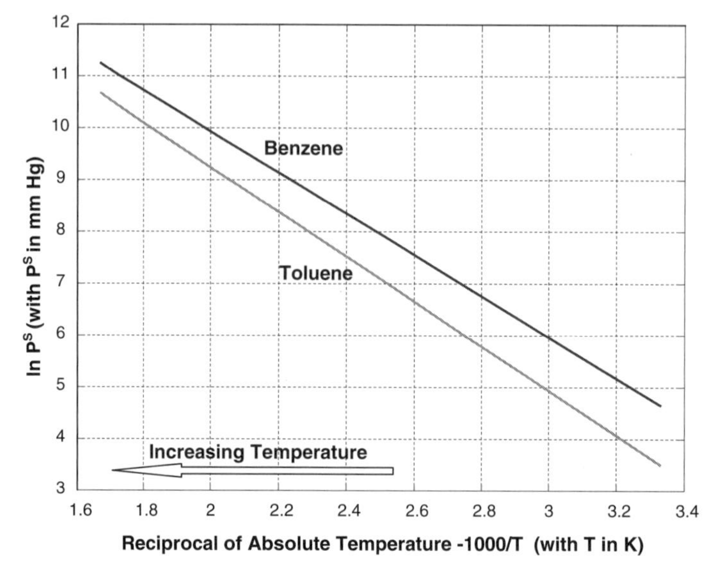

Figure 1.1 Vapor pressures of pure benzene and toluene.

absolute temperature. As temperature increases, we move to the left in the figure, which means a higher vapor pressure. In this particular figure, the vapor pressure P^S of each component is given in units of millimeters of mercury (mmHg). The temperature is given in Kelvin units.

Looking at a vertical constant-temperature line shows that benzene has a higher vapor pressure than does toluene at a given temperature. Therefore benzene is the "lighter" component from the standpoint of volatility (not density). Looking at a constant-pressure horizontal line shows that benzene boils at a lower temperature than does toluene. Therefore benzene is the "lower boiling" component. Note that the vapor pressure lines for benzene and toluene are fairly parallel. This means that the ratio of the vapor pressures does not change much with temperature (or pressure). As discussed in a later section, this means that the ease or difficulty of the benzene/toluene separation (the energy required to make a specified separation) does not change much with the operating pressure of the column. Other chemical components can have temperature dependences that are quite different.

If we have a vessel containing a mixture of these two components with liquid and vapor phases present, the concentration of benzene in the vapor phase will be higher than that in the liquid phase. The reverse is true for the heavier, higher-boiling toluene. Therefore benzene and toluene can be separated in a distillation column into an overhead distillate stream that is fairly pure benzene and a bottoms stream that is fairly pure toluene.

Equations can be fitted to the experimental vapor pressure data for each component using two, three, or more parameters. For example, the two-parameter version is

$$\ln P_j^S = C_j + D_j/T$$

The C_j and D_j are constants for each pure chemical component. Their numerical values depend on the units used for vapor pressure [mmHg, kPa, psia (pounds per square inch absolute), atm, etc.] and on the units used for temperature (K or °R).

1.2 BINARY VLE PHASE DIAGRAMS

Two types of vapor–liquid equilibrium diagrams are widely used to represent data for two-component (binary) systems. The first is a "temperature versus x and y" diagram (Txy). The x term represents the liquid composition, usually expressed in terms of mole fraction. The y term represents the vapor composition. The second diagram is a plot of x versus y.

These types of diagrams are generated at a constant pressure. Since the pressure in a distillation column is relatively constant in most columns (the exception is vacuum distillation, in which the pressures at the top and bottom are significantly different in terms of absolute pressure level), a Txy diagram, and an xy diagram are convenient for the analysis of binary distillation systems.

Figure 1.2 gives the Txy diagram for the benzene/toluene system at a pressure of 1 atm. The abscissa shows the mole fraction of benzene; the ordinate, temperature. The lower curve is the "saturated liquid" line, which gives the mole fraction of benzene in the liquid phase x. The upper curve is the "saturated vapor" line, which gives the mole fraction of benzene in the vapor phase y. Drawing a horizontal line at some temperature and reading off the intersection of this line with the two curves give the compositions of the two phases. For example, at 370 K the value of x is 0.375 mole fraction benzene and the value of y is 0.586 mole fraction benzene. As expected, the vapor is richer in the lighter component.

At the leftmost point we have pure toluene (0 mole fraction benzene), so the boiling point of toluene at 1 atm can be read from the diagram (384.7 K). At the rightmost point we have pure benzene (1 mole fraction benzene), so the boiling point of benzene at 1 atm can be read from the diagram (353.0 K). In the region between the curves, there are two phases; in the region above the saturated vapor curve, there is only a single "superheated" vapor phase; in the region below the saturated liquid curve, there is only a single "subcooled" liquid phase.

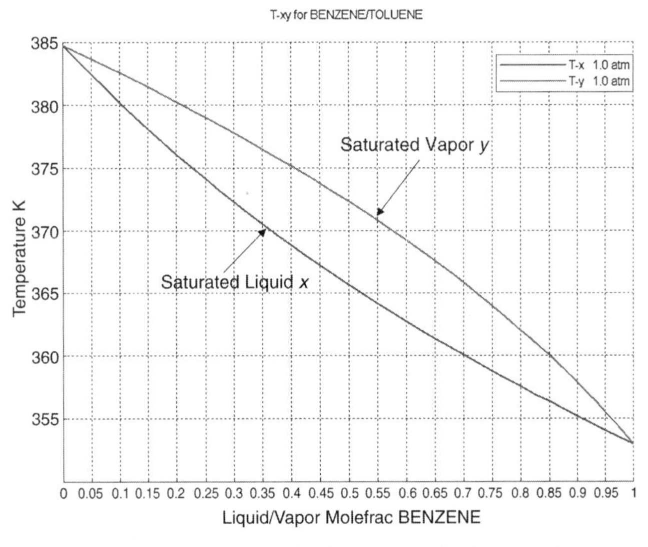

Figure 1.2 Txy diagram for benzene and toluene at 1 atm.

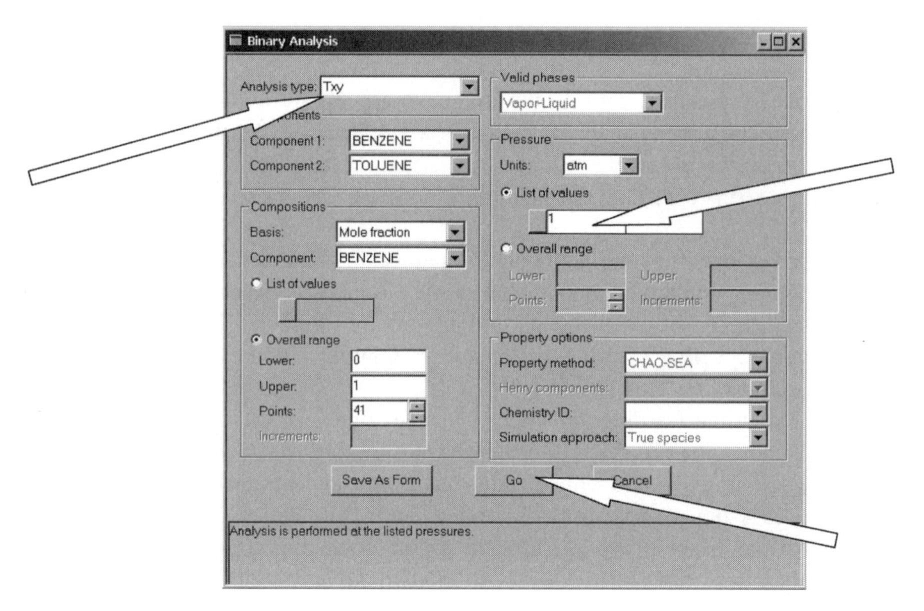

Figure 1.3 Specifying *Txy* diagram parameters.

The diagram is easily generated in Aspen Plus by going to *Tools* on the upper toolbar and selecting *Analysis, Property*, and *Binary*. The window shown in Figure 1.3 opens and specifies the type of diagram and the pressure. Then we click the *Go* button.

The pressure in the *Txy* diagram given in Figure 1.2 is 1 atm. Results at several pressures can also be generated as illustrated in Figure 1.4. The higher the pressure, the higher the temperatures.

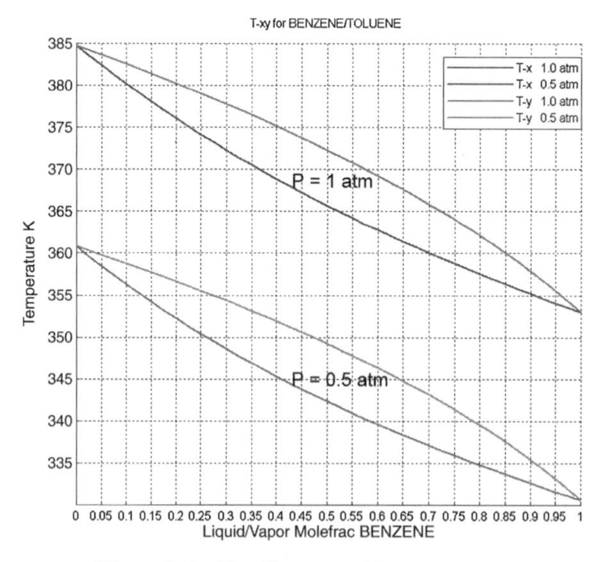

Figure 1.4 *Txy* diagrams at two pressures.

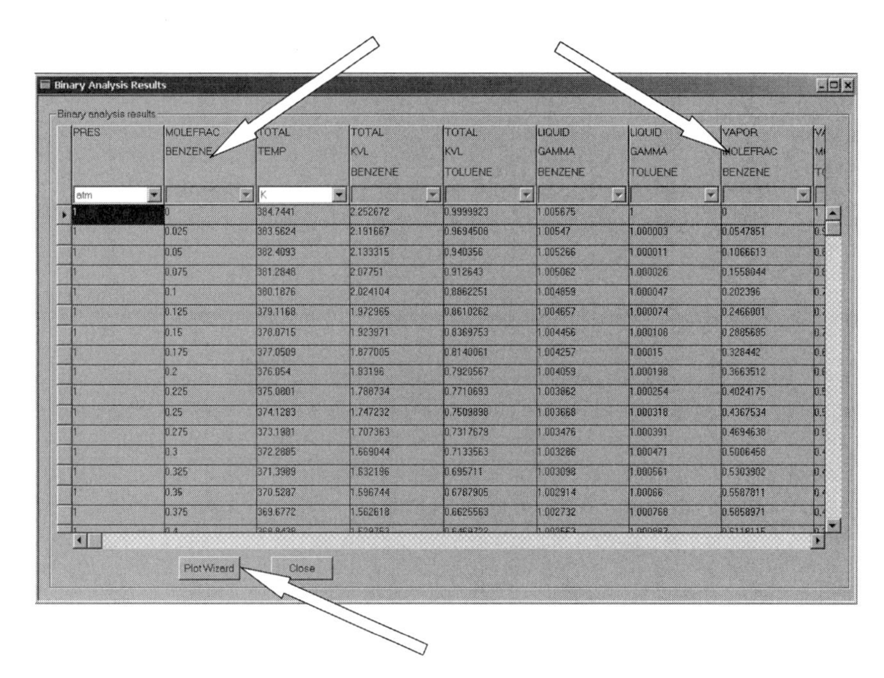

PRES	MOLEFRAC BENZENE	TOTAL TEMP	TOTAL KVL BENZENE	TOTAL KVL TOLUENE	LIQUID GAMMA BENZENE	LIQUID GAMMA TOLUENE	VAPOR MOLEFRAC BENZENE
atm		K					
1	0	384.7441	2.252672	0.9999923	1.005675	1	0
1	0.025	383.5624	2.191667	0.9694508	1.00547	1.000003	0.0547851
1	0.05	382.4093	2.133315	0.940356	1.005266	1.000011	0.1066613
1	0.075	381.2848	2.07751	0.912643	1.005062	1.000026	0.1558044
1	0.1	380.1876	2.024104	0.8862251	1.004859	1.000047	0.202396
1	0.125	379.1168	1.972965	0.8610262	1.004657	1.000074	0.2466001
1	0.15	378.0715	1.923971	0.8369753	1.004456	1.000106	0.2885685
1	0.175	377.0509	1.877005	0.8140061	1.004257	1.00015	0.328442
1	0.2	376.054	1.83196	0.7920567	1.004059	1.000198	0.3663512
1	0.225	375.0801	1.788734	0.7710693	1.003862	1.000254	0.4024175
1	0.25	374.1283	1.747232	0.7509898	1.003668	1.000318	0.4367534
1	0.275	373.1981	1.707363	0.7317679	1.003476	1.000391	0.4694638
1	0.3	372.2885	1.669044	0.7133563	1.003286	1.000471	0.5006458
1	0.325	371.3989	1.632196	0.695711	1.003098	1.000561	0.5303902
1	0.35	370.5287	1.596744	0.6787905	1.002914	1.00066	0.5587811
1	0.375	369.6772	1.562618	0.6625563	1.002732	1.000768	0.5858971

Figure 1.5 Using Plot Wizard to generate *xy* diagram.

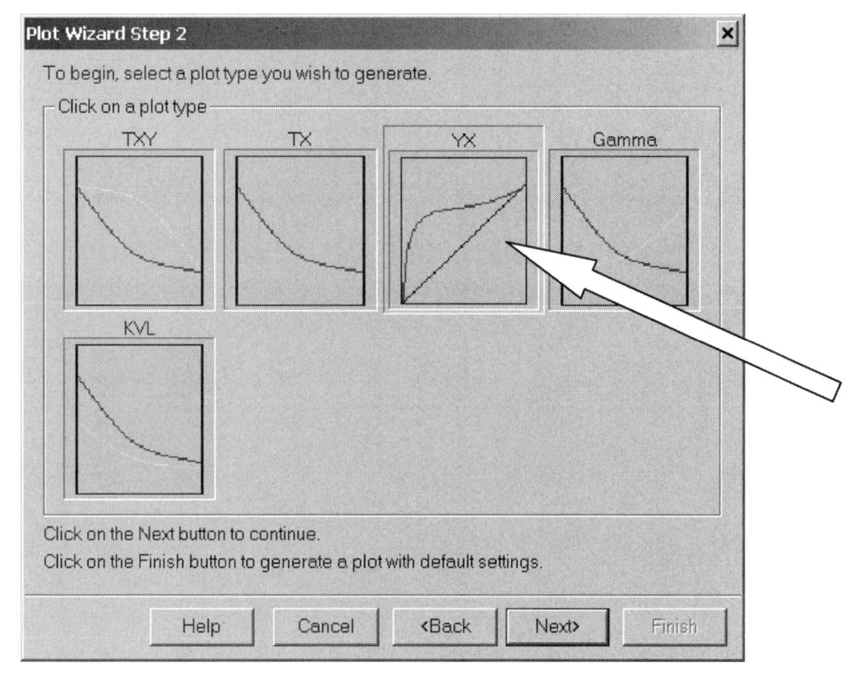

Figure 1.6 Using Plot Wizard to generate *xy* diagram.

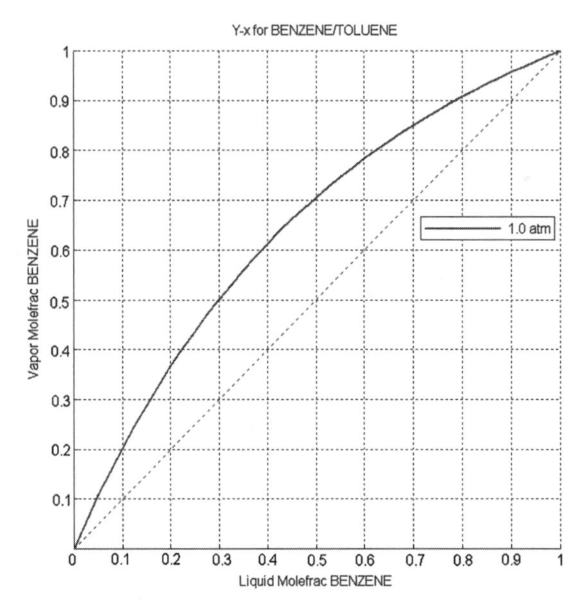

Figure 1.7 *xy* diagram for benzene/toluene.

The other type of diagram, an *xy* diagram, is generated in Aspen Plus by clicking the *Plot Wizard* button at the bottom of the *Binary Analysis Results* window that also opens when the *Go* button is clicked to generate the *Txy* diagram. As shown in Figure 1.5, this window also gives a table of detailed information. The window shown in Figure 1.6 opens, and *YX* picture is selected. Clicking the *Next* and *Finish* buttons generates the *xy* diagram shown in Figure 1.7.

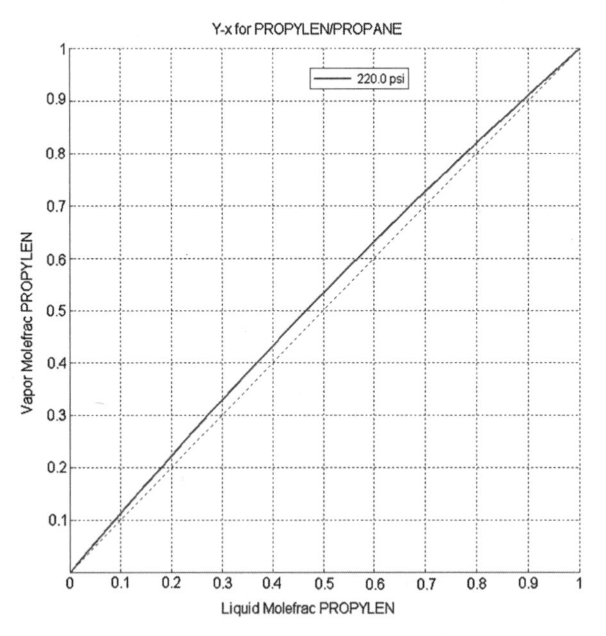

Figure 1.8 *xy* diagram for propylene/propane.

Figure 1.8 gives an xy diagram for the propylene/propane system. These components have boiling points that are quite close, which leads to a very difficult separation.

These diagrams provide valuable insight into the VLE of binary systems. They can be used for quantitative analysis of distillation columns, as we will demonstrate in Chapter 2. Three-component ternary systems can also be represented graphically, as discussed in Section 1.6.

1.3 PHYSICAL PROPERTY METHODS

The observant reader may have noticed in Figure 1.3 that the physical property method specified for the VLE calculations in the benzene/toluene example was "Chao–Seader." This method works well for most hydrocarbon systems.

One of the most important issues involved in distillation calculations is the selection of an appropriate physical property method that will accurately describe the phase equilibrium of the chemical component system. The Aspen Plus library has a large number of alternative methods. Some of the most commonly used methods are Chao–Seader, van Laar, Wilson, Unifac, and NRTL.

In most design situations there is some type of data that can be used to select the most appropriate physical property method. Often VLE data can be found in the literature. The multivolume DECHEMA data books[1] provide an extensive source of data.

If operating data from a laboratory, pilot plant, or plant column are available, they can be used to determine what physical property method fits the column data. There could be a problem in using column data in that the tray efficiency is also unknown and the VLE parameters cannot be decoupled from the efficiency.

1.4 RELATIVE VOLATILITY

One of the most useful ways to represent VLE data is by employing "relative volatility," which is the ratio of the y/x values [vapor mole fraction over (divided by) liquid mole fraction] of two components. For example, the relative volatility of component L with respect to component H is defined in the following equation:

$$\alpha_{LH} \equiv \frac{y_L/x_L}{y_H/x_H}$$

The larger the relative volatility, the easier the separation.

Relative volatilities can be applied to both binary and multicomponent systems. In the binary case, the relative volatility α between the light and heavy components can be used to give a simple relationship between the composition of the liquid phase (x is the mole fraction of the light component in the liquid phase) and the composition of the vapor phase (y is the mole fraction of the light component in the vapor phase):

$$y = \frac{\alpha x}{1 + (\alpha - 1)x}$$

[1]J. Gmehling et al., *Vapor-Liquid Equilibrium Data Collection*, DECHEMA, Frankfurt/Main, 1993.

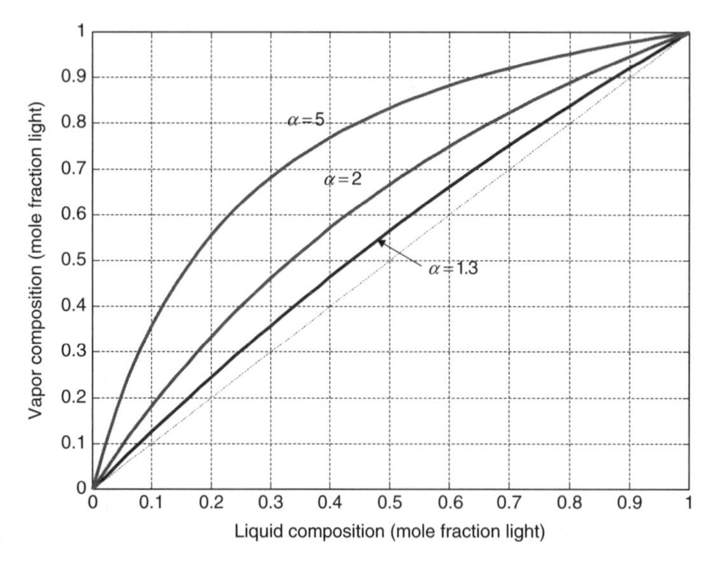

Figure 1.9 *xy* curves for relative volatilities of 1.3, 2, and 5.

Figure 1.9 gives *xy* curves for several values of α, assuming that α is constant over the entire composition space.

In the multicomponent case, a similar relationship can be derived. Suppose that there are *NC* components. Component 1 is the lightest, component 2 is the next lightest, and so forth down to the heaviest of all the components, component *H*. We define the relative volatility of component *j* with respect to component *H* as α_j:

$$\alpha_j = \frac{y_j/x_j}{y_H/x_H}$$

Solving for y_j and summing all the *y* values (which must add to unity) give

$$y_j = \alpha_j x_j \frac{y_H}{x_H}$$

$$\sum_{j=1}^{NC} y_j = 1 = \sum_{j=1}^{NC} \alpha_j x_j \frac{y_H}{x_H}$$

$$1 = \frac{y_H}{x_H} \sum_{j=1}^{NC} \alpha_j x_j$$

Then, solving for y_H/x_H and substituting this into the first equation above give

$$\frac{y_H}{x_H} = \frac{1}{\sum_{j=1}^{NC} \alpha_j x_j}$$

$$y_j = \frac{\alpha_j x_j}{\sum_{j=1}^{NC} \alpha_j x_j}$$

The last equation relates the vapor composition to the liquid composition for a constant relative volatility multicomponent system. Of course, if relative volatilities are not constant, this equation cannot be used. What is required is a "bubblepoint" calculation, which is discussed in the next section.

1.5 BUBBLEPOINT CALCULATIONS

The most common VLE problem is to calculate the temperature and vapor composition y_j that is in equilibrium with a liquid at a known total pressure of the system P and with a known liquid composition (all the x_j values). At phase equilibrium the "chemical potential" μ_j of each component in the liquid and vapor phases must be equal:

$$\mu_j^L = \mu_j^V$$

The liquid-phase chemical potential of component j can be expressed in terms of liquid mole fraction x_j, vapor pressure P_j^S, and activity coefficient γ_j:

$$\mu_j^L = x_j P_j^S \gamma_j$$

The vapor-phase chemical potential of component j can be expressed in terms of vapor mole fraction y_j, the total system pressure P, and fugacity coefficient σ_j:

$$\mu_j^V = y_j P \sigma_j$$

Therefore the general relationship between vapor and liquid phases is

$$y_j P \sigma_j = x_j P_j^S \gamma_j$$

If the pressure of the system is not high, the fugacity coefficient is unity. If the liquid phase is "ideal" (i.e., there is no interaction between the molecules), the activity coefficient is unity. The latter situation is much less common than the former because components interact in liquid mixtures. They can either attract or repulse. Section 1.7 discusses nonideal systems in more detail.

Let us assume that the liquid and vapor phases are both ideal ($\gamma_j = 1$ and $\sigma_j = 1$). In this situation the bubblepoint calculation involves an iterative calculation to find the temperature T that satisfies the equation

$$P = \sum_{j=1}^{NC} x_j P_{j(T)}^S$$

The total pressure P and all the x_j values are known. In addition, equations for the vapor pressures of all components as functions of temperature T are known. The Newton–Raphson convergence method is convenient and efficient in this iterative calculation because an analytical derivative of the temperature-dependent vapor pressure functions P^S can be used.

1.6 TERNARY DIAGRAMS

Three-component systems can be represented in two-dimensional ternary diagrams. There are three components, but the sum of the mole fractions must add to unity. Therefore, specifying two mole fractions completely defines the composition.

A typical rectangular ternary diagram is given in Figure 1.10. The mole fraction of component 1 is shown on the abscissa; the mole fraction of component 2, on the ordinate. Both of these dimensions run from 0 to 1. The three corners of the triangle represent the three pure components.

Since only two compositions define the composition of a stream, the stream can be located on this diagram by entering the appropriate coordinates. For example, Figure 1.10 shows the location of stream F that is a ternary mixture of 20 mol% n-butane (C4), 50 mol% n-pentane (C5), and 30 mol% n-hexane (C6).

One of the most useful and interesting aspects of ternary diagrams is the "ternary mixing rule," which states that if two ternary streams are mixed together (one is stream D with composition x_{D1} and x_{D2} and the other is stream B with composition x_{B1} and x_{B2}), the mixture has a composition (z_1 and z_2) that lies on a **straight** line in a x_1–x_2 ternary diagram that connects the x_D and x_B points.

Figure 1.11 illustrates the application of this mixing rule to a distillation column. Of course, a column *separates* instead of *mixes*, but the geometry is exactly the same. The two products D and B have compositions located at point (x_{D1}–x_{D2}) and (x_{B1}–x_{B2}), respectively. The feed F has a composition located at point (z_1–z_2) that lies on a straight line joining D and B.

This geometric relationship is derived from the overall molar balance and the two overall component balances around the column:

$$F = D + B$$
$$Fz_1 = Dx_{D1} + Bx_{B1}$$
$$Fz_2 = Dx_{D2} + Bx_{B2}$$

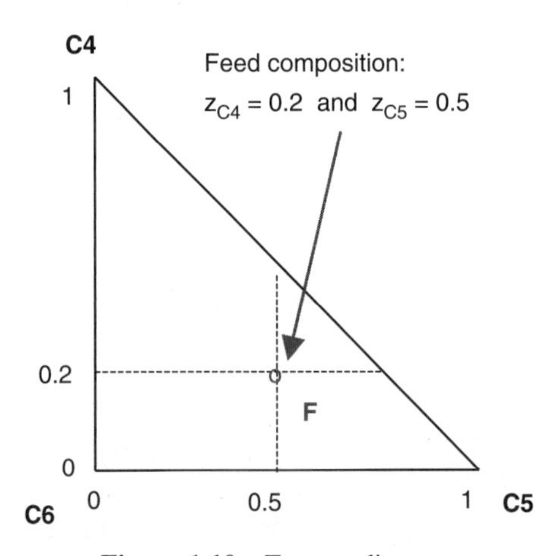

Figure 1.10 Ternary diagram.

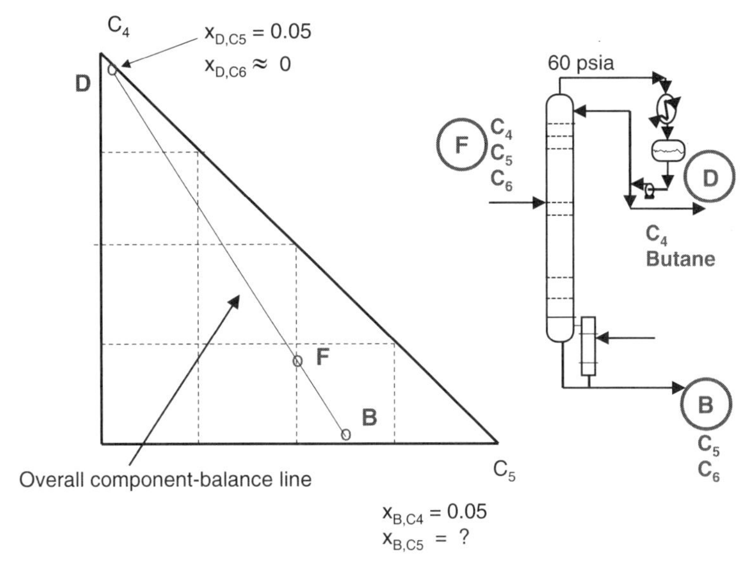

Figure 1.11 Ternary mixing rule.

Substituting the first equation in the second and third gives

$$(D + B)z_1 = Dx_{D1} + Bx_{B1}$$
$$(D + B)z_2 = Dx_{D2} + Bx_{B2}$$

Rearranging these two equations to solve for the ratio of B over D gives

$$\frac{D}{B} = \frac{z_1 - x_{D1}}{x_{B1} - z_1}$$
$$\frac{D}{B} = \frac{z_2 - x_{D2}}{x_{B2} - z_2}$$

Equating these two equations and rearranging give

$$\frac{z_1 - x_{D1}}{x_{B1} - z_1} = \frac{z_2 - x_{D2}}{x_{B2} - z_2}$$
$$\frac{x_{D1} - z_1}{z_2 - x_{D2}} = \frac{z_1 - x_{B1}}{x_{B2} - z_2}$$

Figure 1.12 shows how the ratios given above can be defined in terms of the tangents of the angles θ_1 and θ_2. The conclusion is that both angles must be equal, so the line between D and B must pass through F.

As we will see in subsequent chapters, this straight-line relationship is quite useful in representing what is going on in a ternary distillation system.

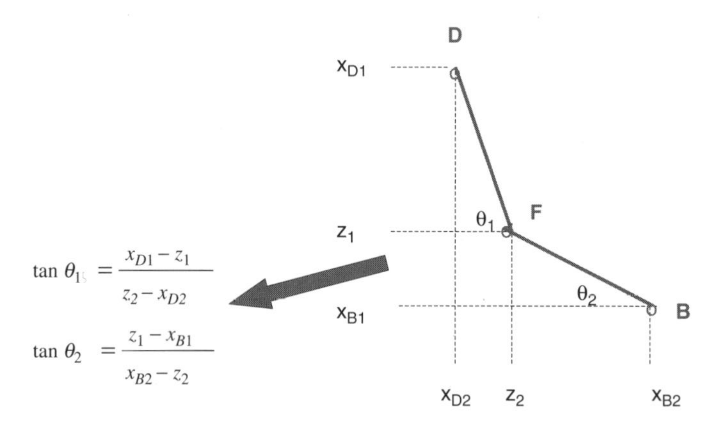

Figure 1.12 Proof of collinearity.

1.7 VLE NONIDEALITY

Liquid-phase ideality (activity coefficients $\gamma_j = 1$) occurs only when the components are quite similar. The benzene/toluene system is a common example. As shown in the sixth and seventh columns in Figure 1.5, the activity coefficients of both benzene and toluene are very close to unity.

However, if components are dissimilar, nonideal behavior occurs. Consider a mixture of methanol and water. Water is very polar. Methanol is polar on the OH end of the molecule, but the CH_3 end is nonpolar. This results in some nonideality. Figure 1.13a gives the *xy* curve at 1 atm. Figure 1.13b gives a table showing how the activity coefficients of the two components vary over composition space. The Unifac physical property method is used. The γ values range up to 2.3 for methanol at the $x = 0$ limit and 1.66 for water at $x = 1$. A plot of the activity coefficients can be generated by selecting the *Gamma* picture when using the *Plot Wizard.* The resulting plot is given in Figure 1.13c.

Now consider a mixture of ethanol and water. The $CH_3{-}CH_2$ end of the ethanol molecule is more nonpolar than the CH_3 end of methanol. We would expect the nonideality to be more pronounced, which is exactly what the *Txy* diagram, the activity coefficient results, and the *xy* diagram given in Figure 1.14 show.

Note that the activity coefficient of ethanol at the $x = 0$ end (pure water) is very large ($\gamma_{EtOH} = 6.75$) and also that the *xy* curve shown in Figure 1.14c crosses the 45° line ($x = y$) at ~90 mol% ethanol. This indicates the presence of an azeotrope. Note also that the temperature at the azeotrope (351.0 K) is lower than the boiling point of ethanol (351.5 K).

An "azeotrope" is defined as a composition at which the liquid and vapor compositions are equal. Obviously, when this occurs, there can be no change in the liquid and vapor compositions from tray to tray in a distillation column. Therefore an azeotrope represents a "distillation boundary."

Azeotropes occur in binary, ternary, and multicomponent systems. They can be "homogeneous" (single liquid phase) or "heterogeneous" (two liquid phases). They can be "minimum boiling" or "maximum boiling." The ethanol/water azeotrope is a minimum-boiling homogeneous azeotrope.

The software Aspen Split provides a convenient method for calculating azeotropes. Go to *Tools* on the top toolbar, then select *Aspen Split* and *Azeotropic Search.* The window shown at the top of Figure 1.15 opens, on which the components and pressure level are specified. Clicking on *Azeotropes* opens the window shown at the bottom of Figure 1.15, which gives the calculated results: a homogeneous azeotrope at 78°C (351 K) with composition 89.3 mol% ethanol.

Up to this point we have been using *Split* as an analysis tool. Aspen Technology plans to phase out *Split* in new releases of their Engineering Suite and will offer another tool called "DISTIL," which has more capability. To illustrate some of the features of DISTIL, let us use it to study a system in which there is more dissimilarity of the molecules

(a)

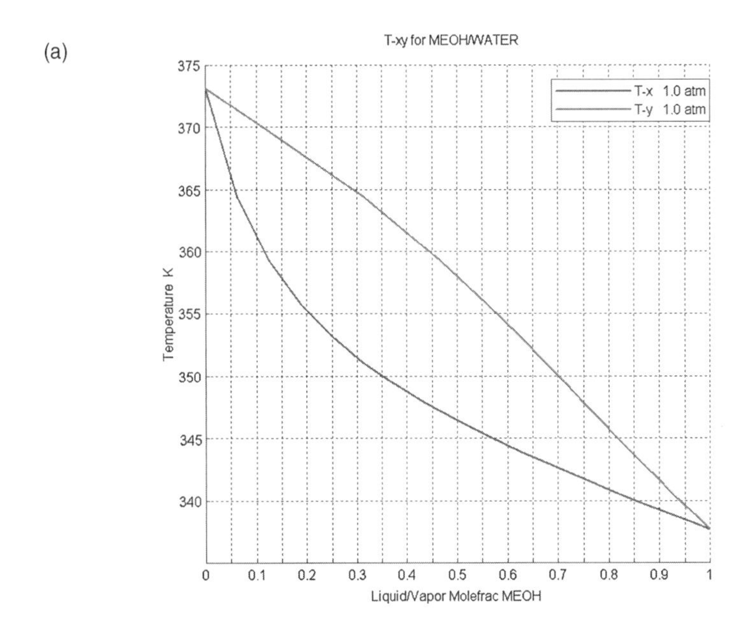

(b) Binary Analysis Results

Binary analysis results

PRES	MOLEFRAC MEOH	TOTAL TEMP	TOTAL KVL MEOH	TOTAL KVL WATER	LIQUID GAMMA MEOH	LIQUID GAMMA WATER	VAPOR MOLEFRAC MEOH	VAPOR MOLEFRAC WATER
atm		K						
1	0	373.1675	7.749335	0.9999874	2.295259	1	0	1
1	0.0625	364.4015	5.043268	0.7304488	1.963214	1.004922	0.3152043	0.6847957
1	0.125	359.2501	3.741208	0.6084334	1.723613	1.01833	0.467599	0.532401
1	0.1875	355.7694	2.986933	0.5415125	1.547374	1.038583	0.5600063	0.4399937
1	0.25	353.1777	2.498613	0.5004891	1.415168	1.064555	0.6246249	0.3753751
1	0.3125	351.1055	2.158348	0.4734803	1.314358	1.095401	0.6744724	0.3255276
1	0.375	349.3581	1.908578	0.4548468	1.236537	1.130547	0.7157214	0.2842786
1	0.4375	347.8249	1.717996	0.4415318	1.175977	1.169571	0.7516415	0.2483585
1	0.5	346.4395	1.568161	0.4317911	1.128672	1.212164	0.7841088	0.2158912
1	0.5625	345.1601	1.447525	0.4245425	1.09176	1.258095	0.8142673	0.1857327
1	0.625	343.9592	1.348499	0.4190841	1.063149	1.307191	0.8428477	0.1571523
1	0.6875	342.8187	1.265891	0.4149424	1.041282	1.359323	0.8703339	0.1296661
1	0.75	341.7258	1.196036	0.4117869	1.024981	1.414393	0.8970556	0.1029444
1	0.8125	340.6716	1.136273	0.4093805	1.013338	1.472327	0.9232424	0.0767576
1	0.875	339.6497	1.084625	0.407549	1.005646	1.533072	0.9490568	0.0509432
1	0.9375	338.6559	1.03961	0.4061703	1.001349	1.596588	0.974614	0.025386
1	1	337.6848	0.9999999	0.4051123	1	1.662845	1	0

Figure 1.13 (a) *Txy* diagram for methanol/water; (b) activity coefficients for methanol/water; (c) activity coefficient plot for methanol/water.

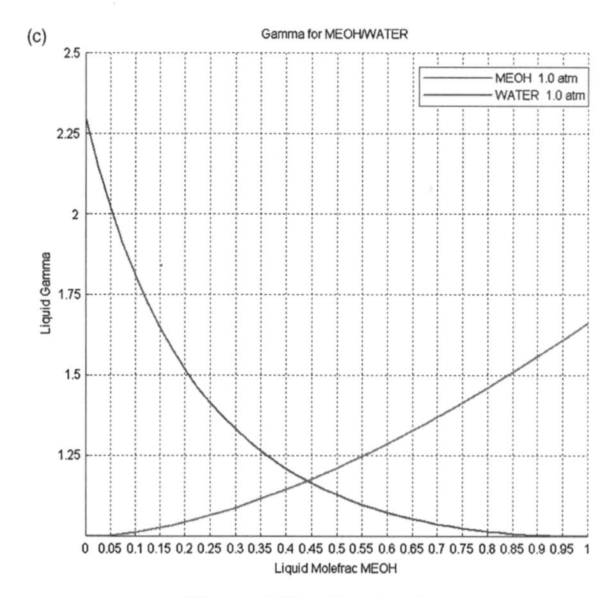

Figure 1.13 *Continued.*

by looking at the *n*-butanol/water system. The normal boiling point of *n*-butanol is 398 K, while that of water is 373 K, so water is the low boiler in this system.

The DISTIL program is opened in the usual way. Clicking the *Managers* button on the top toolbar of the *DISTIL* window and clicking *Fluid Package Manager* opens the window shown at the top of Figure 1.16. Click the *Add* button. The window shown at the bottom of Figure 1.16 opens. On the *Property Package* page tab, the *UNIFAC VLE* package is selected with *Ideal Gas*. Then click the *Select* button on the right side of the window.

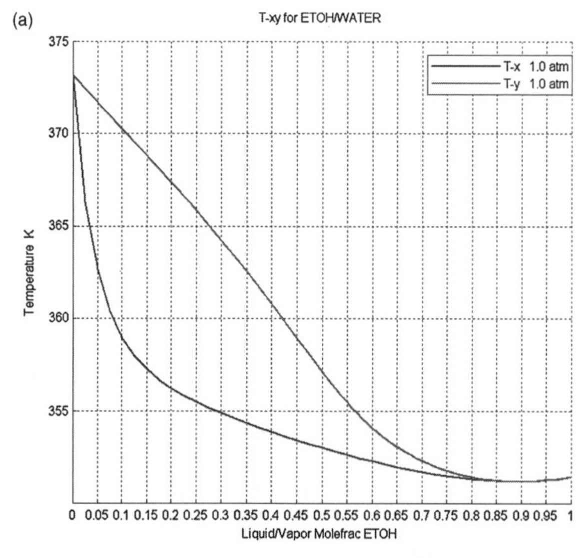

Figure 1.14 *Txy* diagram for ethanol/water (a), activity coefficient plot (b), and *xy* plot (c) for ethanol/water.

(b)

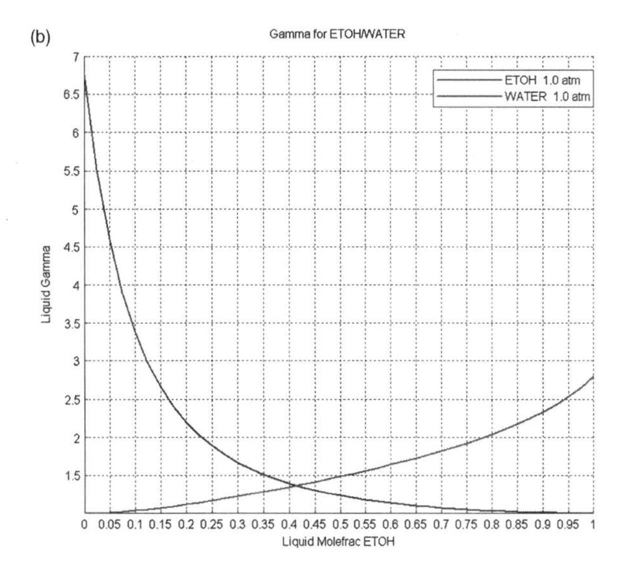

(c)

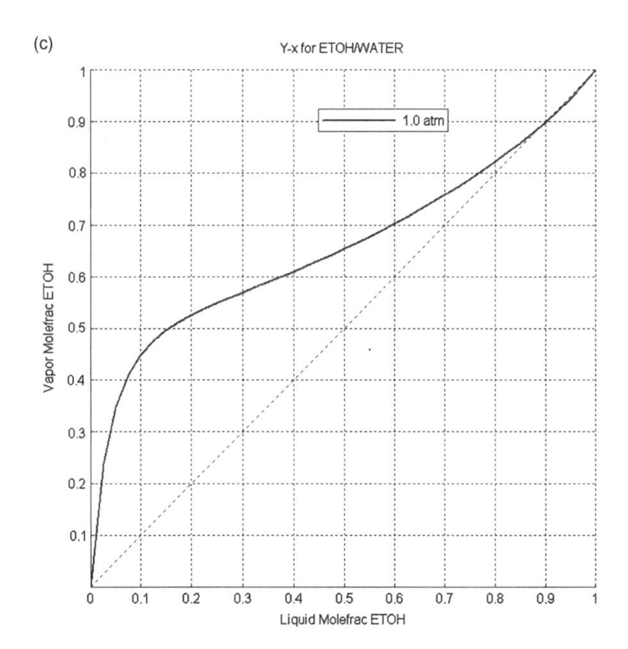

Figure 1.14 *Continued.*

Next click the *Components* page tab. Enter the name of a component in the *Matches* window and click *Select* to add each component to list on the left, as shown at the top of Figure 1.17.

Now click the *Manager* button again on the top toolbar and select *Separation Manager—Separation* and *Azeotrope Analysis*, which opens the window shown at the bottom Figure 1.17. With the *Fluid Package* highlighted under the *Setup* column on the left, the physical property package to be used is selected from the dropdown menu, and the components of interest are selected by clicking under the *Selected* column to

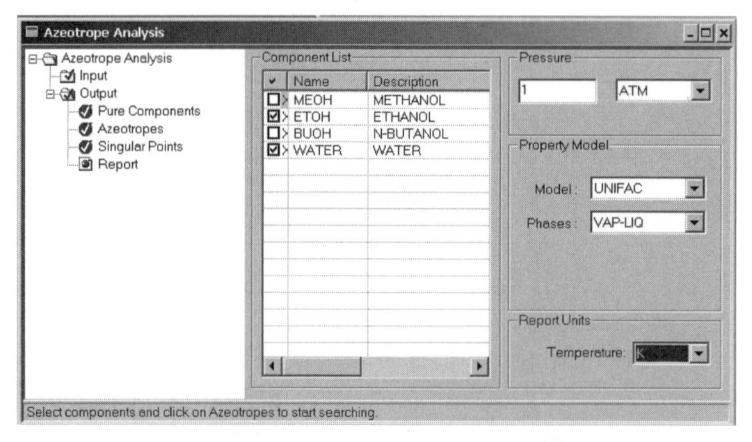

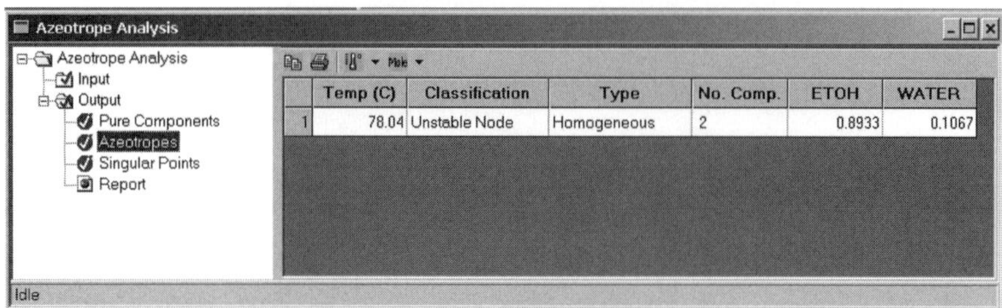

Figure 1.15 Aspen Split ethanol/water.

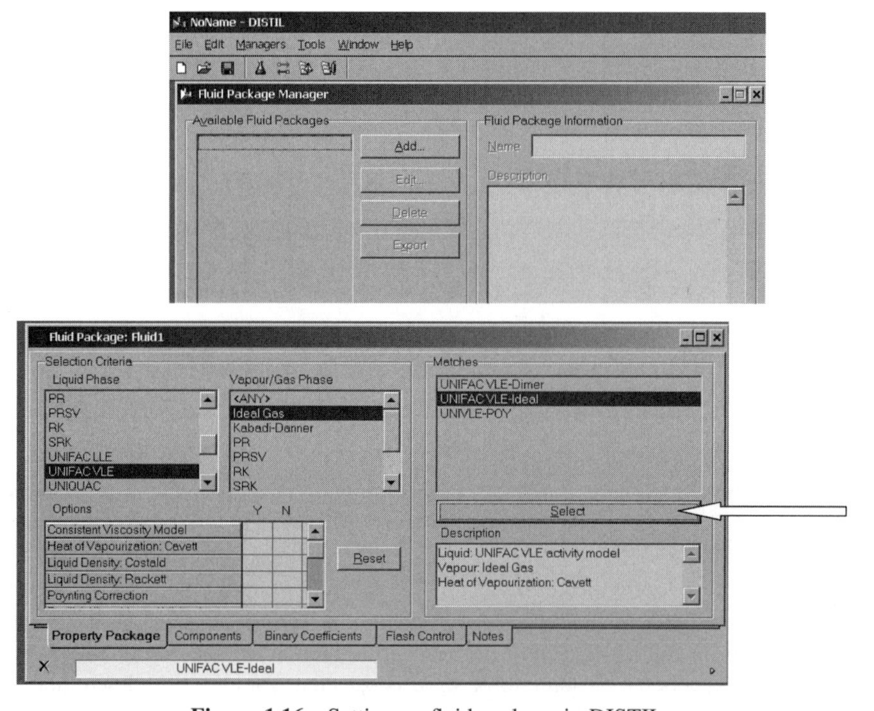

Figure 1.16 Setting up fluid package in DISTIL.

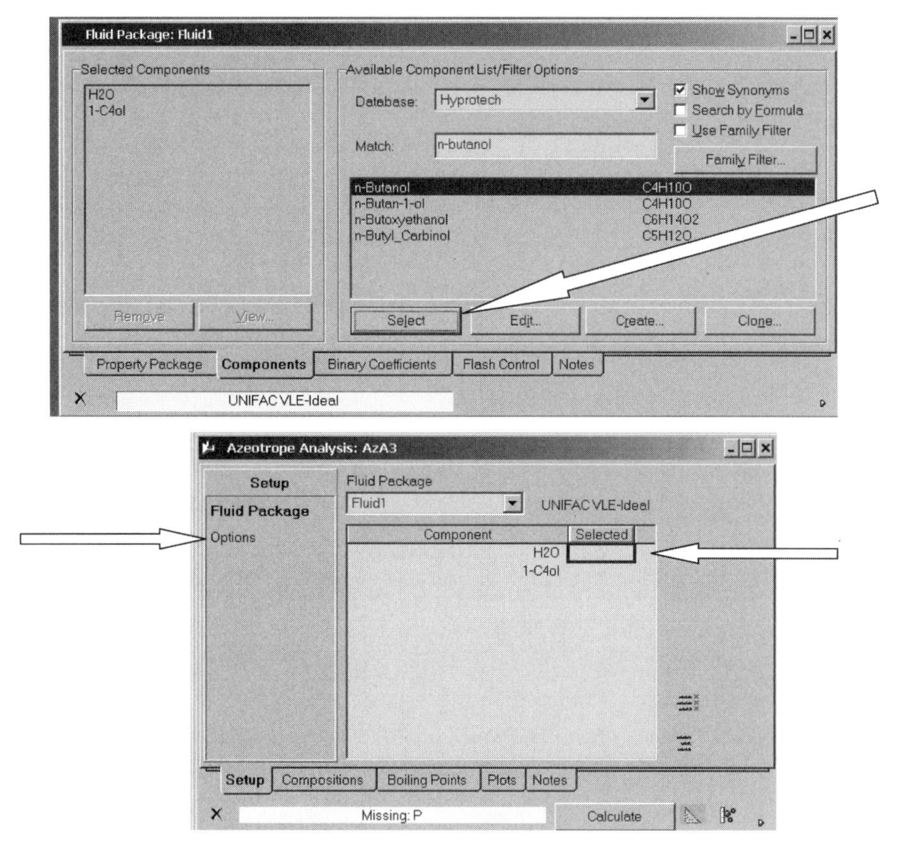

Figure 1.17 Selecting components in DISTIL.

place a green checkmark, not a red "x." Note that there is a message in the yellow box at the bottom of the window stating that the pressure needs to be specified. Clicking *Options* under the *Setup* column on the left opens the window shown at the top of Figure 1.18, where the pressure or a range of pressures is specified. Then click the *Calculate* button at the bottom of the window. Selecting the *Compositions* page tab or the *Boiling Points* page tab gives the results shown in Figure 1.19. Note that the azeotropic temperature (93°C) is lower than the boiling points of pure water (100°C) or butanol (118°C).

Various types of diagrams can be generated by going to the top toolbar and selecting *Managers, Thermodynamic Workbench Manager*, and *Phase Equilibrium Properties*. The window given at the top of Figure 1.20a opens, on which the fluids package and component are selected. Clicking the *Plots* page tab opens the window shown at the bottom of Figure 1.20a, on which pressure is set and the type of plot is specified. A *Txy* diagram is selected. Figure 1.20b gives activity coefficient and *xy* plots.

These results show a huge activity coefficient for butanol ($\gamma_{BuOH} = 40$) at the $x = 1$ point. The horizontal lines in the *Txy* diagram and the *xy* diagram indicate the presence of an heterogeneous azeotrope. The molecules are so dissimilar that two liquid phases are formed. At the azeotrope, the vapor composition is ~76 mol% water and the compositions of the two liquid phases are ~50 and ~97 mol% water.

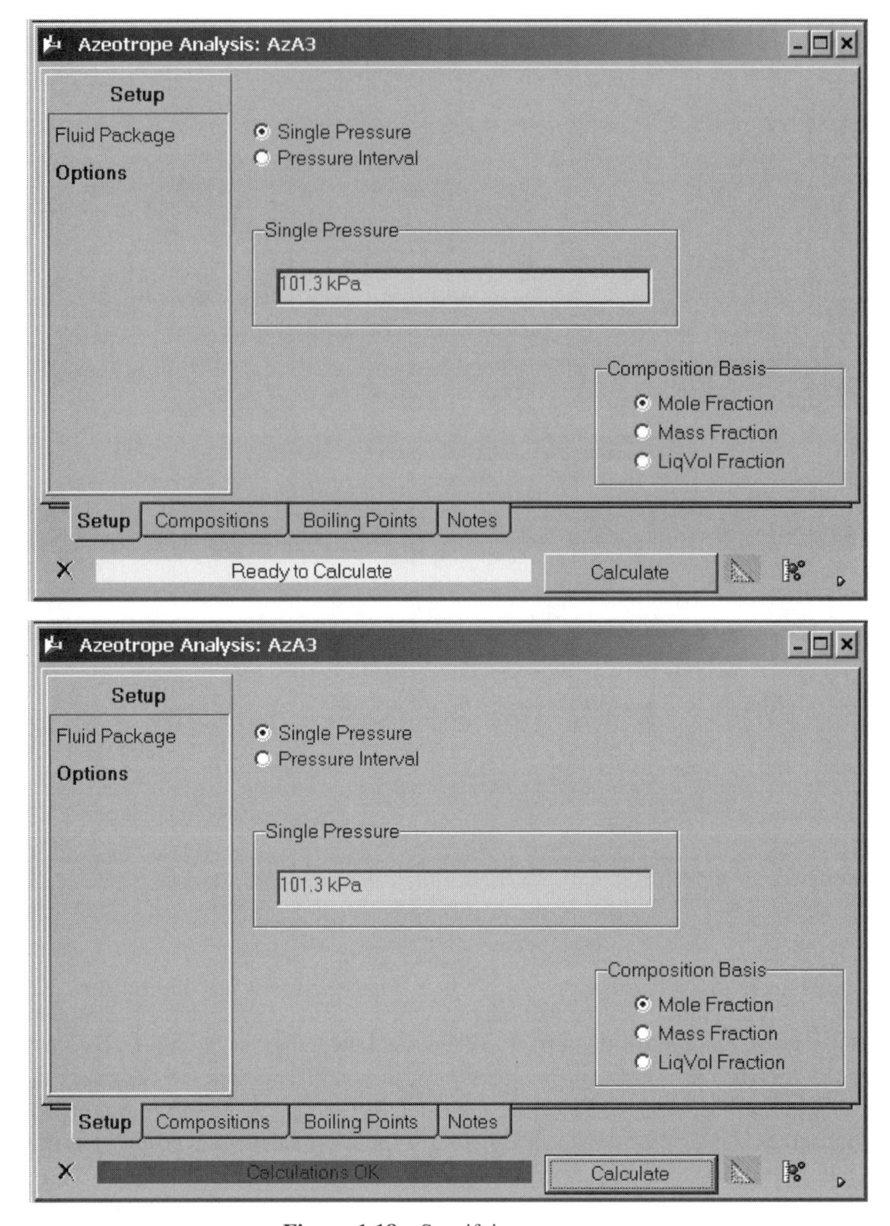

Figure 1.18 Specifying pressure.

Up to this point we have considered only binary systems. In the following section ternary nonideal systems are explored using the capabilities of Aspen Split.

1.8 RESIDUE CURVES FOR TERNARY SYSTEMS

Residue curve analysis is quite useful in studying ternary systems. A mixture with an initial composition $x_{1(0)}$ and $x_{2(0)}$ is placed in a container at some fixed pressure. A

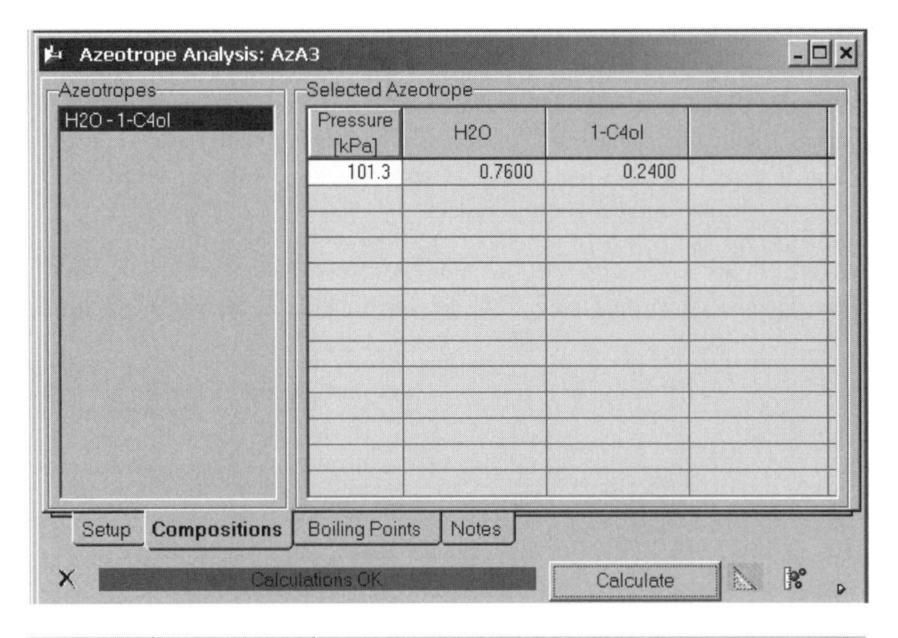

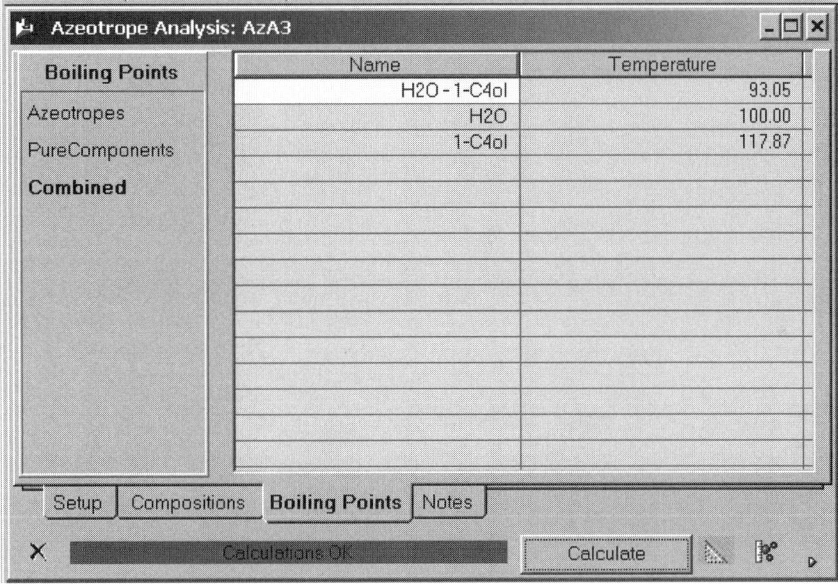

Figure 1.19 Azeotrope composition and temperatures.

vapor stream is continuously removed, and the composition of the remaining liquid in the vessel is plotted on the ternary diagram.

Figure 1.21 gives an example of how the compositions of the liquid x_j and the vapor y_j change with time during this operation. The specific numerical example is a ternary mixture of components A, B, and C, which have constant relative volatilities of $\alpha_A = 4$,

$\alpha_B = 2$, and $\alpha_C = 1$. The initial composition of the liquid is $x_A = 0.5$ and $x_B = 0.25$. The initial amount of liquid is 100 mol, and vapor is withdrawn at a rate of 1 mol per unit of time. Note that component A is quickly depleted from the liquid because it is the lightest component. The liquid concentration of component B actually increases for a while and then drops. Figure 1.22 plots the x_A and x_B trajectories for different initial conditions. These are the "residue curves" for this system.

Residue curves can be easily generated by using Aspen Split. Click on *Tools* in the upper toolbar in the *Aspen Plus* window and select *Aspen Split* and *Ternary Maps*. This opens the window shown in Figure 1.23, on which the three components and pressure

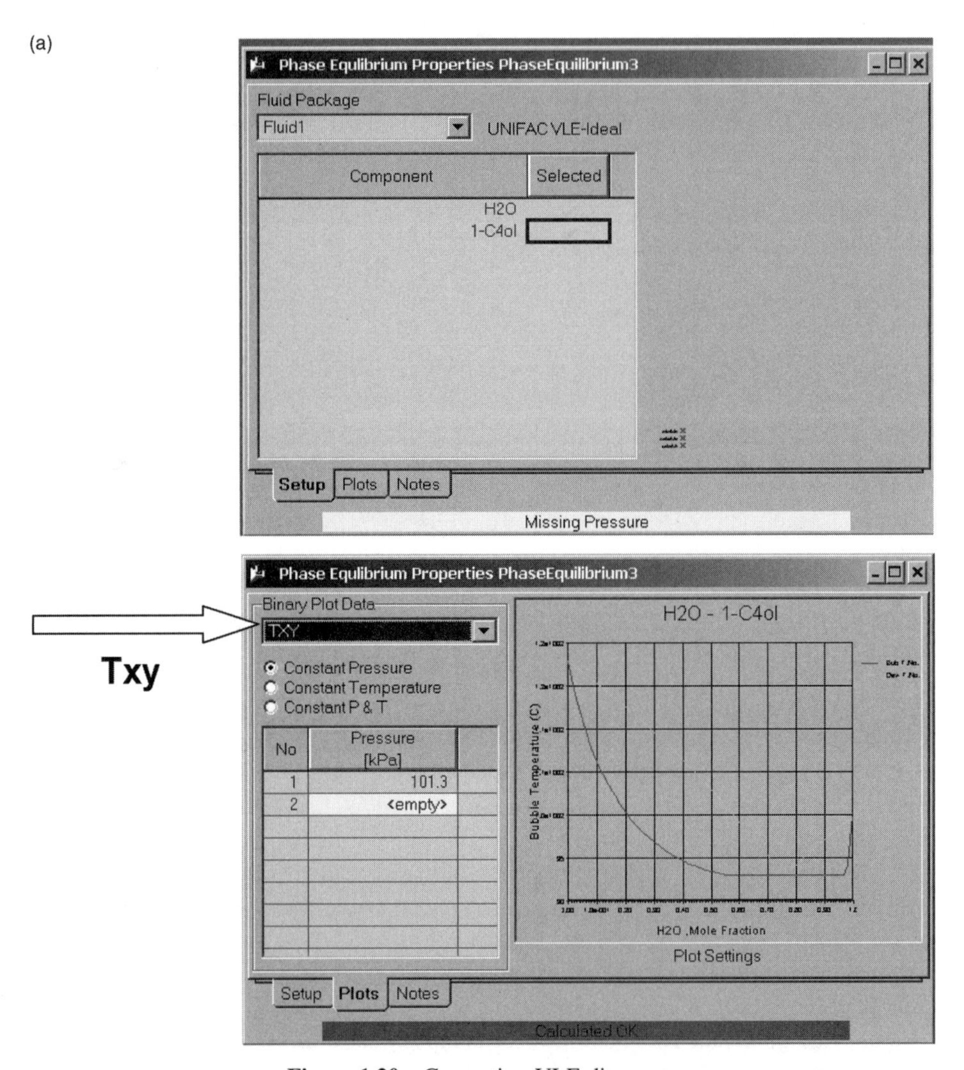

Figure 1.20 Generating VLE diagrams.

(b)

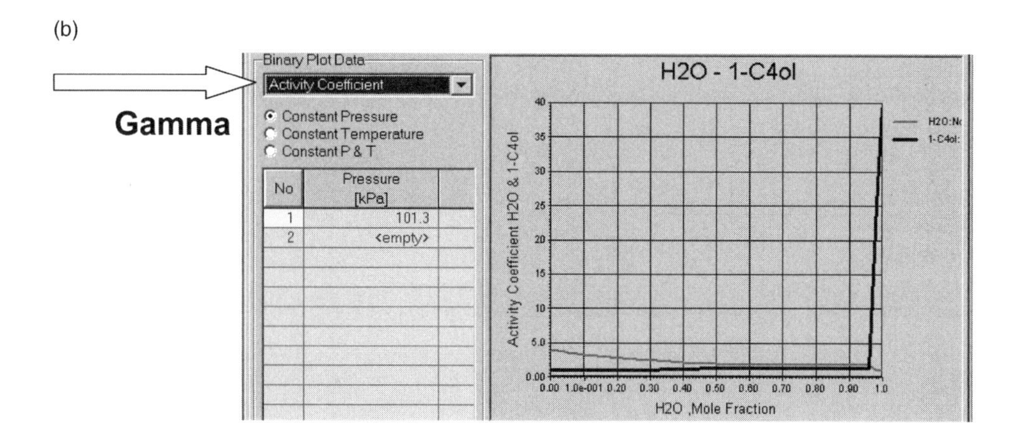

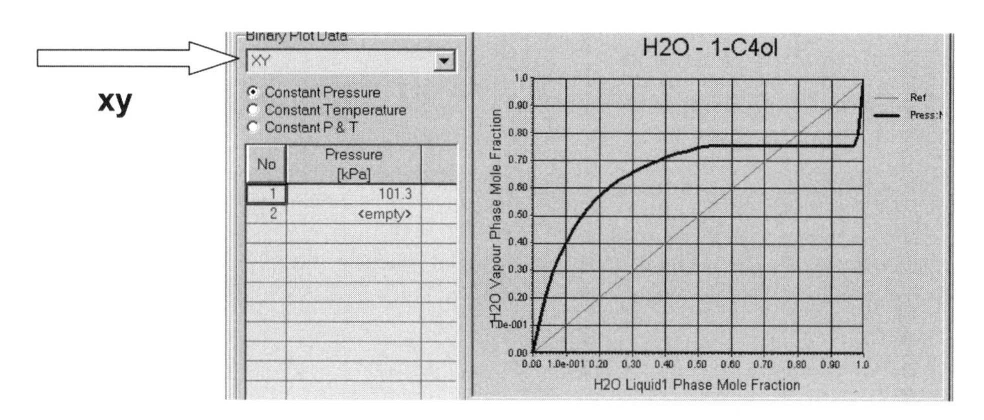

Figure 1.20 *Continued.*

are selected. The numerical example is the ternary mixture of n-butane, n-pentane, and n-hexane. Clicking on *Ternary Plot* opens the window given in Figure 1.24. To generate a residue curve, right-click the diagram and select *Add* and *Curve.* A crosshair appears that can be moved to any location on the diagram. Clicking inserts a residue curve that passes through the selected point, as shown in Figure 1.25a. Repeating this procedure produces multiple residue curves as shown in Figure 1.25b.

Note that all the residue curves start at the lightest component (C4) and move toward the heaviest component (C6). In this sense they are similar to the compositions in a distillation column. The light components go out the top, and the heavy components, go out the bottom. We will show below that this similarity proves to be useful for the analysis of distillation systems.

The generation of residue curves is described mathematically by a dynamic molar balance of the liquid in the vessel M_{liq} and two dynamic component balances for

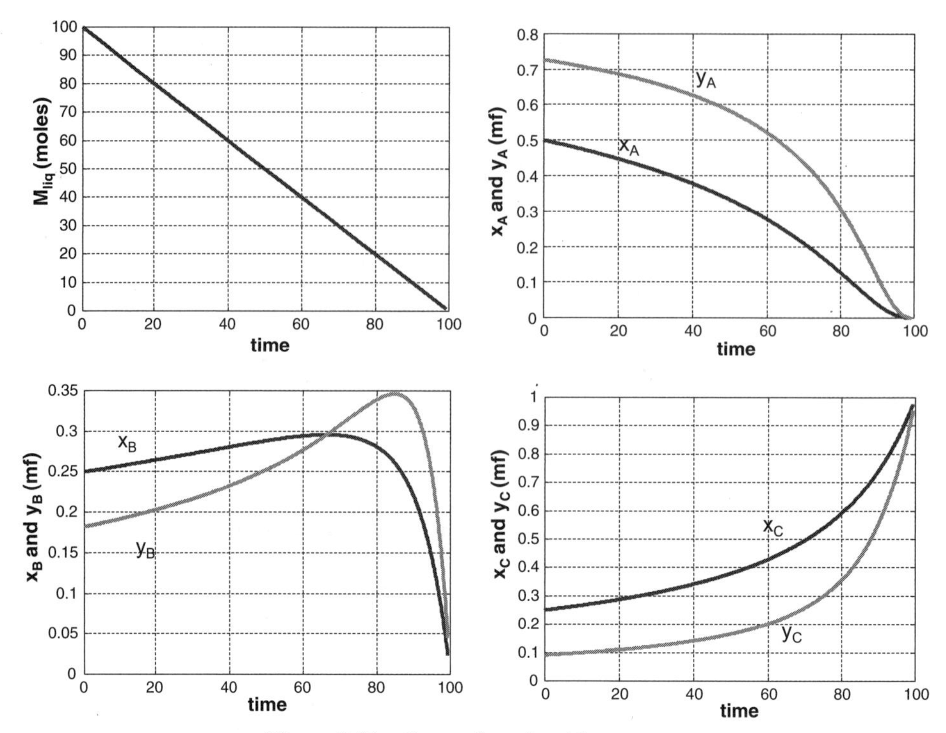

Figure 1.21 Generation of residue curves.

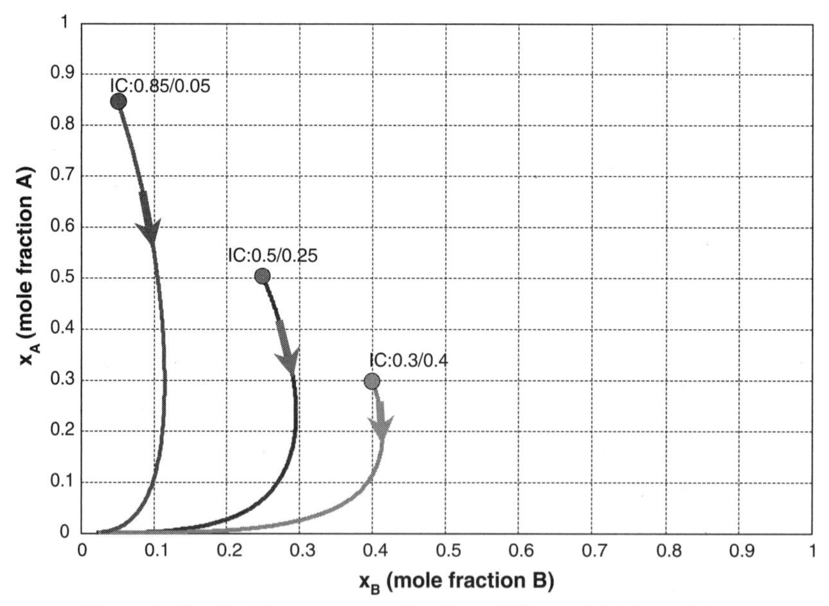

Figure 1.22 Residue curves starting from different initial conditions.

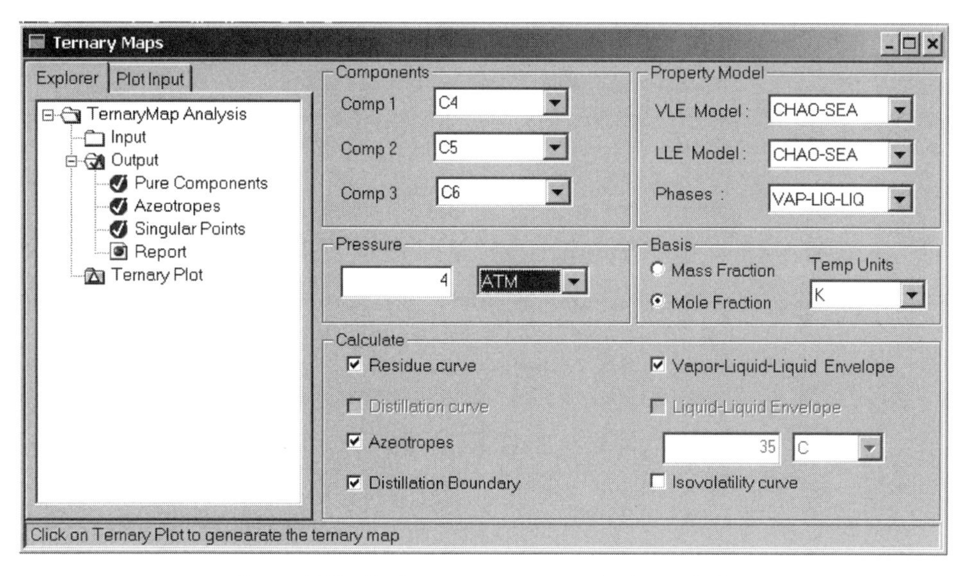

Figure 1.23 Setting up ternary maps in Aspen Split.

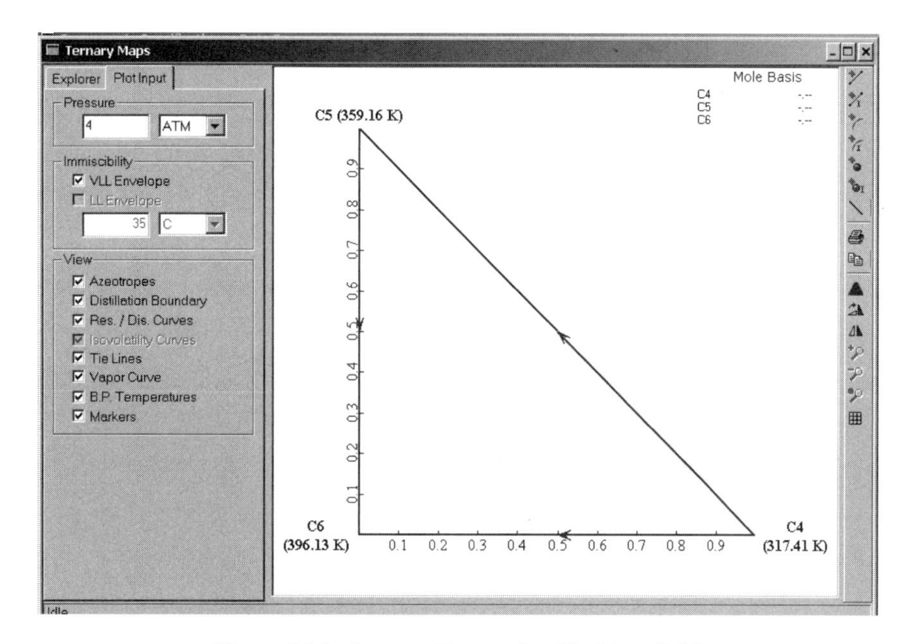

Figure 1.24 Ternary diagram for C4, C5, and C6.

(a)

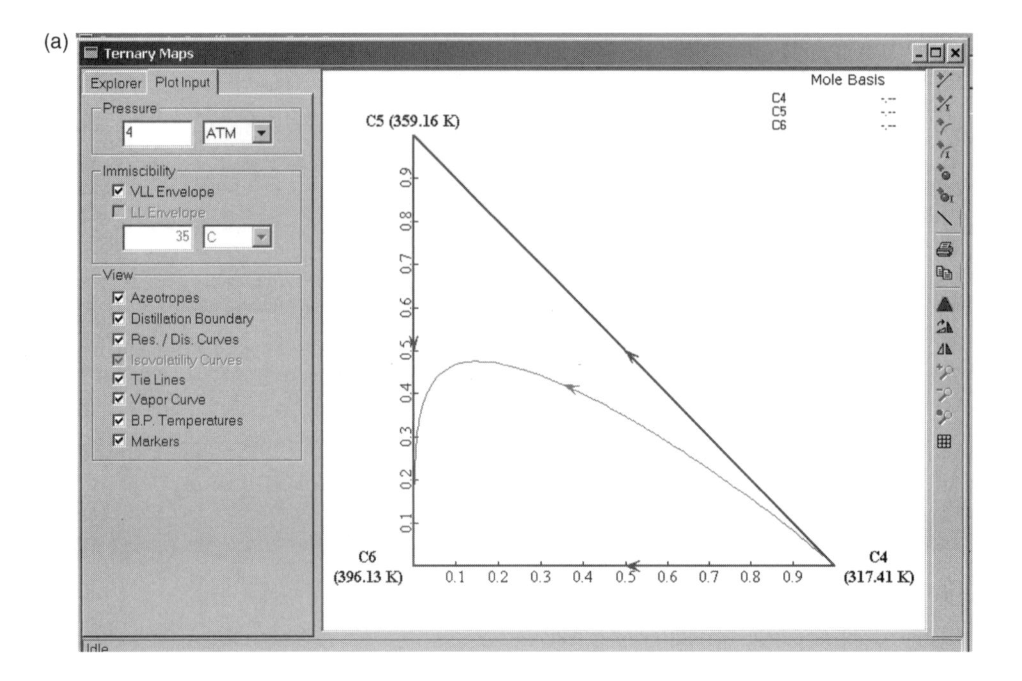

(b)

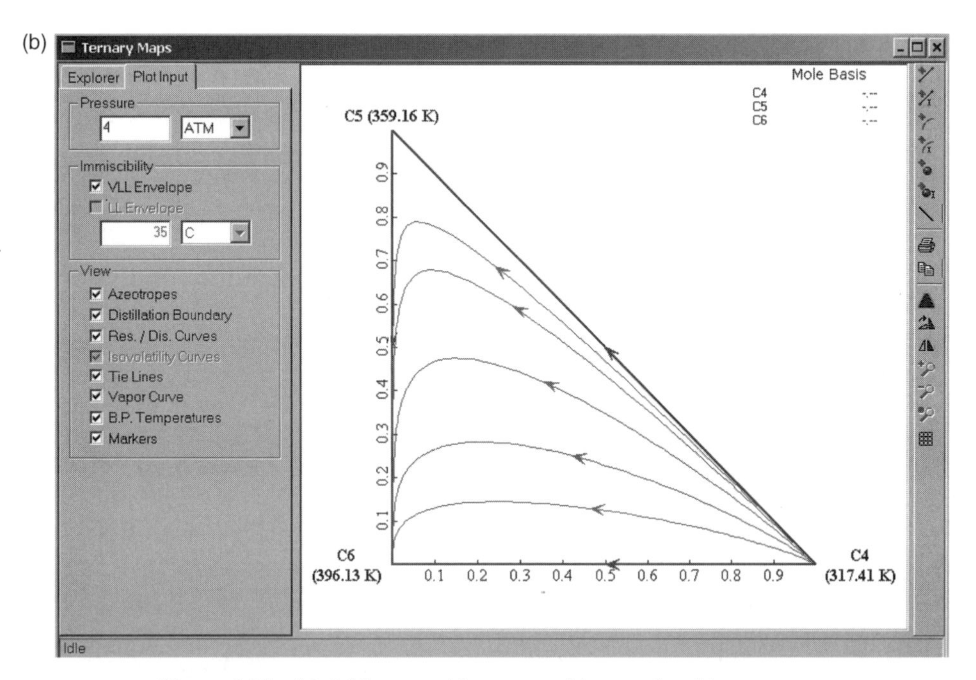

Figure 1.25 (a) Adding a residue curve; (b) several residue curves.

components A and B. The rate of vapor withdrawal is V (moles per unit time):

$$\frac{dM_{\mathrm{liq}}}{dt} = -V$$

$$\frac{d(M_{\mathrm{liq}} x_j)}{dt} = -V y_j$$

Of course, the values of x_j and y_j are related by the VLE of the system. Expanding the second equation and substituting the first equation give

$$M_{\mathrm{liq}} \frac{dx_j}{dt} + x_j \frac{dM_{\mathrm{liq}}}{dt} = -V y_j$$

$$M_{\mathrm{liq}} \frac{dx_j}{dt} + x_j(-V) = -V y_j$$

$$\left(\frac{M_{\mathrm{liq}}}{V}\right) \frac{dx_j}{dt} = x_j - y_j$$

$$\frac{dx_j}{d\theta} = x_j - y_j$$

The parameter θ is a dimensionless time variable. The last equation models how compositions change during the generation of a residue curve. As we describe below, a similar equation expresses the tray-to-tray liquid compositions in a distillation column under total reflux conditions. This relationship permits us to use residue curves to assess which separations are feasible or infeasible in a given system.

Consider the upper section of a distillation column as shown in Figure 1.26. The column is cut at tray n, at which the passing vapor and liquid streams have compositions

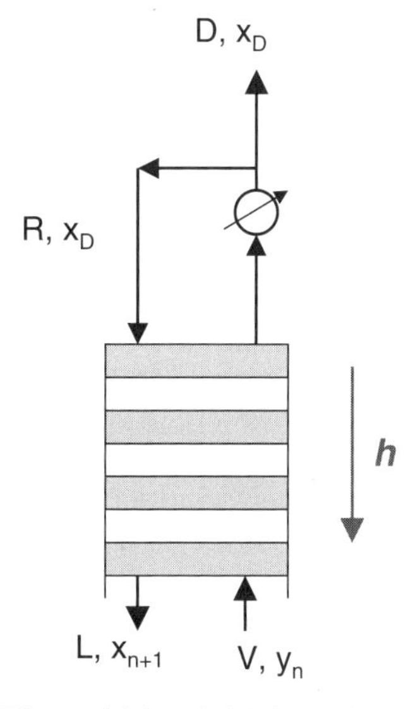

Figure 1.26 Distillation column.

y_{nj} and $x_{n+1,j}$ and flowrates are V_n and L_{n+1}. The distillate flowrate and composition are D and x_{Dj}. The steady-state component balance is

$$V_n y_{nj} = L_{n+1} x_{n+1,j} + D x_{Dj}$$

Under total reflux conditions, D is equal to zero and L_{n+1} is equal to V_n. Therefore y_{nj} is equal to $x_{n+1,j}$.

Let us define a continuous variable h as the distance from the top of the column down to any tray. The discrete changes in liquid composition from tray to tray can be approximated by the differential equation

$$\frac{dx_j}{dh} \approx x_{nj} - x_{n+1,j}$$

At total reflux this equation becomes

$$\frac{dx_j}{dh} = x_{nj} - y_{nj}$$

Note that this is the same equation as developed for residue curves.

The significance of this similarity is that the residue curves approximate the column profiles. Therefore a feasible separation in a column must satisfy two conditions:

1. The distillate compositions x_{Dj} and the bottoms compositions x_{Bj} must lie near a residue curve.
2. They must lie on a straight line through the feed composition point z_j.

We will use these principles in Chapters 2 and 5 for analyzing both simple and complex distillation systems.

1.9 CONCLUSION

The basics of vapor–liquid phase equilibrium have been reviewed in this chapter. A good understanding of VLE is indispensable in the design and control of distillation systems. These basics will be used throughout this book.

CHAPTER 2

ANALYSIS OF DISTILLATION COLUMNS

The major emphasis of this book is the use of rigorous steady-state and dynamic simulation for the design and control of distillation columns. However, several simple approximate methods provide significant insight into how the various design and operating parameters impact separation. Some of these methods employ graphical techniques that give visual pictures of the effects of parameters. Although some of the methods are limited to binary systems, the relationships can be extended to multicomponent systems.

2.1 DESIGN DEGREES OF FREEDOM

The design of a distillation column involves many parameters: product compositions, product flowrates, operating pressure, total number of trays, feed tray location, reflux ratio, reboiler heat input, condenser heat removal, column diameter, and column height. Not all of these variables are independent, so a "degrees of freedom" analysis is useful in pinning down exactly how many independent variables can (and must) be specified to completely define the system.

A rigorous degrees-of-freedom analysis involves counting the number of variables in the system and subtracting the number of equations that describe the system. For a multicomponent, multistage column this can involve hundreds, if not thousands, of variables and equations. Any error in counting is grossly amplified because we are taking the difference between two very large numbers. A simple intuitive approach is used below.

The normal situation in distillation design is that the feed conditions are given: flowrate F [moles per hour (mol/h)], composition z_j (mole fraction component j), temperature T_F and pressure P_F. The desired compositions of the product streams are also typically known. We consider a two-product column, so the normal specifications are to set the heavy-key impurity in the distillate $x_{D,\mathrm{HK}}$ and the light-key impurity in the bottoms $x_{B,\mathrm{LK}}$.

Distillation Design and Control Using Aspen™ Simulation, By William L. Luyben

The design problem is to establish the operating pressure P, the total number of trays N_T, and the feed tray location N_F that produces the desired product purities. All other parameters are then fixed. Therefore, the number of design degrees of freedom is **5**: $x_{D,\mathrm{HK}}$, $x_{B,\mathrm{LK}}$, P, N_T, and N_F. So if the desired product purities and the pressure are given, there are **2** degrees of freedom.

Just to emphasis this point, the five variables that could be specified might be the distillate flowrate D, reflux ratio $RR = R/D$, P, N_T, and N_F. In this case the product compositions cannot be specified but depend on the distillate flowrate and reflux ratio selected.

The steps in the design procedure will be illustrated in subsequent chapters. Our purpose in this chapter is to discuss some of the ways to establish reasonable values of some of the parameters such as the number of stages or the reflux ratio.

2.2 BINARY McCABE–THIELE METHOD

The McCabe–Thiele method is a graphical approach that shows very nicely in pictorial form the effects of VLE, reflux ratio, and number of trays. It is limited to binary systems, but the effects of parameters can be extended to multicomponent systems. The basic effects can be summarized:

1. The easier the separation, the fewer trays required and the lower the required reflux ratio (lower energy consumption).
2. The higher the desired product purities, the more trays required. But the required reflux ratio does not increase significantly as product purities increase.
3. There is an engineering tradeoff between the number of trays and the reflux ratio. An infinite number of columns can be designed that produce exactly the same products but have different heights, different diameters, and different energy consumptions. Selecting the optimum column involves issues of both steady-state economics and dynamic controllability.
4. Minimum values of the number of trays (N_{min}) and of the reflux ratio (RR_{min}) are required for a given separation.

All of these items can be visually demonstrated using the McCabe–Thiele method.

The distillation column considered is shown in Figure 2.1 with the various flows and composition indicated. We assume that the feed molar flowrate F and composition z are given. If the product compositions are specified, the molar flowrates of the two products D and B can be immediately calculated from the overall total molar balance and the overall component balance on the light component:

$$F = D + B$$

$$zF = Dx_D + Bx_B$$

$$\Rightarrow D = F\left(\frac{z - x_B}{x_D - x_B}\right)$$

For the moment let us assume that the pressure has been specified, so the VLE is fixed. Let us also assume that the reflux ratio has been specified, so the reflux flowrate can be calculated

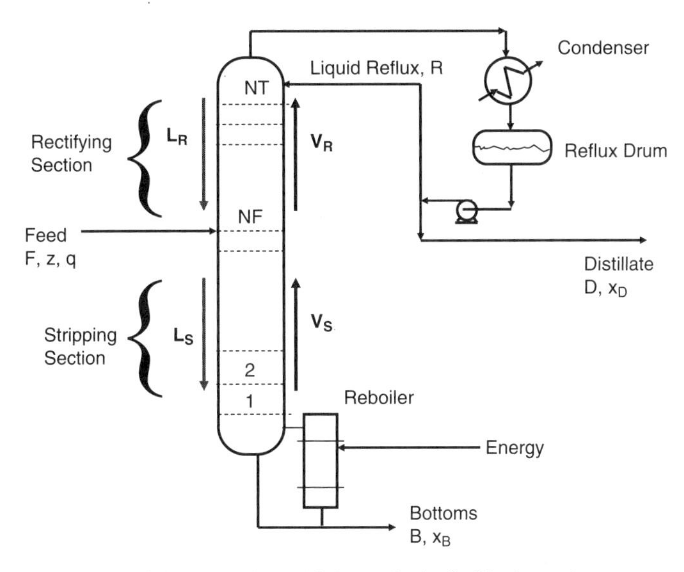

Figure 2.1 McCabe–Thiele method: distillation column.

$R = (RR)(D)$. The "equimolal overflow" assumption is usually made in the McCabe–Thiele method. The liquid and vapor flowrates are assumed to be constant in a given section of the column. For example, the liquid flowrate in the rectifying section L_R is equal to the reflux flowrate R. From an overall balance around the top of the column, the vapor flowrate in the rectifying section V_R is equal to the reflux plus the distillate ($V_R = R + D$).

This method uses an xy diagram whose coordinates are the mole fraction of the light component in the liquid x and the mole fraction of the light component in the vapor phase y. The VLE curve is plotted for the selected pressure. The 45° line is plotted. The specified product compositions x_D and x_B are located on the 45° line, as shown in Figure 2.2.

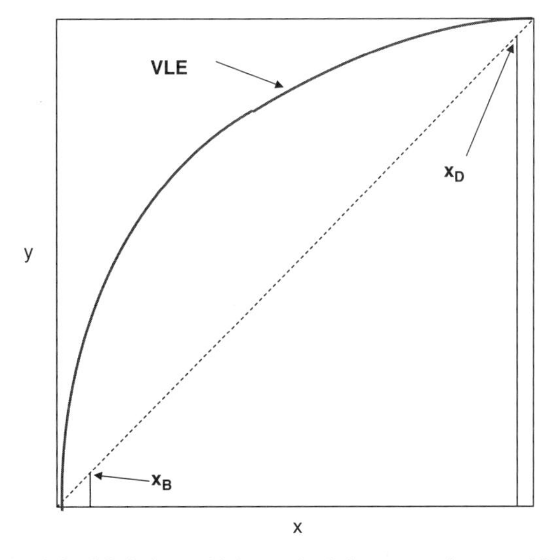

Figure 2.2 McCabe–Thiele method: locate products and VLE.

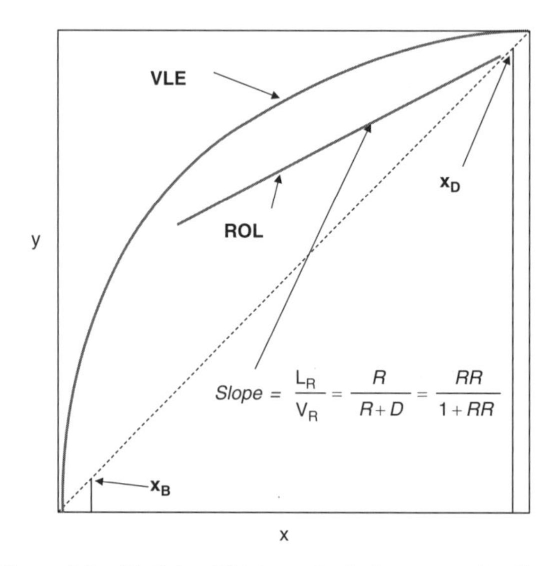

Figure 2.3 McCabe–Thiele method: draw operating lines.

2.2.1 Operating Lines

Next the "rectifying operating line" (ROL) is drawn. This is a straight line with a slope equal to the ratio of the liquid and vapor flowrates in the rectifying section:

$$\text{Slope ROL} = \frac{L_R}{V_R} = \frac{R}{R+D} = \frac{RR}{1+RR}$$

The line intersects the 45° line at the distillate composition x_D, so it is easy to construct (see Fig. 2.3). The proof of this construction can be derived by looking at the top of the column, as shown in Figure 2.4.

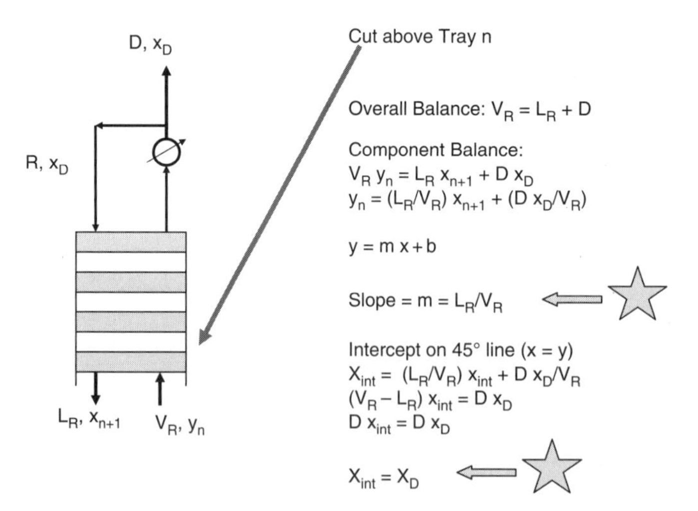

Figure 2.4 ROL construction.

The liquid and vapor flowrates in the stripping section (L_S and V_S) can be calculated if the thermal condition of the feed is known. Since the temperature, pressure, and composition of the feed are given, the fraction of the feed that is liquid can be calculated from an isothermal flash calculation. This fraction is defined as the variable q. Knowing q, we can calculate the liquid and vapor flowrates in the stripping section. If the feed is saturated liquid, q is 1; if the feed is saturated vapor, q is 0:

$$q = \frac{L_S - L_R}{F}$$
$$\Rightarrow L_S = qF + L_R$$
$$V_S = L_S - B$$

The stripping operating line (SOL) can be drawn. It is a straight line with slope L_S/V_S that interects the 45° line at the bottoms composition x_B. The proof of this construction can be derived by looking at the bottom of the column, as shown in Figure 2.5. Figure 2.6 shows the two operating lines.

2.2.2 *q* Line

There is a relationship between the intersection point of the two operating lines and feed conditions. As shown in Figure 2.7, a straight line can be drawn from the location of the feed composition z on the 45° line to this intersection point. As we will prove below, the slope of this line is only a function of the thermal condition of the feed, which is defined in the parameter q. The slope is $-q/(1-q)$, which makes the construction of the McCabe–Thiele diagram very simple:

1. Locate the three compositions on the 45° line (z, x_D, x_B).
2. Draw the ROL from the x_D point with a slope of $RR/(1+RR)$.

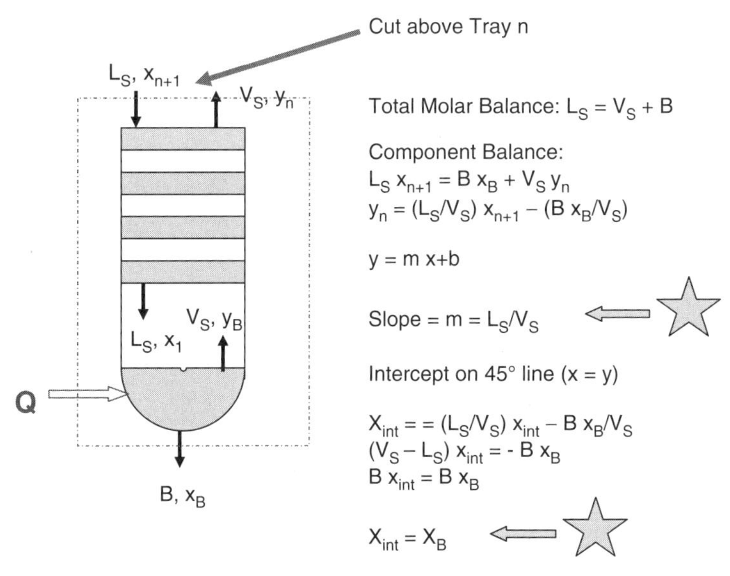

Figure 2.5 SOL construction.

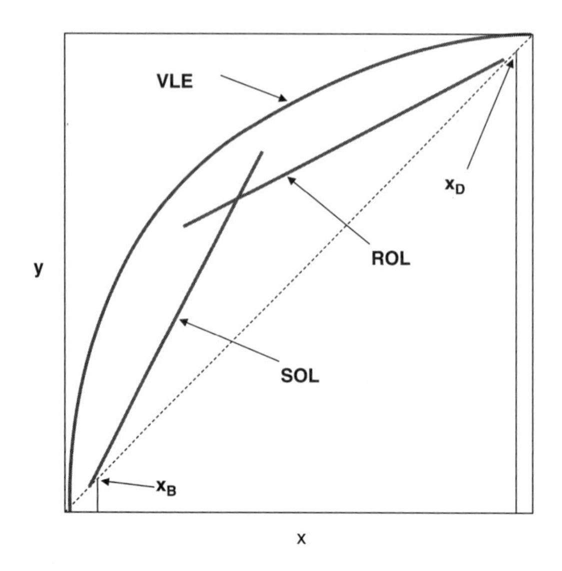

Figure 2.6 Operating lines.

3. Draw the q line from the z point with a slope of $-q/(1-q)$.
4. Draw the SOL from the x_B point to the intersection of the q line and the ROL.

The equations of the rectifying and stripping operating lines are given below in terms of the point of intersection of the two lines at y_{int} and x_{int}:

$$\text{ROL:} \quad y_{\text{int}} = \left(\frac{L_R}{V_R}\right) x_{\text{int}} + \frac{D x_D}{V_R}$$

$$\text{SOL:} \quad y_{\text{int}} = \left(\frac{L_S}{V_S}\right) x_{\text{int}} - \frac{B x_B}{V_S}$$

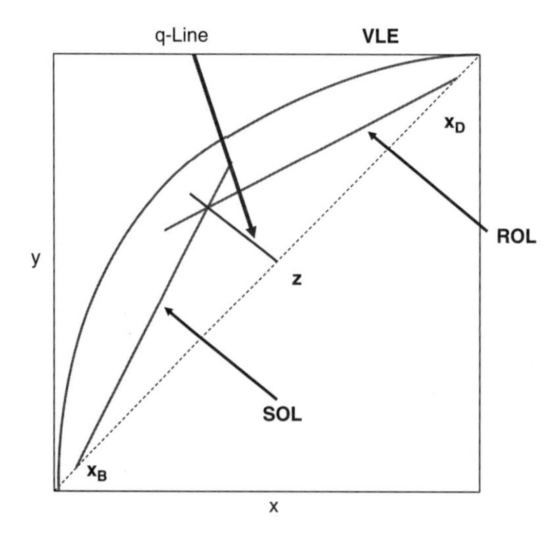

Figure 2.7 q line.

Subtracting the two equations gives

$$(V_R - V_S)y_{\text{int}} = (L_R - L_S)x_{\text{int}} + (Dx_D + Bx_B)$$

The last term on the right is just Fz. Using the definition of q leads to

$$(V_R - V_S) = (1 - q)F$$
$$(L_R - L_S) = -qF$$

Substituting these relationships into the previous equation gives

$$(1 - q)Fy_{\text{int}} = -qFx_{\text{int}} + Fz$$
$$y_{\text{int}} = \left(\frac{-q}{1 - q}\right)x_{\text{int}} + \left(\frac{z}{1 - q}\right)$$

This is the equation of a straight line with slope $-q/(1 - q)$. The q line is vertical for saturated liquid feed ($q = 1$), and it is horizontal for saturated vapor feed ($q = 0$). On the 45° line, x_{int} is equal to y_{int}. We can define this as x_{45}:

$$(1 - q)x_{45} = -qx_{45} + z$$
$$x_{45} = z$$

Thus the q line intersects the 45° line at the feed composition z.

2.2.3 Stepping off Trays

The number of trays is determined by moving vertically from the x_B point on the 45° line to the VLE line. This is the composition of the vapor y_B leaving the partial reboiler. Then we move horizontally over to the SOL. This step represents the partial reboiler. The value of x on the SOL is the composition of liquid x_1 leaving tray 1 (if we are numbering from the bottom of the column up). This stepping is repeated, moving vertically to y_1 and horizontally to x_2. Stepping continues until we cross the intersection of the operating lines. This is the feed tray. Then the horizontal line is extended to the ROL. Continuing to step until the x_D value is crossed gives the total number of trays. A numerical example is given below.

2.2.4 Effect of Parameters

We know enough now about the McCabe–Thiele diagram to make several observations, which can be applied to any distillation system, not just a binary separation:

1. The farther the VLE curve is from the 45° line, the smaller the slope of the rectifying operation line. This means a smaller reflux ratio and therefore lower energy consumption. A "fat" VLE curve corresponds to large relative volatilities and an easy separation.
2. The easier the separation, the fewer trays is takes to make a given separation.
3. The higher the product purities, the more trays it takes to make a given separation.

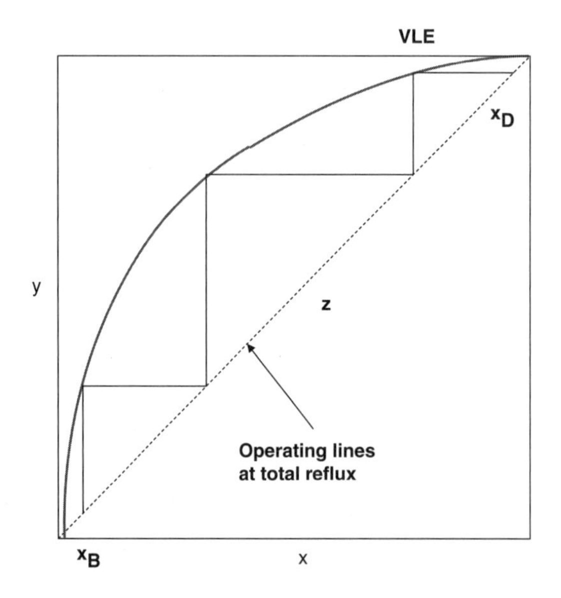

Figure 2.8 Minimum number of trays.

4. Increasing product purities does not have a significant effect on the required reflux ratio.
5. Increasing the liquid to vapor ratio in a section of a column increases the separation that occurs in that section.

These effects apply to all types of separations and distillation columns.

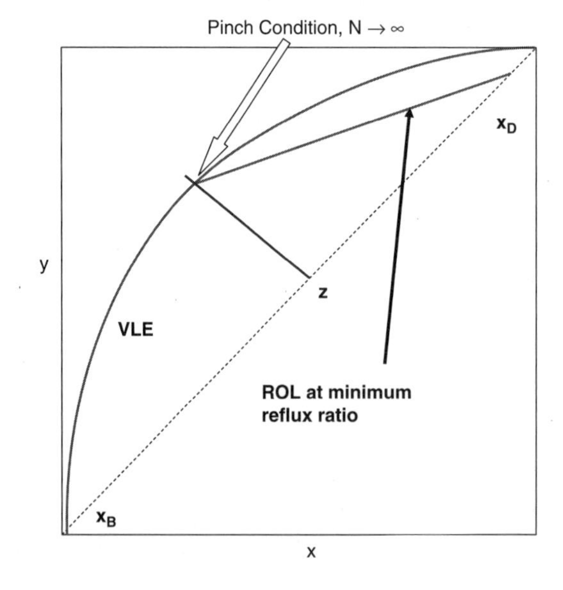

Figure 2.9 Minimum reflux ratio.

2.2.5 Limiting Conditions

Before we go through a specific example, we need to discuss some of the limiting conditions. The minimum number of trays for a specified separation corresponds to total reflux operation. If the column is run under total reflux conditions, the distillate flowrate is zero. Therefore the reflux ratio is infinite, and the slope of the operating lines is unity. This is the 45° line. Thus the minimum number of trays can be determined by simply stepping up between the 45° line and the VLE curve (see Fig. 2.8).

The minimum reflux ratio for a specified separation corresponds to having an infinite number of trays. This usually occurs when the operating lines and the q line intersect exactly on the VLE curve. This is a "pinch" condition. It would take an infinite number of trays to move past this point. This is illustrated in Figure 2.9. The minimum reflux ratio is calculated from the slope of this limiting operating line.

2.2.6 Numerical Example using DISTIL

To illustrate this method on a numerical example, consider the separation of benzene and toluene. The column operates at atmospheric pressure. The feed is 100 kmol/h of saturated liquid ($q = 1$) with composition $z = 0.40$ mole fraction (mf) benzene. The desired product compositions are $x_D = 0.95$ and $x_B = 0.05$. The flowrates of the two products are

$$D = F\left(\frac{z - x_B}{x_D - x_B}\right) = 100\left(\frac{0.40 - 0.05}{0.95 - 0.05}\right) = 38.9\,\text{kmol/h}$$

$$B = F - D = 100 - 38.9 = 61.1\,\text{kmol/h}$$

The software DISTIL can perform the McCabe–Thiele analysis in a very convenient way. Start the program and use the *Fluid Package Manager* to define fluids property package and the components using the method described in Chapter 1. We use the Wilson fluids property method for the benzene/toluene system. Select the *Separation Manager* and *Graphical Column Design* on the top toolbar. The window shown at the top of Figure 2.10 opens with the *Setup* page tab selected, on which the fluid package and components are selected. Clicking *Options* under the *Setup* column on the left opens the window shown at the bottom of Figure 2.10, on which the pressure is specified.

Clicking the *Configuration* page tab opens the window shown at the top of Figure 2.11. We want a standard two-product column, so there is nothing to specify on this window. Clicking the *Spec Entry* page tab opens the window given at the bottom of Figure 2.11. This view is used to specify the feed conditions and the bottoms and distillate specification. First click the *Feed* button and give the feed composition (0.40 mf benzene) and the feed thermal condition ($q = 1$). Under the *Internal Flow Specifications*, we set a preliminary value of the reflux ratio to be 2. This will be changed later after the program calculates the minimum reflux ratio for us.

Clicking the *Distillate* button opens the window shown at the top of Figure 2.12, on which the distillate purity (0.95 mf benzene) is specified. Clicking the *Bottoms* button opens the window shown at the bottom of Figure 2.12, on which the bottoms impurity (0.05 mf benzene) is specified. Now the yellow message at the bottom of the window tells us that the program is *Ready to Calculate*.

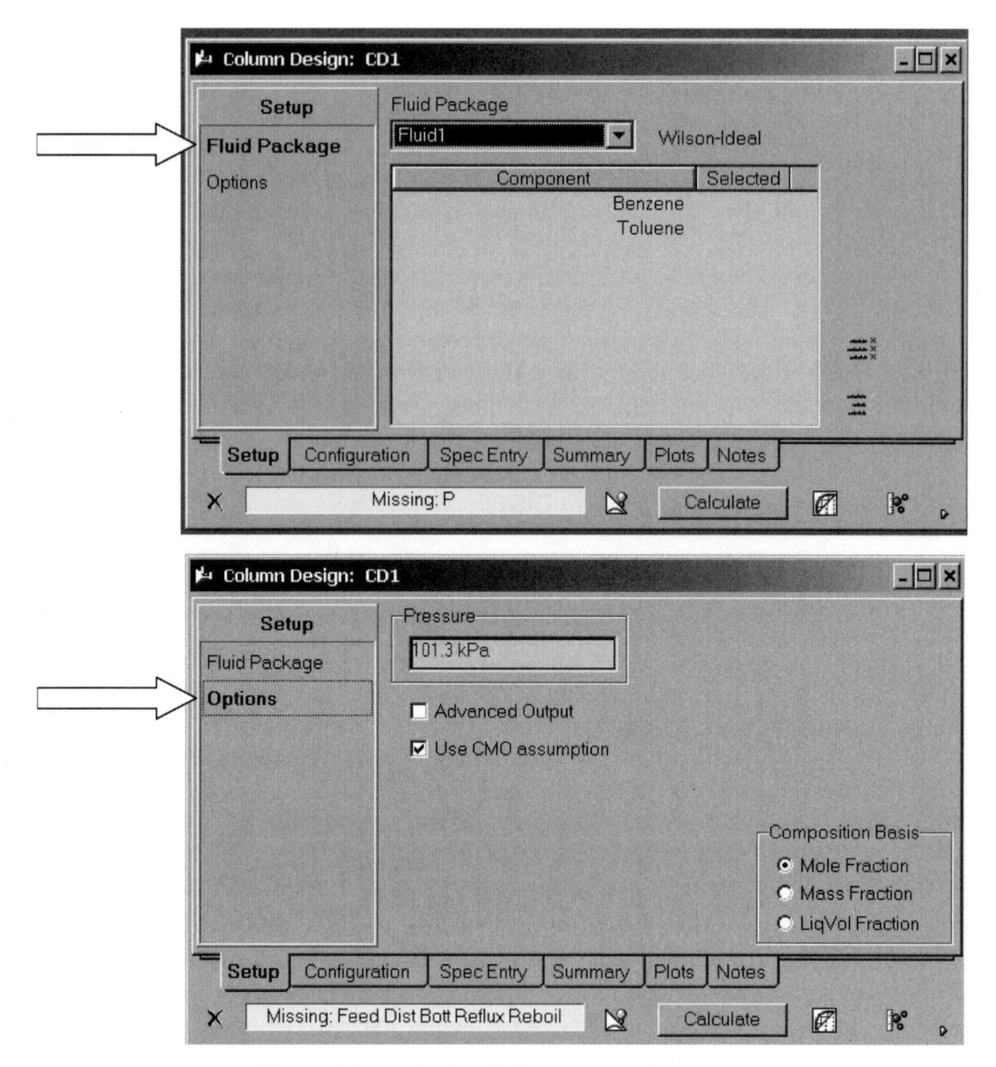

Figure 2.10 Selecting fluid package and pressure.

Click *Calculate* to the right of the message. The message window turns green and says *Calculations OK*, as shown at the top of Figure 2.13. Note that the minimum reflux ratio has been calculated ($RR_{\min} = 1.489$). The reflux ratio that has been specified is 2, which gives a ratio of the actual reflux ratio to the minimum of $2/1.489 = 1.34$. The economic optimization of a distillation column is discussed in Chapter 4. Typical values of the *RR* to $RR_{\min}$ ratio are around 1.2, so our initial guess is a little high.

Clicking the *Summary* page tab gives the window shown at the bottom of Figure 2.13. The DISTIL program uses the stage numbering convention of starting at the top. The reflux drum is stage 0, and the partial reboiler is the last stage. For the specified reflux ratio and product purities, 13 stages plus a partial reboiler are required, with the feed

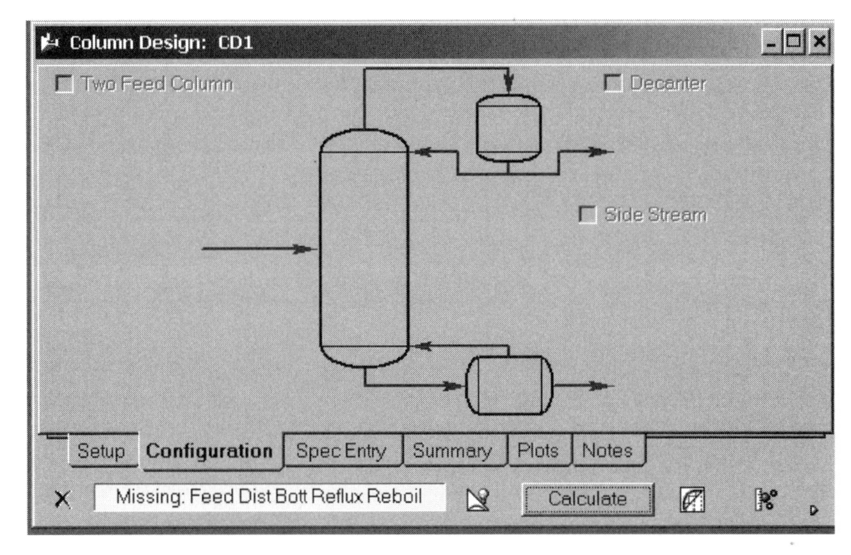

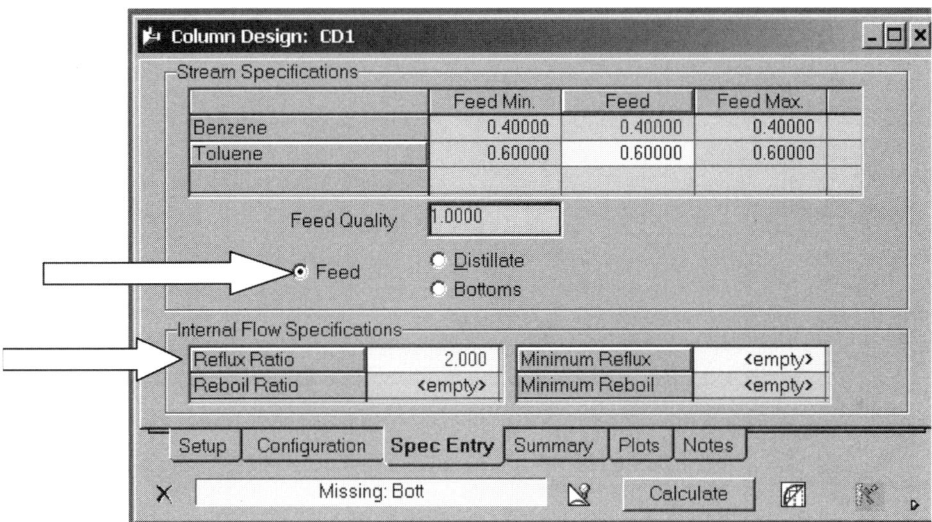

Figure 2.11 Selecting configuration, feed conditions, and reflux ratio.

fed on stage 7. If the reflux ratio is reduced to 1.2 $RR_{\min}$ [$RR = (1.2)(1.389) = 1.67$] and the calculations are redone, the number of stages increases to 17 with feed introduced on stage 9.

Various types of plots can be generated by clicking the *Plots* page tab and selecting one of the three types of plots listed under *Plot Generation*. Figure 2.14 gives the McCabe–Thiele xy diagram with the operating lines, the q line, the VLE curve, and the stages stepped off. Some of the features of this plot can be modified by right-clicking the plot and selecting *Chart Control*. Figure 2.15 gives the temperature profile and the benzene liquid and vapor mole fraction on all stages.

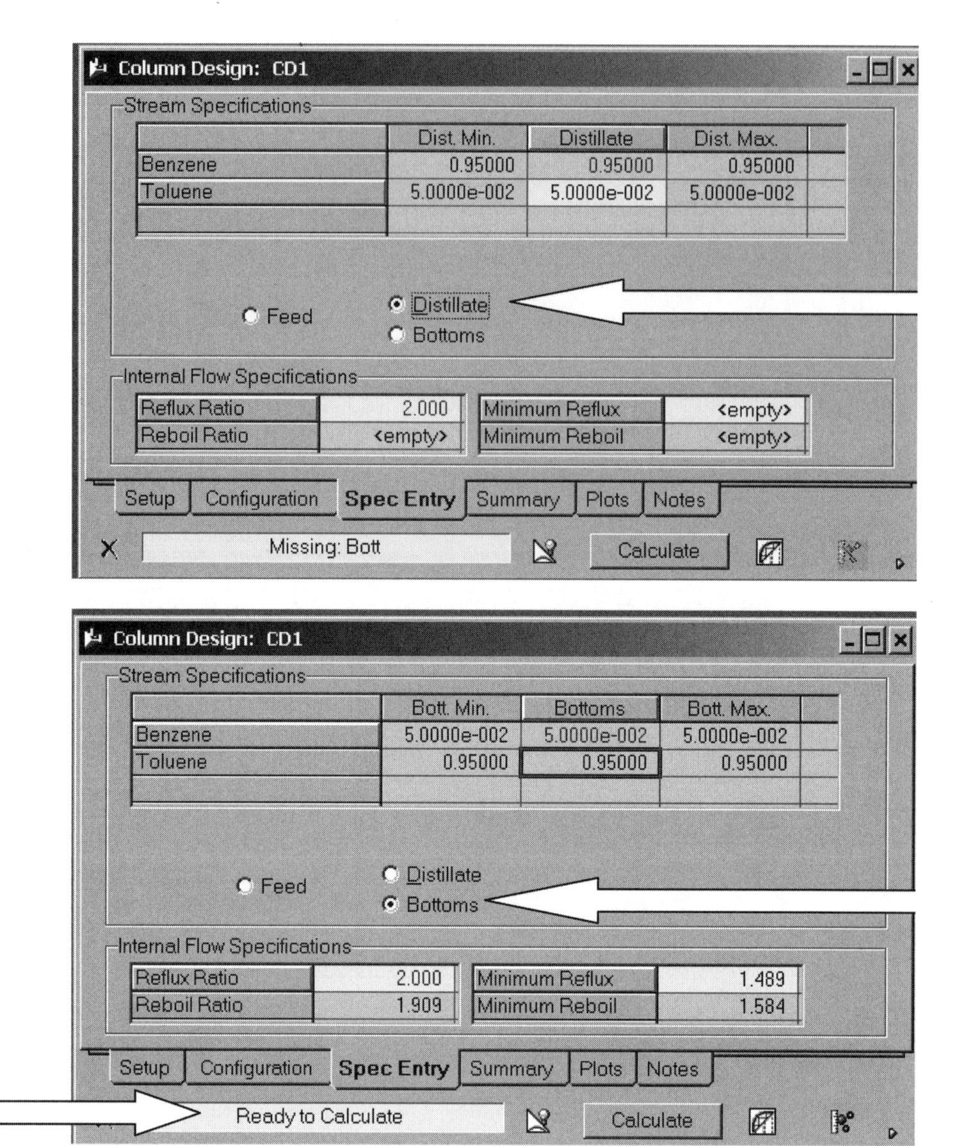

Figure 2.12 Specifying distillate and bottoms.

The DISTIL program can also be used to design columns in ternary systems, as we illustrate later in this chapter.

2.3 APPROXIMATE MULTICOMPONENT METHODS

Many years before the availability of computers for rigorous analysis, several simple approximate methods were developed for analyzing multicomponent systems. These methods are still quite useful for getting quick estimates of the size of a column

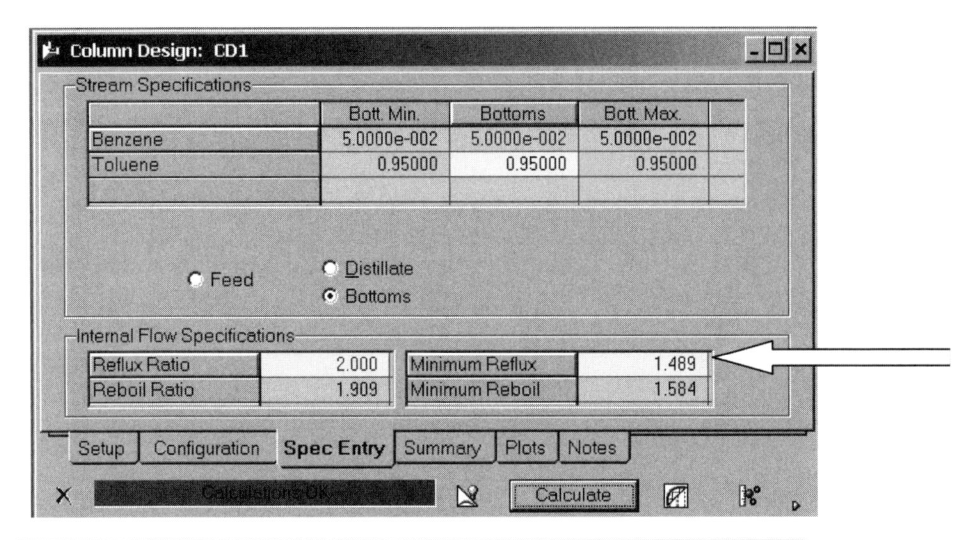

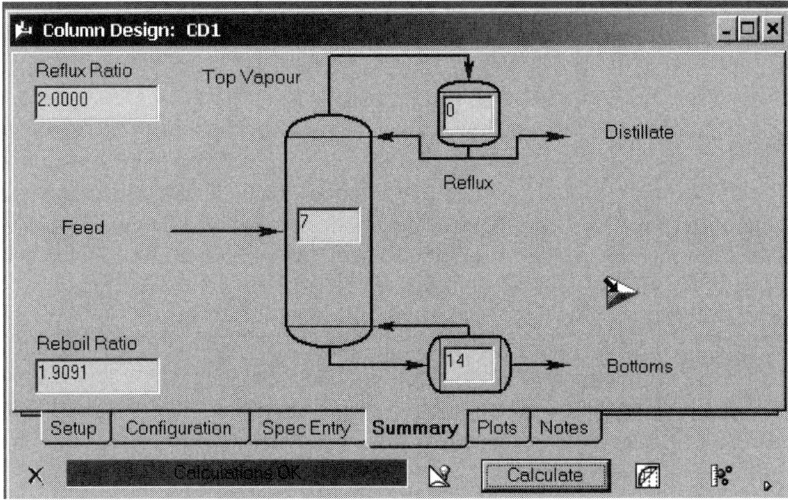

Figure 2.13 Summary of calculated results.

(number of trays) and the energy consumption (reflux ratios and the corresponding vapor boilup and reboiler heat input).

2.3.1 Fenske Equation for Minimum Number of Trays

The minimum number of trays corresponds to total reflux operation (an infinite reflux ratio). The Fenske equation relates the compositions at the two ends of a column to the number of stages in the column under this limiting condition:

$$N_{\min} + 1 = \frac{\log\left[\left(\frac{x_{D,\mathrm{LK}}}{x_{D,\mathrm{HK}}}\right)\left(\frac{x_{B,\mathrm{HK}}}{x_{B,\mathrm{LK}}}\right)\right]}{\log\left(\alpha_{\mathrm{LK,HK}}\right)}$$

where $N_{\min}$ is the minimum number of stages, $x_{D,\mathrm{LK}}$ is the mole fraction of the light-key

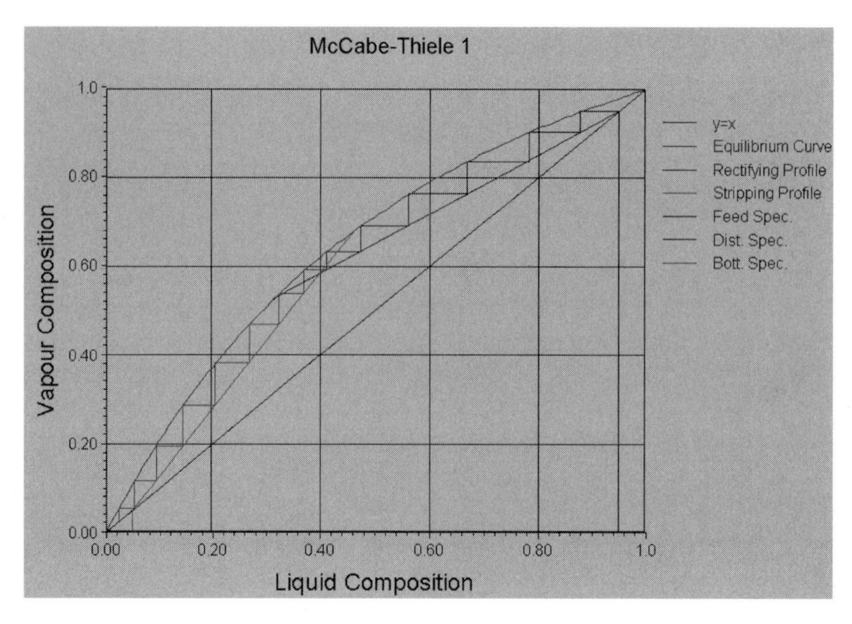

Figure 2.14 *xy* diagram.

component at the top of the column, $x_{D,\mathrm{HK}}$ is the mole fraction of the heavy-key component at the top of the column, $x_{B,\mathrm{HK}}$ is the mole fraction of the heavy-key component at the bottom of the column, $x_{B,\mathrm{LK}}$ is the mole fraction of the light-key component at the bottom of the column, and $\alpha_{\mathrm{LK,HK}}$ is the relative volatility between the LK and HK components.

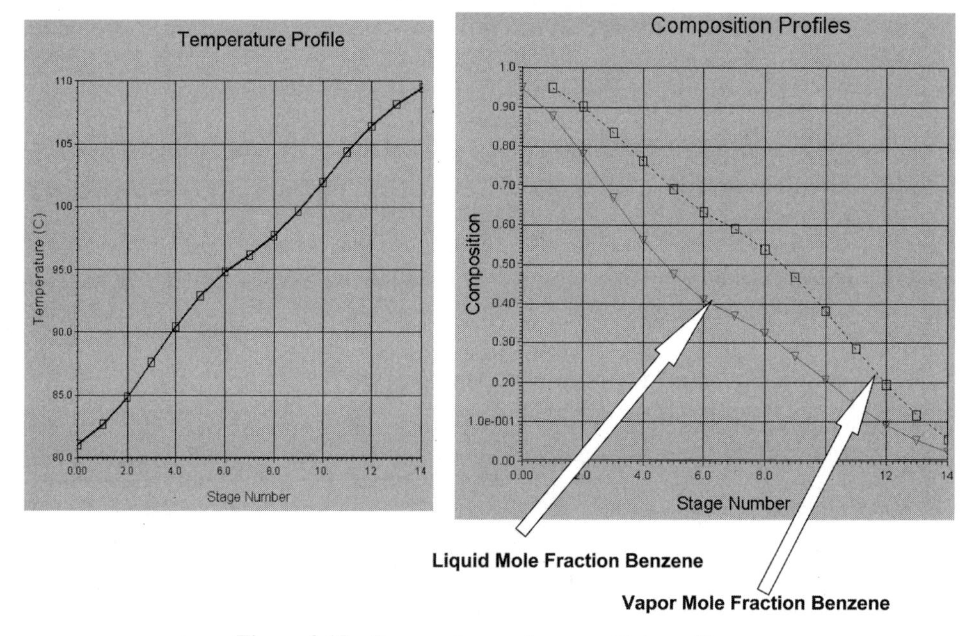

Figure 2.15 Temperature and composition profiles.

This equation is applicable to multicomponent systems, but it assumes a constant relative volatility between the two components considered.

An example of the use of the Fenske equation is given in Chapter 4. Results of this approximate method will be compared with the results found from rigorous simulation.

2.3.2 Underwood Equations for Minimum Reflux Ratio

The Underwood equations can be used to calculate the minimum reflux ratio in a multicomponent system if the relative volatilities of the components are constant. There are two equations:

$$\sum_{j=1}^{NC} \frac{\alpha_j z_j}{\alpha_j - \theta} = 1 - q$$

$$\sum_{j=1}^{NC} \frac{\alpha_j x_{Dj}}{\alpha_j - \theta} = 1 + RR_{\min}$$

The feed composition z_j (mole fractions $j = 1, NC$), the desired distillate composition x_{Dj} ($j = 1, NC$), and the feed thermal condition q are specified. The relative volatilities α_j ($j = 1, NC$) of the multicomponent mixture are known.

The first equation contains one unknown parameter θ. However, expanding the summation of NC terms and multiplying through by all the denominator terms $(\alpha_j - \theta)$ give a polynomial in θ whose order is NC; therefore there are NC roots of this polynomial. One of these roots lies between the two relative volatility values α_{LK} and α_{HK}. This is found using some iterative solution method. It is substituted into the second equation, which can then be solved explicitly for the minimum reflux ratio.

An example of the use of the Underwood equations is given in Chapter 4. The results of this approximate method will be compared with the results found from rigorous simulation.

2.4 ANALYSIS OF TERNARY SYSTEMS USING DISTIL

The graphical column design feature of DISTIL can also be used for ternary systems. As an example, consider a column operating at 4 atm with a feed that is 30 mol% *n*-butane, 30 mol% *n*-pentane, and 40 mol% *n*-hexane. The Unifac fluids package is set up, the components are specified and the pressure is set (see the top of Fig. 2.16). The feed composition is specified on the *Spec Entry* page tap, and a reflux ratio of 2 is specified, as shown in the lower picture in Figure 2.16. Three compositions of the two product streams can be specified (see Fig. 2.17). We set the C4 in the bottoms at 1 mol% and the C5 in the distillate at 1 mol%. The remaining specification is the small amount of C6 in the distillate (0.001 mol%).

The required column configuration (number of stages and feed tray) are given on the *Summary* page tab, shown in Figure 2.18.

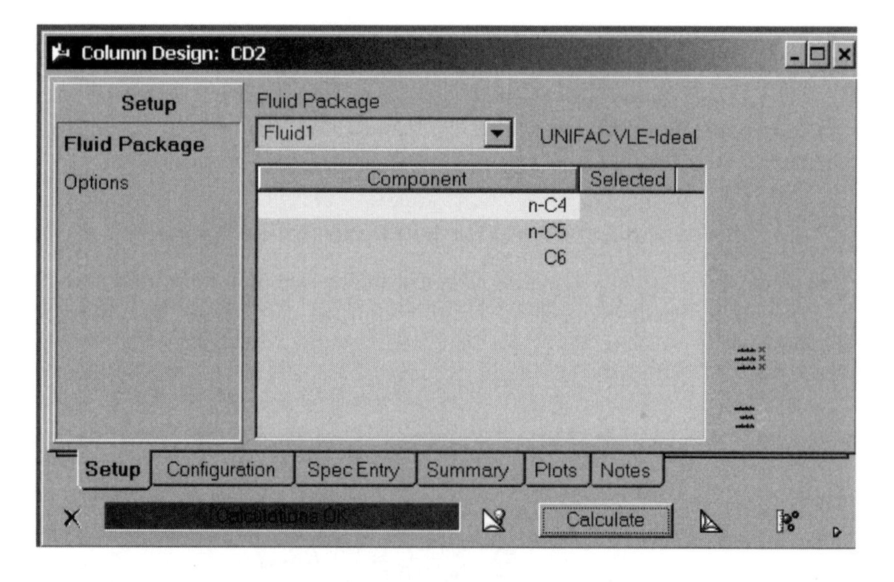

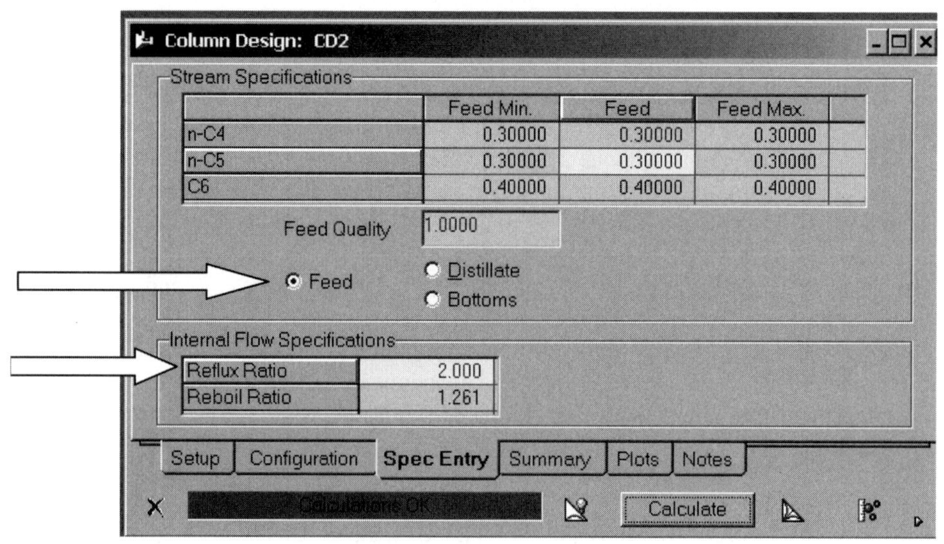

Figure 2.16 Ternary analysis using DISTIL; setting feed and reflux ratio.

Detailed results are plotted on the ternary diagram given in Figure 2.19. The feed, distillate, and bottoms points have been labeled. They lie on a straight line (the heavy dashed line), as required by the overall component balances. The liquid compositions on all the trays in the stripping and rectifying trays are shown. Residue curves are added by clicking any point on the diagram.

Note that a residue curve passes through (or near) the bottoms and distillate point, so the specified product compositions should yield a feasible design. Composition profiles (Fig. 2.20) and temperature profiles can also be plotted.

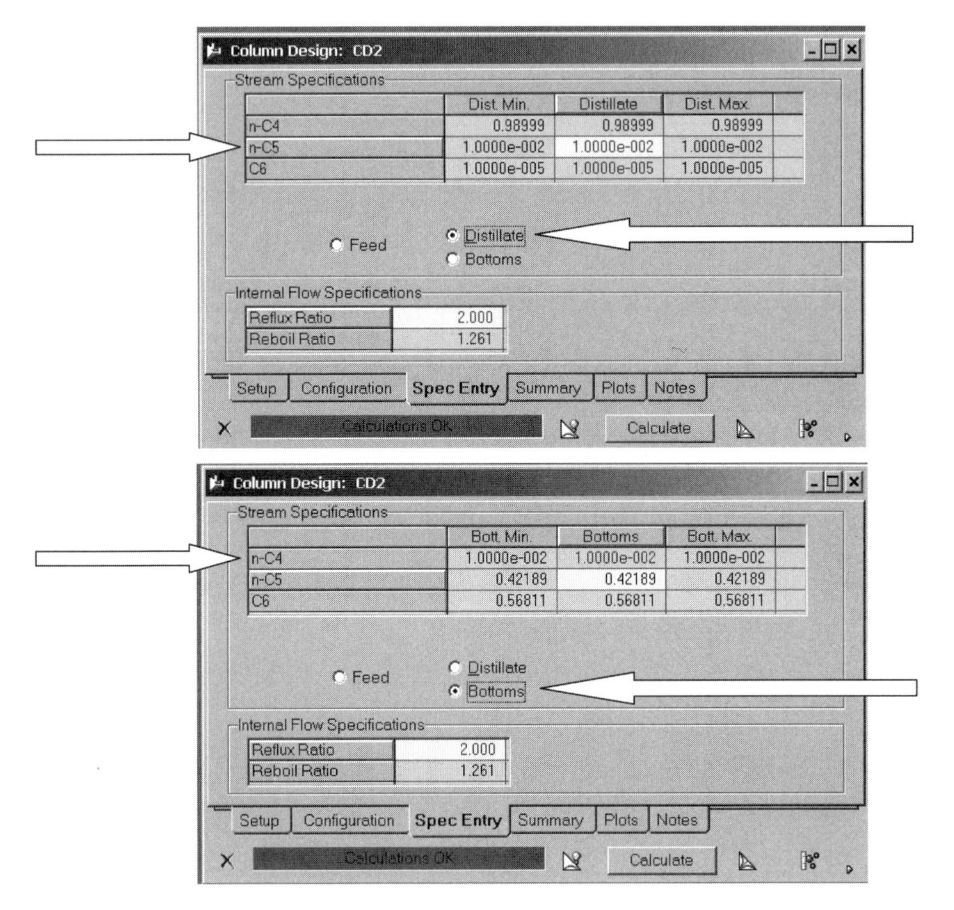

Figure 2.17, Specifying distillate and bottoms composition.

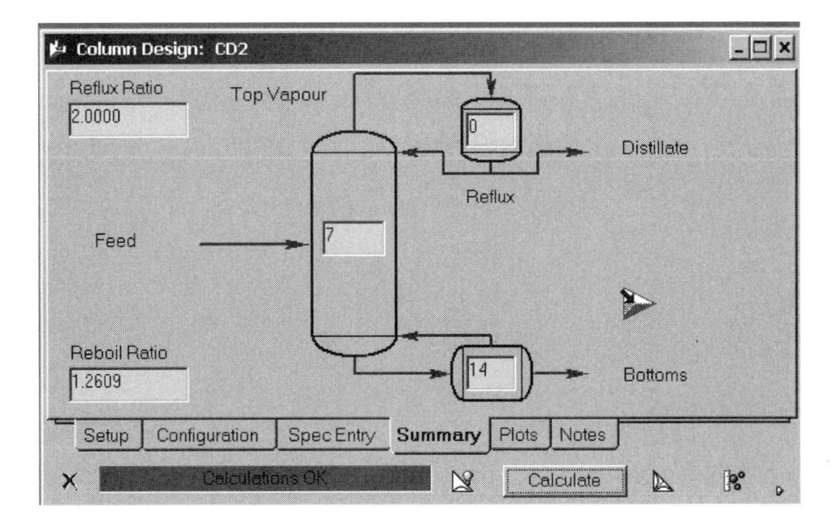

Figure 2.18 Column configuration.

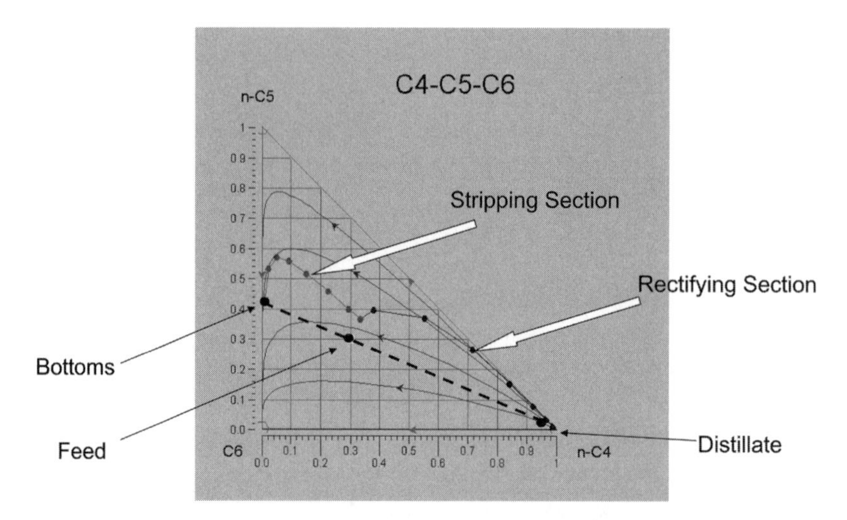

Figure 2.19 Ternary plot.

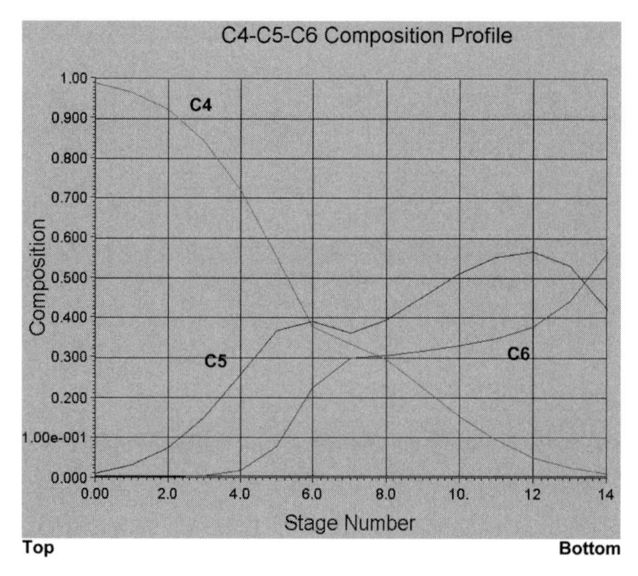

Figure 2.20 Composition profiles.

The value of these graphical ternary diagrams will be demonstrated more thoroughly in Chapter 5.

2.5 CONCLUSION

Several methods for analyzing distillation columns have been presented in this chapter. Graphical methods provide valuable insight into how various parameters affect separations in distillation.

CHAPTER 3

SETTING UP A STEADY-STATE SIMULATION

In this chapter we begin at the beginning. We take a simple binary separation and go through all the details of setting up a simulation of this system in Aspen Plus using the rigorous distillation column simulator *RadFrac*.

All the pieces of a distillation column will be specified (column, control valves, and pumps) so that we can perform a dynamic simulation after the steady-state simulation is completed. If we were interested only in a steady-state simulation, pumps and control valves would not have to be included in the flowsheet. But if we want the capability to do "simultaneous design" (steady state and dynamic), these items must be included to permit a "pressure-driven" dynamic simulation.

3.1 CONFIGURING A NEW SIMULATION

Open up a blank flowsheet by going to *Start* and *Programs* and then clicking sequentially on *Aspen Tech*, *Aspen Engineering Suite*, *Aspen Plus 12.1*, and *Aspen Plus User Interface*. The window shown in Figure 3.1 opens up. Selecting the *Blank Simulation* button and clicking *OK* opens up the blank flowsheet shown in Figure 3.2. The page tabs along the bottom let us choose which unit operations to place on the flowsheet. We are going to need a distillation column, two pumps, and three control valves.

Clicking the *Columns* page tab and clicking the arrow just to the right of *RadFrac* opens the window shown in Figure 3.3, which contains several types of columns, including full columns, strippers (with a reboiler but no condenser), rectifiers (with a condenser but no reboiler), and absorbers (with neither). Click the full-column button on the top row, second from the left, and move the cursor to the blank flowsheet. The cursor becomes a cross. If we click on the flowsheet, a column icon appears, as shown in Figure 3.4.

Distillation Design and Control Using Aspen™ Simulation, By William L. Luyben

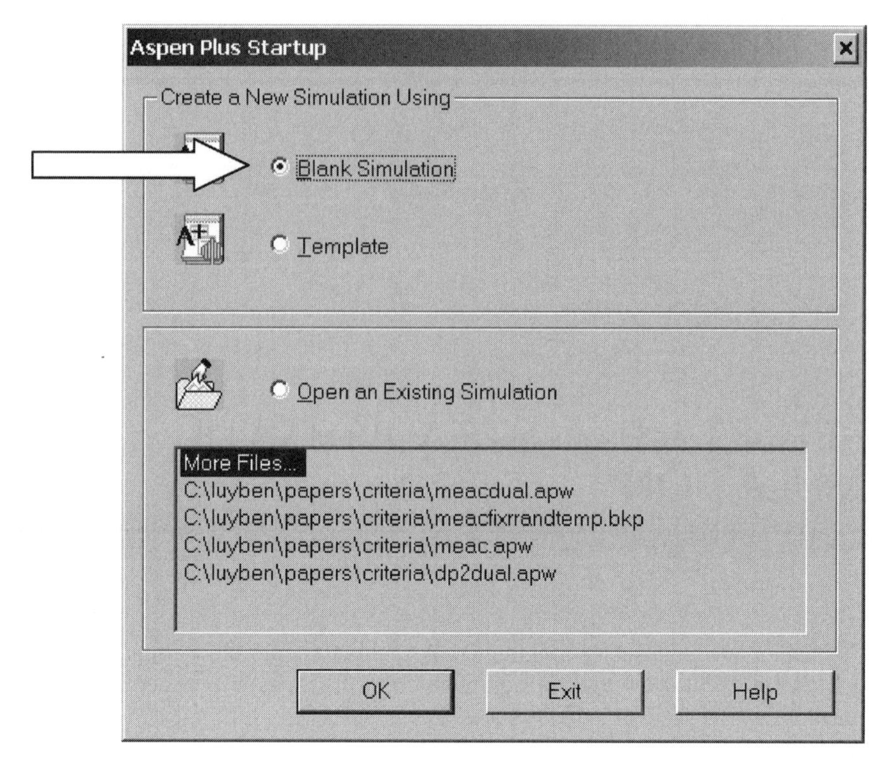

Figure 3.1 Aspen Plus startup.

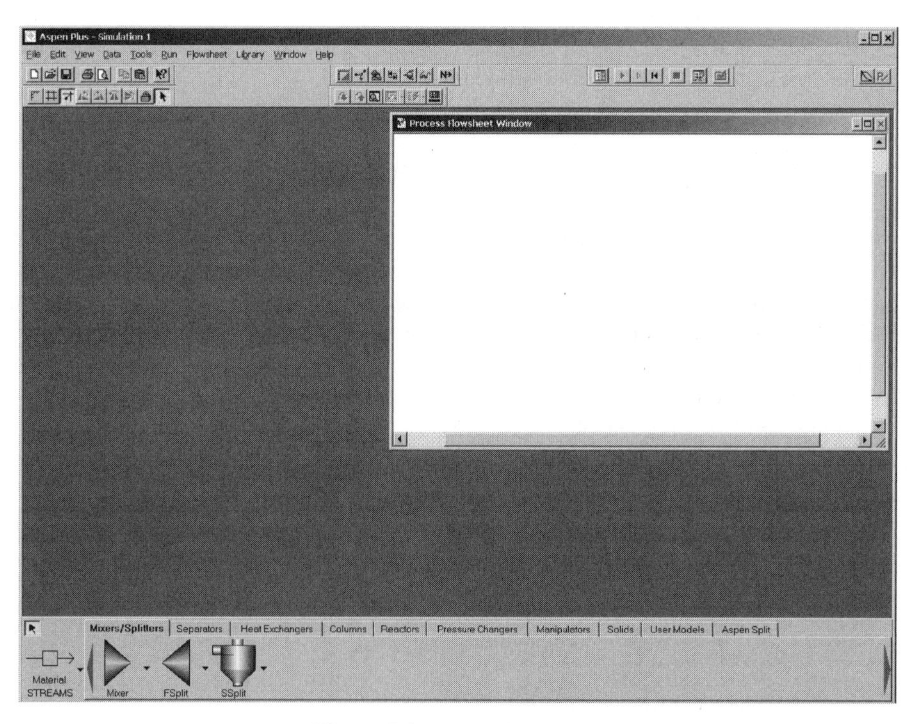

Figure 3.2 Blank flowsheet.

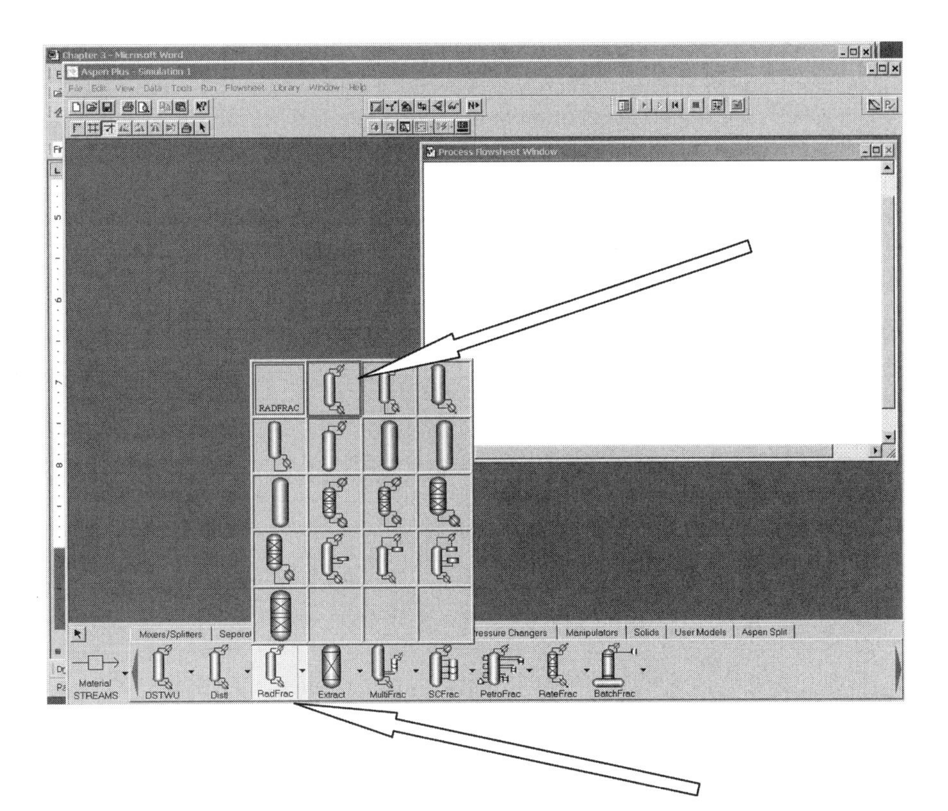

Figure 3.3 Selecting type of column.

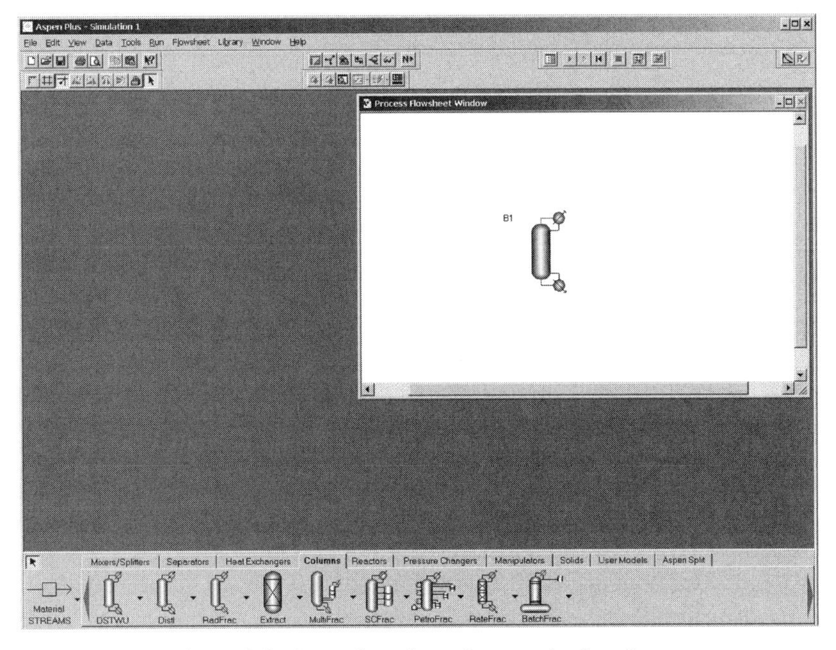

Figure 3.4 Paste the column icon on the flowsheet.

To add the pumps and control valves to the flowsheet, click the page tab on the bottom of the window labeled *Pressure Changers*. This opens the window shown in Figure 3.5a, on which we can select the *Pump* button or the *Valve* button. Clicking the arrow just to the right of the *Valve* button lets us select a valve icon (Fig. 3.5b). Move the cursor to the flowsheet and paste a valve to the left of the column (Fig. 3.5c). The cursor remains a cross on the flowsheet, and we can paste as many additional valves as needed. Two more are inserted to the right of the column in Figure 3.5d. In a similar way, we click the *Pump* button, select an icon, and paste two pumps on the flowsheet as shown in Figure 3.6.

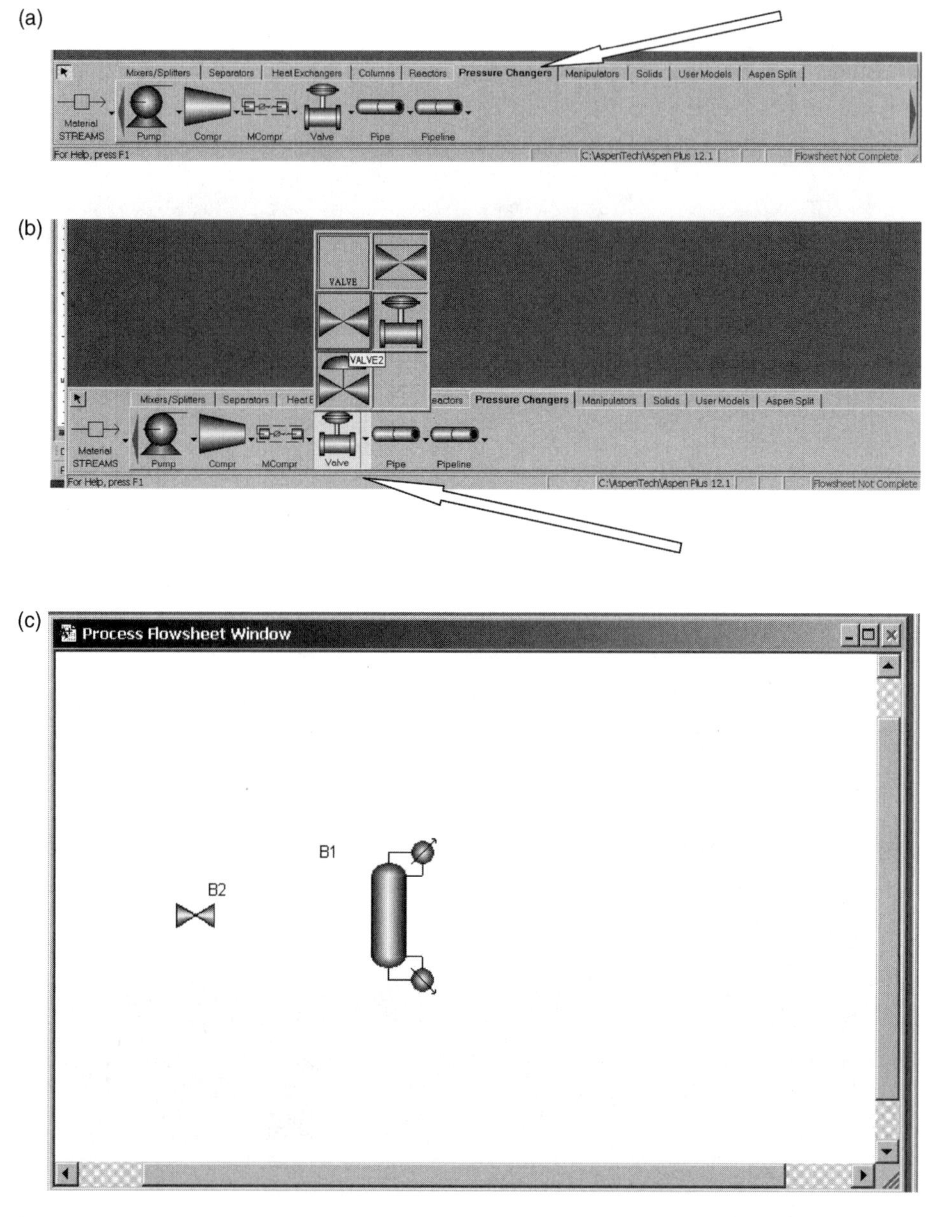

Figure 3.5 (a) Pressure changers; (b) selecting a valve icon; (c) paste valve on flowsheet; (d) flowsheet with three valves added.

(d)

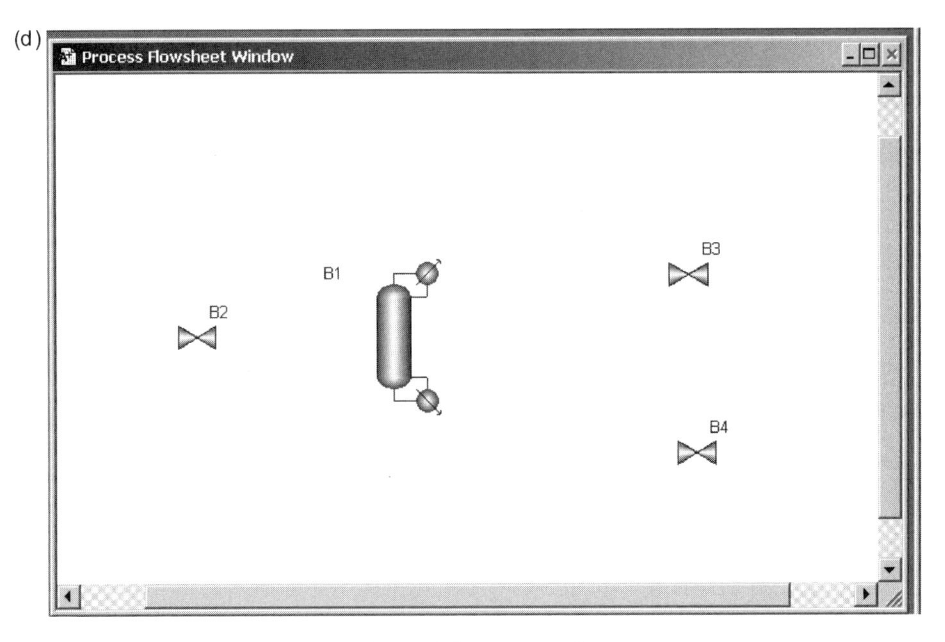

Figure 3.5 *Continued.*

The next job is to add streams to connect all the pieces in the flowsheet. This is achieved by moving the cursor all the way to the left at the bottom of the window and clicking the *Material STREAMS* button as shown in Figure 3.7. Click the *Material* button and move the cursor to the flowsheet. A number of arrows appear (Fig. 3.8) that show all the possible

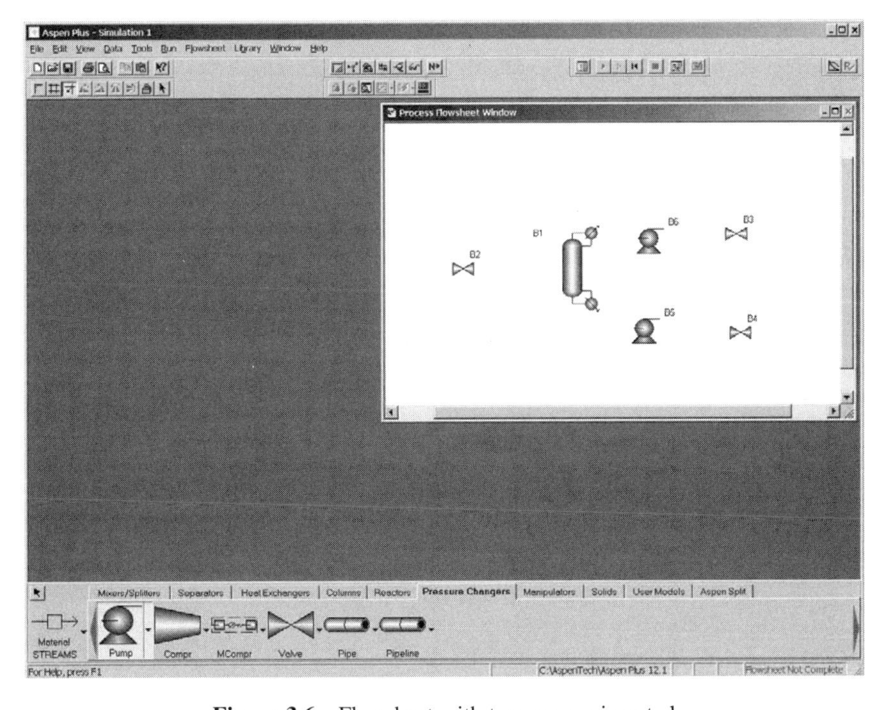

Figure 3.6 Flowsheet with two pumps inserted.

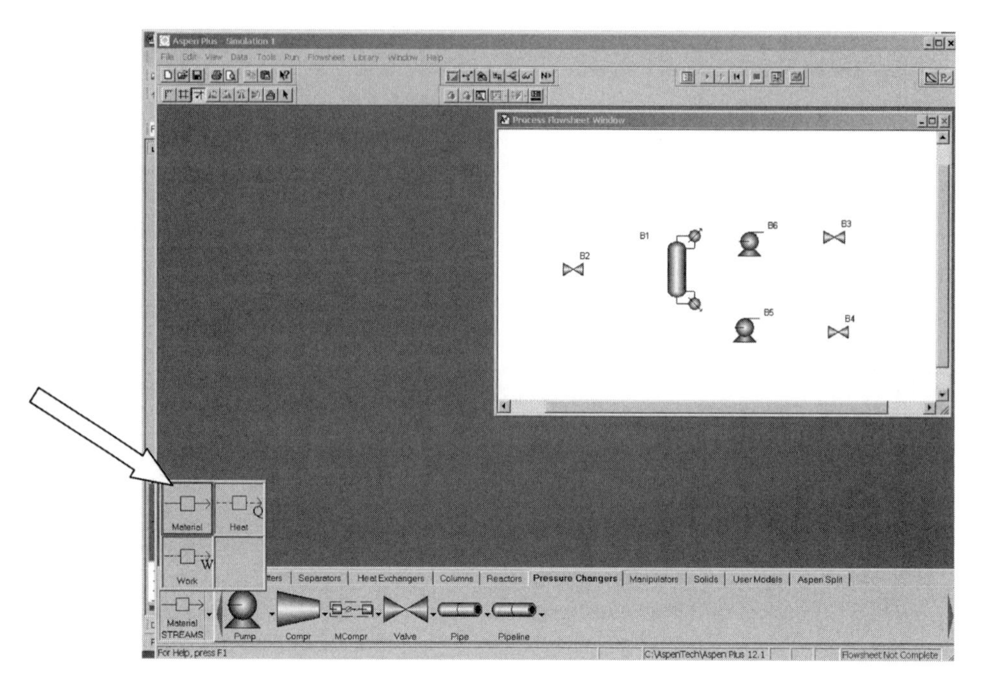

Figure 3.7 Adding streams to flowsheet.

places where a material stream can be located as an input stream or an output stream from each unit. Place the cursor on one arrow and left-click. Then place the cursor on a second arrow where you want to connect the stream and click. Figure 3.9 shows a stream 1 that connects valve block B2 with column block B1. To connect a stream to the inlet of valve

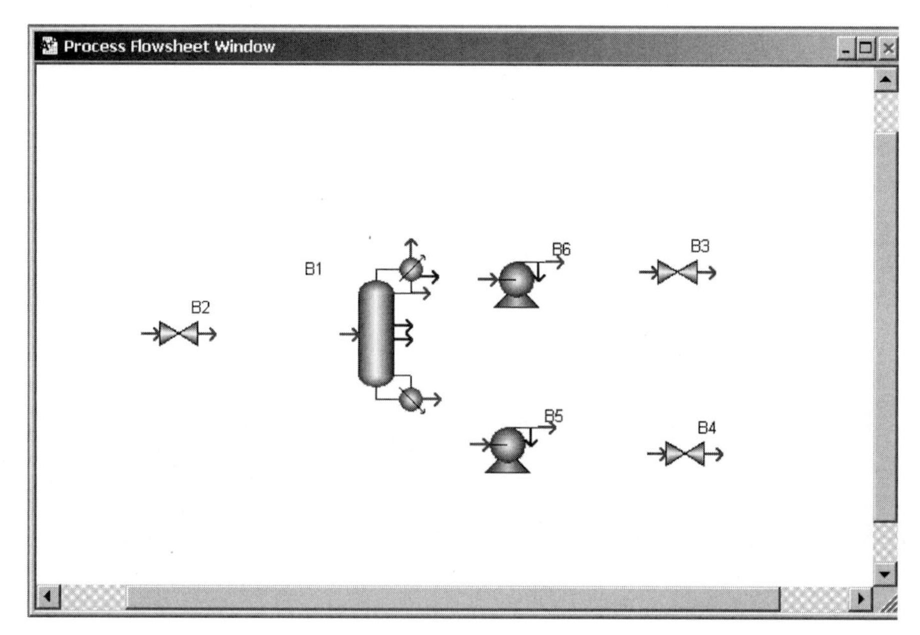

Figure 3.8 Flowsheet with possible connections displayed.

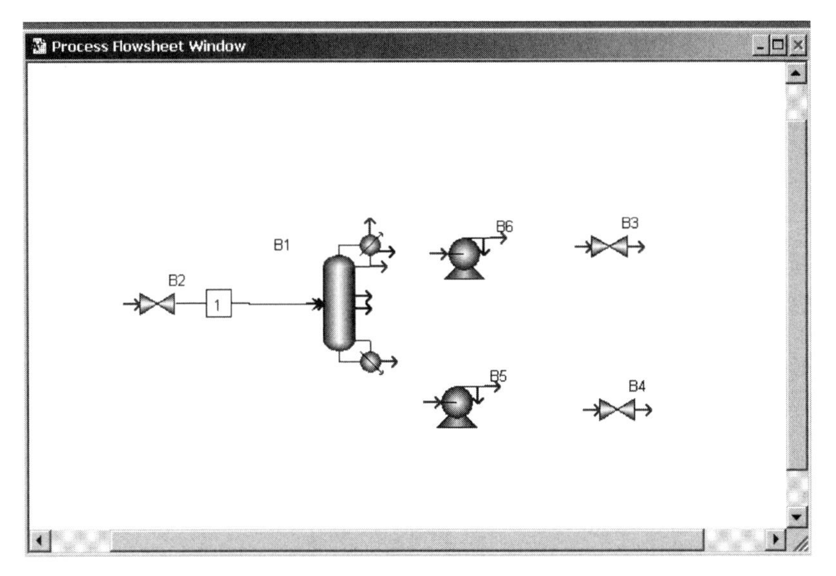

Figure 3.9 Connecting a valve to the column.

block B2, we click the inlet red arrow and then click on the flowsheet to the left of the valve (Fig. 3.10). This inserts stream 2.

All the other material streams are connected in a manner similar to that shown in Figure 3.11. Note that the stream 3 from the top of the column has been connected at the arrow that is below the condenser symbol. This gives a liquid distillate product. If the stream had been connected to the arrow coming out the top of the condenser, the distillate would be a vapor and the condenser would be a partial condenser. When you are

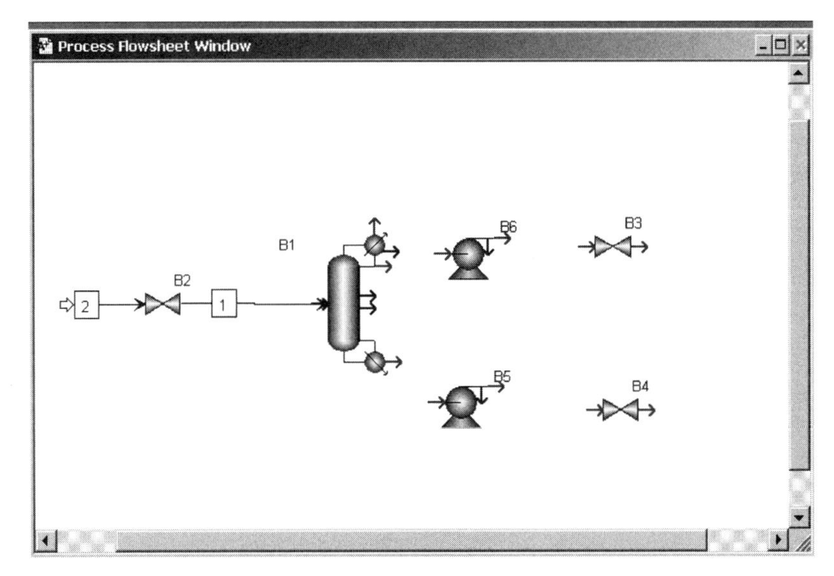

Figure 3.10 Connecting stream to inlet valve.

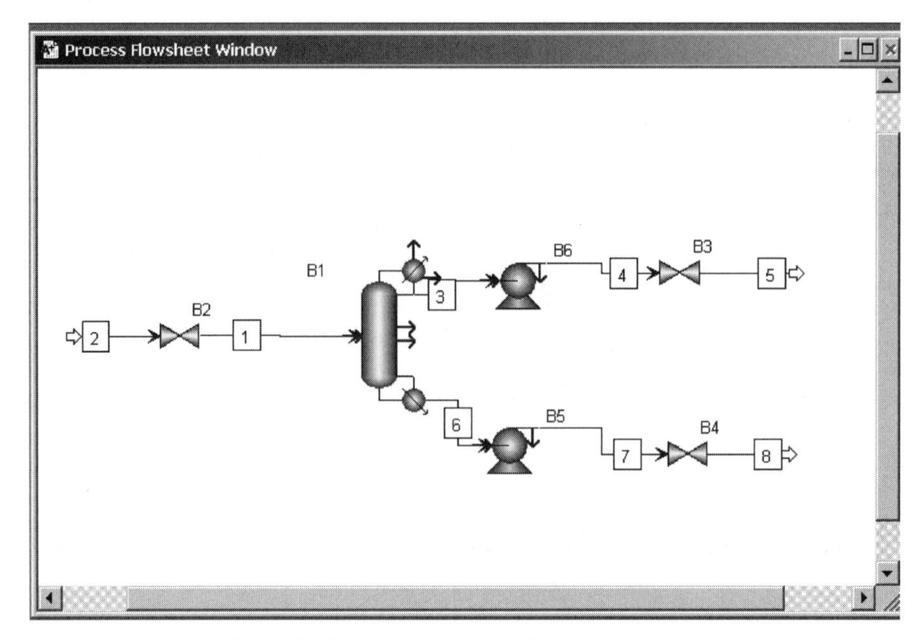

Figure 3.11 Flowsheet with all streams attached.

finished adding streams and units, click the arrow that is immediately above the *Material STREAMS* block at the bottom left of the window (see Fig. 3.6). This is the *Cancel Insert Mode* button.

At this point, the flowsheet has been configured. It is a good idea to rename the various pieces of equipment and the streams to help keep track of the identity of each of these components. To rename a block, left-click its icon and right-click to get a dropdown menu. Selecting *Rename* gives the window shown in Figure 3.12. Type in the desired new name. The same procedure is used to rename streams. Figure 3.13 gives the flowsheet with all blocks and streams renamed to correspond to conventional distillation terminology. The feedstream is F1, the distillate stream is D1, and the bottoms stream is B1. Note that the pumps and valves have been renamed for easy association with this

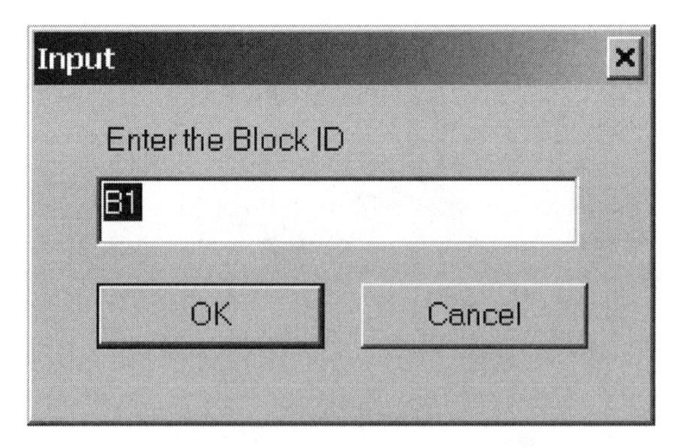

Figure 3.12 Renaming a block.

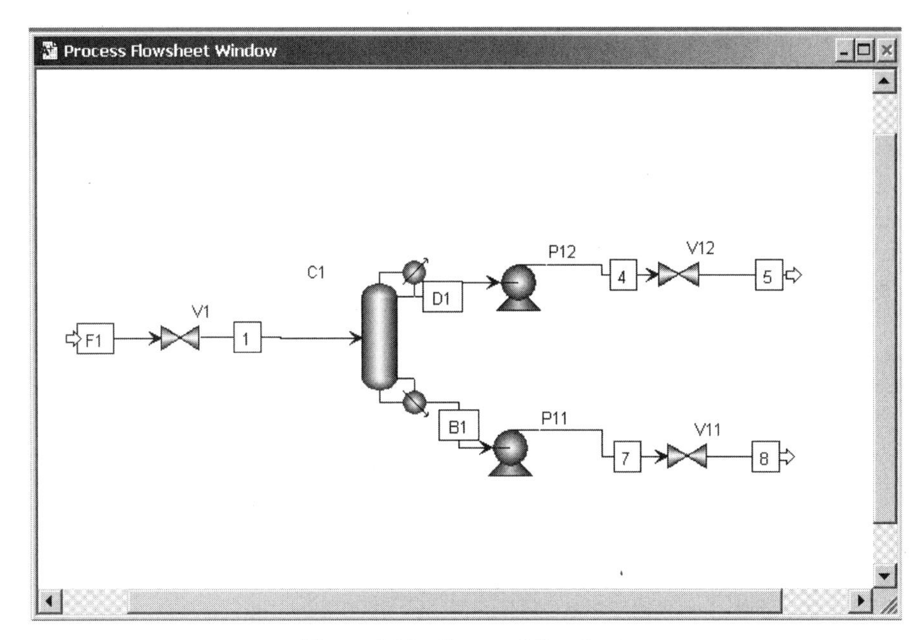

Figure 3.13 Renamed flowsheet.

column C1. Some logical scheme for renaming units and streams is essential in a large plantwide simulation with many units and many streams.

3.2 SPECIFYING CHEMICAL COMPONENTS AND PHYSICAL PROPERTIES

The structure of the flowsheet is now completely specified. Next we must define the chemical components involved in the separation and specify what physical property package is to be used. Chapter 1 discussed many aspects of choosing a physical property relationship to ensure that an accurate representation of reality is used.

A simple binary separation of propane from isobutane is used in the numerical example considered in this chapter. The VLE relationships for most hydrocarbon systems are well handled by the Chao–Seader correlation, so we select that package.

To start specifying chemical components, go to the toolbar at the top of the window and click the fourth item from the left *Data*. Select the top item *Setup*. This opens the *Data Browser* window shown in Figure 3.14. This is the window that is used to look at all aspects of the simulation. It is used to define components, set physical properties, specify the parameters of the equipment (e.g., the number of stages in the column and the pressure), and specify properties of various streams (e.g., flowrate, composition, temperature and pressure).

A couple of preliminary items should be done first. In the middle of the window there are two boxes in which we can specify the units to be used in the simulation. Figure 3.15 shows the standard three alternatives: *ENG* (English engineering), *MET* (metric) and *SI* (Système International). We will use SI units in most of the examples in this book because of their increasing importance in our global economy. However, we will make one departure from regular SI units. In the SI system, pressures are expressed in pascals

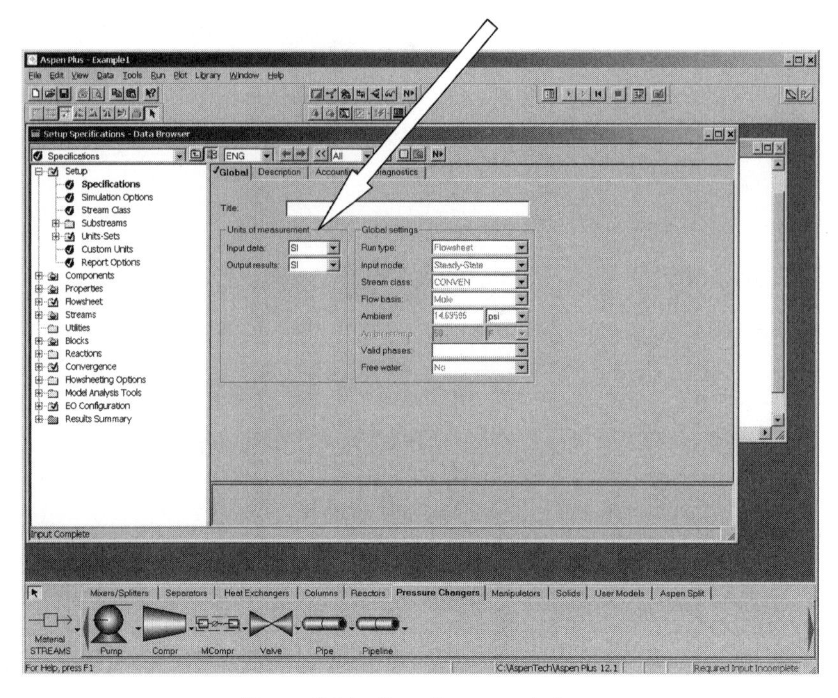

Figure 3.14 *Data Browser* window.

(N/m^2), which are quite inconvenient for most chemical processes because typical pressures are very large numbers in pascals (1 atm = 101,325 Pa). Therefore we will use pressures in atmospheres in most of the examples. However, make sure that you select the correct units when you enter data.

The second preliminary item is to indicate what properties we want to see for all the streams. The defaults do not include compositions in mole fractions, which are very

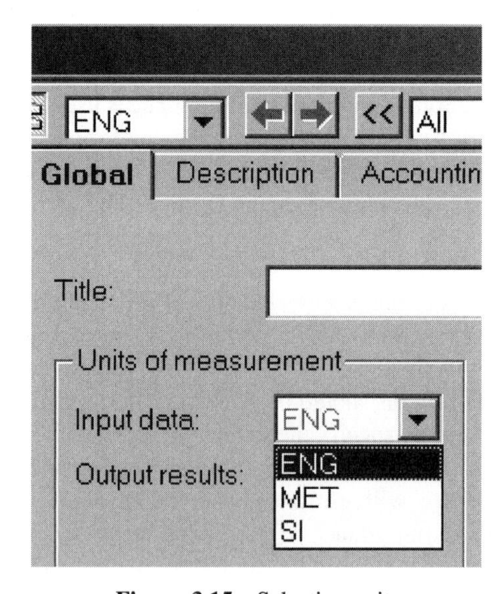

Figure 3.15 Selecting units.

useful in distillation calculations. To include mole fractions in the list of stream properties, click on the last item on the *Setup* list at the left of the window, which is labeled *Report Options*. This opens the window shown in Figure 3.16a. Click the page tab labeled *Stream*. This opens the window shown in Figure 3.16b. Select *Mole* under the *Fraction basis* column in the middle of the window.

With these bookkeeping issues out of the way, we can select the chemical components by clicking *Components* on the left of the *Data Browser* window. This opens the window shown in Figure 3.17a. Clicking the *Find* button near the bottom of the window opens another window shown in Figure 3.17b. Type in *propane* and click *Find now* as shown in Figure 3.18. A list of components is opened at the bottom of the window. Click

(a)

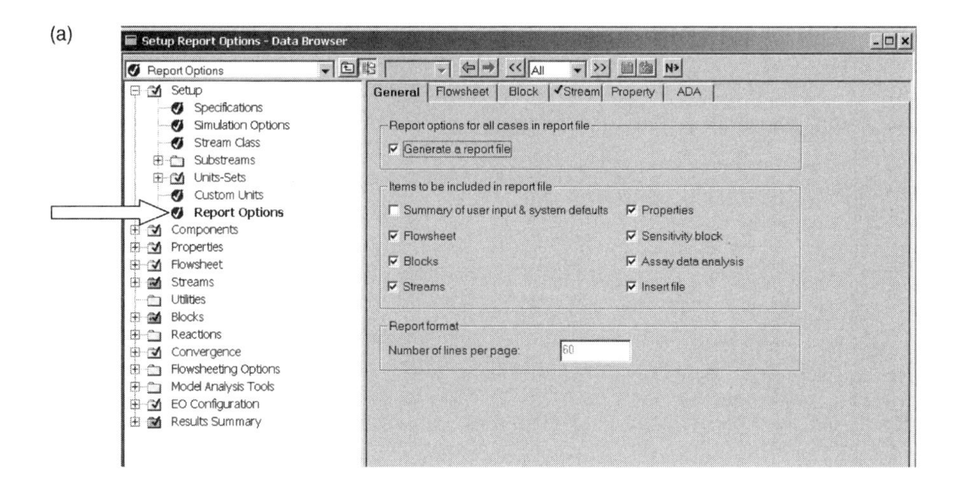

(b)

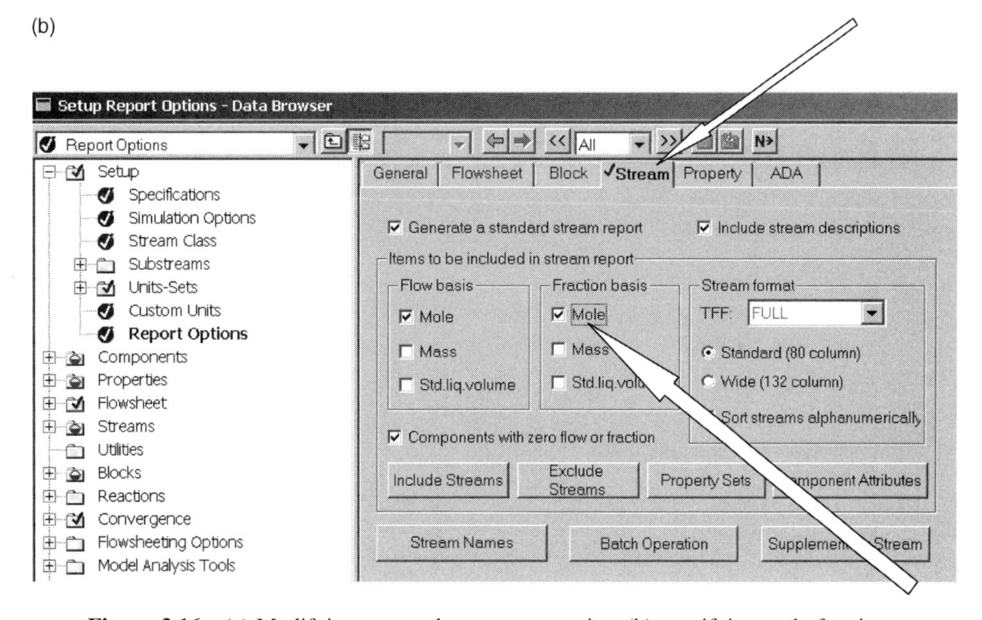

Figure 3.16 (a) Modifying reported stream properties; (b) specifying mole fractions.

PROPANE and click the *Add* button at the bottom of the window. Repeat for *ISOBUTANE* and click *Close*.

The two components have now been selected (see Fig. 3.19). It is often desirable to change the name of a component. For example, suppose that we want to use "C3" for propane and "IC4" for isobutane. This can be accomplished by highlighting the name listed under *Component ID* and typing the desired name. Click anyplace on the window, and the message shown in Figure 3.20 will appear. Click *Rename*. Repeat for isobutane.

Now we are ready to select a physical property package. Click *Properties* and *Specifications* on the left side of the window. Figure 3.21 shows the window that opens. Under

(a)

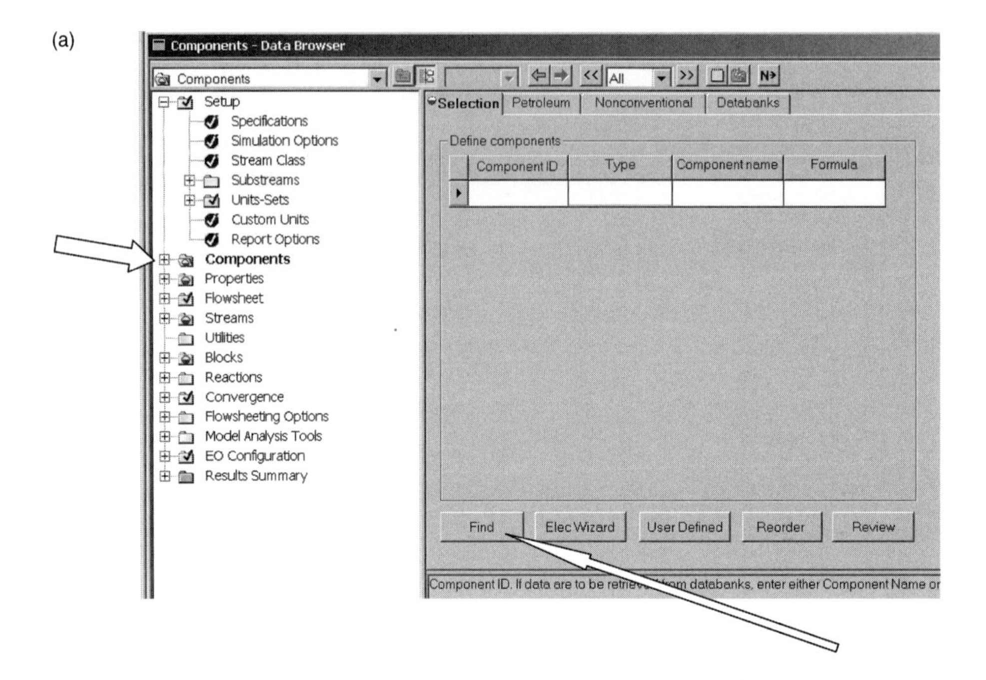

(b)

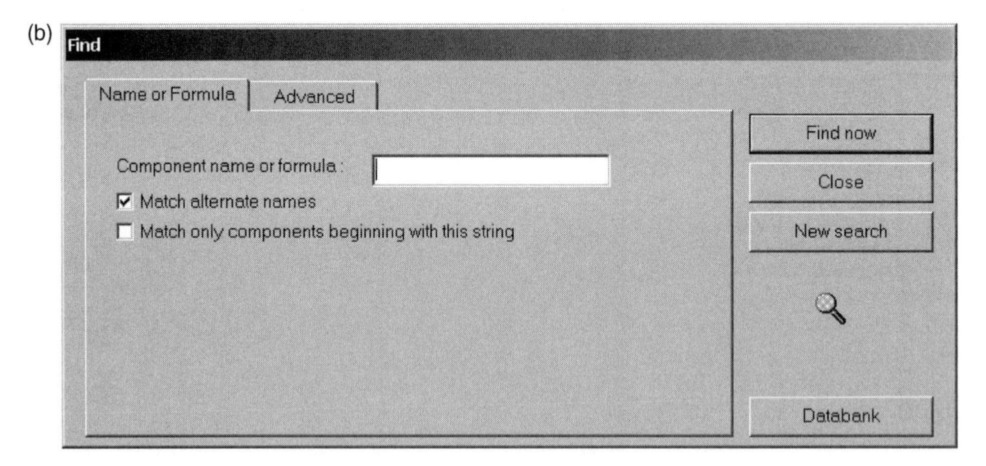

Figure 3.17 (a) Specifying chemical components; (b) finding components.

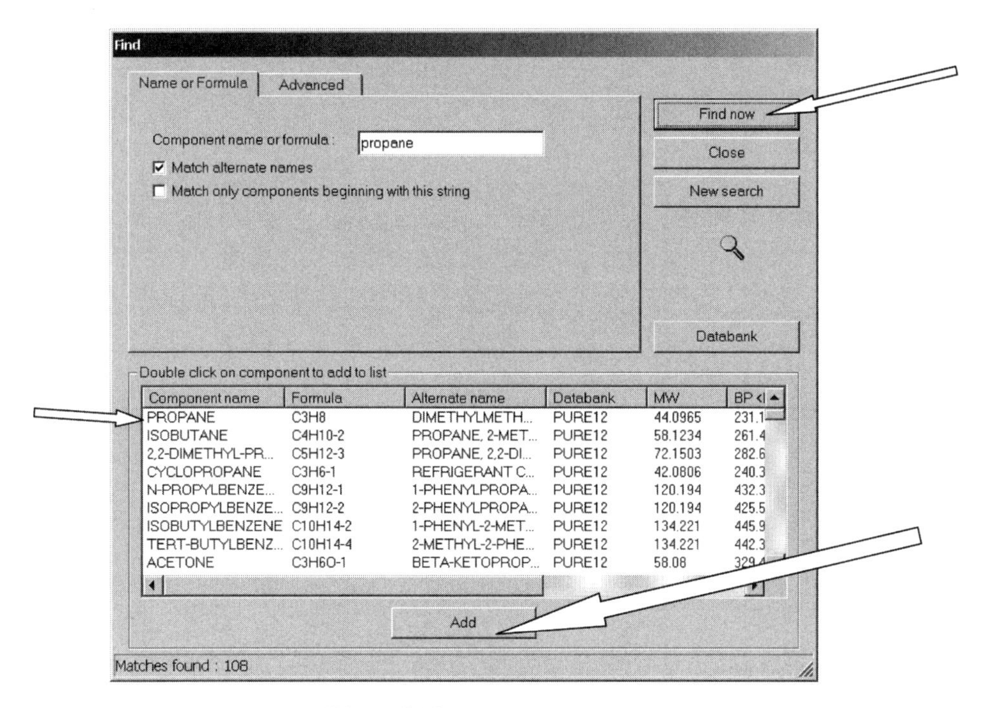

Figure 3.18 Type in components.

Property methods & models click the arrow on the right of *Base method*. A long list of alternatives appears as shown in Figure 3.22. Scroll down and select *CHAO–SEAD*.

It should be noted that different physical property packages can be used in different unit operations in a flowsheet. For example, we may be simulating a distillation column and a

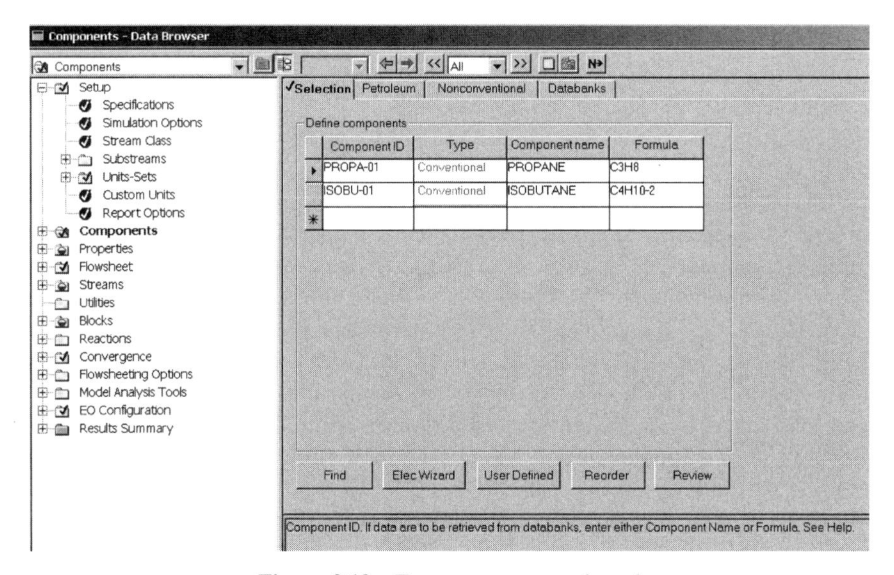

Figure 3.19 Two components selected.

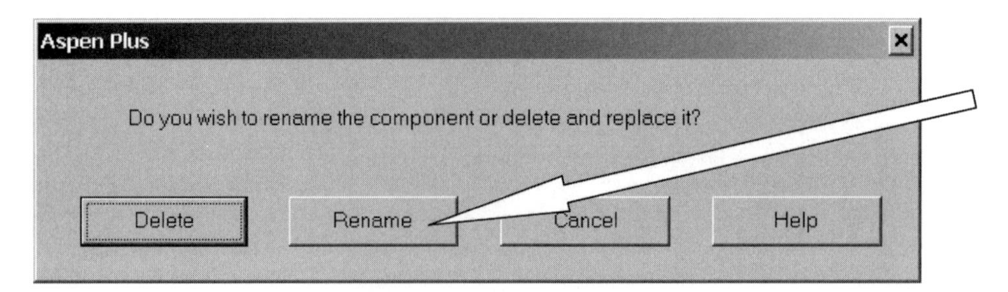

Figure 3.20 Message to rename.

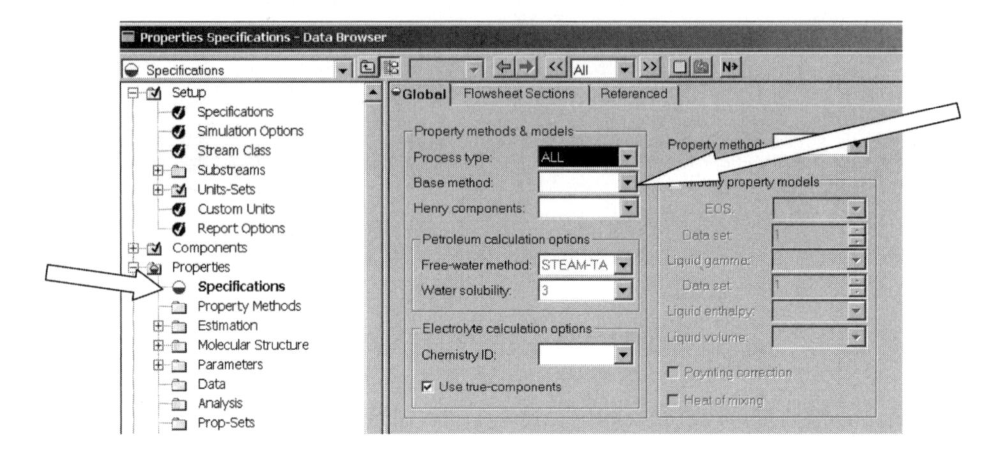

Figure 3.21 Specifying physical properties.

decanter. The best VLE package should be used in the column, while the best LLE package should be used in the decanter.

3.3 SPECIFYING STREAM PROPERTIES

The input streams to the process must be specified. In this example, there is only one input stream, the feedstream F1. The flowrate, composition, temperature, and pressure of this stream must be specified. Clicking *Streams* and *F1* and then *Input* opens the window shown in Figure 3.23.

In distillation calculations, molar flowrates, and compositions are usually employed. Let us assume that the feed flowrate is 1 kmol/s and the feed temperature is 322 K (120°F). These are entered in the middle of the window. The feed composition is 40 mol% propane and 60 mol% isobutane. The composition can be entered in terms of mole or mass fractions, or it can be entered in terms of molar or mass flowrates. In our example, we use the dropdown arrow to change to *Mole-Frac* and enter the appropriate values (see Fig. 3.24).

Specification of the feed pressure takes a little thought. We will discuss the selection of column pressure in more detail later in this chapter. We know that the distillate product is propane. We will want to use cooling water in the condenser since it is an inexpensive

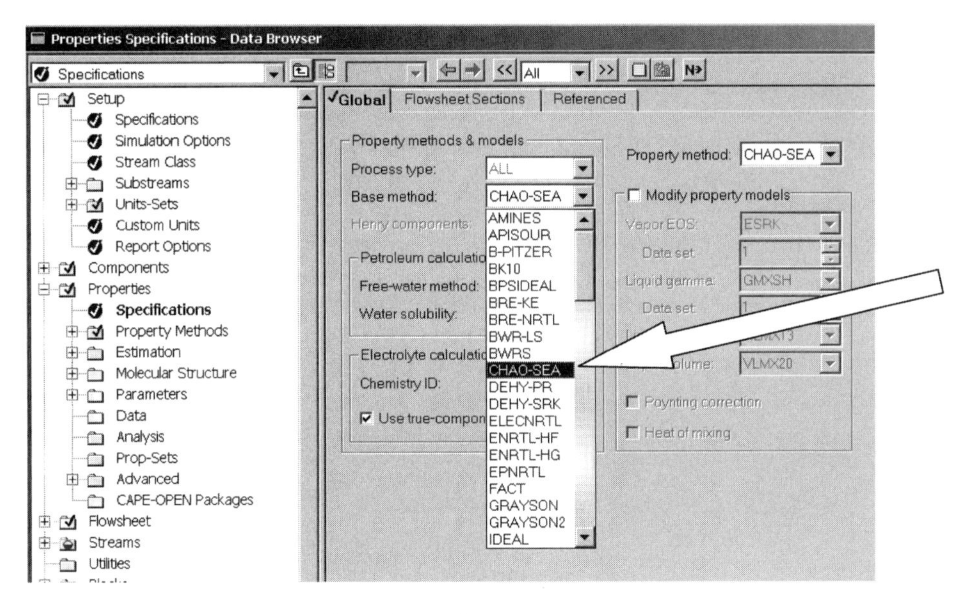

Figure 3.22 Specify base method.

heat sink compared to refrigeration. Cooling water is typically available at about 305 K. A reasonable temperature difference for heat transfer in the condenser is 20 K. Therefore the reflux drum temperature will be ~325 K. The vapor pressure of propane at 325 K is ~14 atm (206 psia). Therefore, the column will have a pressure at the feed tray of something a little higher than 14 atm.

In addition, we need to allow for some pressure drop over the control valve on the feedstream. The subject of selecting control valve pressure drops will be discussed later

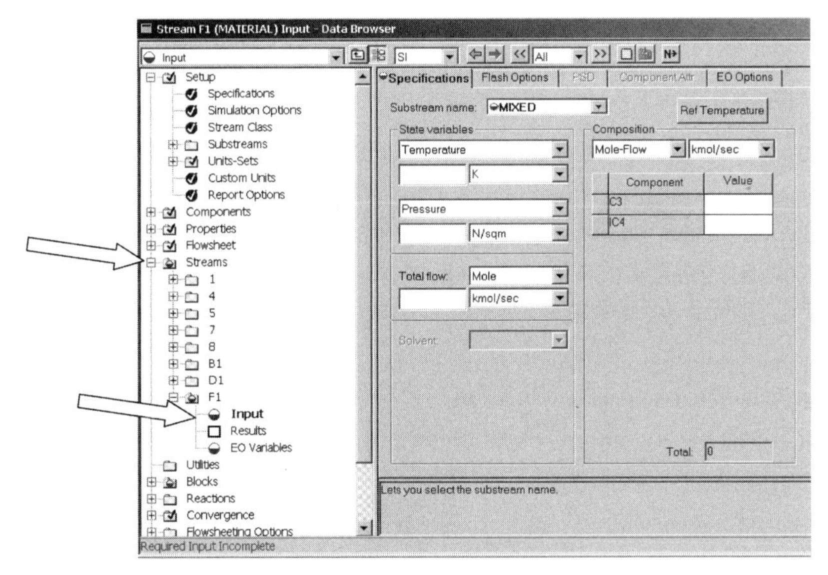

Figure 3.23 Input stream data.

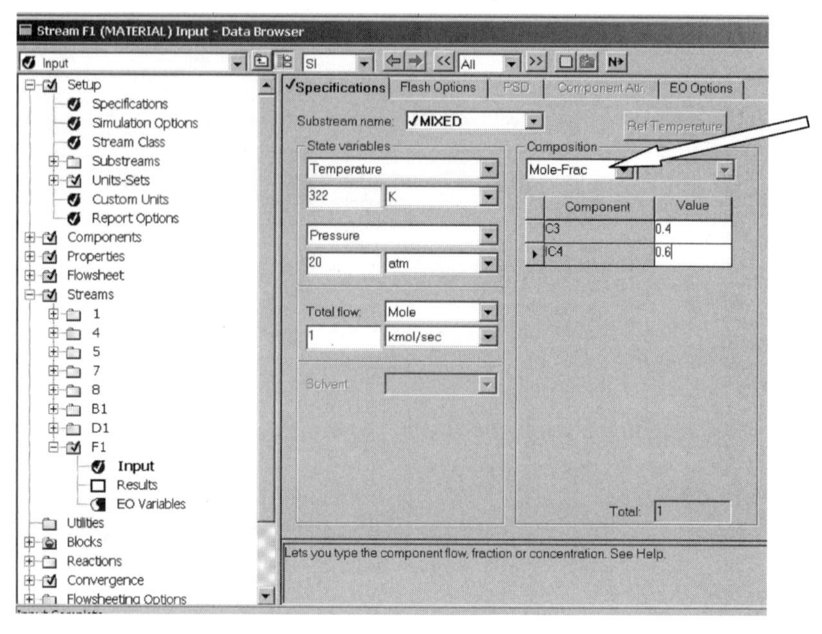

Figure 3.24 Feed input conditions.

in this book. For the moment, let us assume that we need a 5 atm pressure drop. Thus the feed pressure should be set at ~20 atm. Be careful to specify "atm" instead of the default "N/m^2" for the pressure units. The final conditions of the feed are shown in Figure 3.24.

3.4 SPECIFYING EQUIPMENT PARAMETERS

The parameters for all the equipment must be specified. Clicking on *Blocks* on the left side of the *Data Browser* window produces a list of all the blocks that must be handled. Any block with a red color is not completely specified. The column is the most complex and has the most parameters to fix. So we will start with the column C1.

3.4.1 Column C1

Clicking on the block labeled *C1* opens a window with a long list of items. The top subitem is labeled *Setup*. Clicking it opens the window shown in Figure 3.25. There are several page tabs. The first is *Configuration*, on which the number of total stages, the type of condenser, the type of reboiler, the numerical convergence method, and two other variables are specified. We consider each of these below.

1. *Number of Stages.* The rigorous way to select the number of stages is to perform an economic optimization. We discuss this in detail in Chapter 4. For the moment let us select a column with 32 stages. Aspen uses the tray numbering convention of defining the reflux drum as stage 1. The top tray is stage 2 and so forth on down the column. The base of the column in this example is stage 32. Therefore this column has 30 trays.

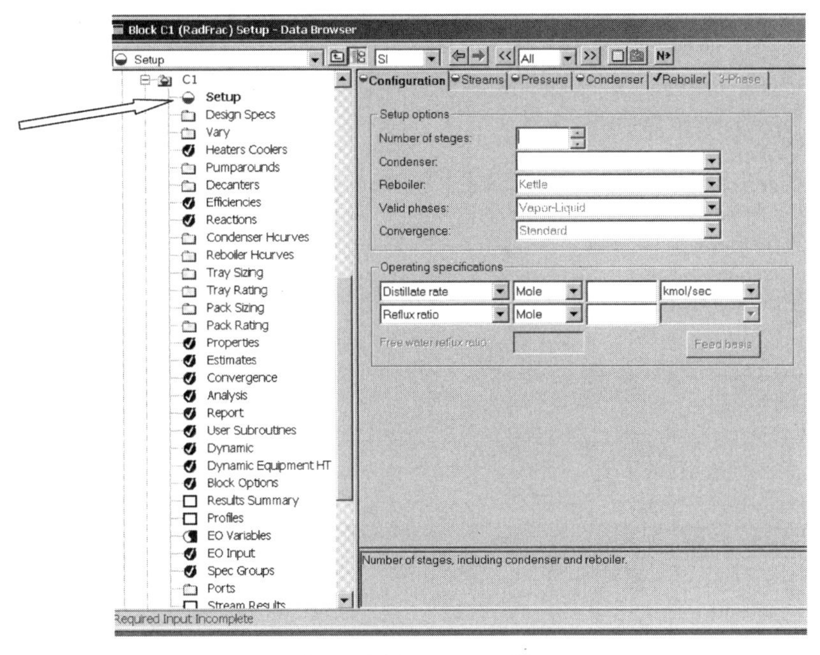

Figure 3.25 Column C1.

2. *Condenser.* Use the dropdown menu to select *Total.* If the distillate were removed as a vapor, *Partial-Vapor* should be selected.
3. *Reboiler.* Both the kettle and the thermosyphon reboilers are partial reboilers (the vapor from the reboiler is in equilibrium with the liquid bottoms product withdrawn), so it does not matter which you select.
4. *Convergence.* The *standard* method works well in hydrocarbon systems. Alternative methods must be use in highly nonideal systems. Examples in latter chapters will illustrate this.
5. *Operating Specifications.* As discussed in Chapter 2, a distillation column has 2 degrees of freedom once the feed, pressure, number of trays, and feed tray location have been fixed. There are several alternative ways to select these 2 degrees of freedom, as shown in Figure 3.26. At this stage in our simulation, the usual approach is to fix the distillate flowrate and the reflux ratio. Later, once we obtain a converged solution, we will change the specified variable so that the product specifications are met. For now let us fix the distillate flowrate at 0.4 kmol/s since we know that this is the molar flowrate of propane in the feed. In addition, let us select a reflux ratio of 2 since the propane/isobutane separation is neither very difficult nor very easy. Figure 3.27 shows the *Configuration* page with all these data inserted. Note that the red dot on the C1 block becomes a blue checkmark when all the required input data have been provided.

Now click the *Streams* page tab. A window opens, on which the location of the feed tray must be given. For the moment we set this in the middle of the column on stage 16 (see Fig. 3.28). Later we will return to this question and determine the "optimum" feed tray location by finding the tray that minimizes reboiler heat input.

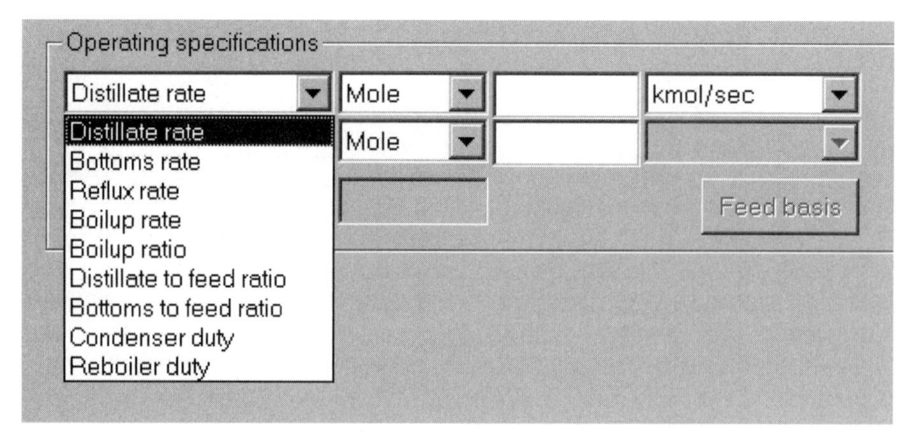

Figure 3.26 Alternative choices of operating specifications.

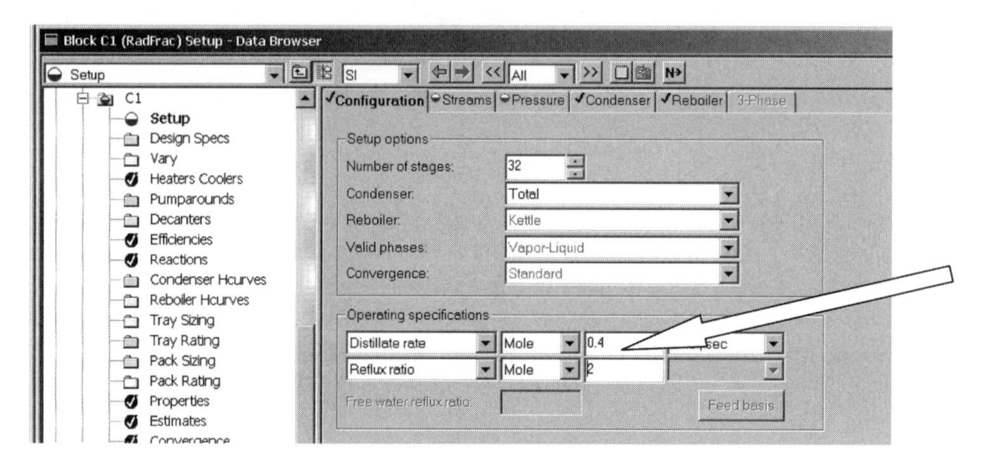

Figure 3.27 Configuration page with all data.

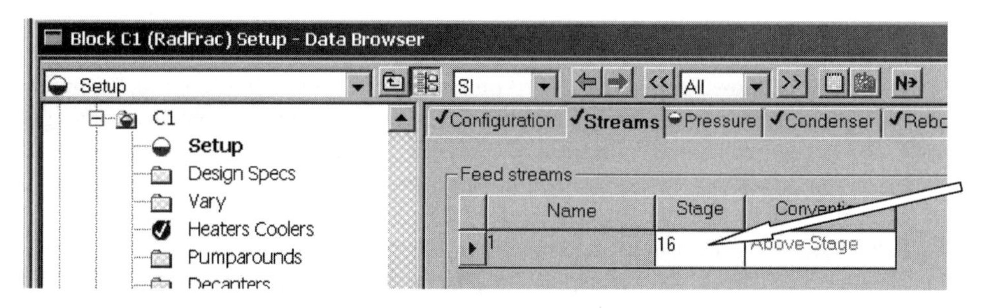

Figure 3.28 Specifying the feed stage.

The last page tab is *Pressure*. Clicking it opens the window shown in Figure 3.29, in which we specify the pressure in the reflux drum (condenser) and the pressure drop through each of the trays in the column. As discussed above, we set the reflux drum pressure at 14 atm (be careful to change from N/m^2). A reasonable tray pressure drop is about 0.0068 atm per tray (0.1 psi per tray).

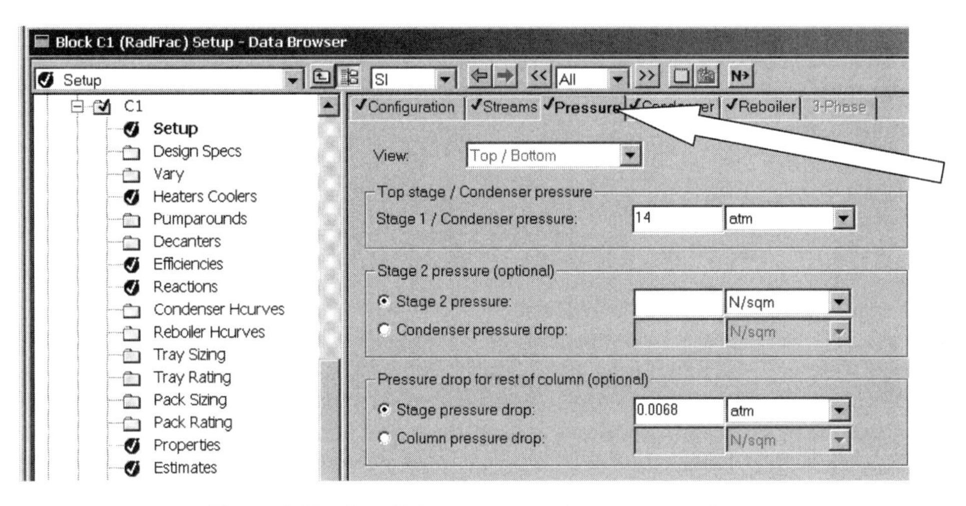

Figure 3.29 Specifying pressure and tray pressure drop.

All the items in the C1 block are now blue, so the column is completely specified. Next, the design parameters of all the valves and pumps must be specified.

3.4.2 Valves and Pumps

We assume that both pumps generate a pressure difference of 6 atm between suction and discharge. Click *Setup* under pump P11 and enter these data. Figure 3.30 shows the input

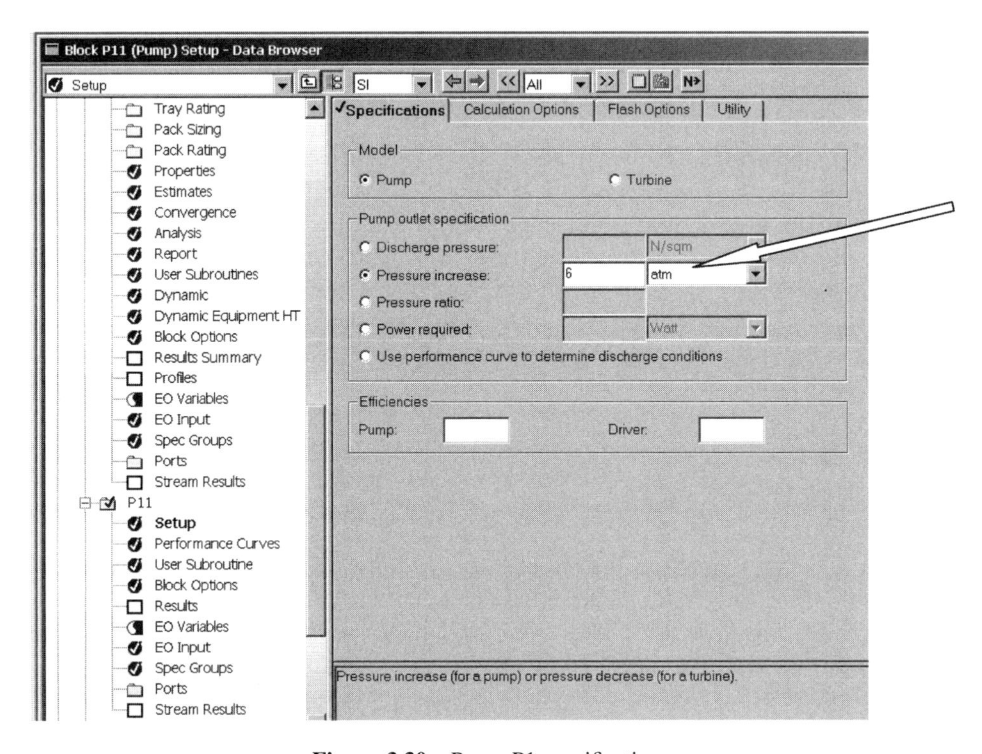

Figure 3.30 Pump P1 specifications.

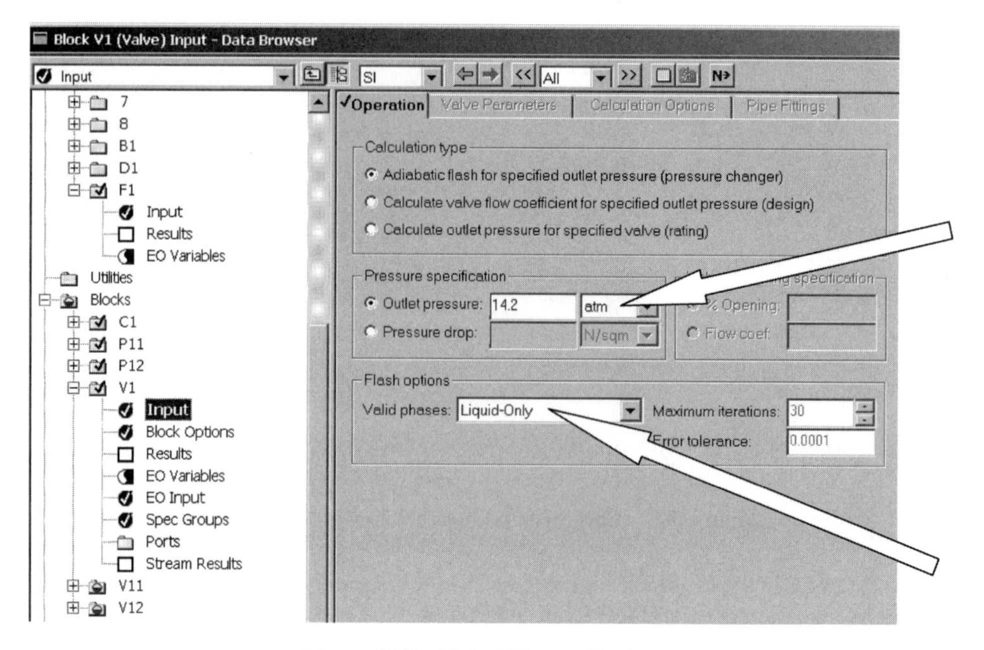

Figure 3.31 Valve V1 specifications.

form for pump P11 with the *Pressure increase* button selected and 6 atm pump head entered. Pump P12 is handled in the same way.

The pressure at the exit of the feed valve (V1) must be equal to the pressure on the feed tray (stage 16). We do not know exactly what this is at this point, but we guess it to be about 14.2 atm. Clicking *Setup* for the V1 block opens the window shown in Figure 3.31. The outlet pressure is set at 14.2 atm. We will come back and adjust this pressure after the flowsheet has been converged and when we know exactly what the pressure on the feed tray is.

Note that under the *Flash options* the *Valid phases* has been set to *Liquid-Only*. This is not necessary for a steady-state simulation, but it will become useful when we move into dynamic simulation in Chapter 7. The other two control valves are each given a pressure drop of 3 atm, as shown in Figure 3.32, and the *Flash options* for *Valid phases* are set to *Liquid-Only*.

The flowsheet is fully specified at this point. All the read buttons are blue, and we are ready to run the simulation.

3.5 RUNNING THE SIMULATION

The blue *N* button ("next") at the top of the *Data Browser* window on the far right is clicked to run the simulation. If any information is needed, the program will go to that location on the window and display a red symbol. If everything is ready to calculate, the message shown in Figure 3.33 will appear, and you should click *OK*. The *Control Panel* window shown in Figure 3.34 opens and indicates that the column was successfully converged. It took four iterations to converge the column.

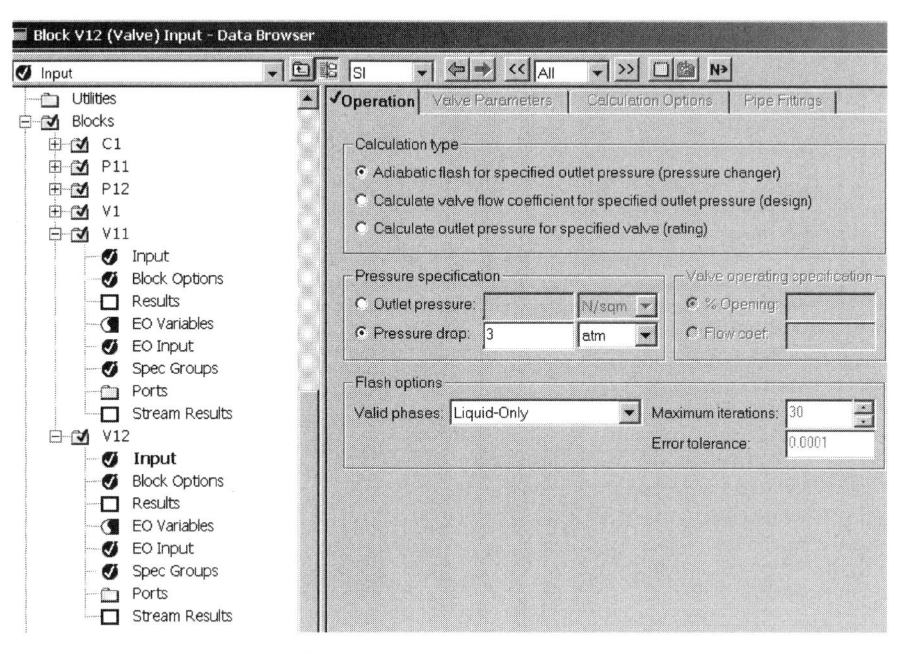

Figure 3.32 Valves V11 and V12 specifications.

Now we want to look at the compositions of the product streams leaving the column to see if they satisfy their desired purities. We assume that the specification of the heavy impurity in the distillate (isobutane = iC_4) is 2 mol% and that the specification of the light impurity in the bottoms (propane = C_3) is 1 mol%. To look at the properties of these streams, we open the C1 block in the *Data Browser* window and click the item *Stream Results*, which is at the very bottom of the list.

There are three streams in the table shown in Figure 3.35. Stream 1 is the feed inlet to the column. Stream D1 is the liquid distillate leaving the reflux drum. Stream B1 is the liquid bottoms leaving the base of the column. We can see that there is about 12 mol% iC_4 in the distillate and 8 mol% C_3 in the bottoms. The purities are too low, so we need to increase the reflux ratio or add more stages to get a better separation.

If we go back to *Setup* in the C1 block, change the reflux ratio to 3, and click the *N* button, the simulation converges with the results shown in Figure 3.36. The distillate impurity has decreased to about 2 mol% iC_4, and the bottoms now has about 1.5 mol% C_3. So we are getting pretty close to the desired purities. We could continue to manually change the reflux ratio and the distillate flowrate to attempt to achieve the desired product purities by trial and error. However, there is a much easier way, as discussed in the next section.

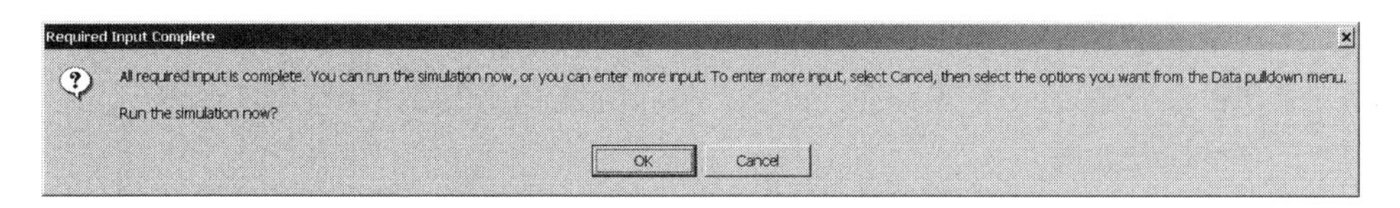

Figure 3.33 Ready-to-run message.

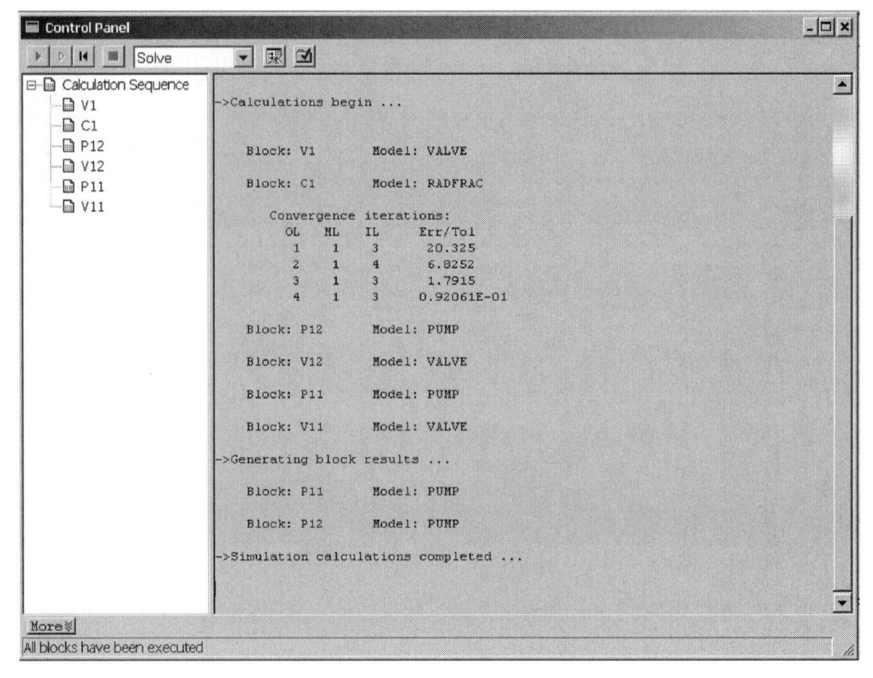

Figure 3.34 Control panel.

3.6 USING "DESIGN SPEC/VARY" FUNCTION

The specifications for product impurities are 1 mol% propane in the bottoms and 2 mol% isobutane in the distillate. To achieve these precise specifications, Aspen Plus uses the

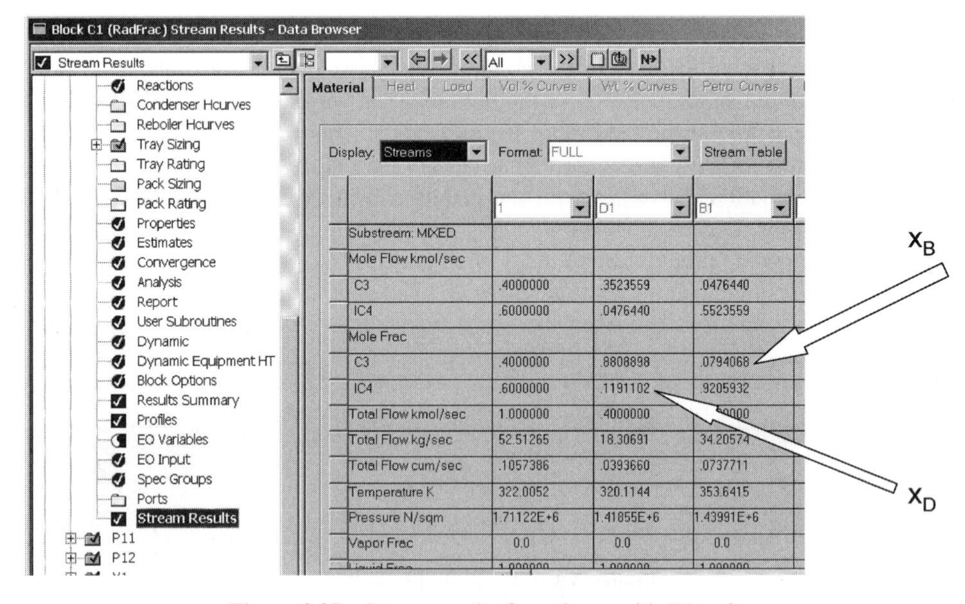

Figure 3.35 Stream results for column with $RR = 2$.

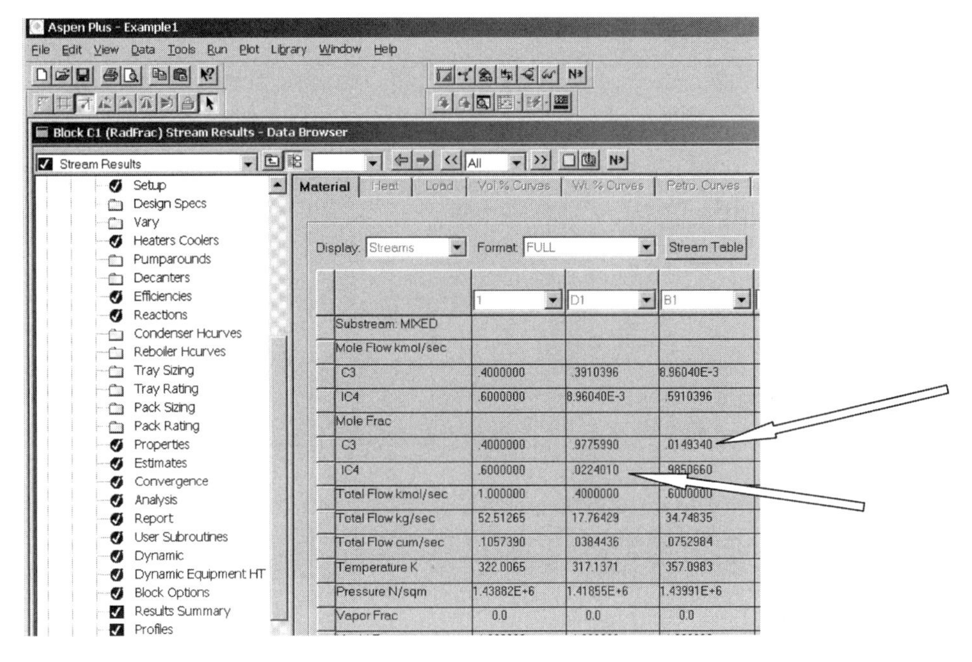

Figure 3.36 Stream results for column with $RR = 3$.

"Design Spec/Vary" function. A desired value of some "controlled" variable is specified, and the variable to be manipulated is specified. The simulation attempts to adjust the manipulated variable in such a way that the specified value of the controlled variable is achieved.

In the example under study, we want to find the values of distillate flowrate and reflux ratio that drive the distillate composition to 2 mol% isobutane and the bottoms composition to 1 mol% propane.

A word of caution might be useful at this point. The solution of a large set of simultaneous nonlinear algebraic equations is very difficult. There is no guarantee that a solution will be found because of numerical problems. In addition, if good engineering judgment is not used in selecting the target values, there may be no physically realizable solution. For example, if the specified number of stages is less than the minimum required for the specified separation, there is no value of the adjusted variable that can produce the desired result.

Another possible complication is multiplicity. Because the equations are nonlinear, there may be multiple solutions. Sometimes the program will converge to one solution and at other times it will converge to another solution, depending on the initial conditions.

It is usually a good idea to start by converging only one variable at a time instead of trying to handle several simultaneously. In our example, we will converge the distillate specification first by adjusting distillate flowrate. Then, with this specification active, we will converge the bottoms specification by adjusting the reflux ratio. The order of this sequential approach is deliberately selected to use the distillate first because the effect of distillate flowrate on compositions throughout the column is much larger than the effect of the reflux ratio.

To set up the Design Spec/Vary function, click on *Design Spec* under the C1 block in the *Data Browser* window. The window shown in Figure 3.37a opens up. Clicking the *New* button opens the window shown in Figure 3.37b. Click *OK* and another window opens

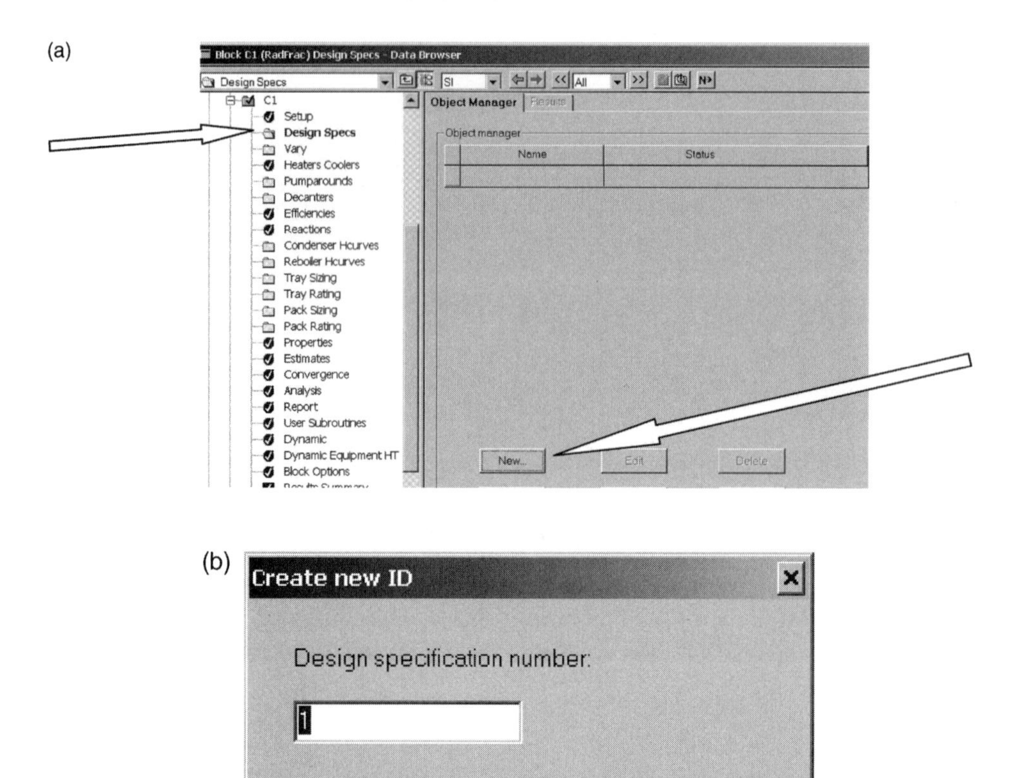

Figure 3.37 Setting up the design spec (a) and the design spec number (b).

(Fig. 3.38a), which has several page tabs. On the first one, *Specifications*, you can specify the type of variable and what its desired value is. Clicking the dropdown menu under *Design specification* and *Type* opens a long list of possible types of specifications (Fig. 3.38b). Select *mole purity*. Go down to *Target* and type in "0.02." This is the desired mole fraction of isobutane in the distillate. Then click the second page tab *Components*. Click the *IC4* in the left column under *Available* components. Clicking the ">" button moves IC4 over to the right *Selected components* column (Fig. 3.39b).

Click the third page tab *Feed/Product Streams*, select *D1* in the left column, and click the ">" button to move it to the right column. The Design Spec is now completed. Note that the number "1" in Figure 3.39c is blue. Now we must specify what variable to adjust. Clicking the *Vary* item under the C1 block opens the window shown in Figure 3.40a. Clicking the *New* button and specifying the number as "1" opens the window shown in Figure 3.40b, where the manipulated variable is defined.

Opening the dropdown menu under *Adjusted variable* and *Type* produces a long list of possible variables. We select *Distillate rate*, which opens several boxes (Fig. 3.40c) in which the range of changes in the distillate flowrate can be restricted. We set the lower bound at 0.2 and the upper bound at 0.6 kmol/s.

(a)

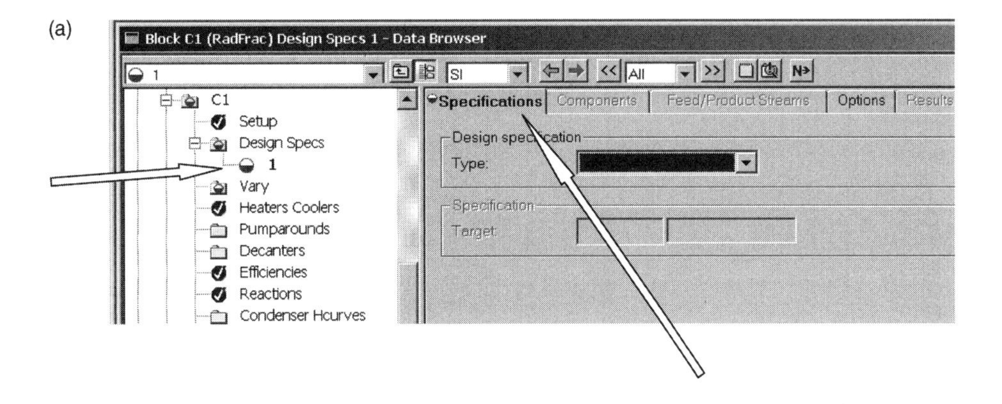

(b)

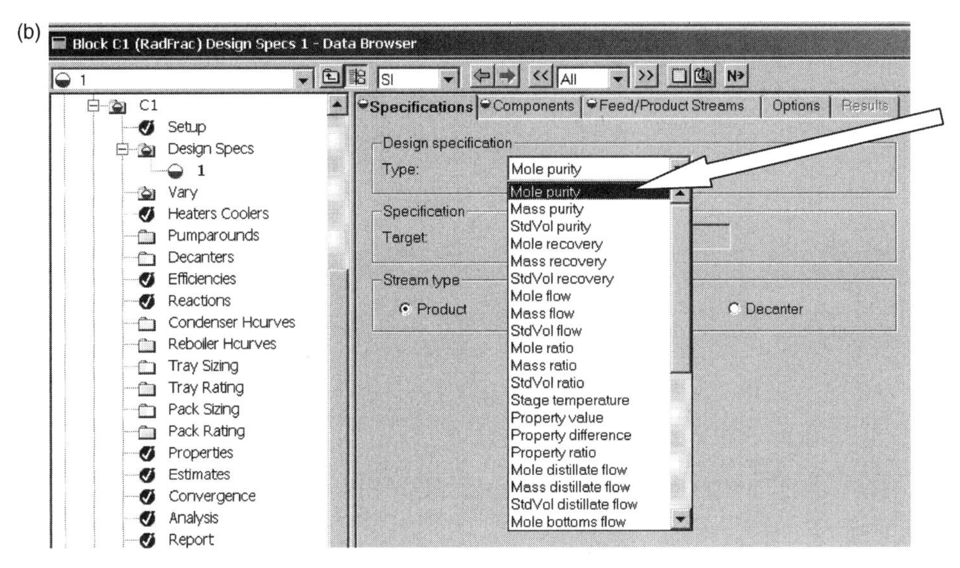

Figure 3.38 Specifying the controlled variable (a) and selecting the type of variable (b).

Note that all the items in the *Data Browser* window are blue, so the simulation is ready to run. We click the blue *N* button and run the program. The *Control Panel* window opens and tells us that it has taken three iterations to converge (Fig. 3.41a). Going down to *Stream Results* at the bottom of the list under the C1 block lets us look at the new values of the stream properties. Figure 3.41b shows that mole fraction of IC4 in D1 is 0.01999713, which is within the error tolerance of the 0.02 mole fraction (mf) desired. Note that the flowrate of D1 has changed to 0.39753743 kmol/s.

The second Design Spec/Vary is set up in the same way. Clicking *Design Spec* opens a window on which you specify a new design spec ("2"). Then the mole purity of the bottoms B1 is specified to be 0.01 mf propane. See Figures 3.42a–3.42c for the three steps on the three page tabs.

Next a second *Vary* ("2") is set up as shown in Figure 3.43 with reflux ratio selected. The upper and lower bounds are set at 1 and 5, respectively, since we know a reflux ratio of $\sim$3 gives results that are close to the desired.

Everything is ready to run again. Clicking the blue *N* button executes the program. The simulation converges in three iterations.

Figure 3.44 shows the new stream results. The mole fraction of iC_4 in the distillate is 0.0200027, and the mole fraction of the C_3 in the bottoms is 0.0100008. Both are now very close to their specified values. Of course, the distillate flowrate and the reflux ratio have

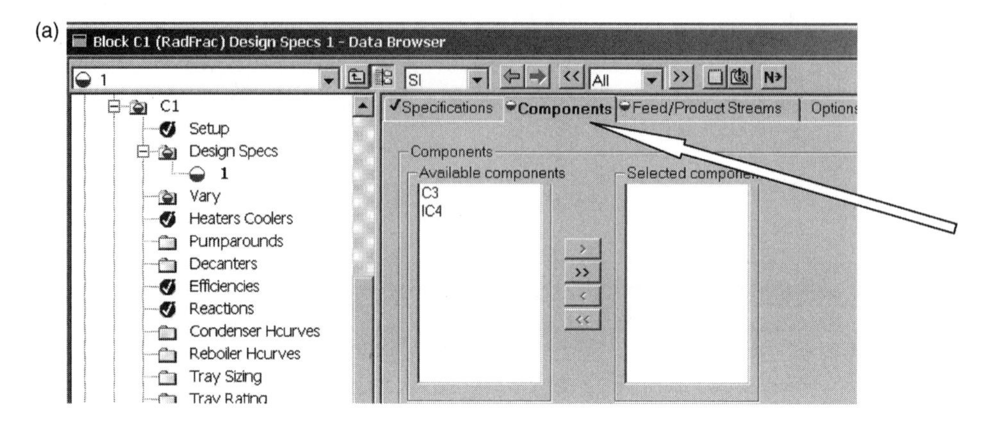

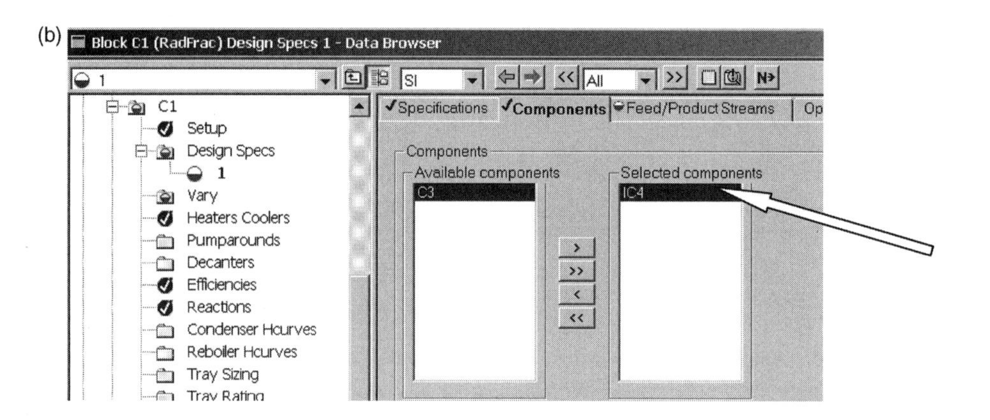

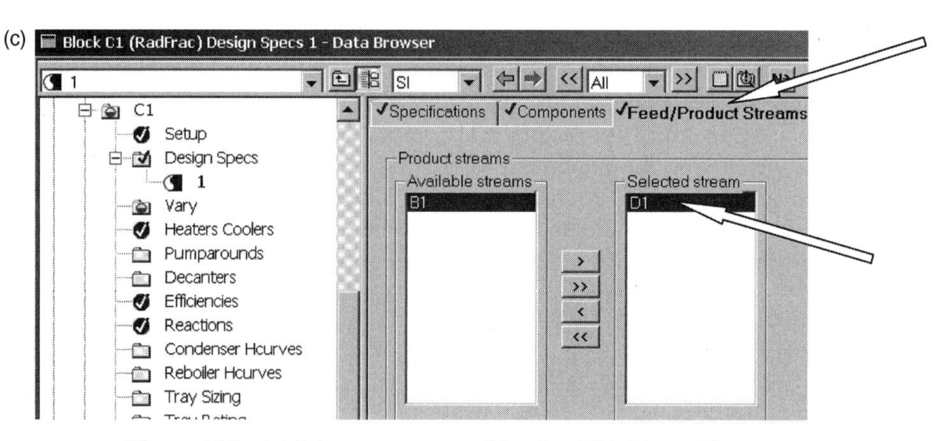

Figure 3.39 (a) Select components; (b) select IC4; (c) specify stream.

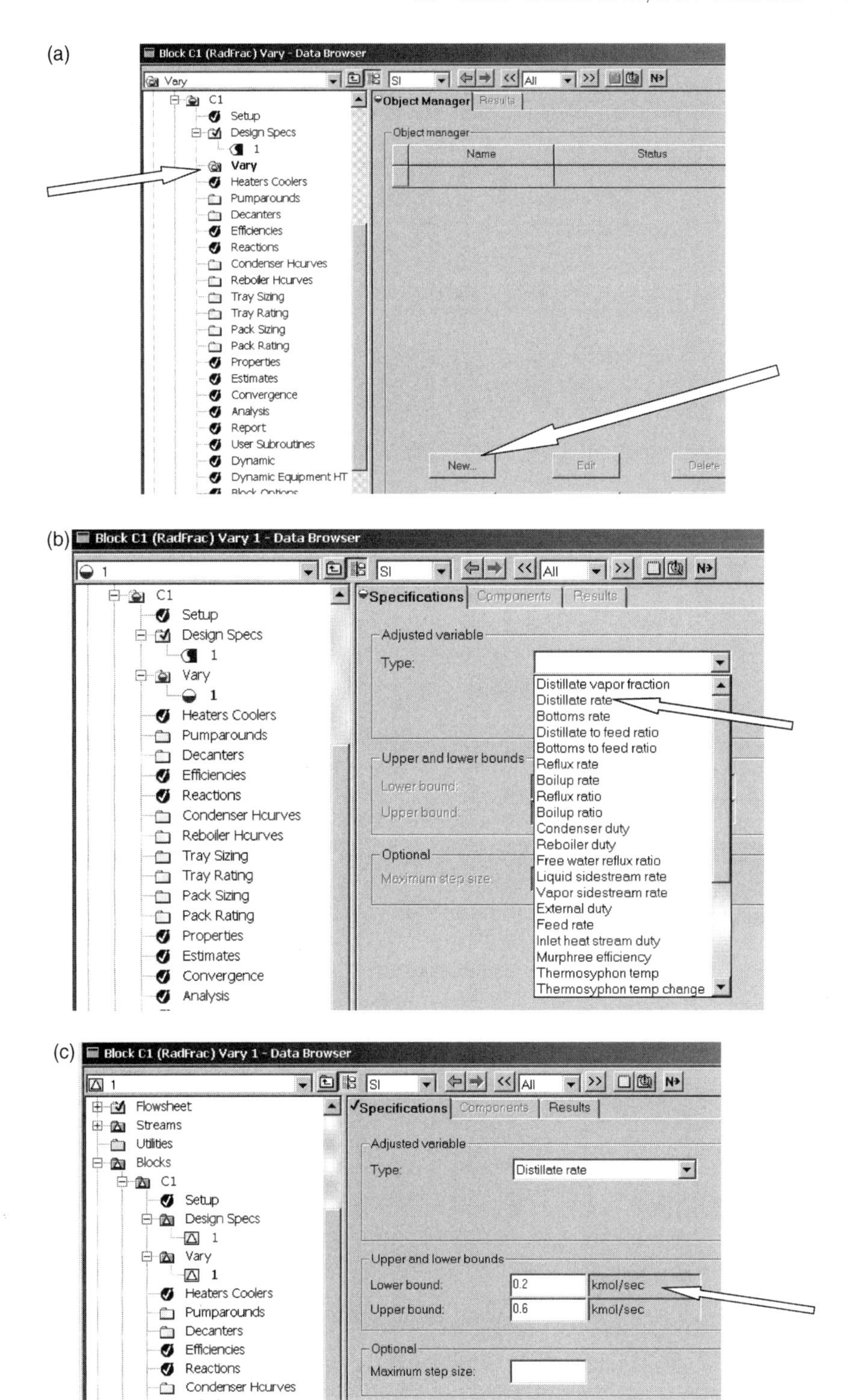

Figure 3.40 (a) Opening the *Vary* window; (b) defining manipulated variable; (c) setting limits on distillate flowrate.

(a)

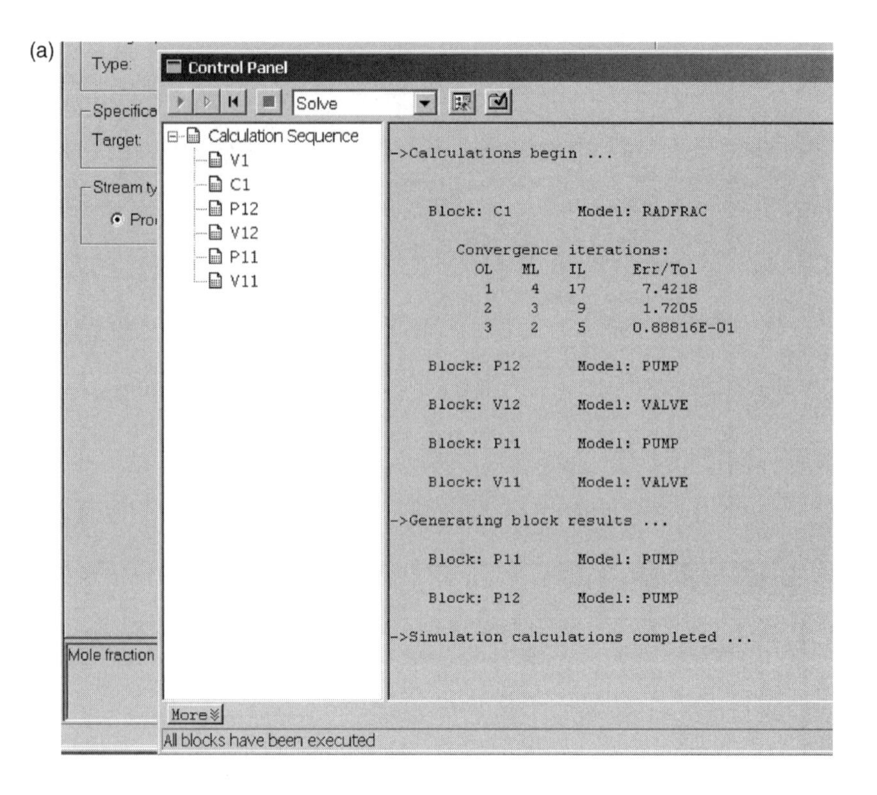

(b)

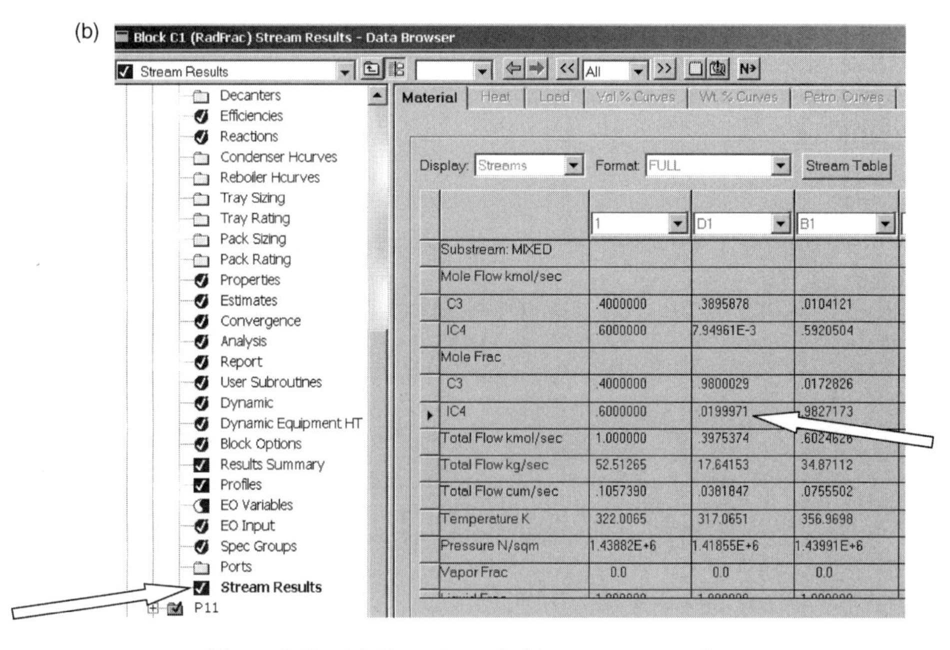

	1	D1	B1
Substream: MIXED			
Mole Flow kmol/sec			
C3	.4000000	.3895878	.0104121
IC4	.6000000	7.94961E-3	.5920504
Mole Frac			
C3	.4000000	.9800029	.0172826
IC4	.6000000	.0199971	.9827173
Total Flow kmol/sec	1.000000	.3975374	.6024626
Total Flow kg/sec	52.51265	17.64153	34.87112
Total Flow cum/sec	.1057390	.0381847	.0755502
Temperature K	322.0065	317.0651	356.9698
Pressure N/sqm	1.43882E+6	1.41855E+6	1.43991E+6
Vapor Frac	0.0	0.0	0.0

Figure 3.41 (a) Control panel, (b) new stream results.

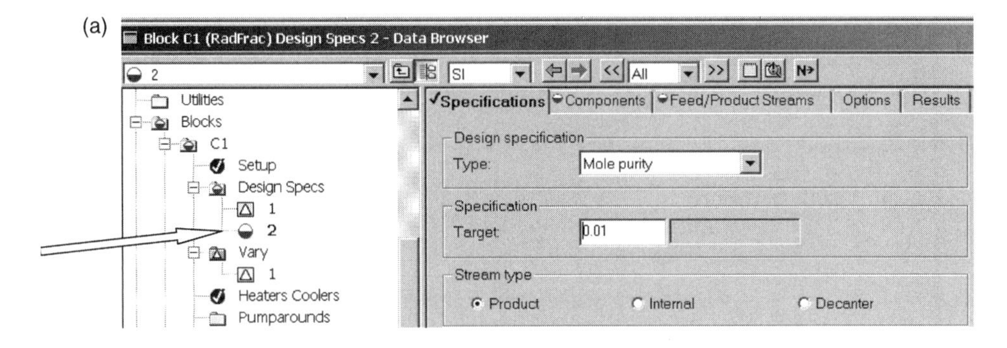

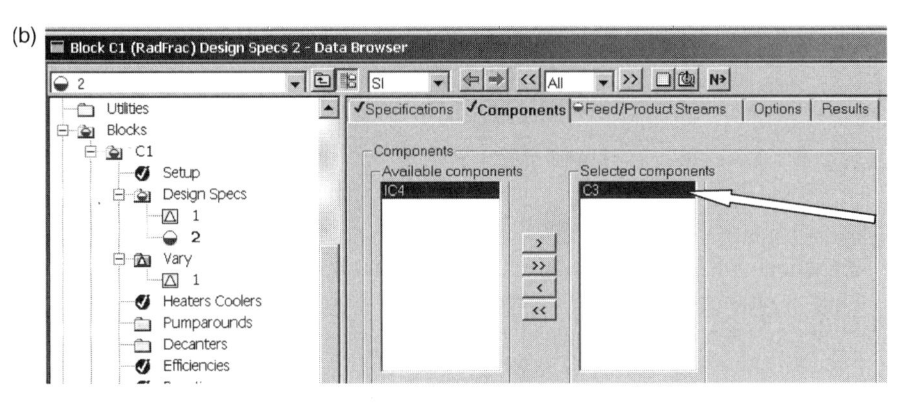

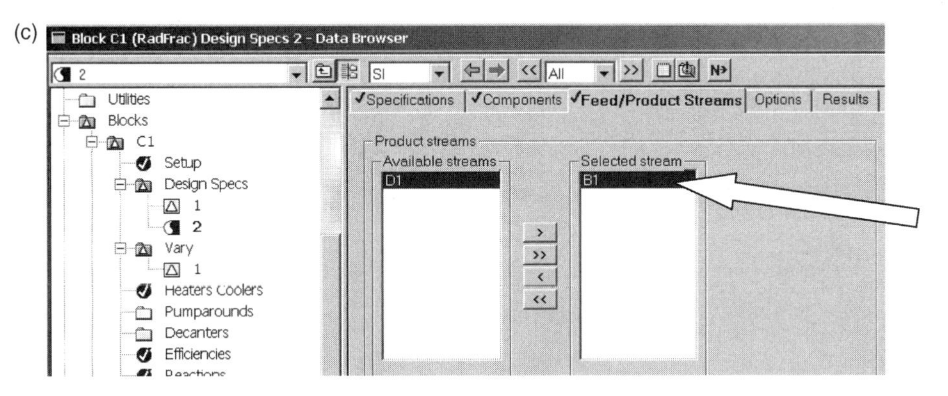

Figure 3.42 (a) Setting up second design spec; (b) selecting *Propane*; (c) selecting *Bottoms*.

been changed to produce the desired product purities. The stream results show that the flowrate of D1 is 0.4021 kmol/s. To ascertain the reflux ratio, click on *Results Summary* under the C1 block. The window shown in Figure 3.45 opens, on which the conditions at the top of the column are given. The reflux ratio is 3.095.

The other important pieces of information in the window are the condenser heat removal [−22.29 MW (megawatts)] and the reflux drum temperature (317.06 K) at the 14 atm pressure that we specified. If you recall, we guessed that a pressure of 14 atm would give us a reflux drum temperature of ~325 K so cooling water could be used in the

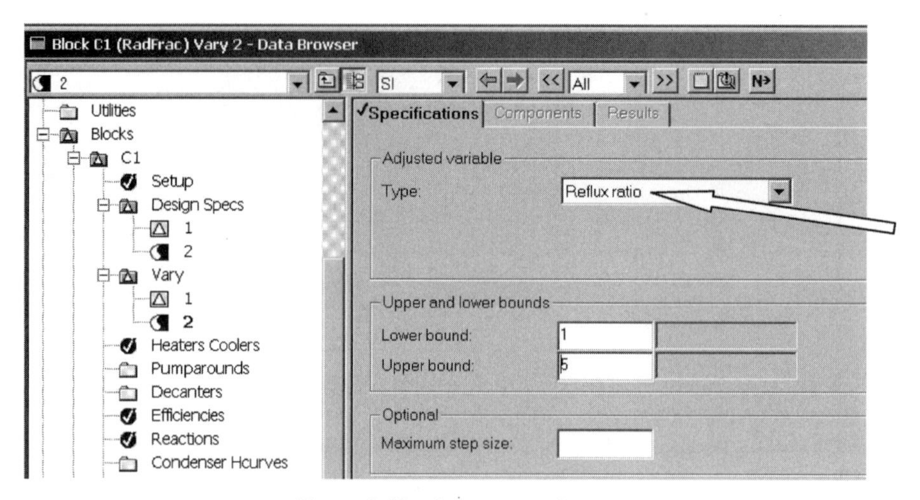

Figure 3.43 Set up second *Vary*.

condenser. To attain the desired 325 K, the pressure should be increased a little. If we rerun the simulation with a pressure of 16.8 atm, the reflux drum temperature will be 325.06 K.

Of course, at this new pressure the required reflux ratio changes. It increases from 3.095 to 3.511 (see Fig. 3.46). This shows the adverse effect of pressure on relative volatilities that occurs in most hydrocarbon systems. The column should be operated at as low a pressure as possible to save energy.

To find the conditions at the base of the column, we use the dropdown menu that is next to *View* on the *Results Summary* window and select *Reboiler/Column base*. Figure 3.47 shows the information obtained. The most important piece of information is the reboiler heat input 27.409 MW. The base temperature is 366.11 K. This will dictate the pressure

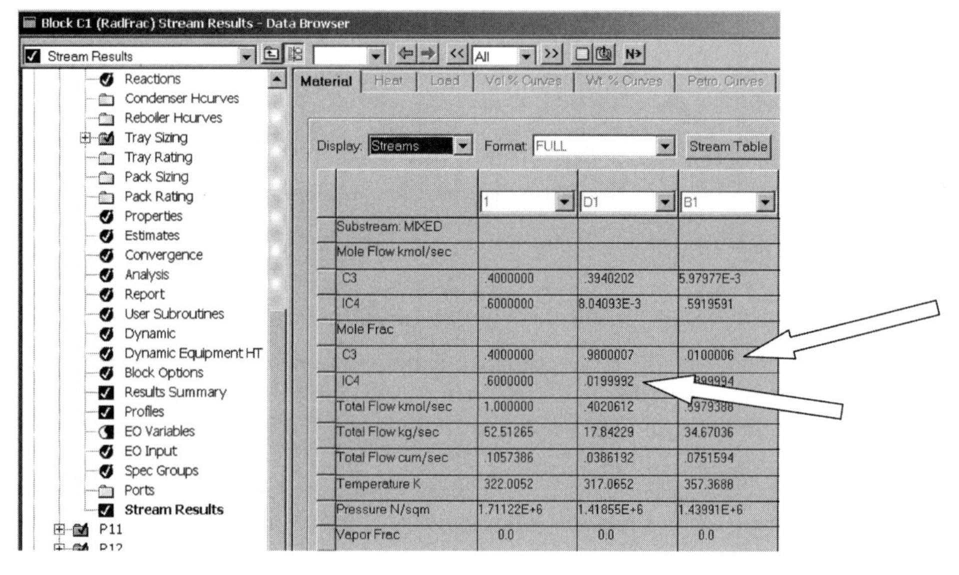

Figure 3.44 Stream results.

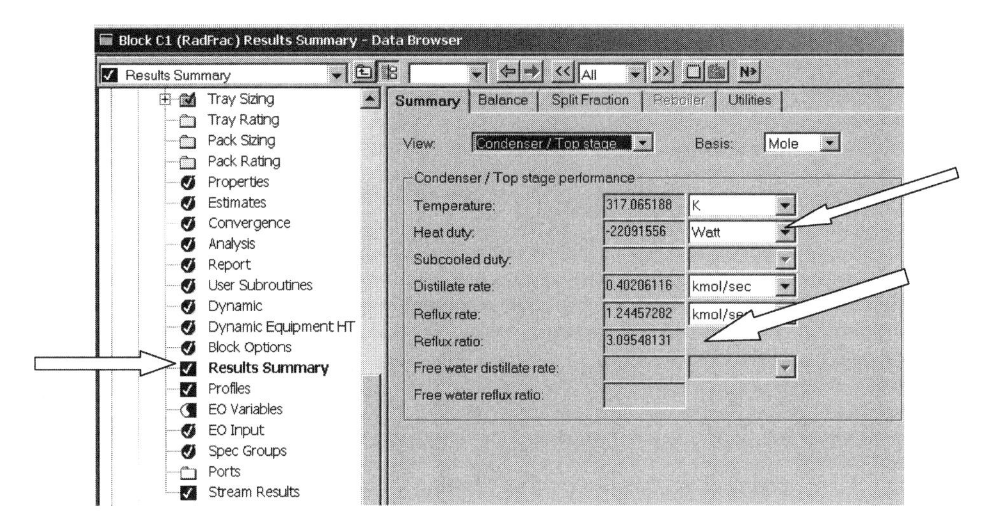

Figure 3.45 Results for top of column at 14 atm pressure.

of the steam used in the reboiler. A reasonable differential temperature is 40 K, which corresponds to a saturated steam pressure of ~3 atm at 406 K. So, if steam is available in the plant at ~6 atm, it can be used to supply the heat required in the reboiler, assuming a 3 atm pressure drop over the steam control valve.

The stream conditions at the 16.8 atm column pressure are shown in Figure 3.48. The column temperature and composition profiles can be obtained by selecting *Profiles* in the C1 block. The window that opens is shown in Figure 3.49. There are several page tabs. The first *TPFQ* (temperature, pressure, flow, heat) gives the temperature and pressure on each stage. Selecting the second page tab *Compositions* opens the window shown in Figure 3.50, in which *Liquid* has been selected from the dropdown menu in the *View* box.

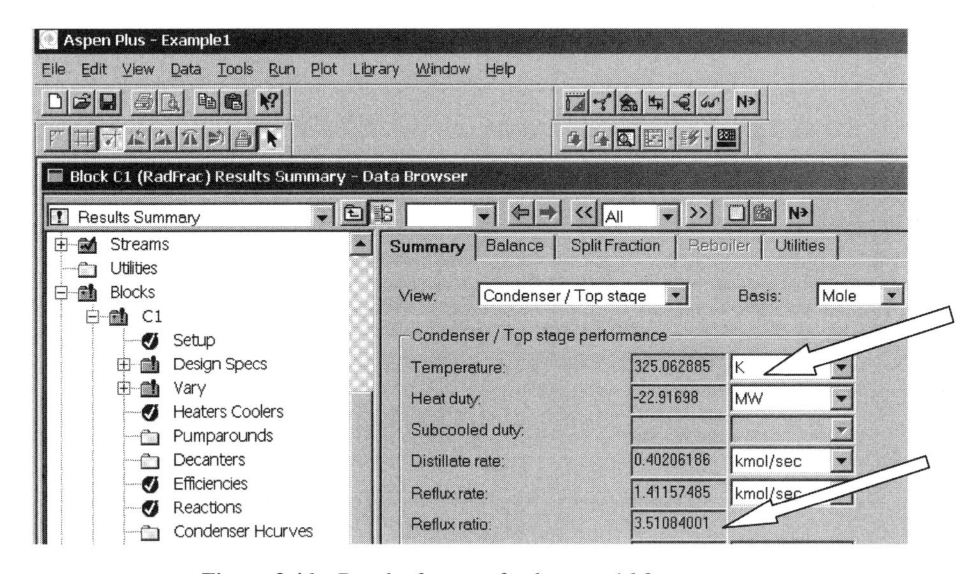

Figure 3.46 Results for top of column at 16.8 atm pressure.

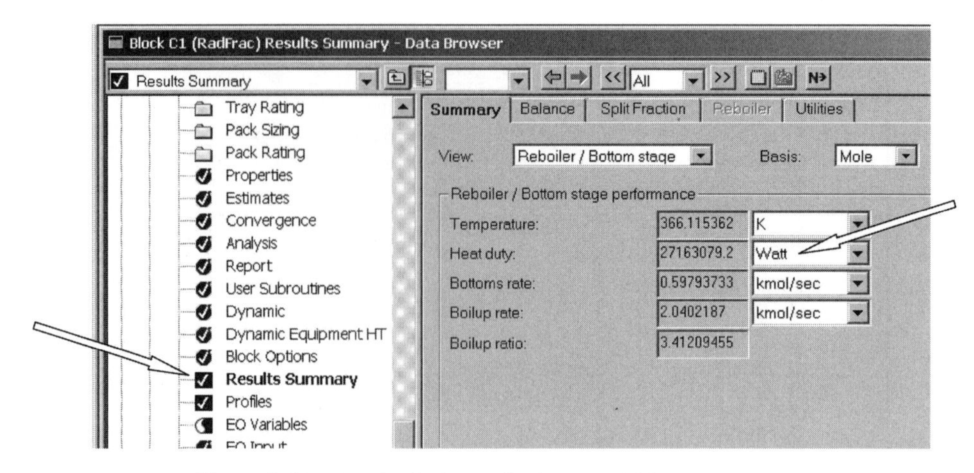

Figure 3.47 Results for base of column at 16.8 atm pressure.

Using the Plot Wizard program makes generating plots of these profiles quite easy. Click on *Plot* at the top toolbar of the Aspen Plus simulation window. Then click *Plot Wizard*. This opens the window shown in Figure 3.51a. Clicking *Next* opens the window shown in Figure 3.51b. Clicking on the upper left picture labeled *Temp* produces the temperature profile plot given in Figure 3.51c. Clicking on the picture labeled *Comp* and then clicking *Next* opens the window on which you can select what components to plot and what phase (liquid or vapor compositions). Figure 3.52a shows the selections, and Figure 3.52b gives the composition profile.

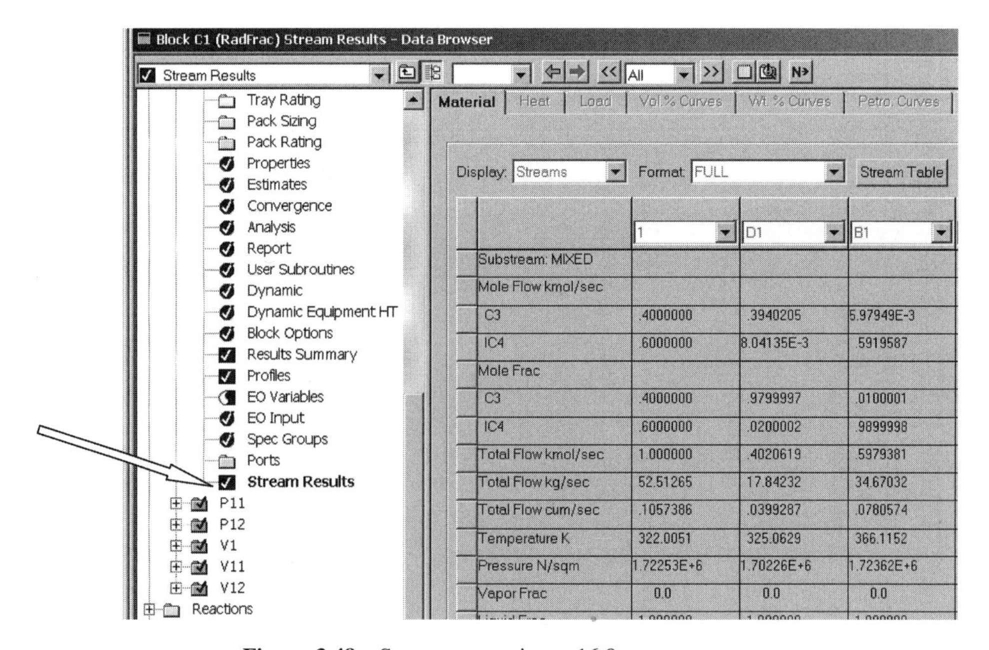

Figure 3.48 Stream properties at 16.8 atm pressure.

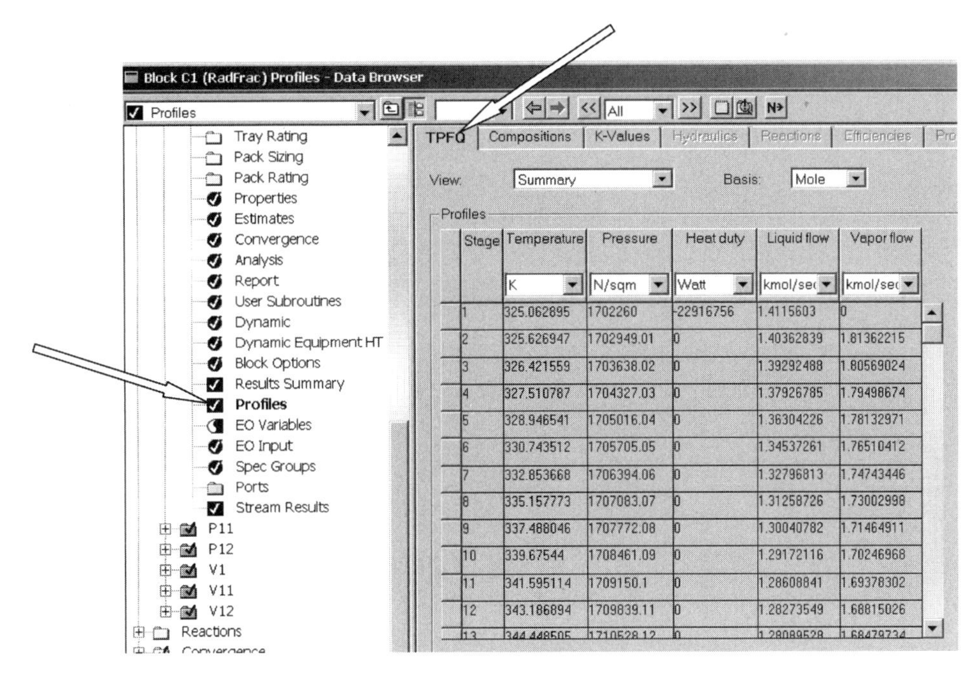

Stage	Temperature K	Pressure N/sqm	Heat duty Watt	Liquid flow kmol/sec	Vapor flow kmol/sec
1	325.062895	1702260	-22916756	1.4115603	0
2	325.626947	1702949.01	0	1.40362839	1.81362215
3	326.421559	1703638.02	0	1.39292488	1.80569024
4	327.510787	1704327.03	0	1.37926785	1.79498674
5	328.946541	1705016.04	0	1.36304226	1.78132971
6	330.743512	1705705.05	0	1.34537261	1.76510412
7	332.853668	1706394.06	0	1.32796813	1.74743446
8	335.157773	1707083.07	0	1.31258726	1.73002998
9	337.488046	1707772.08	0	1.30040782	1.71464911
10	339.67544	1708461.09	0	1.29172116	1.70246968
11	341.595114	1709150.1	0	1.28608841	1.69378302
12	343.186894	1709839.11	0	1.28273549	1.68815026

Figure 3.49 Temperature and pressure profiles.

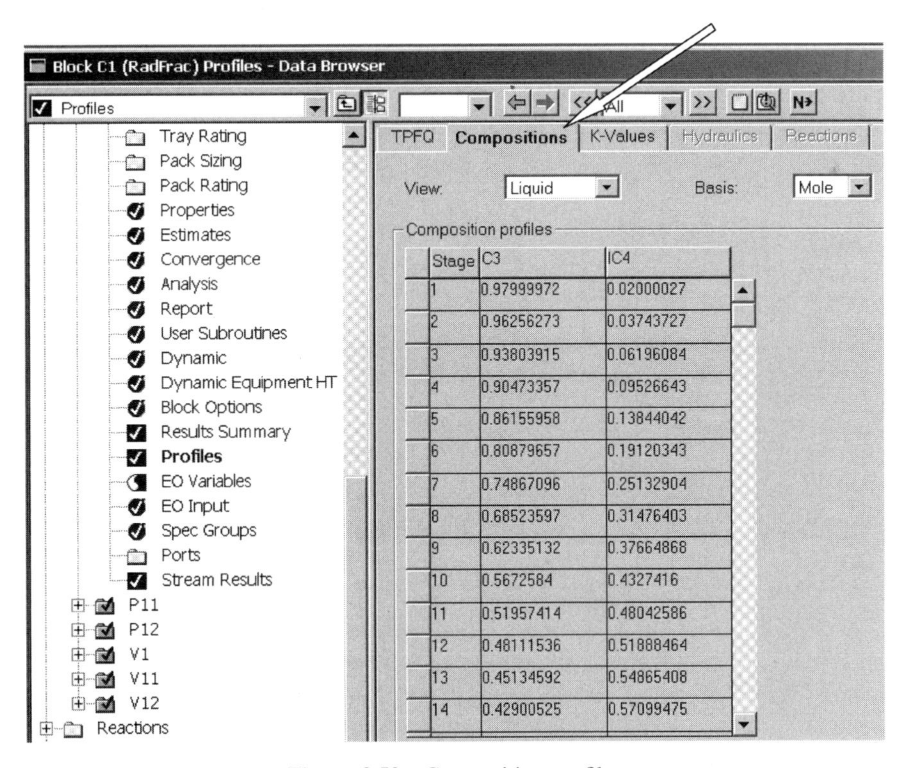

Stage	C3	IC4
1	0.97999972	0.02000027
2	0.96256273	0.03743727
3	0.93803915	0.06196084
4	0.90473357	0.09526643
5	0.86155958	0.13844042
6	0.80879657	0.19120343
7	0.74867096	0.25132904
8	0.68523597	0.31476403
9	0.62335132	0.37664868
10	0.5672584	0.4327416
11	0.51957414	0.48042586
12	0.48111536	0.51888464
13	0.45134592	0.54865408
14	0.42900525	0.57099475

Figure 3.50 Composition profiles.

(a)

(b)

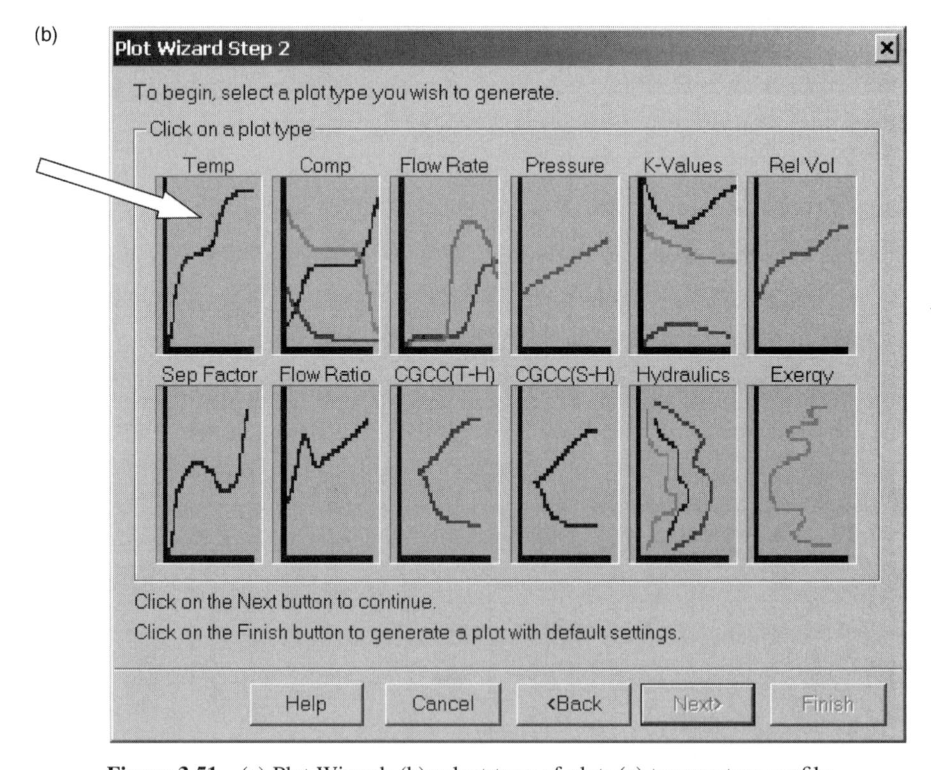

Figure 3.51 (a) Plot Wizard; (b) select type of plot; (c) temperature profile.

(c)

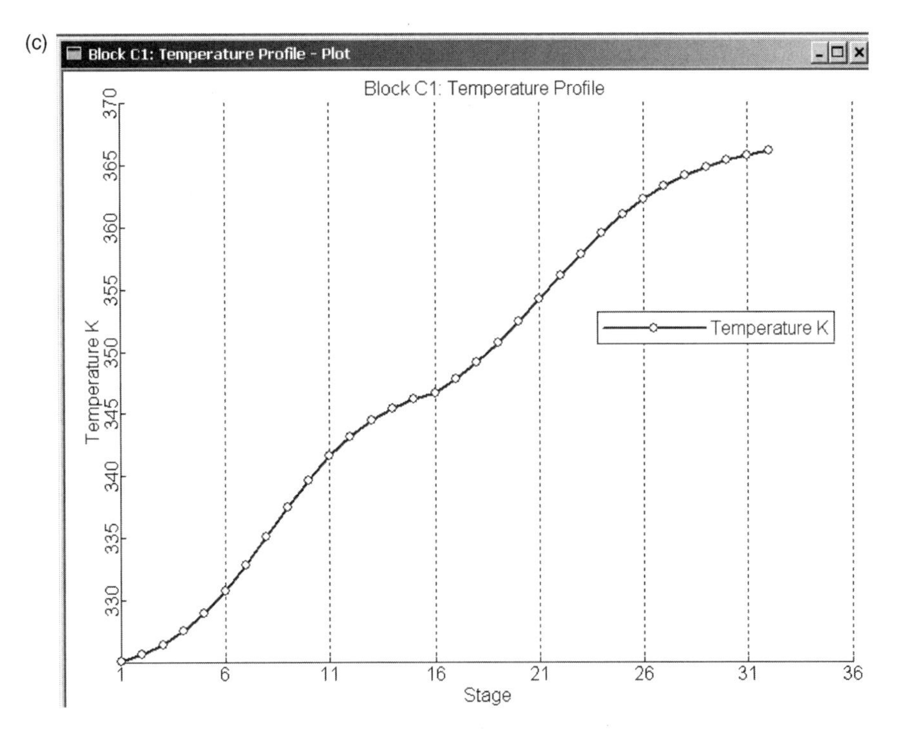

Figure 3.51 *Continued.*

(a)

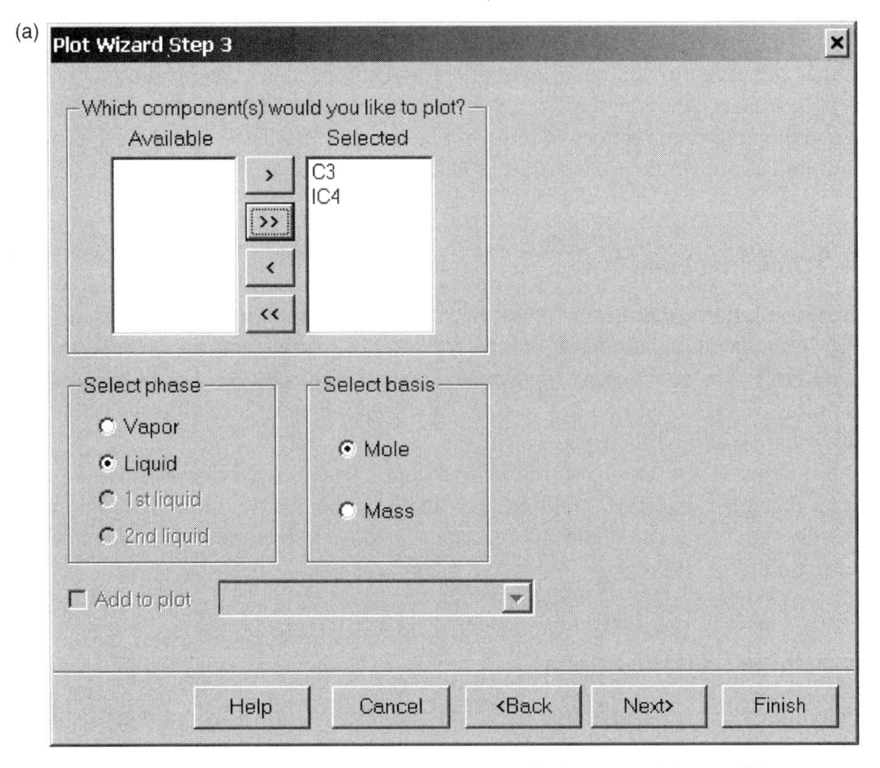

Figure 3.52 (a) Selecting components; (b) liquid composition profile.

(b)

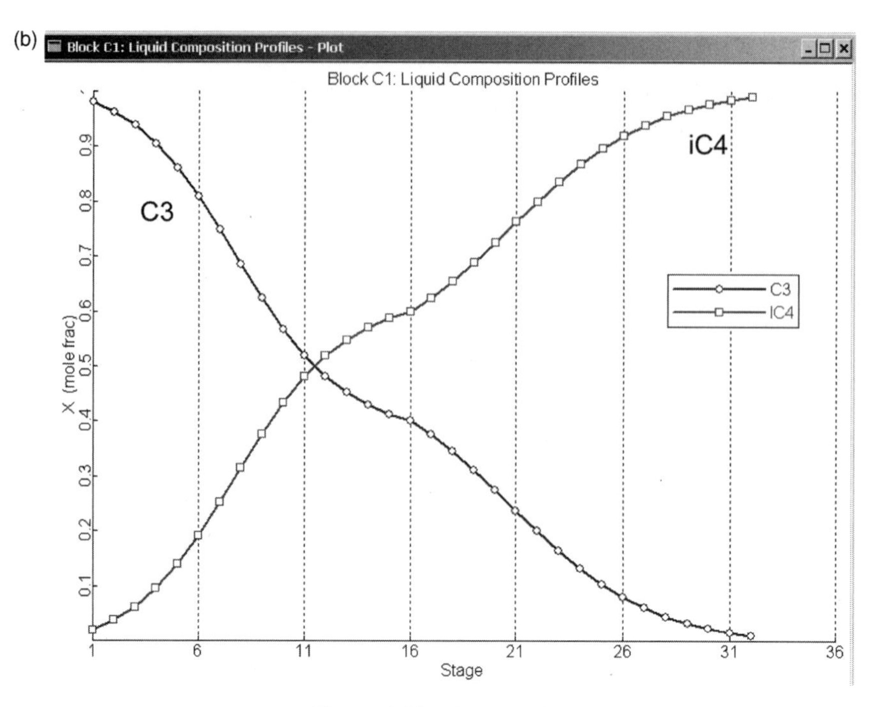

Figure 3.52 *Continued.*

3.7 FINDING THE OPTIMUM FEED TRAY AND MINIMUM CONDITIONS

Now that the pressure has been determined and the product specifications attained, we need to go back and find the "optimum" feed tray. In addition, the minimum reflux ratio and the minimum number of trays can be determined. These will be useful for heuristic optimization, which is discussed in detail in Chapter 4.

3.7.1 Optimum Feed Tray

In most distillation columns, the major operating cost is reboiler energy consumption. Of course, if refrigeration were used in the condenser, this heat removal expense would also be quite large. For our propane/isobutane example, the pressure was deliberately set so that cooling water could be used in the condenser. Therefore, reboiler heat input is the quantity that should be minimized.

The simulation is run using different feed stages. The purities of both products are held constant. The feed stage that minimizes reboiler heat input is the optimum. Table 3.1 gives the results of these calculations. Feeding on stage 14 gives the minimum energy consumption.

3.7.2 Minimum Reflux Ratio

The simulation can be used to find the minimum reflux ratio by increasing the number of stages until there is no further reduction in the reflux ratio. Product purities are held

TABLE 3.1 Effect of Feed Stage on Reboiler Heat Input

Feed Stage Number	Reboiler Heat Input (MW)	Condenser Heat Removal (MW)	Reflux Ratio
12	27.94	23.45	3.616
13	27.39	22.90	3.508
14*	**27.17**	22.68	3.463
15	27.18	22.64	3.465
16	27.41	22.92	3.511

*Optimum.

TABLE 3.2 Minimum Reflux Ratio

Total Stages	Feed Stage	Reflux Ratio
32	14	3.463
48	21	2.959
64	28	2.912
96	42	2.908

constant. It is assumed that the feed stage is a fixed ratio of the total number of stages. The results, given in Table 3.2, show that the minimum reflux ratio is about 2.9.

3.7.3 Minimum Number of Trays

The simulation can also be run to find the minimum number of trays by decreasing the number of stages until the required reflux ratio becomes very large. Product purities are held constant. It is assumed that the feed stage is a fixed ratio of the total number of stages. Results are given in Table 3.3 and show that the minimum number of stages is 15.

3.8 COLUMN SIZING

The last topic to discuss in this chapter before going into economic optimization of column design is how to determine the diameter and length of the vessel.

TABLE 3.3 Minimum Number of Stages

Total Stages	Feed Stage	Reflux Ratio
32	14	3.463
22	10	6.021
20	9	8.100
18	8	13.56
17	8	20.59
16	7	21.35
15	7	160.8

3.8.1 Length

Calculating the height of the column is fairly easy if the number of trays is given. The typical distance between trays (tray spacing) is 0.61 m (2 ft). If there are N_T stages, the number of trays is $N_T - 2$ (one stage for the reflux drum and one for the reboiler).

In addition to the trays, some space is needed at the top where the reflux piping enters the vessel and at the feed tray for feed distribution piping. More significantly, space is needed at the base to satisfy two requirements: (1) liquid holdup is needed for surge capacity and (2) the liquid height in the base of the column must be high enough above the elevation of the bottoms pump to provide the necessary net positive suction head (NPSH) requirements for this pump.

Therefore, a design heuristic is to provide an additional 20% more height than that required for just the trays. So the length of the vessel can be estimated from the following equation.

$$L = 1.2(0.61)(N_T - 2)$$

3.8.2 Diameter

The diameter of a distillation column is determined by the maximum vapor velocity. If this velocity is exceeded, the column liquid and vapor hydraulics will fail and the column will flood. Reliable correlations are available to determine this maximum vapor velocity.

Since the vapor flowrates change from tray to tray in a nonequimolal overflow system, the tray with the highest vapor velocity will set the minimum column diameter. If the vapor mass flowrate and the vapor density are known, the volumetric flowrate of the vapor can be calculated. Then, if the maximum allowable velocity is known, the cross-sectional area of the column can be calculated.

Aspen Plus has an easy-to-use tray sizing capability. Click the subitem *Tray Sizing* under the C1 block, and then click *New* and *OK* for the identification number. A window will open, where the column sections to be sized and the type of tray can be entered. Figure 3.53a shows the parameter values used in the example. The stages run from stage 2 (the top tray) to stage 31 (the bottom tray). Sieve trays are specified.

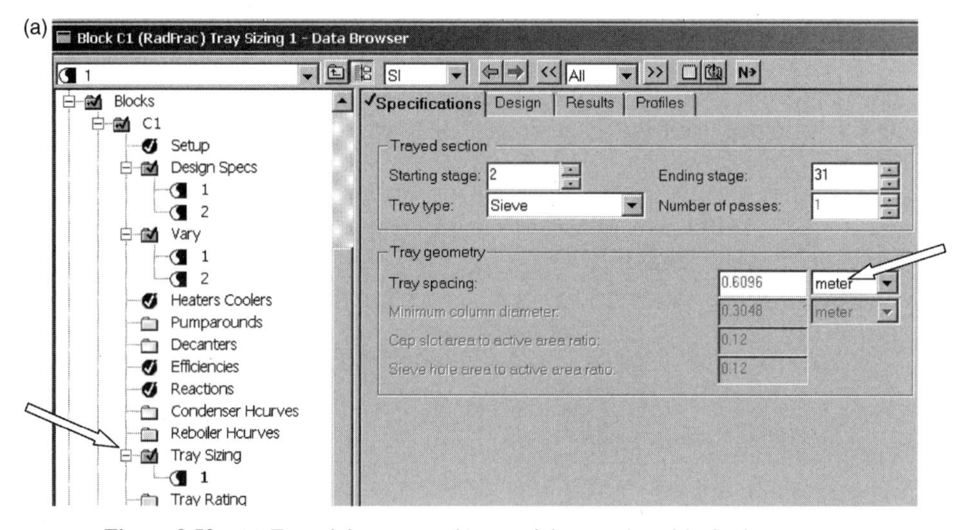

Figure 3.53 (a) Tray sizing setup; (b) tray sizing results with single-pass trays.

(b)

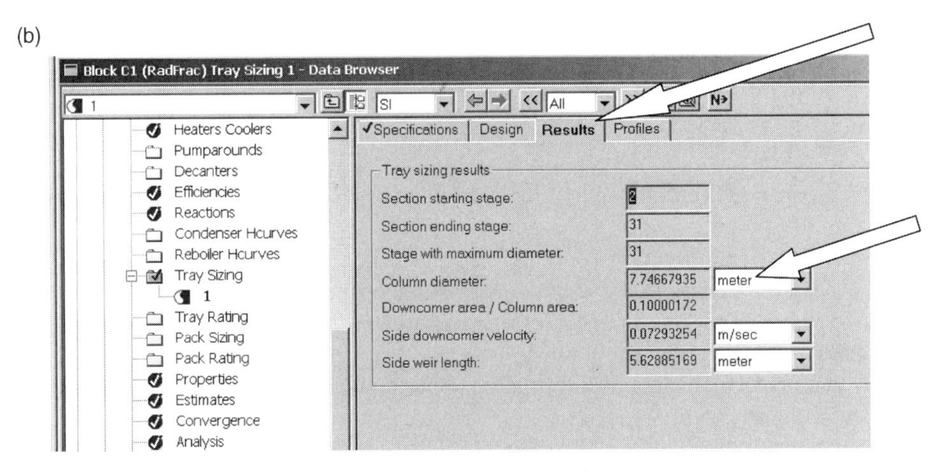

Figure 3.53 *Continued.*

The simulation must be run by clicking the *N* button. Then the page tab *Results* is clicked (see Fig. 3.53b), and the column diameter is seen to be 7.75 m. This is a very large distillation column, and therefore a single liquid pass would produce very large liquid gradients across the tray and liquid heights over the weir. A column this large would use at least 2-pass trays. Changing the number of passes to 2 on the *Specifications* page tab produces a large change in the calculated diameter, dropping it from 7.73 to 5.91 m.

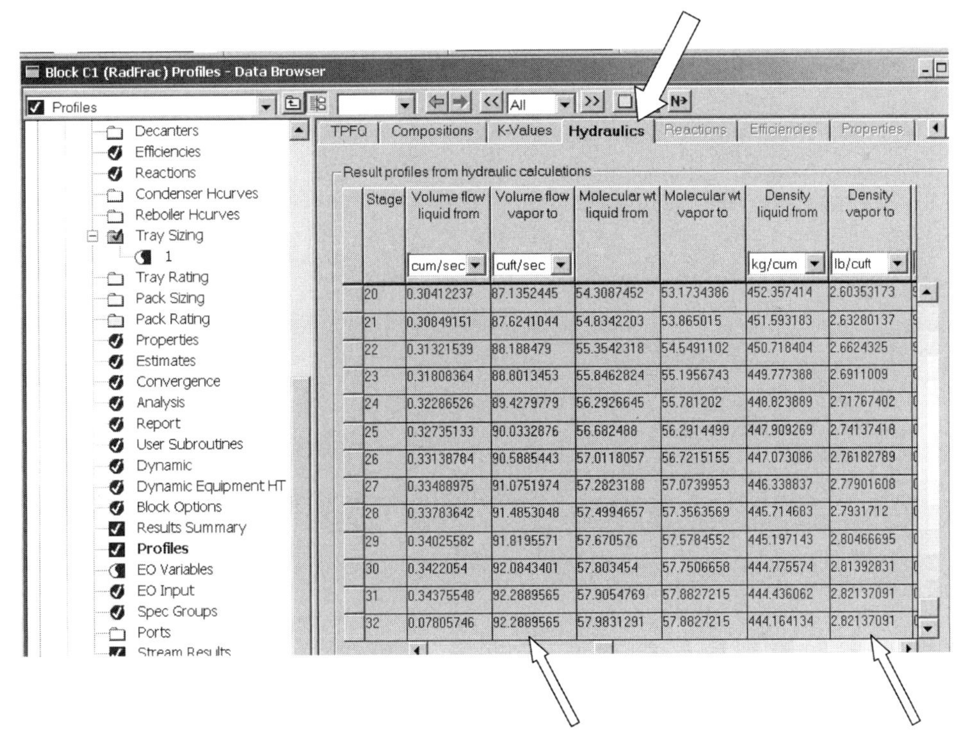

Stage	Volume flow liquid from (cum/sec)	Volume flow vapor to (cuft/sec)	Molecular wt liquid from	Molecular wt vapor to	Density liquid from (kg/cum)	Density vapor to (lb/cuft)
20	0.30412237	87.1352445	54.3087452	53.1734386	452.357414	2.60353173
21	0.30849151	87.6241044	54.8342203	53.865015	451.593183	2.63280137
22	0.31321539	88.188479	55.3542318	54.5491102	450.718404	2.6624325
23	0.31808364	88.8013453	55.8462824	55.1956743	449.777388	2.6911009
24	0.32286526	89.4279779	56.2926645	55.781202	448.823889	2.71767402
25	0.32735133	90.0332876	56.682488	56.2914499	447.909269	2.74137418
26	0.33138784	90.5885443	57.0118057	56.7215155	447.073086	2.76182789
27	0.33488975	91.0751974	57.2823188	57.0739953	446.338837	2.77901608
28	0.33783642	91.4853048	57.4994657	57.3563569	445.714683	2.7931712
29	0.34025582	91.8195571	57.670576	57.5784552	445.197143	2.80466695
30	0.3422054	92.0843401	57.803454	57.7506658	444.775574	2.81392831
31	0.34375548	92.2889565	57.9054769	57.8827215	444.436062	2.82137091
32	0.07805746	92.2889565	57.9831291	57.8827215	444.164134	2.82137091

Figure 3.54 Hydraulic results.

The diameter given by Aspen Plus can be checked by using the approximate heuristic that the F factor should be equal to 1 (in English engineering units)

$$F = V_{\max}\sqrt{\rho_V}$$

where $V_{\max}$ is the maximum vapor velocity in units of ft/s and ρ_V is the vapor density in units of lb/ft^3.

All the detailed information about the vapor and liquid flows throughout the column can be accessed by clicking the subitem *Report* under the C1 block and under *Property Options* checking the box in front of *Include hydraulic parameters*. Then, after the program is run, click the subitem *Profiles* and click the *Hydraulics* page tab. The window that opens gives lots of information about liquid and vapor rates and properties, as shown in Figure 3.54.

The maximum vapor volumetric flowrate is 9.23 ft^3/s and occurs on stage 32. The vapor density on this stage is 2.82 lb/ft^3. Using an F factor of 1, the maximum velocity is 0.595 ft/s, which give a cross-sectional area of 155 ft^2. This corresponds to a diameter of 14.0 ft or 4.28 m, which is somewhat lower than the Aspen Plus result (5.91 m).

In either case, this is a very large distillation column, and, as we will see in the next chapter, it is very expensive to buy.

3.9 CONCLUSION

All the details of setting up and running a steady-state simulation of a simple distillation column have been presented in this chapter. These methods are applied to a variety of columns in later chapters.

CHAPTER 4

DISTILLATION ECONOMIC OPTIMIZATION

In Chapter 3 we studied how to design a distillation column given the feed conditions, the desired product specifications, and the total number of stages. The calculated design parameters included the operating pressure of the column, the reboiler and condenser heat duties, and the length and diameter of the column vessel.

In this chapter the steady-state economic optimization of a distillation column is discussed. Basically, we need to find the optimum number of total stages. There are some simple approaches, and there are more rigorous approaches. The simple methods use heuristics such as setting the total number of trays equal to twice the minimum. The rigorous methods determine how the capital and energy costs change with the number of trays and find the minimum total annual cost design.

4.1 HEURISTIC OPTIMIZATION

There are two widely used heuristics for optimum distillation design. We will discuss both of them in this section and compare the designs that result from applying each. It should be emphasized that they cannot **both** be used simultaneously for rigorous design because fixing one of the two completely specifies the design of the column.

4.1.1 Set Total Trays to Twice Minimum Number of Trays

We discussed the use of the Fenske equation to find the minimum number of trays in Chapter 2 for constant-relative-volatility systems. We found the minimum number of trays more rigorously in Chapter 3 by using the simulator to find the number of stages where the required reflux ratio became very large. In the numerical example, the

Distillation Design and Control Using Aspen™ Simulation, By William L. Luyben

minimum number of trays is 15. Taking twice this number and adding two stages for the reflux drum and reboiler give a 32-stage column, which is the column we designed in Chapter 3.

It is interesting to compare this rigorous number with what the Fenske equation predicts for the same system. The first issue is to find an appropriate value for the relative volatility for the propane/isobutane system at the operating pressure of the column (16.8 atm). The usual approach is to find the relative volatility at the temperature at the top of the column (stage 2 temperature is 325.6 K) and at the temperature in the reboiler (stage 32 temperature is 366.1 K).

By definition, relative volatility α_{LH} is the ratio of the vapor and liquid compositions of component L divided by the same ratio of component H. The propane and isobutane compositions on the top tray (stage 2) can be seen in the subitem *Profiles* under the column C1 block in the *Data Browser* window:

$$\text{Propane:} \quad x = 0.96256322; \quad y = 0.98$$

Of course, in this binary system, the isobutane compositions are simply 1 minus the propane mole fractions. Therefore the relative volatility between propane and isobutane on the top tray is

$$\alpha = \frac{y_{C_3}/x_{C_3}}{y_{iC_4}/x_{iC_4}} = \frac{0.98/0.96356322}{(1-0.98)/(1-0.96356322)} = 1.8529$$

Repeating this calculation for conditions in the reboiler gives

$$\text{Propane:} \quad x = 0.01; \quad y = 0.01715812$$

$$\alpha = \frac{y_{C_3}/x_{C_3}}{y_{iC_4}/x_{iC_4}} = \frac{0.01715812/0.01}{(1-0.01715812)/(1-0.01)} = 1.7633$$

The geometric average of these relative volatilities is

$$\alpha_{\text{average}} = \sqrt{(1.8529)(1.7633)} = 1.8076$$

The distillate and bottoms propane compositions are $x_D = 0.98$ and $x_B = 0.01$. Substituting these values in the Fenske equation gives the minimum number of trays $N_{\min}$:

$$N_{\min} + 1 = \frac{\log\left[\left(\frac{x_D}{1-x_D}\right)\left(\frac{1-x_B}{x_B}\right)\right]}{\log(\alpha_{\text{average}})} = \frac{\log\left[\left(\frac{0.98}{1-0.98}\right)\left(\frac{1-0.01}{0.01}\right)\right]}{\log(1.8076)} = 14.34$$

Therefore the minimum number of trays is 13.43, which is close to that found from the simulation (15). Using this heuristic would lead us to set the actual number of tray equal to 2 times 15 = 30. This would give 32 stages.

4.1.2 Set Reflux Ratio to 1.2 Times Minimum Reflux Ratio

The other common distillation economic heuristic is to select a reflux ratio that is 20% larger than the minimum reflux ratio. The minimum reflux ratio found in Chapter 3 from the simulation was 2.9. Multiplying this by 1.2 gives an actual reflux ratio of 3.48. This is very close to the reflux ratio (3.49) that we found was necessary in Chapter 3 for a 32-stage column.

It is interesting to compare the minimum reflux ratio found in the simulation with that predicted by the Underwood equations. These equations are derived assuming constant relative volatilities. As we have seen in the previous section, the relative volatility between propane and isobutane is almost constant. It varies from 1.85 to 1.76. Therefore, the Underwood equations should predict the minimum reflux ratio quite well.

As discussed in Chapter 2, there are two equations. The first is solved for a parameter θ that is one of the roots of this equation

$$\sum_{j=1}^{NC} \frac{\alpha_j z_j}{\alpha_j - \theta} = 1 - q$$

where NC is the number of components, α_j is the relative volatility of component j, and q is the thermal condition of the feed ($q = 1$ for saturated liquid feed, $q = 0$ for saturated vapor feed).

Applying this to the binary propane/isobutane system for saturated liquid feed with composition 40 mol% propane and using the average relative volatility of 1.8076 for propane and 1 for isobutane give

$$\sum_{j=1}^{NC} \frac{\alpha_j z_j}{\alpha_j - \theta} = \frac{(1.8076)(0.40)}{1.8076 - \theta} + \frac{(1)(0.60)}{1 - \theta} = 1 - 1$$

$$0.72304(1 - \theta) + (0.6)(1.8076 - \theta) = 0$$

$$\theta = 1.3662$$

This value of θ is substituted into the second Underwood equation, using the the distillate composition $x_D = 0.98$ mf propane:

$$\sum_{j=1}^{NC} \frac{\alpha_j x_{Dj}}{\alpha_j - \theta} = 1 + RR_{\min}$$

$$\frac{(1.8076)(0.98)}{1.8076 - 1.3662} + \frac{(1)(0.02)}{1 - 1.3662} = 1 + RR_{\min}$$

$$RR_{\min} = 2.959$$

As expected, this is very close to the 2.9 value found in the simulation.

4.2 ECONOMIC BASIS

Equations used to calculate the capital cost of all the equipment and the energy cost of the heat added to the reboiler are needed to perform economic optimization calculations. The major pieces of equipment in a distillation column are the column vessel (of length L and diameter D, both with units of meters) and the two heat exchangers (reboiler and condenser with heat transfer areas A_R and A_C, respectively, with units of m^2). Smaller items such as pumps, valves, and the reflux drum are seldom significant at the conceptual design stage. The cost of the trays themselves is usually small compared to the costs of the vessel and heat exchangers unless expensive internals are used such as structured packing. Table 4.1 gives the economic parameter values and the sizing relationships and parameters used.

The sizing of the column vessel has been discussed in Chapter 3. The condenser and reboiler heat duties are determined in the simulation, but we need to have an overall heat transfer coefficient and a differential temperature driving force in each heat exchanger to be able to calculate the required area. The values of these parameters given in Table 4.1 are typical of condensing and boiling hydrocarbon systems. Note that the overall heat transfer coefficient of the condenser is larger than that of the reboiler. Reboilers have a greater tendency to foul because of the higher temperature (more coking or polymerization) and because any heavy material in the feed drops to the bottom of the column.

Various objective functions are used for economic optimization. Some are quite elegant and incorporate the concept of the "time value of money." Examples are "net present value" and "discounted cash flow." These methods are preferred by business majors, accountants, and economists because they are more accurate measures of profitability over an extended time period. However, many assumptions must be made in applying these methods, and the accuracy of these assumptions is usually quite limited. The prediction of future sales, prices of raw materials and products, and construction schedule is usually a guessing game made by marketing and business managers whose track record for predicting the future is almost as poor as that of the weather forecaster (meteorologist).

Therefore, the use of some simple economic objective function usually serves the purpose of optimizing a distillation column design. We will use the total annual cost (TAC). As shown in Table 4.1, this measure incorporates both energy cost and the annual cost of capital. The units of TAC are \$/year. The units of capital investment are U.S. dollars (\$). The units of annual cost of capital are \$/year, and it is obtained by dividing the cost of capital by a suitable payback period.

TABLE 4.1 Basis of Economics

Parameter	Value
Condensers	
Heat transfer coefficient	0.852 kW/K·m^2
Differential temperature	13.9 K
Capital cost	7296 (area in m^2)$^{0.65}$
Reboilers	
Heat transfer coefficient	0.568 kW/K·m^2
Differential temperature	34.8 K
Capital cost	7296 (area in m^2)$^{0.65}$
Column vessel capital cost	17,640 [diameter (D) in meters]$^{1.066}$ [length (L) in meters]$^{0.802}$
Energy cost	\$4.7/10^6 kJ

$$\text{TAC} = \frac{\text{capital cost}}{\text{payback period}} + \text{energy cost}$$

Payback period	3 years

The cost of energy varies quite a bit from plant to plant. In some locations energy sources are plentiful and inexpensive. For example, in Saudi Arabia gas coming from an oil well is sometimes simply flared (burned). In other locations, fuel is quite expensive because it must be transported long distances. For example, in Japan some of the natural gas is shipped in from Indonesia on liquefied natural gas tankers (LNG), which are very expensive. Therefore energy costs depend on location. A value of \$4.7 per kilojoule is used in the results presented below.

4.3 RESULTS

The equations to calculate the capital cost of all the equipment and the energy cost of the energy are given in the Matlab program shown in Table 4.2. The numerical example is for the 32-stage column studied in Chapter 3.

Table 4.3 gives results for a range of values for the total number of stages. The 32-stage case is shown in the second column. The capital cost of the column shell, which is 5.91 m in diameter and 1.2(0.61)(30) = 22 m in length, is \$1,400,000. The capital cost of the two large heat exchangers at \$1,790,000 is more than the vessel. The total annual cost is \$5,090,000 per year. Note that most of this is energy (\$4,030,000 per year).

The other columns in Table 4.3 give results for columns with other total stages. If the number of stages is reduced to 24, which gives a shorter column, reboiler heat input increases. This increases column diameter and heat exchanger areas. This results in an increase in both capital and energy costs.

If the number of stages is increased, the column becomes taller, but its diameter becomes smaller because reboiler heat input decreases. This decreases heat exchanger costs and energy costs. However, the cost of the vessel increases because it is longer.

So the effect of increasing the number of stages is to increase the capital cost of the shell and to decrease the capital cost of the heat exchangers and energy costs. As more and more

TABLE 4.2 Matlab Program to Evaluate Economics

```
% Program ''economics.m''
% economics for distillation column Example 1 (depropanizer)
% Given Qr, Qc and number of trays, calculate TAC
% for standard column
% using SI units (m, K, MW)
% Cost of energy=$4.7 per kJ
ur=0.568;uc=.852;dtr=34.8;dtc=13.9;costenergy=4.7;
% 32 stage column
nt=30;d=5.91;qr=27.17;qc=22.68;
l=nt*2*1.2/3.281;
shell=17640*(d^1.066)*(l^0.802);
ar=qr*1.055*2.54e6/3600/.7457/(dtr*ur);ac=qc*1.055*2.54e6/3600/
  .7457/(dtc*uc);
hx=7296*(ar^0.65 +ac^0.65);
energy=qr*costenergy*3600*24*365/1000;
capital=shell+hx;
tac=energy+capital/3;
nt+2,ac,ar,shell,hx,energy,capital,tac
```

TABLE 4.3 Rigorous Optimization Results

Stages	24	32	36	42	44	48
N_F	10	14	16	18	19	21
D (m)	6.82	5.91	5.77	5.67	5.65	5.63
Q_C (MW)	39.0	22.7	21.4	20.5	20.3	20.1
RR	5.10	3.46	3.21	3.04	3.00	2.96
Q_R (MW)	35.5	27.2	25.9	25.0	24.8	24.6
A_C (m^2)	3280	1910	1800	1730	1710	1700
A_R (m^2)	1800	1370	1310	1260	1260	1240
Shell (10^6 \$)	1.27	1.40	1.50	1.68	1.74	1.87
HX (10^6 \$)	2.36	1.79	1.73	1.69	1.68	1.67
Energy (10^6/year)	5.26	4.03	3.84	3.71	3.68	3.65
Capital (10^6 \$)	3.62	3.18	3.23	3.37	3.42	3.53
TAC (10^6 \$/year)	6.46	5.09	4.92	4.83	**4.82***	4.83

*Optimum.

stages are added, the incremental decrease in reboiler heat input gets smaller and smaller. The cost of the shell continues to increase (to the 0.802 power as shown in Table 4.1). Figure 4.1 shows how the variables change with the number of stages.

The total annual cost reaches a minimum of \$4,823,000 per year for a column with 44 stages. Thus in this numerical case, the optimum ratio of the actual number of trays to the minimum is $42/15 = 2.8$ instead of the heuristic 2. The reflux ratio is 3 at the optimum 44-stage design, which gives a ratio of actual to minimum of $3/2.9 = 1.04$ instead of the heuristic 1.2.

These differences may seem quite large and indicate that the heuristics are not very good. However, good engineers always build in some safety factors in their designs.

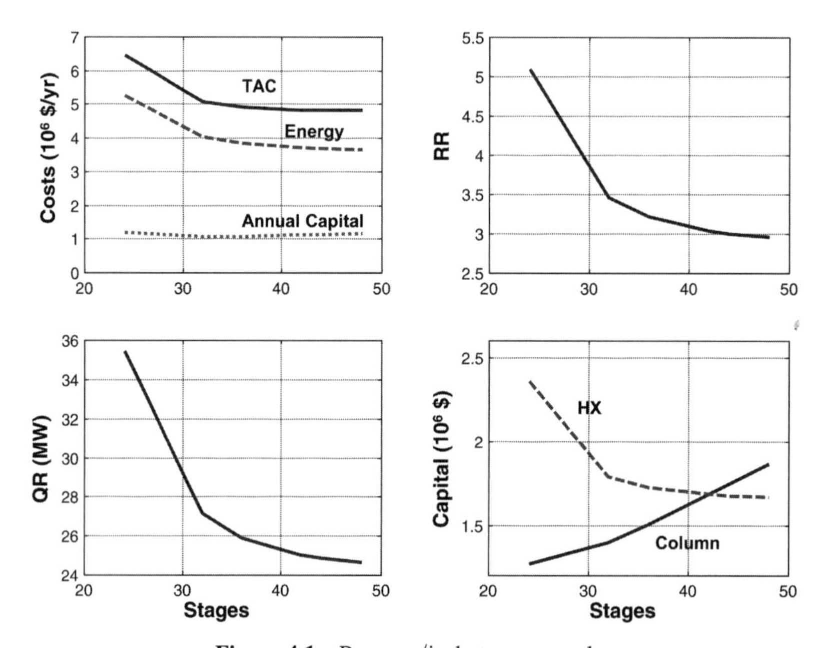

Figure 4.1 Propane/isobutane example.

Building a column that is larger in diameter and has more heat exchanger area than the real economic optimum is good conservative engineering. The number of trays in a column can sometimes be increased by going to smaller tray spacing or installing more efficient contacting devices. But changing the diameter requires a completely new vessel. Therefore, the heuristics give a pretty good design.

It should also be noted that the optimum is quite flat. The TAC decreases only from 5.09 to 4.82×10^6 \$/year as the number of stages is increased from the heuristic 34 stages to the optimum 44 stages. This is only 5%.

If the cost of energy is reduced, the optimum number of stages becomes smaller. Using an energy cost of half that assumed above, the optimum number of stages is 42 instead of 44 and the TAC drops from \$4,823,000 to \$2,980,000 per year. It is clear that energy costs dominate the design of distillation columns.

Stainless steel is used in the cost estimates given in Table 4.1. If the materials of construction were more exotic, the optimum number of stages would decrease.

4.4 OPERATING OPTIMIZATION

In the discussion up to this point, we have been considering the "design problem," namely, finding the optimum number of stages. A second type of optimization problem of equal importance is the "rating problem": finding the optimum operating conditions for a given column with a fixed number of stages.

There are several types of rating problems. One of the most common is finding the product purities that maximize profit. In the design problem considered in previous sections, we assume that the product purities were given. In many columns the purity of one product may be fixed by a maximum impurity specification, but the other product may have no set purity. For example, suppose that the propane product is more valuable than the isobutane and has a maximum impurity specification of 2 mol% isobutane. We know that distillate flowrate should be maximized and that as much isobutane as possible should be included in this stream, up the impurity constraint. This can be achieved by minimizing the concentration of propane that is lost in the bottoms. But reducing x_B requires an increase in reboiler heat input, which increases energy cost. Therefore, there is some value of x_B that maximizes profit. The optimization must take into account the value of the propane product compared to the bottoms and the cost of energy.

The steady-state simulator can be used to find this optimum operating condition. The distillate composition is held constant using a Design Spec/Vary. A value of the bottoms composition is specified in a second Design Spec/Vary, and the simulation is run to find the corresponding reboiler heat input, the distillate flowrate, and the bottoms flowrate. The profit is calculated for this value of x_B by multiplying the price of each product (\$/kg) by its mass flowrate (kg/s), multiplying the price of the feed by its mass flowrate and multiplying the reboiler heat input (MW) by the cost of energy (\$/MW · s). Profit (\$/s) is defined as the income from the two products minus the cost of the feed minus energy cost. Then a new value of bottoms composition is specified and the calculations are repeated.

Figure 4.2 shows the results of these calculations using the following parameter values:

1. Value of distillate = 0.528 \$/kg
2. Value of bottoms = 0.264 \$/kg

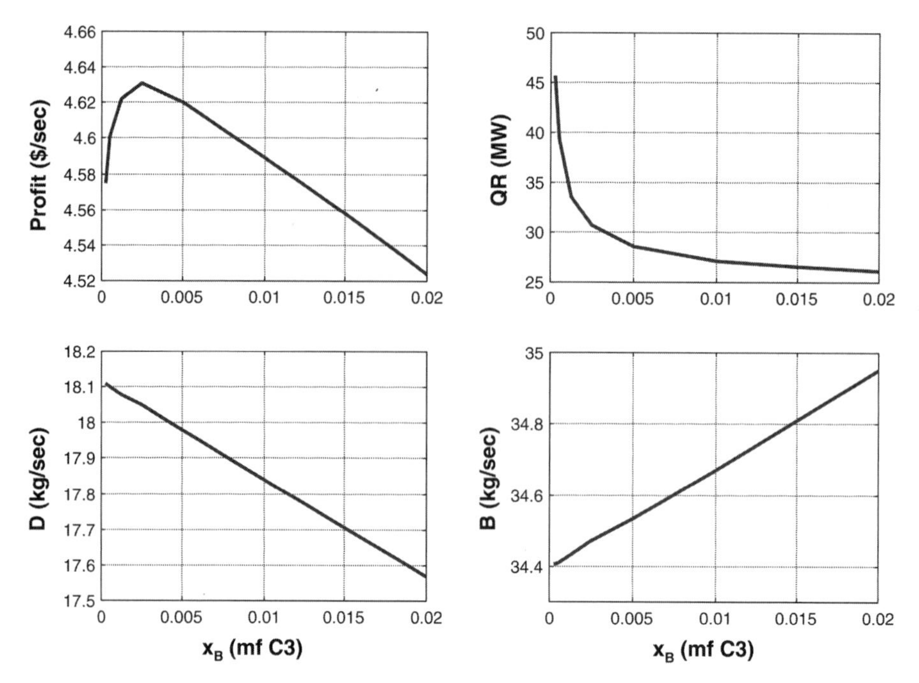

Figure 4.2 Optimum bottoms purity.

3. Cost of feed = 0.264 \$/kg
4. Cost of energy = \$ 4.7/10^6 kJ

As the bottoms composition decreases, the reboiler heat input and the distillate flowrate increase. There is a rapid rise in reboiler heat input below 0.2 mol% propane. The maximum profit is obtained with a bottoms composition of 0.25 mol% propane.

This type of optimization is a "nonlinear programming" problem (NLP), which can be performed automatically in Aspen Plus. Click *Model Analysis Tools* on the *Data Browser* window and select *Optimization*. Click the *New* button and then *OK* to create an ID. The window shown in Figure 4.3 opens, which has a number of page tabs.

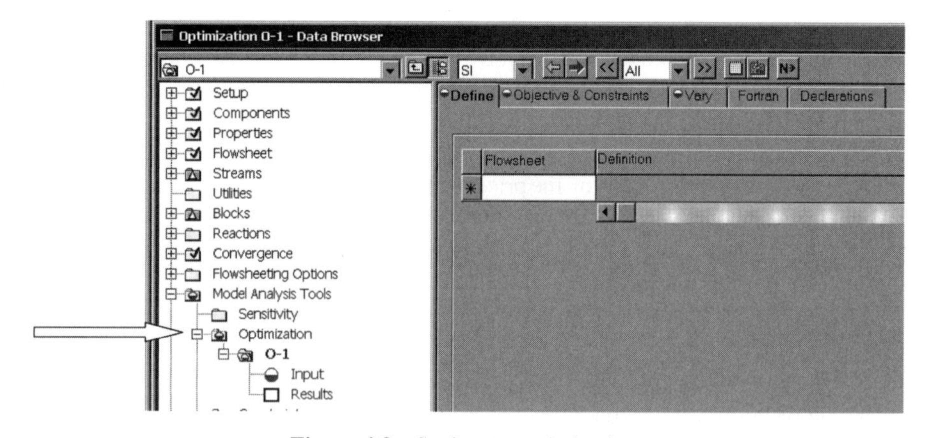

Figure 4.3 Setting up optimization.

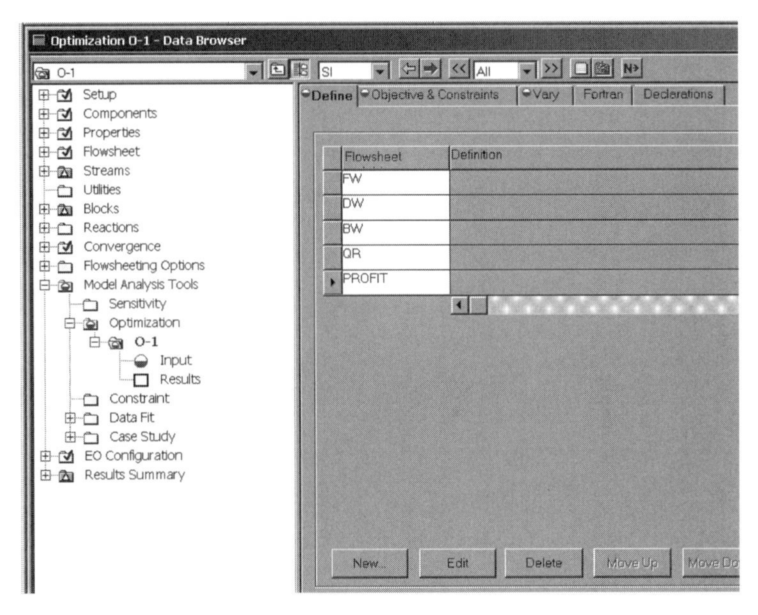

Figure 4.4 Define variables.

On the *Define* page the variables to be used in calculating the profit are defined. Type a variable name under the *Flowsheet* label. Figure 4.4 shows that several variables have been entered. The mass flowrates of feed, distillate, and bottoms are *FW*, *DW*, and *BW* (in *Flowsheet* column) in kilograms per second. Reboiler heat input is *QR* in watts.

Placing the cursor on one of the lines and clicking the *Edit* button open the windows shown in Figure 4.5, where the information about that variable is specified. For example, *FW* is edited in Figure 4.5a. Under the *Category* heading, *Streams* is selected. Under the *Reference* heading, the type is *Stream-Var*, the stream is *F1*, the variable is *Mass-Flow*.

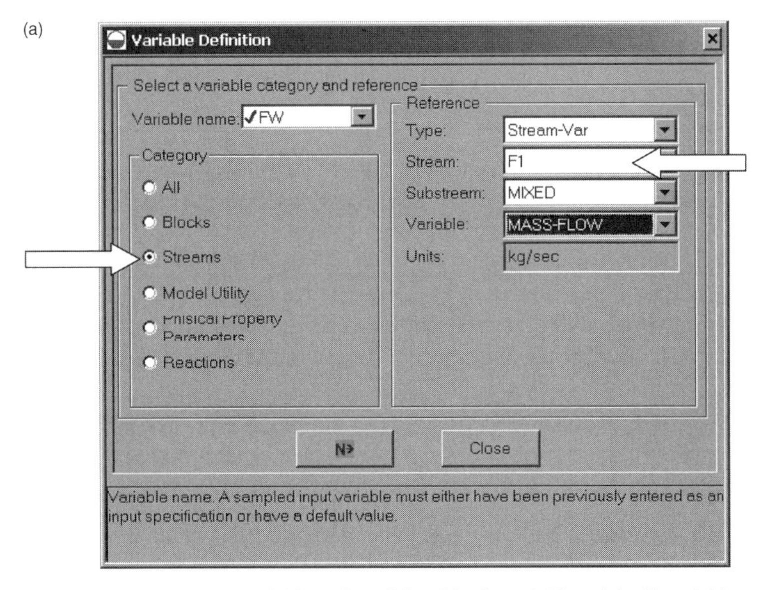

Figure 4.5 (a) Editing stream variables; (b) editing block variables; (c) all variables specified.

(*b*)

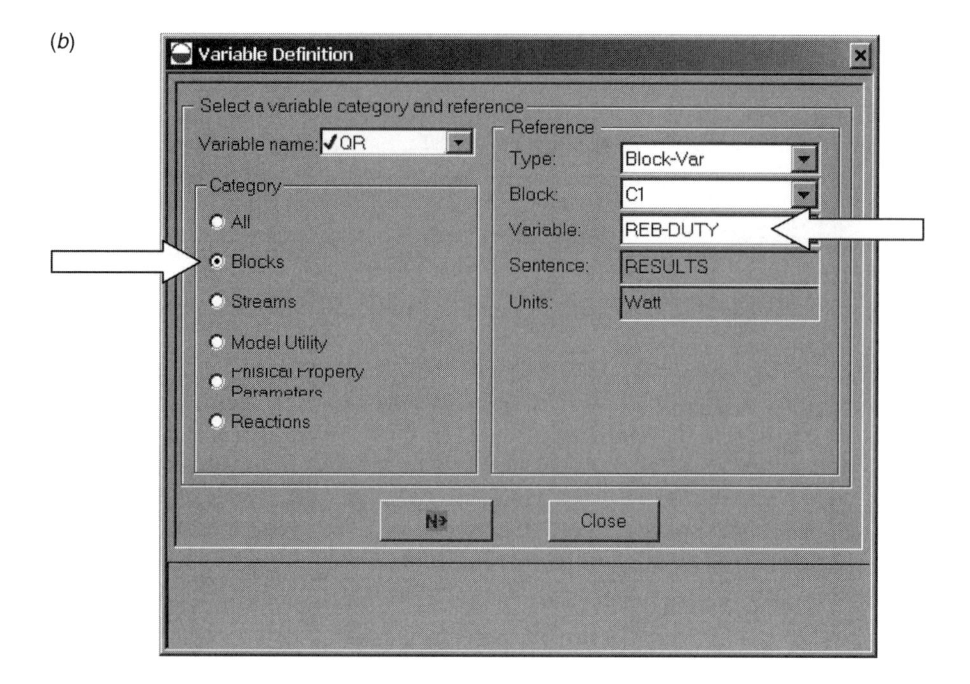

(*c*)

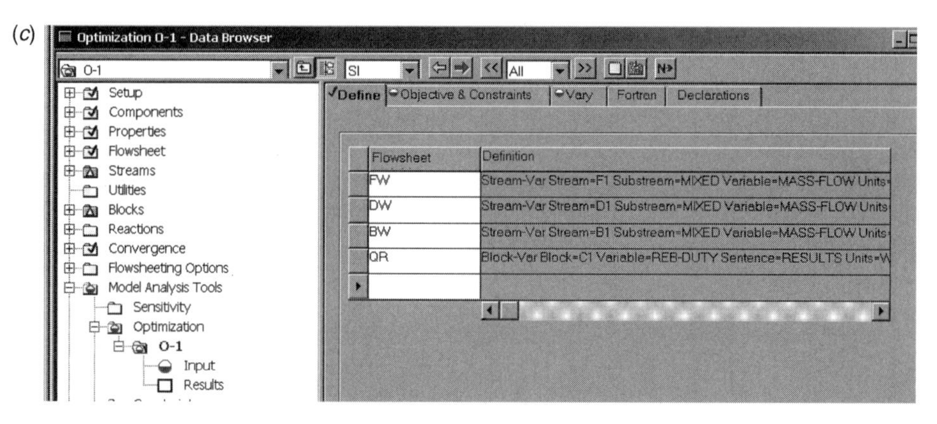

Figure 4.5 *Continued.*

Figure 4.5b shows the editing for the reboiler heat input. Since it is in the C1 block, *Blocks* under the *Category* heading is selected. Figure 4.5c shows that all variables have been defined. Clicking the *Objectives & Constraints* page tab opens the window shown in Figure 4.6, on which *PROFIT* is specified to be maximized. This variable is defined by clicking the *Fortran* page tab and entering the equation for profit as shown in Figure 4.7:

```
PROFIT=DW*0.528+BW*0.264-FW*0.264-QR*4.7e-9
```

(where `4.7e-9` $= 4.7 \times 10^{-9}$). Selecting the final page tab *Vary* opens the window shown in Figure 4.8, in which the variable to be manipulated is defined. The distillate composition is being held constant by manipulating distillate flow using a Design Spec/Vary.

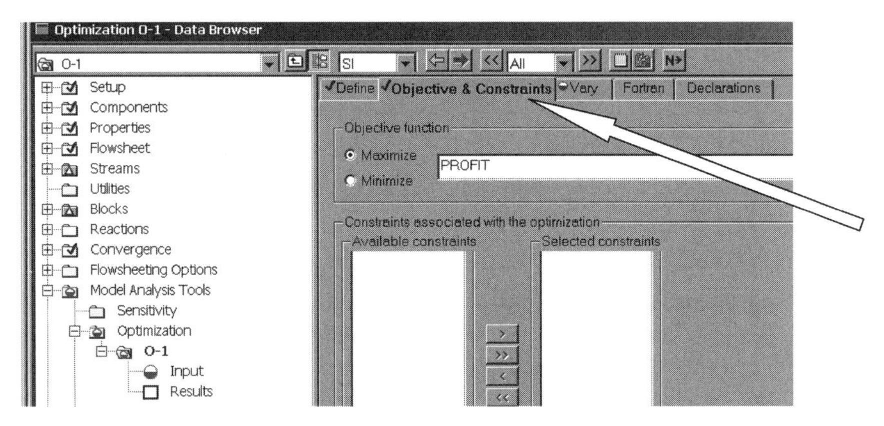

Figure 4.6 Defining the objective function.

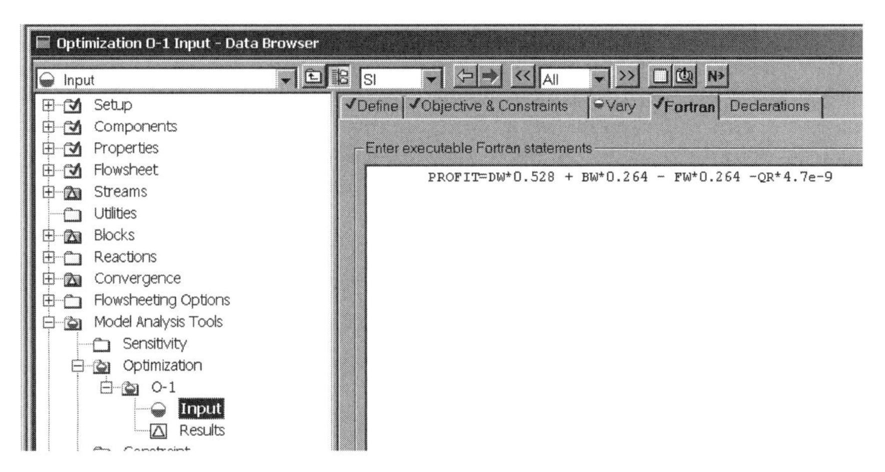

Figure 4.7 Equation for profit.

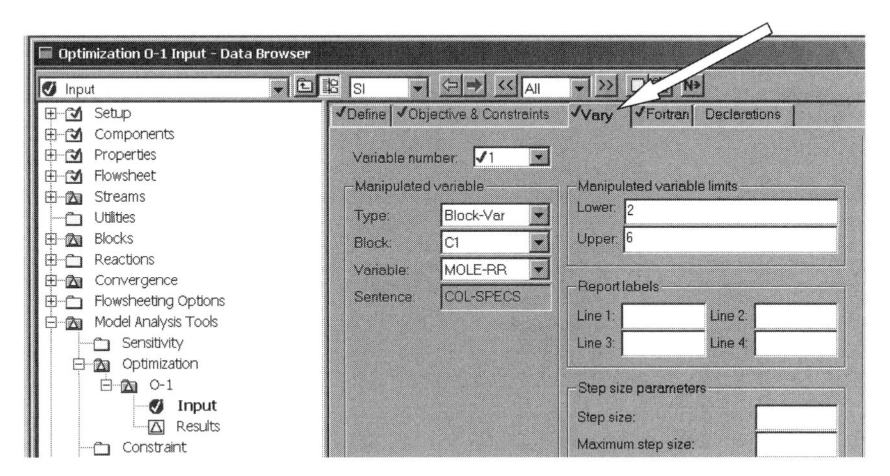

Figure 4.8 Specifying reflux ratio to Vary.

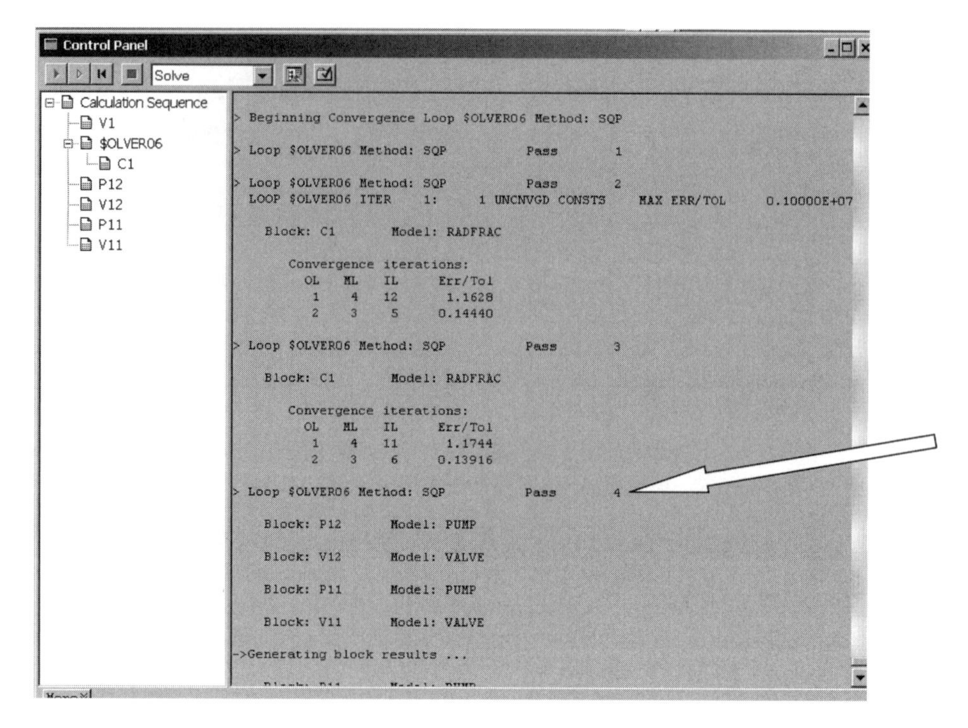

Figure 4.9 Control panel results.

The variable selected to vary in order to find the maximum profit is the reflux ratio. When the window first opens, the box to the right of *Variable number* is blank. Right-clicking opens a little window with *Create* that can be selected. Then the variable *MOLE-RR* is selected in block C1 and limits on its possible range are inserted.

The optimizer is now ready to run. Clicking the *N* button executes the program. The *Control Panel* window (Fig. 4.9) shows that the optimizing algorithm is sequential quadratic programming (SQP), and it took four iterations to find the maximum profit.

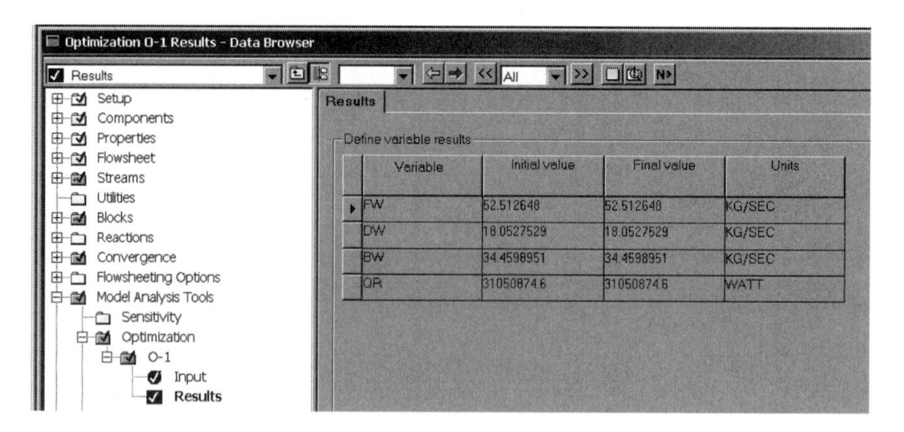

Variable	Initial value	Final value	Units
FW	52.512648	52.512648	KG/SEC
DW	18.0527529	18.0527529	KG/SEC
BW	34.4598951	34.4598951	KG/SEC
QR	31050874.6	31050874.6	WATT

Figure 4.10 NLP optimization results.

The resulting values of the variables can be seen by clicking *Results* and by looking at the stream results in the C1 block (Fig. 4.10).

The optimum value of bottoms composition is 0.2246 mol% propane, which gives distillate and bottoms flowrates of 18.05 and 34.46 kg/s, respectively. The reboiler heat input is 27.17 MW. The profit is $4.637/s.

4.5 CONCLUSION

Several types of distillation optimizations have been considered in this chapter. The approaches presented are simple and practical. There are many advanced techniques in the optimization area that are beyond the scope of this book.

CHAPTER 5

MORE COMPLEX DISTILLATION SYSTEMS

In the example distillation system considered in Chapters 3 and 4 we studied the binary propane/isobutane separation in a single distillation column. This is a fairly ideal system from the standpoint of vapor–liquid equilibrium, and it has only two components, a single feed and two product streams. In this chapter we will show that the steady-state simulation methods can be extended to multicomponent nonideal systems and to more complex column configurations.

Other methods of analysis can also be applied to these more complex systems. In particular, we will find that the use of ternary diagrams provides very useful insight into the design of these complex systems. In addition, the effects of various design parameters in these nonideal systems are sometimes counterintuitive and significantly different from those in an ideal system.

5.1 METHYL ACETATE/METHANOL/WATER SYSTEM

We start with a ternary system that is fairly nonideal. Mixtures of methyl acetate, methanol, and water are generated in the production of polyvinyl alcohol. Both the methyl acetate and the methanol must be recovered for recycling or for further processing.

5.1.1 Vapor–Liquid Equilibrium

Methyl acetate and methanol form a homogeneous minimum-boiling azeotrope at 1.1 atm with a composition of 66.4 mol% methyl acetate and a temperature of 329 K. This means that the overhead product from a distillation column separating this binary mixture cannot have a methyl acetate composition greater than this azeotropic composition. In addition,

Distillation Design and Control Using Aspen™ Simulation, By William L. Luyben

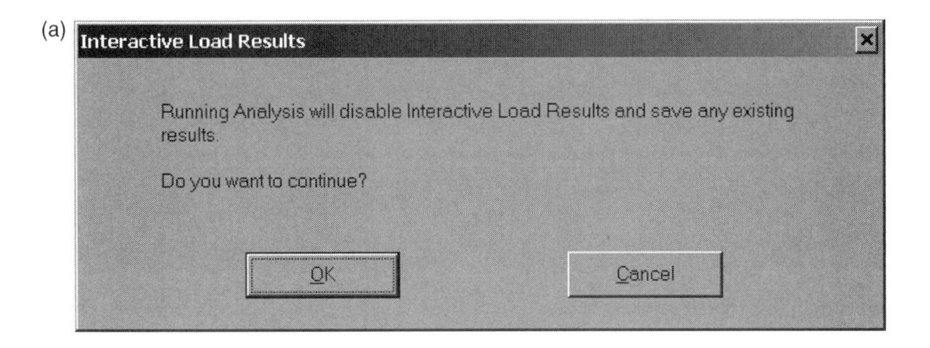

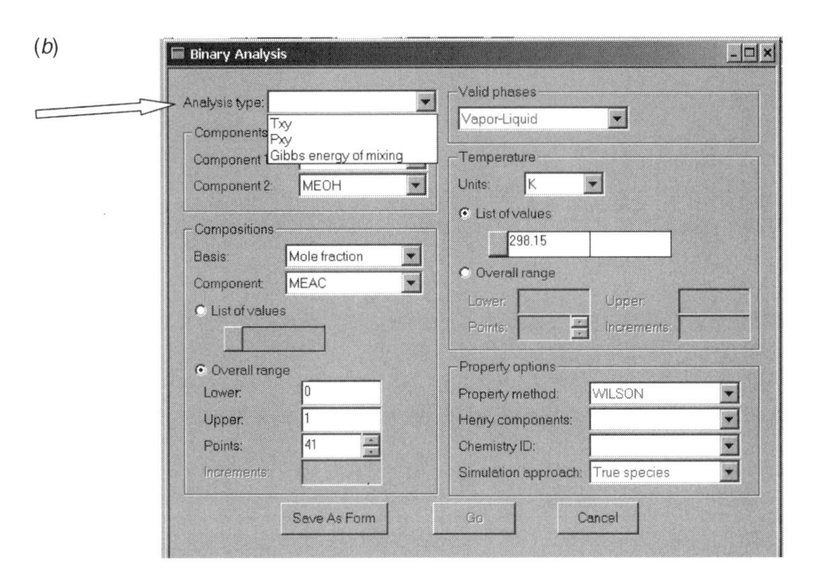

Figure 5.1 (a) Opening binary analysis; (b) selecting type of analysis.

methyl acetate and water form a homogeneous minimum-boiling azeotrope at 1.1 atm with a composition of 88.84 mol% methyl acetate and a temperature of 333.8 K.

As discussed in Chapter 1, Aspen Plus has several nice analysis tools. To look at binary VLE, go to the top toolbar and click *Tools, Analysis, Property*, and *Binary*. The window shown in Figure 5.1a opens, and clicking *OK* opens the window shown in Figure 5.1b, on which we can specify the type of analysis. The dropdown menu at the top left gives three choices. Since distillation columns run at fixed pressures, the *Txy* diagram is the most appropriate.

We set the pressure at 1.1 atm (see Fig. 5.2) because, as we will find out shortly, this pressure will give a reflux drum temperature (325 K) that is high enough to permit the use of inexpensive cooling water in the condenser. Next we select the two components of interest, MEAC and MEOH, by using the dropdown menu arrows. Note that the Wilson physical property package has been specified. Clicking the *Go* button at the bottom of the window produces the *Txy* diagram shown in Figure 5.3. The minimum-boiling homogeneous azeotrope at 66.4 mol% methyl acetate and a temperature of 329 K is shown. Repeating for the other two binary pairs produces the *Txy* diagrams shown in

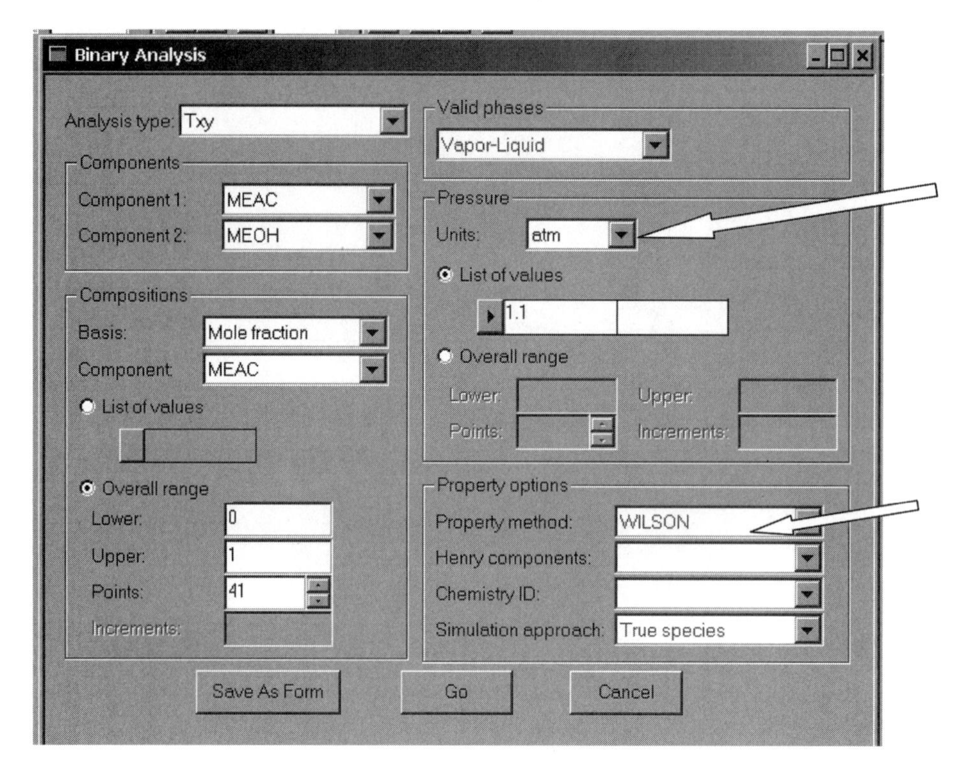

Figure 5.2 Specifying pressure.

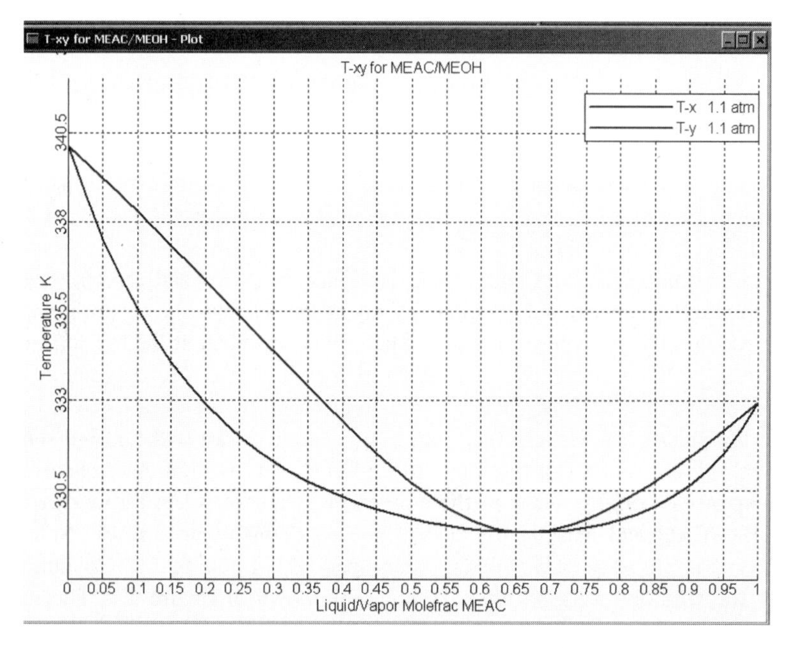

Figure 5.3 *Txy* diagram for methyl acetate/methanol.

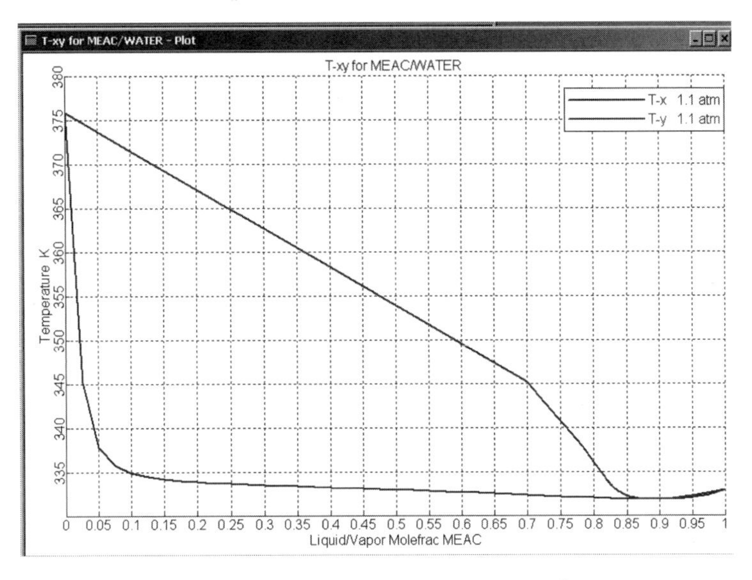

Figure 5.4 *Txy* diagram for methyl acetate/water.

Figures 5.4 and 5.5. The pure component boiling points at 1.1 atm can be found by referring to the table in *Binary Analysis Results*. Methyl acetate is 332.13 K, methanol is 340.13 K and water is 375.86 K.

Vapor–liquid equilibrium information for the ternary system can be seen by going up to the top toolbar and clicking *Tools*, *Analysis*, *Properties*, and *Residue*. The window shown in Figure 5.6a opens, and pressure and the three components of interest are specified. Clicking the *Go* button at the bottom of the window produces an equilateral ternary

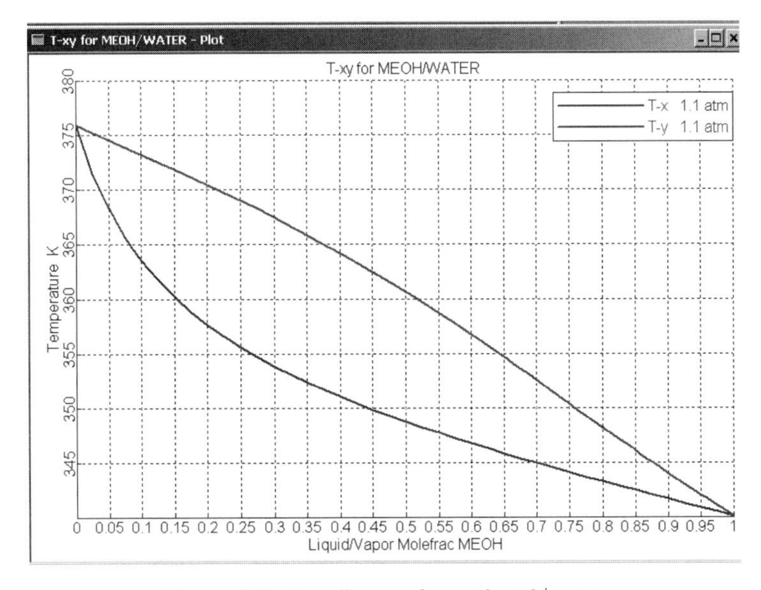

Figure 5.5 *Txy* diagram for methanol/water.

(a)

Residue Curves

Ternary system
Component 1: MEAC
Component 2: MEOH
Component 3: WATER

Pressure
1.1 atm

Valid phases
Vapor-Liquid

Number of curves
3 - 5 curves
10 - 15 curves
15 - 20 curves

Property options
Property method: WILSON
Henry components:
Chemistry ID:

Save as form | Go | Cancel

(b)

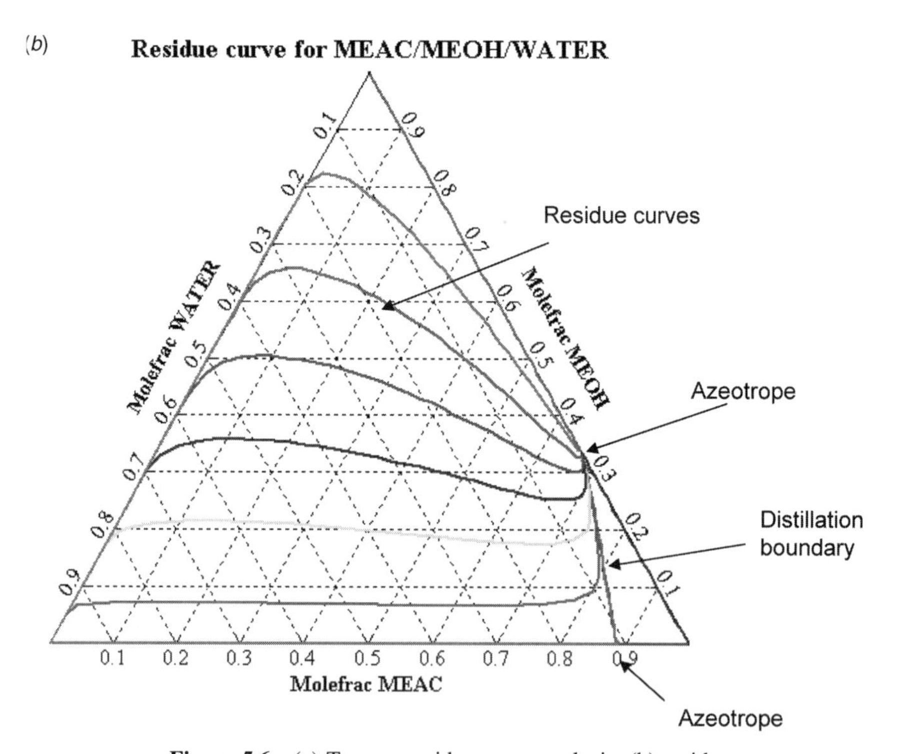

Figure 5.6 (a) Ternary residue curve analysis; (b) residue curves.

diagram that shows the residue curves (see Fig. 5.6b). The residue curve lines lead from the minimum-boiling azeotrope to the highest-boiling component, which is water. Note the distillation boundary that runs between the two azeotropes.

As discussed in Chapter 1, Aspen Split provides additional insight. Click on *Tools, Analysis*, and *Aspen Split*. Two choices can be made: *Azeotropic Search* and *Ternary Diagrams*. Selecting the first opens the window shown in Figure 5.7a, where the pressure and components are specified. Clicking on *Azeotropes* in the list on the far left of the window under *Output* produces the results shown in Figure 5.7b. Clicking *Report* gives information about the two azeotropes (Fig. 5.7c).

The second alternative in Aspen Split is to select *Ternary Diagrams*. This opens the window shown in Figure 5.8a, on which the pressure and the three components are specified. Then clicking *Ternary Plot* at the bottom of the list on the left side of the window opens a right triangle, which displays boiling points of the pure components, the compositions and temperatures of the azeotropes, and the distillation boundaries (see Fig. 5.8b).

To add the residue curve lines, right-click on the graph, select *Add* and *Curve*. The cursor becomes a cross. Move it to a spot on the diagram and left-click. A residue curve through that point is drawn. Figure 5.8c gives the ternary diagram with four residue curves added. Note that a grid has also been added. This is achieved by clicking the bottom icon on the far right that is *Toggle Grid*.

(a)

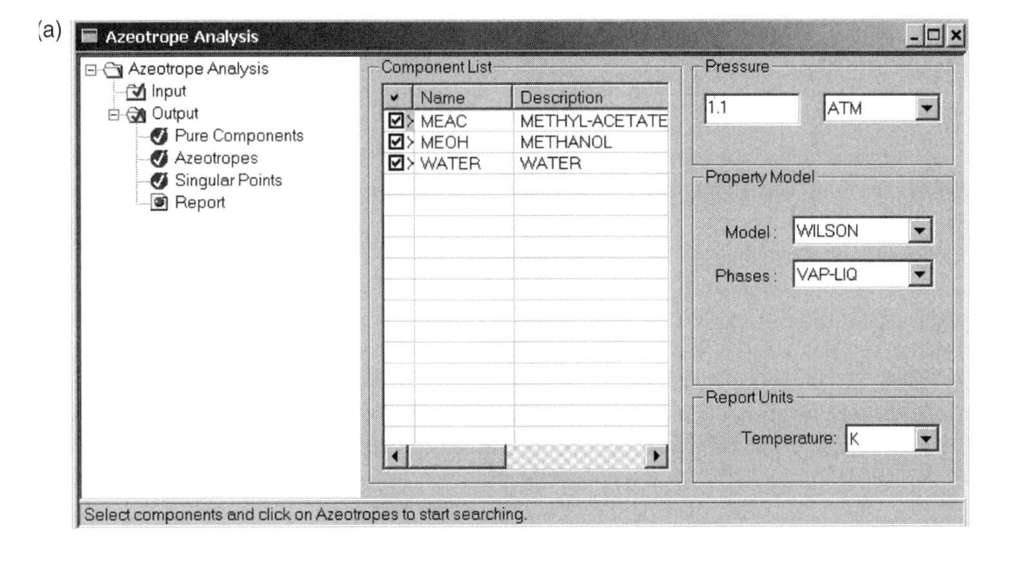

(b)

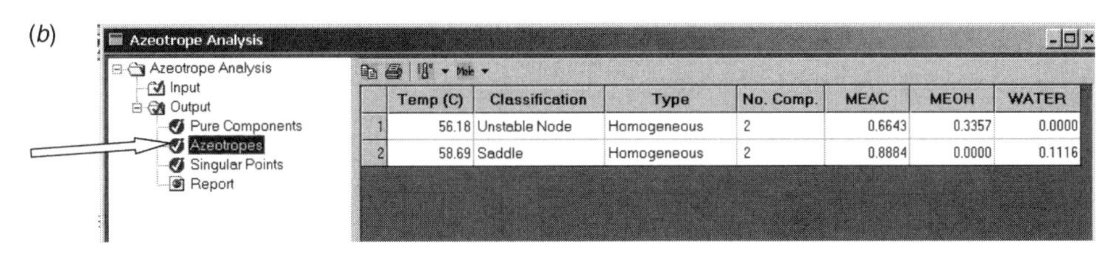

Figure 5.7 (a) Azeotropic search; (b) azeotropic results; (c) report.

(c) **Physical Property Model:** WILSON **Valid Phase:** VAP-LIQ

Mixture Investigated For Azeotropes At A Pressure Of 1.1 ATM

Comp ID	Component Name	Classification	Temperature
MEAC	METHYL-ACETATE	Stable Node	332.96 K
MEOH	METHANOL	Saddle	340.13 K
WATER	WATER	Stable Node	375.86 K

2 Azeotropes Sorted by Temperature

01
Number Of Components: 2 — Temperature 329.33 K
Homogeneous — Classification: Unstable Node

	MOLE BASIS	MASS BASIS
MEAC	0.6643	0.8207
MEOH	0.3357	0.1793

02
Number Of Components: 2 — Temperature 331.84 K
Homogeneous — Classification: Saddle

	MOLE BASIS	MASS BASIS
MEAC	0.8884	0.9704
WATER	0.1116	0.0296

Figure 5.7 *Continued.*

A number of parameters of the plot can be changed by right-clicking and selecting *Properties*. The color and thickness of lines and curves can be specified, and the plot can be rotated if desired.

Another useful feature is the ability to add a marker at some desired location. This is done by clicking the sixth icon from the top on the right side of the window. A window opens on which you can enter the composition coordinates at which you want to place a mark. In Figure 5.9a the composition of the feed to a distillation column is located at 30 mol% MeAc, 50 mol% MeOH, and 20 mol% H_2O. Then click *OK* to place the mark at the correct location (Fig. 5.9b). These markers can be used to show the location of the feed (F), distillate (D), and bottoms (B), as illustrated in Figure 5.10. Note that both the D and B points lie inside the upper region of the diagram and are **not** in two regions separated by a distillation boundary. Also note that these points lie near a residue curve. Therefore this should be a feasible design.

5.1.2 Ternary Column Design

Let us consider a numerical case in which there are several design objectives:

1. The concentration of the methyl acetate in the distillate should be fairly close to the azeotropic composition because this stream is sent for further processing. It may be sent to another distillation column operating at a higher pressure that shifts

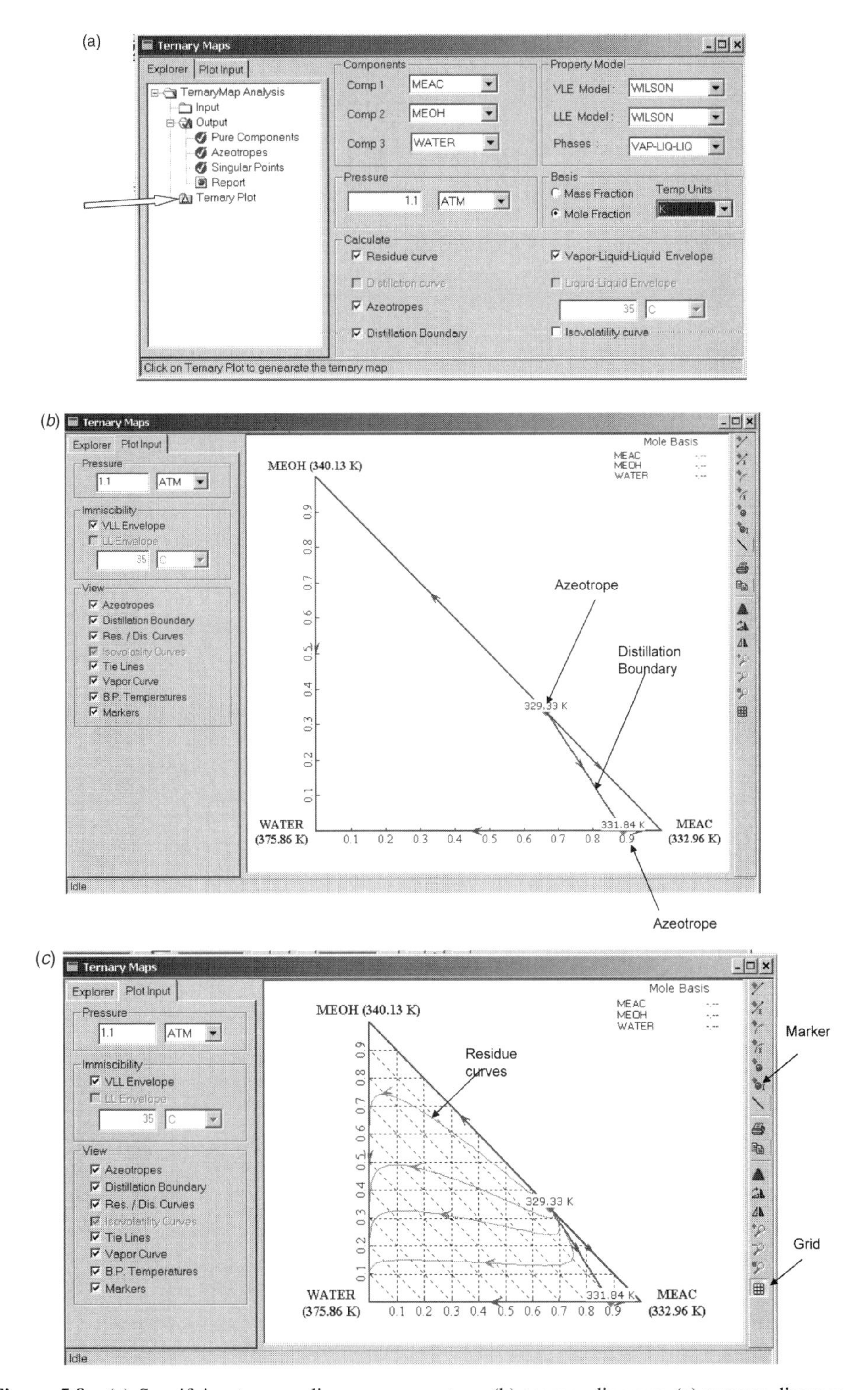

Figure 5.8 (a) Specifying ternary diagram parameters; (b) ternary diagram; (c) ternary diagram with residue curves.

(a)

Add Item By Value

User Points

Basis: Mole fraction

MEAC: 0.3

MEOH: 0.5

WATER: 0.2

OK | Cancel

(b)

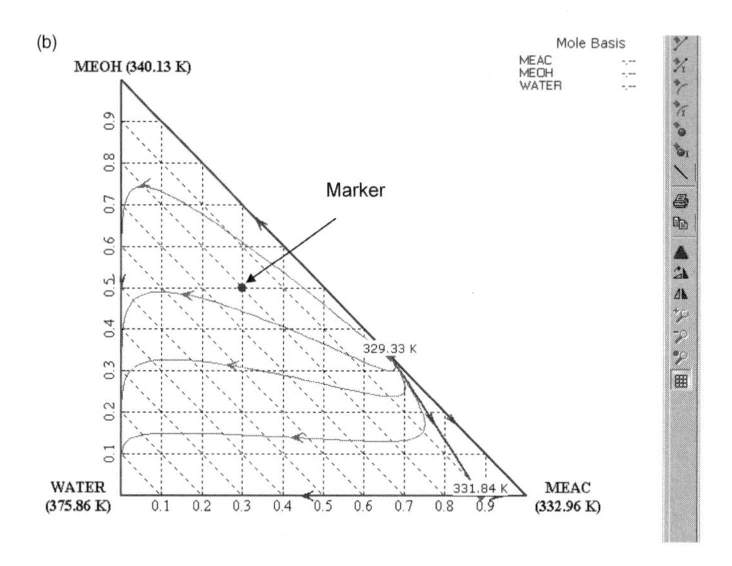

Figure 5.9 (a) Specifying composition point; (b) adding marker.

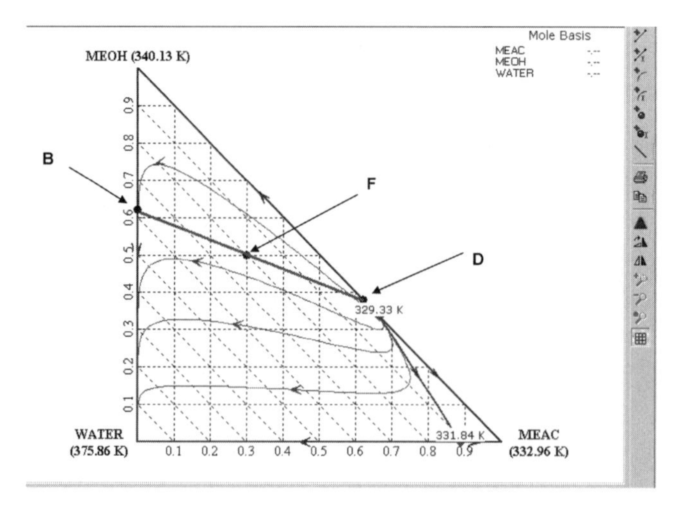

Figure 5.10 F, D, and B points marked.

the azeotropic composition enough to permit the production of methyl acetate out the bottom and the high-pressure azeotrope out the top of this second column. In other plants the stream may be fed to a reactor where it reacts with water to form acetic acid and methanol. In any event, we want as high a concentration of methyl acetate as possible in the distillate.

2. The concentration of water in the distillate should be 0.1 mol%.
3. The concentration of methyl acetate in the bottoms should be 0.1 mol%.

The feed flowrate is 0.1 kmol/s, and the design feed composition is 30 mol% methyl acetate (MeAc), 50 mol% methanol (MeOH), and 20 mol% water. Column pressure is initially set at 1.1 atm, but we will check the reflux drum temperature to make sure that cooling water can be used in the condenser.

For a preliminary design, a column with 32 stages is selected. We will vary the feed tray location to minimize reboiler heat input. Initially stage 16 in the middle of the column is specified. Figure 5.11 shows the flowsheet with valves and pumps installed.

Because of the methyl acetate azeotrope, the distillate cannot be richer in methyl acetate than 66.4 mol%. To guess an initial distillate flowrate, let us assume that the distillate contains all the methyl acetate and that its composition is 62 mol% MeAc. Therefore the distillate flowrate is calculated from a methyl acetate balance:

$$
\begin{aligned}
F z_{\mathrm{MeAc}} &= D x_{D,\mathrm{MeAc}} \\
(0.1)(0.30) &= D(0.62) \\
D &= 0.0484 \text{ kmol/s}
\end{aligned}
$$

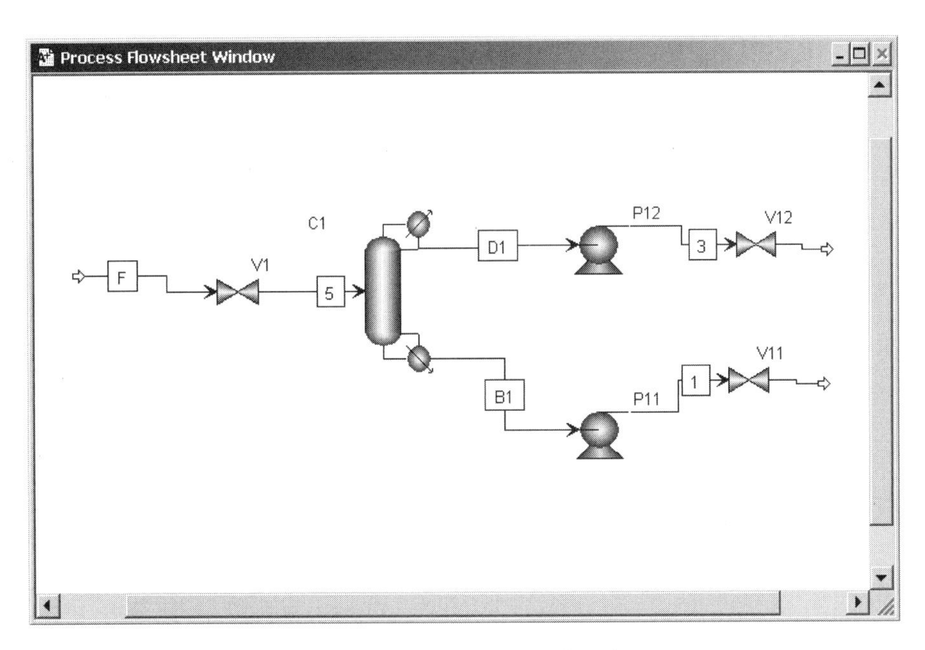

Figure 5.11 Methyl acetate flowsheet.

A guess for the reflux ratio of 2 is made, and the simulation is run. The resulting product compositions are 2.475×10^{-3} mol% H_2O and 61.98 mol% MeAc in the distillate and 3.356×10^{-7} mol% MeAc in the bottoms. To drive the bottoms methyl acetate composition to the specified 0.1 mol%, we set up a Design Spec/Vary manipulating distillate flowrate and click the *N* button. **The column does not converge** after the default 25 iterations. Clicking *Convergence* under the C1 block lets us change the number of iterations to 50 and try again. The column still does not converge.

The problem is that we have been using the *standard* method of convergence. This must be changed to either *Azeotropic* or *Strongly nonideal liquid* to achieve convergence in this nonideal system. This is done on the *Configuration* page tab under *Setup*, as shown in Figure 5.12.

It is sometimes necessary to switch back and forth between *Azeotropic* and *Strongly nonideal liquid* to achieve convergence. It is also sometimes necessary to temporarily *Hide* one or both of the Design Spec/Vary functions to achieve convergence. This is done by right-clicking on the individual Design Spec and selecting *Hide*. The same procedure is followed for the corresponding *Vary*. After the simulation has converged without the Design Spec and Vary active, they can be reactivated by right-clicking *Design Spec* or *Vary* and selecting *Reveal*.

The new distillate flowrate to achieve the bottoms methyl acetate of 0.1 mol% is 0.04424 kmol/s. The water composition of the distillate is now 0.565 mol% instead of the desired 0.1 mol%. A second Design Spec/Vary is set up manipulating the reflux ratio. The simulation is run, and the required reflux ratio is 4.833.

Now the effect of the feed tray location is explored to find the optimum where reboiler heat input is at a minimum. However, we also need to keep track of what is happening to the methyl acetate concentration in the distillate. Figure 5.13 gives results for varying the feed stage number. Unlike the ideal system explored in Chapter 3, energy consumption and product compositions are very sensitive to feed tray location. The feed stage that minimizes reboiler heat input is stage 27, which is very near the bottom of the column. However, the methyl acetate composition in the distillate drops drastically for feed stages greater than stage 26. Therefore a design tradeoff must be made between the cost of energy in this column and the cost of feeding a lower-purity methyl acetate stream

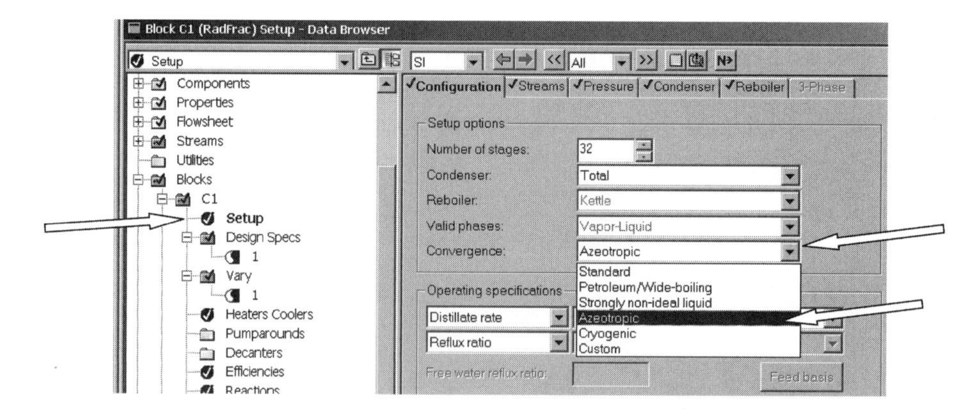

Figure 5.12 Changing convergence method.

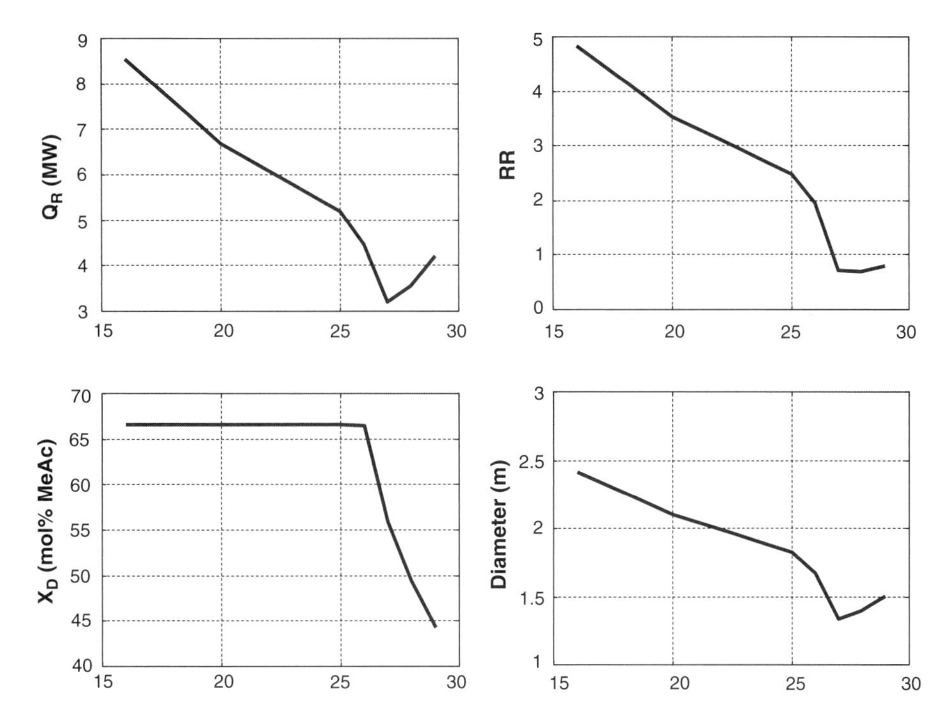

Figure 5.13 Effect of feed stage; 32 total stages.

to the downstream unit. This is a good example of the typical multiunit tradeoffs that occur in complex chemical plants.

This also illustrates that the simple heuristics cannot always be applied in complex nonideal systems. The basic complexity in this example is that we have three design specifications instead of the normal two. Therefore another degree of freedom is required, which is the feed tray location.

The next step is to see the effect of using different numbers of total stages. Figure 5.14 gives results for four different total stages: 17, 22, 27, and 32. The breakpoints where the distillate methyl acetate purity drops precipitously in the four cases are stages 12, 17, 21, and 26, respectively. At these four points, the reboiler heat inputs are 6.611, 3.318, 3.199, and 4.452 MW, respectively. So adding more stages reduces energy consumption up to about 27 stages and then it actually increases. This indicates that either the 22-stage design or the 27-stage design is the economic optimum.

Evaluating the total annual cost of these four designs shows that the 27-stage design gives the lowest TAC. Table 5.1 gives details of the design and economic parameters.

It is interesting to check the rigorous simulation developed above with the ternary analysis using the graphical column design method in DISTIL. Figure 5.15a gives the *Spec Entry* page tab with the feed composition and thermal condition ($q = 1$) given. The reflux ratio is specified to be 1, which would correspond to that found in the optimum 27-stage design. Figure 5.15b gives the bottoms specification (0.001 mf MeAc). Figure 5.15c gives the two specifications made for the distillate (0.001 mf water and 0.63 mf MeAc). Note that the latter is close to the azeotropic composition.

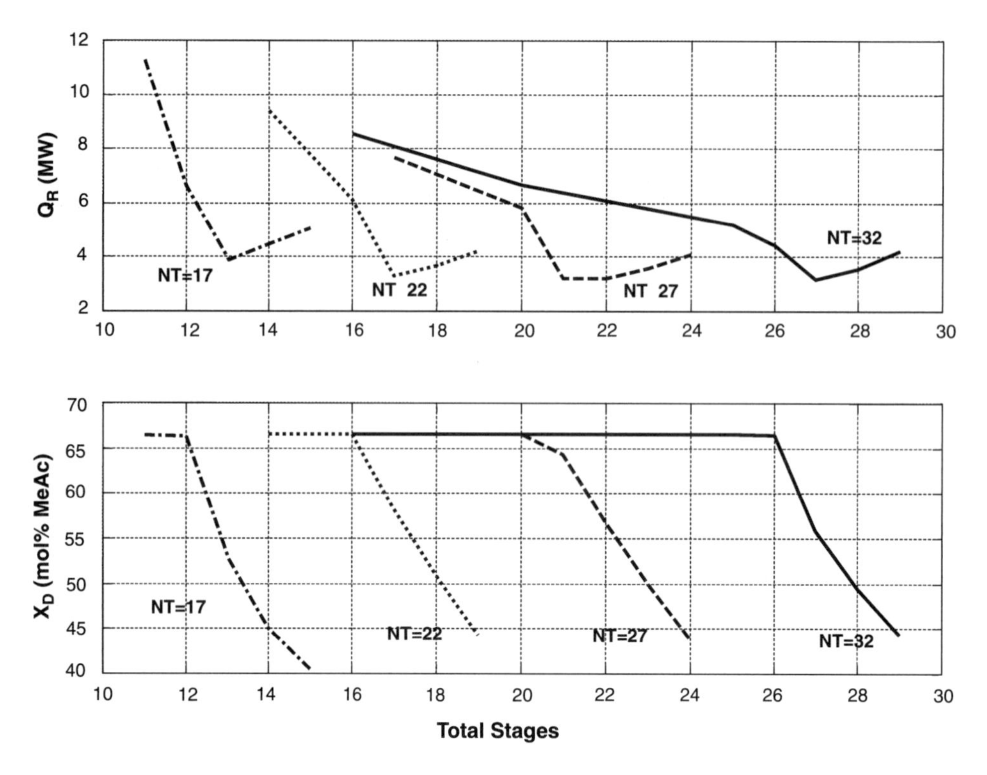

Figure 5.14 Effect of total stages.

Figure 5.15d gives the results of the DISTIL analysis. The total number of stages is 30 with the feed introduced on stage 25. These results are in good agreement with the rigorous simulations. It should be noted that the results are very sensitive to the methyl acetate specification in the distillate. Figure 5.16 shows the ternary diagram with the stream points and the different sections of the column labeled.

TABLE 5.1 Economics of Four Designs

Stages	17	22	27	32	42
N_F	12	17	21	26	35
D (m)	2.09	2.00	1.38	1.67	1.71
Q_C (MW)	6.40	3.10	2.98	4.23	4.40
RR	3.47	0.878	1.01	1.96	2.08
Q_R (MW)	6.61	3.32	3.20	4.45	4.63
A_C (m^2)	539	262	251	356	370
A_R (m^2)	334	168	161	225	234
Shell (10^6 \$)	0.263	0.317	0.255	0.363	0.468
HX (10^6 \$)	0.755	0.476	0.464	0.579	0.594
Energy (10^6/year)	0.980	0.492	0.474	0.660	0.686
Capital (10^6 \$)	1.02	0.793	0.719	0.942	1.06
TAC (10^6 \$/year)	1.32	0.756	**0.714**	0.974	1.04

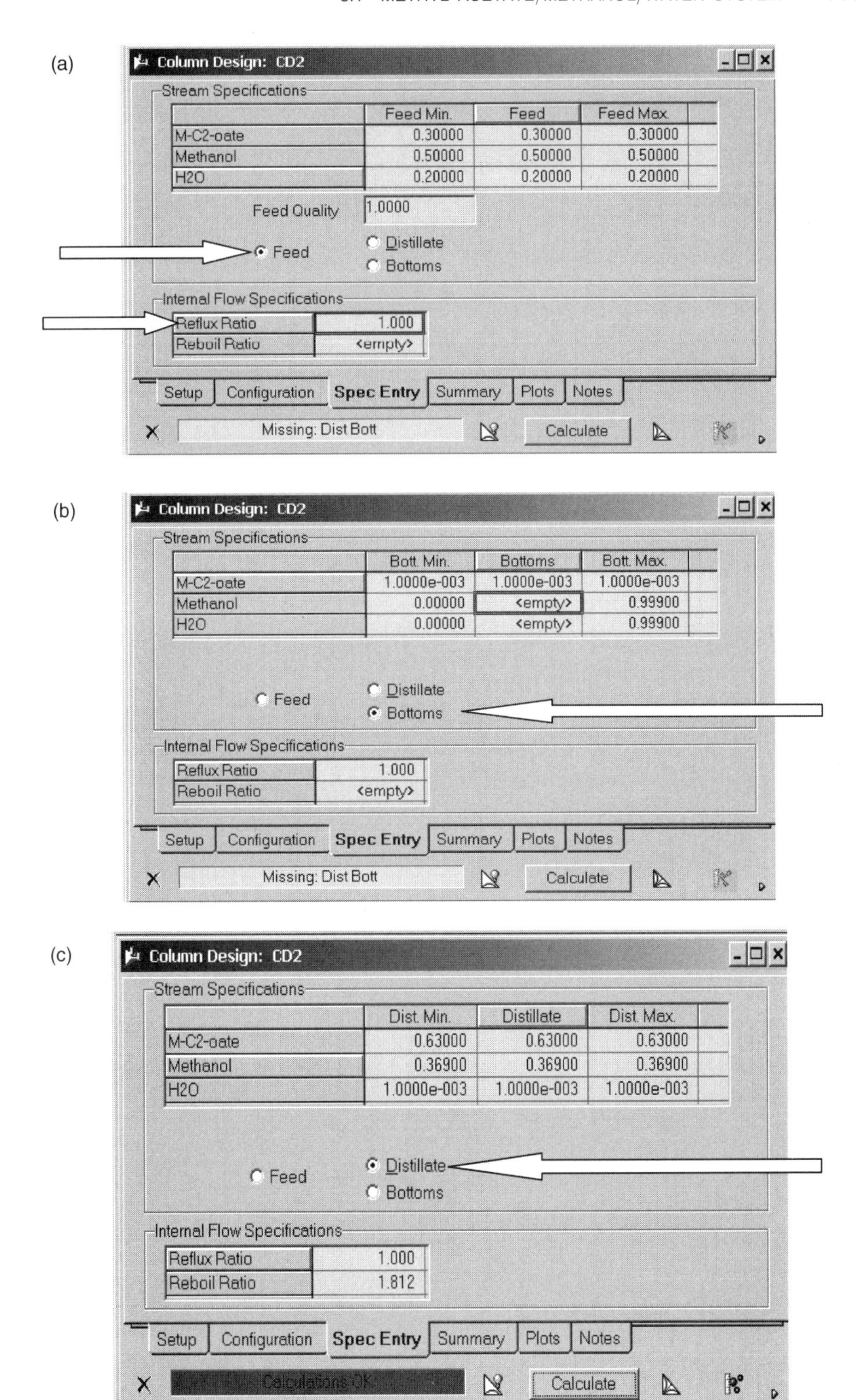

Figure 5.15 (a) Setting feed composition and quality and reflux ratio in DISTIL; (b) setting bottoms composition; (c) setting distillate composition; (d) results.

(d)

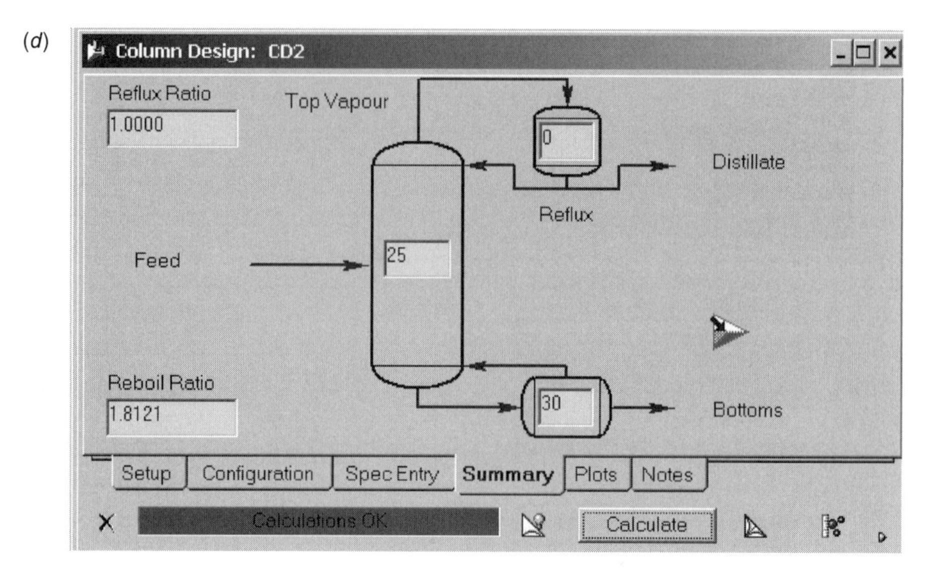

Figure 5.15 *Continued.*

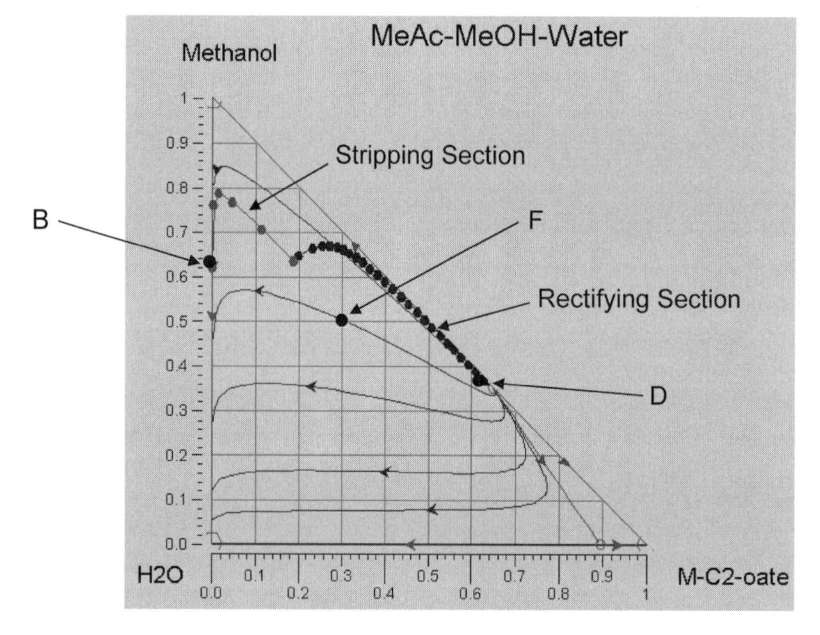

Figure 5.16 Ternary diagram.

5.2 ETHANOL DEHYDRATION

For our second nonideal system we study a process that has extremely nonideal VLE behavior and has a complex flowsheet. The components involved are ethanol, water, and benzene. Ethanol and water at atmospheric pressure form a minimum-boiling homogeneous azeotrope at 351 K of composition 90 mol% ethanol. Much more complexity is introduced by the benzene/water system, which forms two liquid phases with partial miscibility. The flowsheet contains two distillation columns and a decanter. There are two recycle streams, which create very difficult convergence problems, as we will see. So this complex example is a challenging simulation case.

The origins of the example go back over a century when a process to produce high-purity ethanol was needed. Ethanol is widely produced in fermentation processes. A typical mixture from a fermentation process has very low ethanol concentrations (4–6 mol%). If this mixture is fed to a distillation column operating at atmospheric pressure, high-purity water can be produced out the bottom but the ethanol purity of the distillate cannot exceed 90 mol% because of the azeotrope.

Some ingenious engineers came up with the idea of running the fermentation liquid through a conventional "preconcentrator" distillation column that takes most of the water out the bottom and produces a distillate that is perhaps 84 mol% ethanol and 16 mol% water. This binary stream is then fed into a second distillation column. Also fed to the top of this column as reflux is a stream that contains a high concentration (80 mol%) of benzene. The benzene serves as a "light entrainer" that goes overhead and preferentially takes water with it because the large dissimilarity between water and benzene renders the water very volatile. The ethanol goes out the bottom of this column despite the fact that water is the "heavier" component (normal boiling point of ethanol is 351.5 K, while that of water is 373.2 K).

The overhead vapor is a ternary mixture of all three components. When it is condensed, the repulsion between the water molecules and the organic benzene molecules is so great that two liquid phases form. The reflux drum becomes a decanter. The lighter organic liquid phase is pumped back to the column as organic reflux. The heavier aqueous phase contains significant amounts of ethanol and benzene, so it is fed to a second distillation column in which the water is removed in the bottoms stream. The distillate stream is recycled back to the first column.

The first item to explore is the complex VLLE of this heterogeneous vapor–liquid–liquid system.

5.2.1 VLLE Behavior

The phase equilibrium is reported to be well described by the Uniquac physical property package. Two binary *Txy* diagrams are given in Figures 5.17a and 5.17b. The two homogeneous minimum-boiling azeotropes are clearly shown. The ternary diagram (Fig. 5.17c) is generated using Aspen Split as described earlier in this chapter. A large part of the composition space has two liquid phases. The liquid–liquid equilibrium tielines are shown. The aqueous phase is on the left, and the organic phase is on the right.

Figure 5.17d gives a report of all of the azeotropes in this very nonideal VLLE system. Note that the ternary azeotrope is heterogeneous and has the lowest boiling point (337.17 K) of any of the azeotropes or of the pure components. This means that the overhead vapor from the first column will have a composition that is similar to that of this azeotrope. Note also that the diagram is split up into three regions by the distillation boundaries that connect the four azeotropes. As you may recall from Chapter 2, these boundaries limit the separation that is attainable in a single column. The bottoms and distillate points must lie in the same region.

(a)

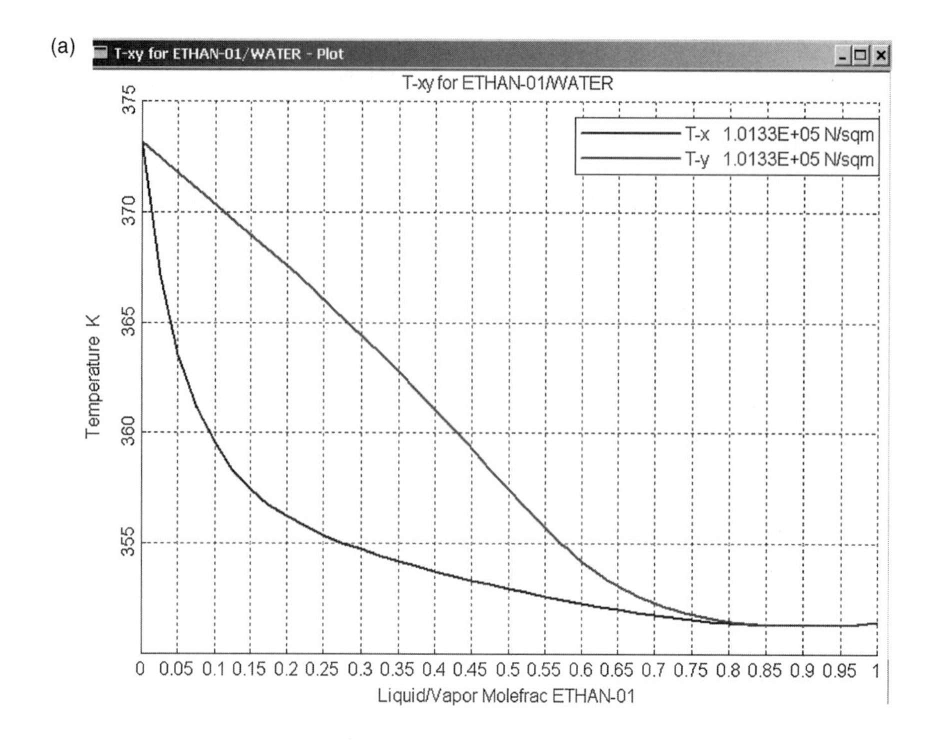

(b)

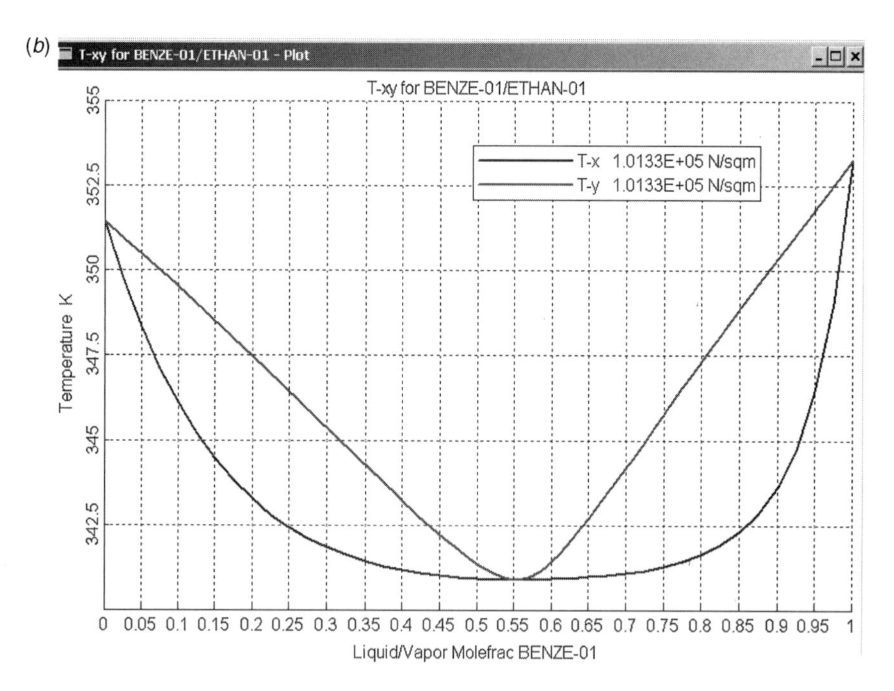

Figure 5.17 (a) *Txy* diagram for ethanol/water; (b) *Txy* diagram for ethanol benzene; (c) ternary diagram; (d) report of azetropes.

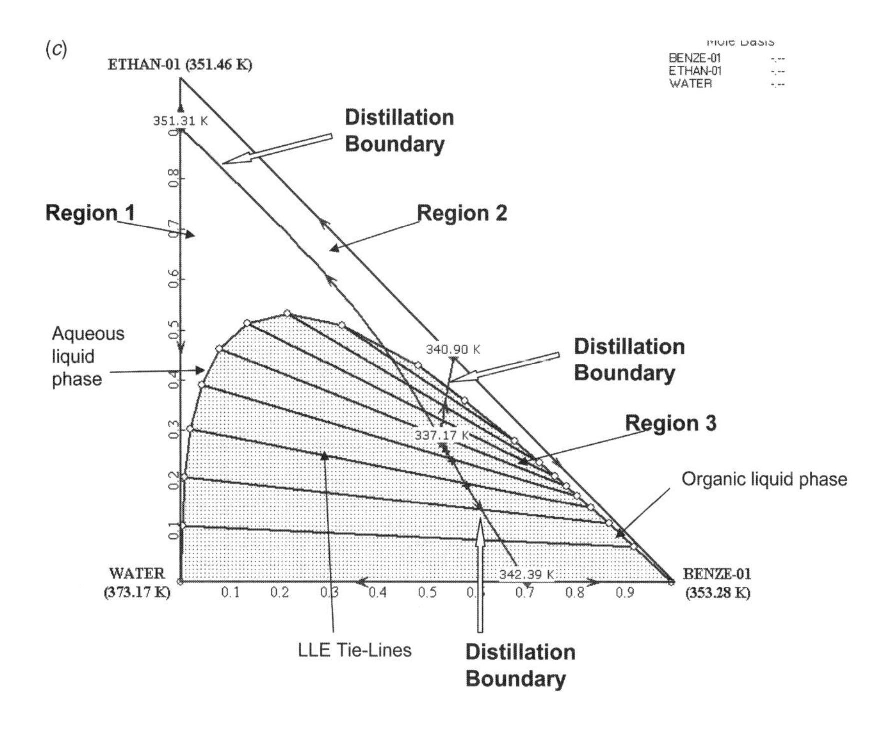

(d)

01 — Number Of Components: 2, Homogeneous; Temperature 340.90 K; Classification: Saddle

	MOLE BASIS	MASS BASIS
BENZE-01	0.5543	0.6783
ETHAN-01	0.4457	0.3217

02 — Number Of Components: 3, Heterogeneous; Temperature 337.17 K; Classification: Unstable Node

	MOLE BASIS	MASS BASIS
BENZE-01	0.5306	0.7194
ETHAN-01	0.2749	0.2198
WATER	0.1945	0.0608

03 — Number Of Components: 2, Heterogeneous; Temperature 342.39 K; Classification: Saddle

	MOLE BASIS	MASS BASIS
BENZE-01	0.7024	0.9110
WATER	0.2976	0.0890

04 — Number Of Components: 2, Homogeneous; Temperature 351.31 K; Classification: Saddle

	MOLE BASIS	MASS BASIS
ETHAN-01	0.8999	0.9583
WATER	0.1001	0.0417

Figure 5.17 *Continued.*

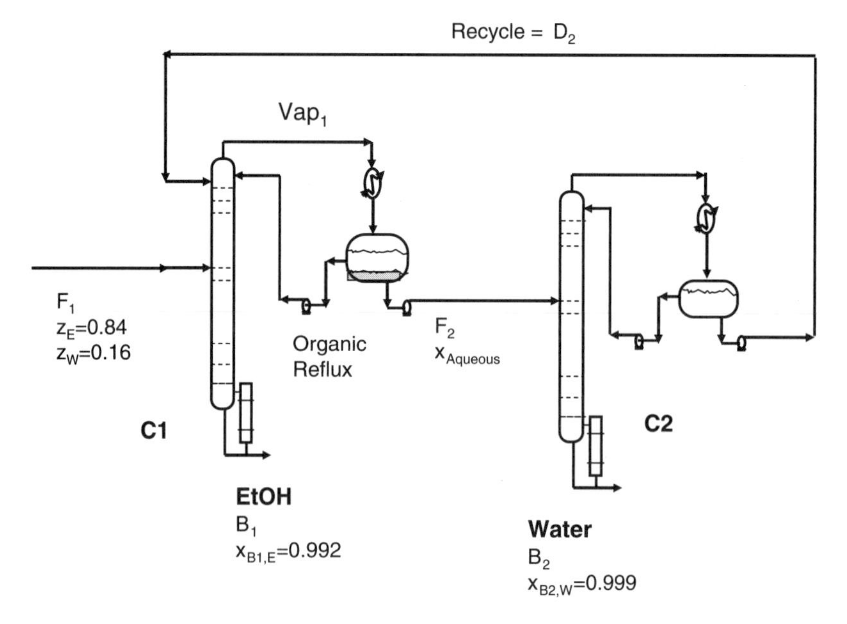

Figure 5.18 Conceptual flowsheet.

It may be useful at this point to locate on the ternary diagram several points to get a preliminary feel for the design problem that we are facing. The feed has a composition 0/84/16 mol% B/E/W (benzene/ethanol/water). So the F point is located on the ordinate in region labeled region 1 in Figure 5.17c. One of the desired products is very pure water, which is located at the bottom left corner in region 1. However, the other desired product is very pure ethanol, which is located at the top corner in region 2. This is on the other side of a distillation boundary, so the separation cannot be achieved in a single simple distillation column. The decanter and the second column are added so that the distillation boundary can be crossed.

Now that the complexity of the VLLE is apparent, let us develop a simulation of a flowsheet to produce high-purity ethanol and water. The flowsheet will have two distillation columns and a decanter. There are two recycle streams back to the first column: organic phase from the decanter and distillate from the second column. A simple conceptual flowsheet is given in Figure 5.18.

5.2.2 Process Flowsheet Simulation

The first column does not have a condenser, so the appropriate "stripper" icon is selected from the many possible types under *RadFrac*, as shown in Figure 5.19. The three streams fed to this column are added with control valves. The stream *Feed* is specified to be 0.06 kmol/s with a composition of 84 mol% ethanol. The other two streams fed to column C1 are unknown. We must make some reasonable guesses of what the flowrates and the compositions of the organic reflux and the recycle from the top of the second column will be.

One way to estimate these compositions is to recognize that the overhead vapor from the first column will have a composition that is close to that of the ternary azeotrope: 53.06/27.49/19.45 mol% benzene/ethanol/water (B/E/W). We set up a simulation

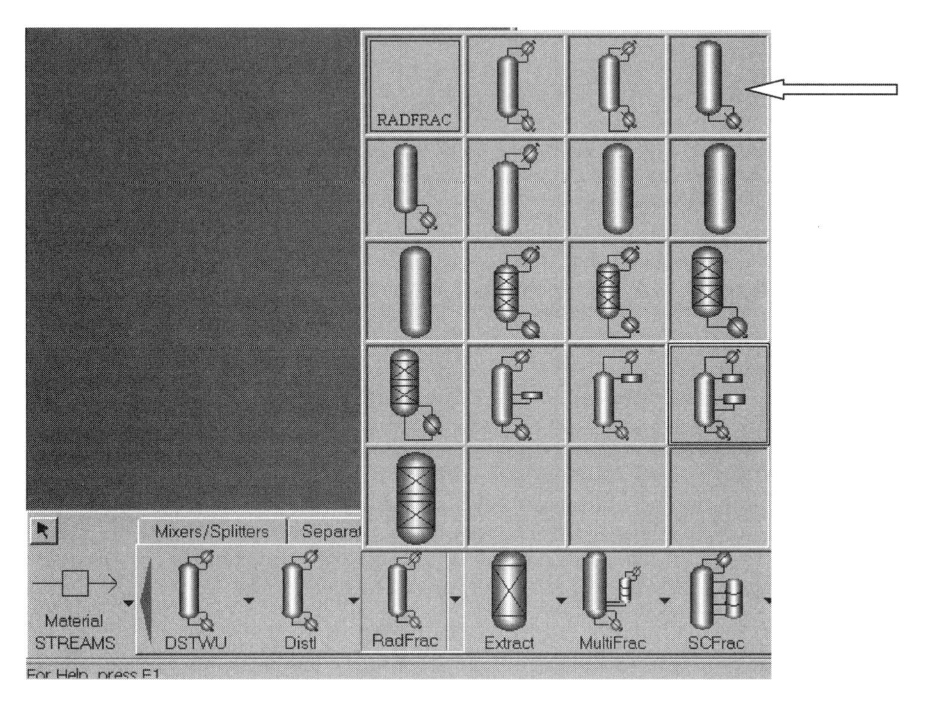

Figure 5.19 Selecting a stripper column.

with a stream with this composition feeding a decanter operating at 313 K. The predicted compositions of the organic and aqueous liquid phases are tabulated as follows:

Composition (mol%)	Organic	Aqueous
Benzene	84.35	7.24
Ethanol	14.14	47.04
Water	1.51	45.72

The composition of the organic reflux should be close to the composition of the organic liquid phase. The composition of the feed to the second column should be close to that of the aqueous liquid phase. Since essentially all the water in the feed comes out the bottom of the second column at a high purity, the amount of water removed from the feed is only (0.06 kmol/s)(0.16) = 0.0096 kmol of water/s (per second). Therefore, as a first estimate, we can use the composition of the aqueous liquid phase for a guess of the recycle composition.

The next issue is guessing the flowrates of the reflux and the recycle. One brute-force way to do this is to guess a recycle flowrate and then find the flowrate of organic reflux to column C1 that is required to keep water from leaving in the bottoms. When this is achieved, the resulting aqueous phase is fed to the second column, and the calculated distillate D2 is compared with the guessed value of recycle (both in flowrate and composition). The composition of the organic phase from the decanter is also compared to the guessed composition. Compositions are adjusted and a new guess of the recycle is made.

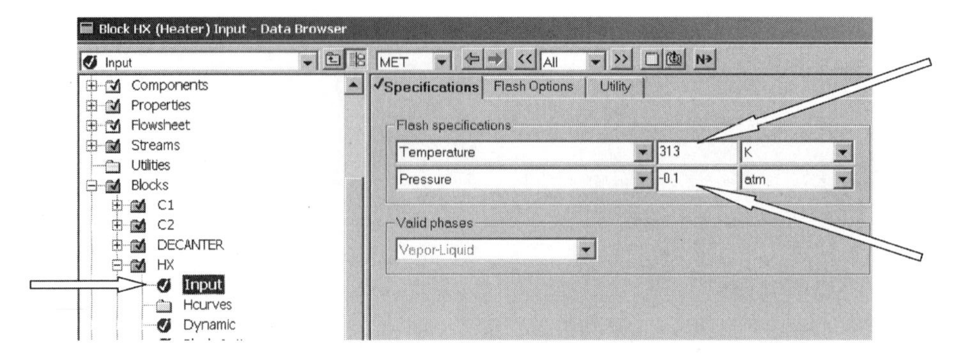

Figure 5.20 Heat exchanger specifications.

The simulation of the first column is extremely tricky, as we demonstrate below. A very small change in the organic reflux can produce a drastic change in the product compositions. Multiple steady states also occur; the same reflux flowrate can give two different column profiles and product compositions.

The number of stages in column C1 is set at 31. Note that since there is no condenser, the top tray is stage 1. Organic reflux is fed at the top. Recycle is fed at stage 10. Fresh feed is fed at stage 15. Column pressure is set at 2 atm because we need a control valve on the overhead vapor line.

The vapor from column C1 goes through a valve V12 and to a heat exchanger HX. The conditions specified in the *HX* block on the *Input* item (Fig. 5.20) are the exit temperature of 313 K and a 0.1 atm pressure drop (entering a negative number for *Pressure* means a pressure drop).

A decanter is then inserted on the flowsheet by clicking on *Separator* at the bottom of the window and selecting *Decanter*, as shown in Figure 5.21a. The operation of the decanter is specified by clicking the *Input* item under the *Decanter* block. The pressure is set at 1 atm and adiabatic operation is selected (heat duty is zero as shown in Fig. 5.21b). Under the item *Key components to identify 2nd liquid phase*, the benzene component is specified by moving benzene into the right *Key components* window (see Fig. 5.21b).

A very small amount of benzene will be lost in the two product streams, so a small makeup stream of fresh benzene is added to the organic phase from the decanter before it is fed to the first column as reflux.

Finally, a second column C2 is added in the normal way. A 22-stage column is specified with feed on stage 11 and operating at 1 atm. The final flowsheet with all the pumps and valve installed is shown in Figure 5.22.

5.2.3 Converging the Flowsheet

The fresh feed is 0.06 kmol/s with a composition of 84 mol% ethanol and 16 mol% water. Essentially all the ethanol must come out in the bottoms B1 from the first column. So in the setup of this column, a bottoms flowrate is fixed at $(0.06)(0.84) = 0.0504$ kmol/s. This column has only 1 degree of freedom because it has no condenser or internal reflux. The organic reflux will eventually be adjusted to achieve the desired purity of the ethanol bottoms product (99.92 mol% ethanol). Note that both benzene and water can appear in the bottoms as impurities.

(a)

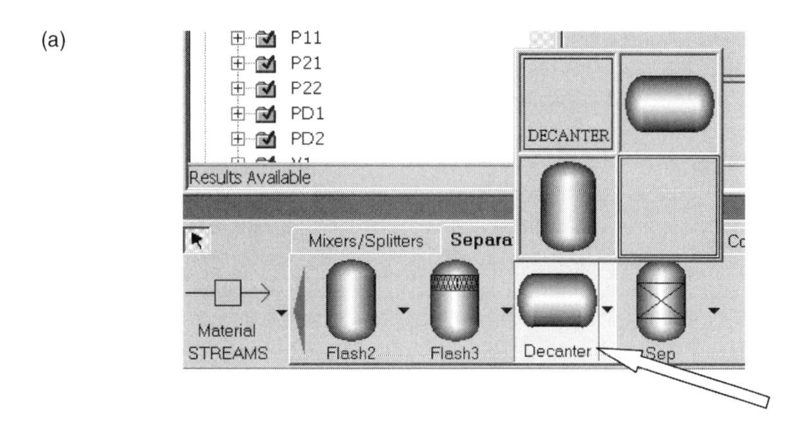

(b)

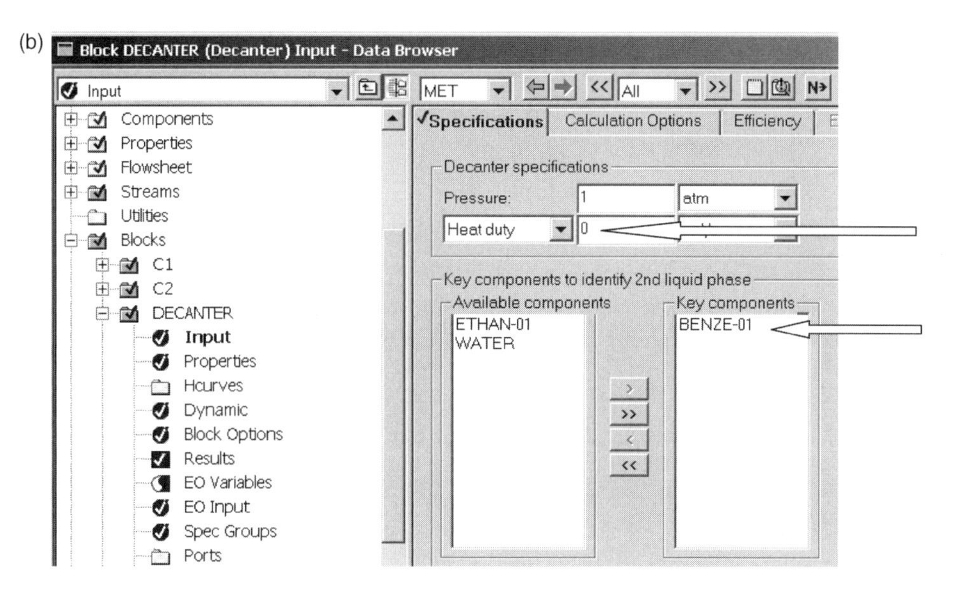

Figure 5.21 (a) Inserting a decanter; (b) specifying conditions in the decanter.

Likewise, essentially all the water must come out in the bottoms B2 of the second column. So in the setup of this column, the bottoms is fixed at (0.06)(0.14) = 0.0096 kmol/s. Initially the reflux ratio is fixed at 2 as the other degree of freedom. This will be adjusted later to achieve the desired purity of the water product (99.9 mol% water).

The first guesses of the compositions of the recycle and reflux are inserted in the *Input* of these streams. First guesses of reflux and recycle flowrates are made of 0.12 and 0.06 kmol/s, respectively. The simulation is run giving a bottoms composition of 21 mol% benzene and 7×10^{-4} mol% water in the first column. The water is driven overhead, but there is too much benzene in the bottoms because the organic reflux flowrate is too large. Reflux flowrate is reduced from 0.12 to 0.1 to 0.09 kmol/s (as shown in Table 5.2), which reduces the benzene impurity in the bottoms. However, when the reflux flowrate is reduced to 0.08 kmol/s, there is a drastic change in the bottoms

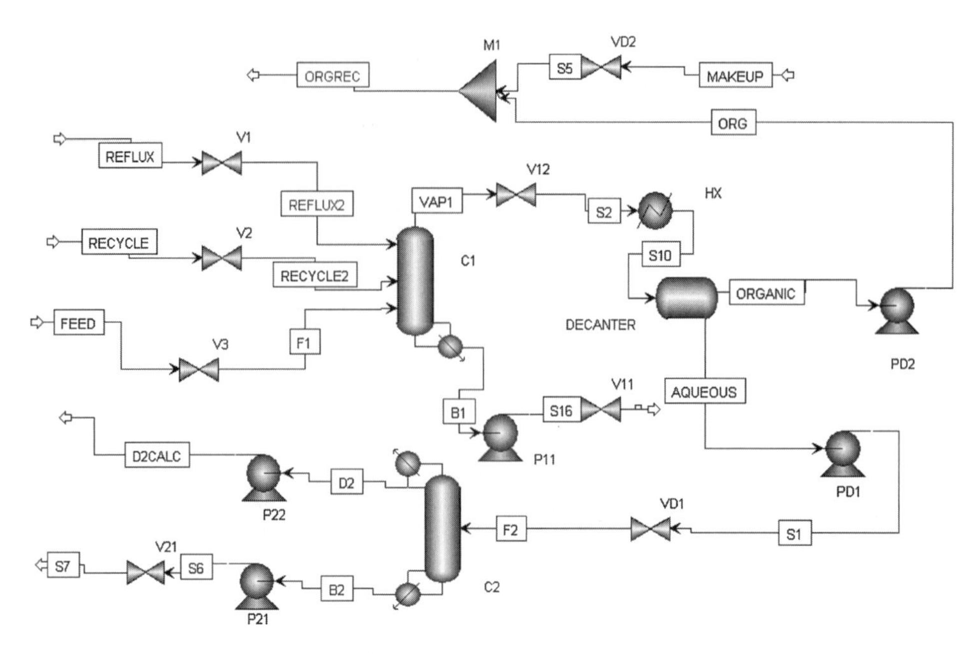

Figure 5.22 Full flowsheet.

composition. Now the water is not driven out in the overhead. It comes out in the bottoms because there is not enough benzene to entrain the water overhead.

Now, if the reflux is increased back to 0.09 kmol/s, the column does not converge to the same steady state that it had previously at this flowrate. The flowrate must be increased to about 0.11 kmol/s to reestablish the desired low water content in the bottoms. This multiple steady-state phenomenon is one of the severe complexities that simulations of distillation columns experience when highly nonideal VLLE relationships are involved.

Figure 5.23 compares temperature and composition profiles at two different steady states. The reflux flowrate is 0.082 kmol/s in both cases. The fresh feed and recycle are identical. The bottoms flowrate is the same. However, the bottoms composition is drastically different.

Obviously the steady state indicated by the solid lines in Figure 5.23 is the desired one. The bottoms purity with this steady state is 99.27 mol% ethanol. The impurities are

TABLE 5.2 Effect of Changing Reflux Flowrate

Reflux (kmol/s)	Bottoms Composition (mol% B)	Bottoms Composition (mol% W)	Notes
0.12	20.9	7×10^{-4}	
0.10	10.1	8×10^{-3}	
0.09	4.70	0.04	
0.08	3×10^{-17}	43.6	Jump to high water B_1
0.09	3×10^{-17}	43.9	
0.10	3×10^{-17}	43.8	
0.11	15.5	2×10^{-3}	Jump back to low water B_1

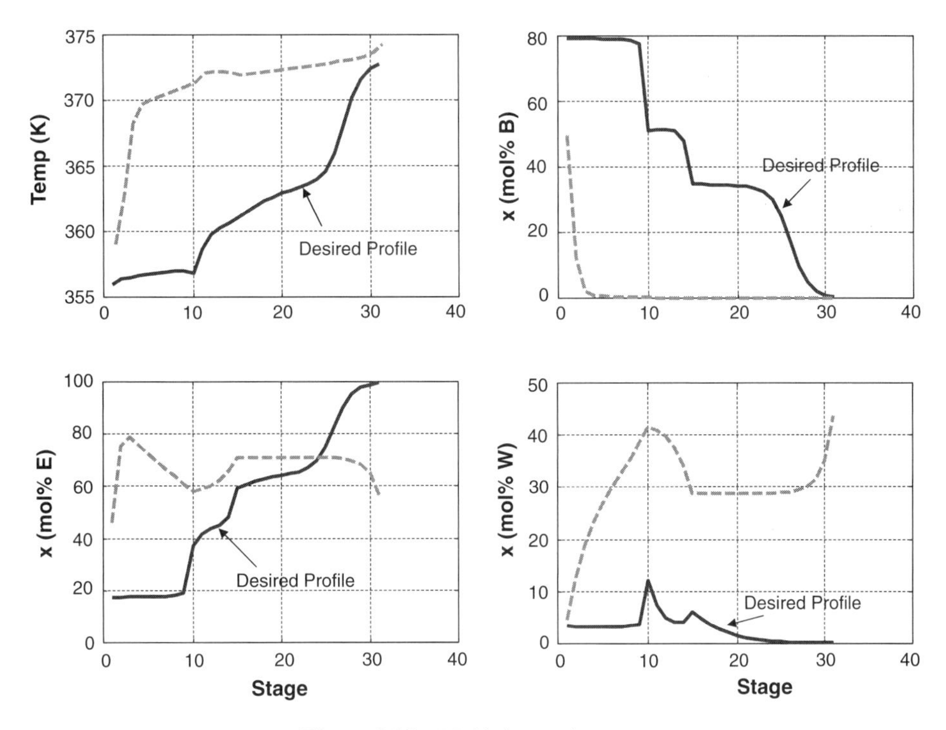

Figure 5.23 Multiple steady states.

0.50 mol% benzene and 0.23 mol% water. Getting the simulation to converge to this steady state is quite difficult.

The calculated compositions of the reflux and recycle are compared to the guessed values. The reflux composition is quite close: 84.4/14.0/1.6 mol% B/E/W calculated versus 84.4/14.1/1.5 mol% assumed. The recycle composition is somewhat different: 4.7/41.6/53.7 mol% calculated versus 7.2/47.1/45.7 mol% assumed. Changing these compositions to the calculated values and rerunning the program give a new B_1 composition of 0.926/98.92/0.157 mol% B/E/W. The calculated flowrate of the distillate D_2 under these conditions is 0.0704 kmol/s versus the 0.06 kmol/s assumed.

Now that we have some reasonable guesses for the values of the recycle streams, the Design Spec/Vary capability can be used to drive the compositions of the two product streams to their desired values. The key feature in the first column is to keep enough benzene in the column to entrain out the water so that the bottoms is high-purity ethanol. On the other hand, if too much benzene reflux is fed to the column, it will go out the bottom and drive the bottoms off specification. A Design Spec/Vary is set up to maintain the benzene composition of the bottoms at 0.5 mol% by manipulating the REFLUX stream, which is consider a *Feed rate* on the list of choices given in the *Vary, Specifications, Adjusted variable*, and *Type*.

The initial guessed value of the reflux ratio in the second column was 2. The bottoms purity was very high. The reflux ratio was reduced to $RR \sim 0.2$ without affecting the bottoms purity significantly. A second Design Spec/Vary is set to maintain the ethanol composition of the bottoms of the second column at 0.1 mol% by varying the bottoms flowrate B_2.

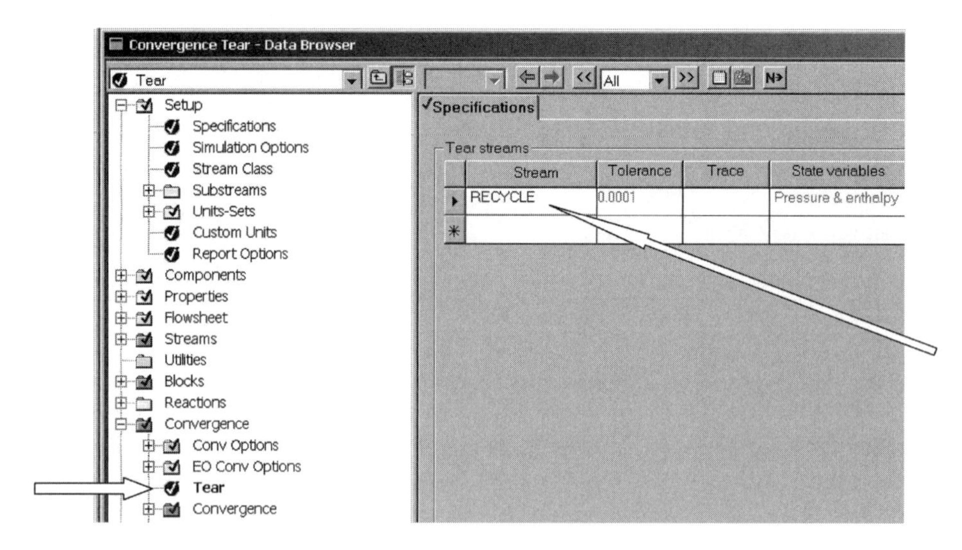

Figure 5.24 Specifying tear stream.

Several runs are made in which the guessed compositions of the reflux and the recycle are compared with those of the calculated organic stream from the decanter and the distillate D_3 from the second column. When these variables are fairly close, the recycle/D_2 loop is closed. The procedure for doing this involves three steps:

1. Delete the stream labeled "D2CALC" on the lower left in Figure 5.22.
2. Click the stream labeled "REYCLE" and reconnect it to the valve labeled "V2."
3. Go down near the bottom of the list of item on the *Data Browser* window and click *Convergence* and then *Tear*. This opens the window shown in Figure 5.24, where the dropdown menu is used to select *RECYCLE*.

When the program is rerun, it converges to the values shown on the flowsheet given in Figure 5.25. Figure 5.26 gives the ternary diagram with the locations of all the streams marked.

In theory, the next and final step is to close the organic reflux loop. The stream labeled "ORGREF" is deleted, the stream "REFLUX" is connected to the mixer "M1," and "REFLUX" is defined as a *TEAR* stream. Unfortunately, this loop does not converge even though the initial values of the guessed and calculated values are very close in both composition and flowrate. An alternative way to converge this system using dynamic simulation will be discussed in Chapter 8 after we have discussed the details of dynamic simulation.

5.3 HEAT-INTEGRATED COLUMNS

In the first two examples we studied columns with nonideal phase equilibrium, and the second example involved a more complex flowsheet with two columns. The last example discussed in this chapter has fairly simple phase equilibrium but a complex process structure. Two columns are operated at two different pressures so that the

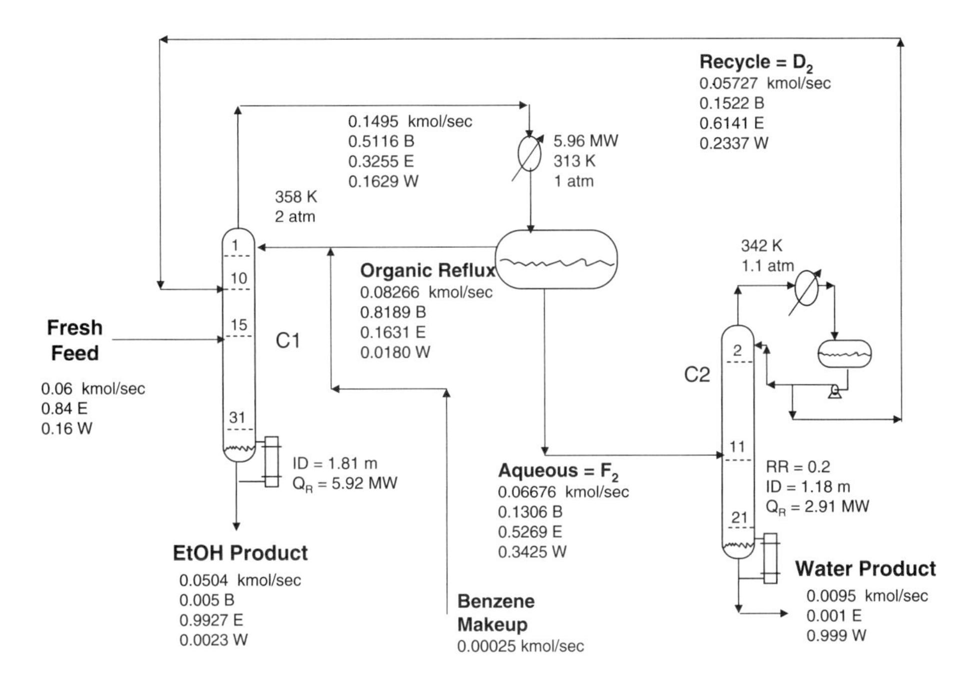

Figure 5.25 Flowsheet with recycle/D_2 loop closed.

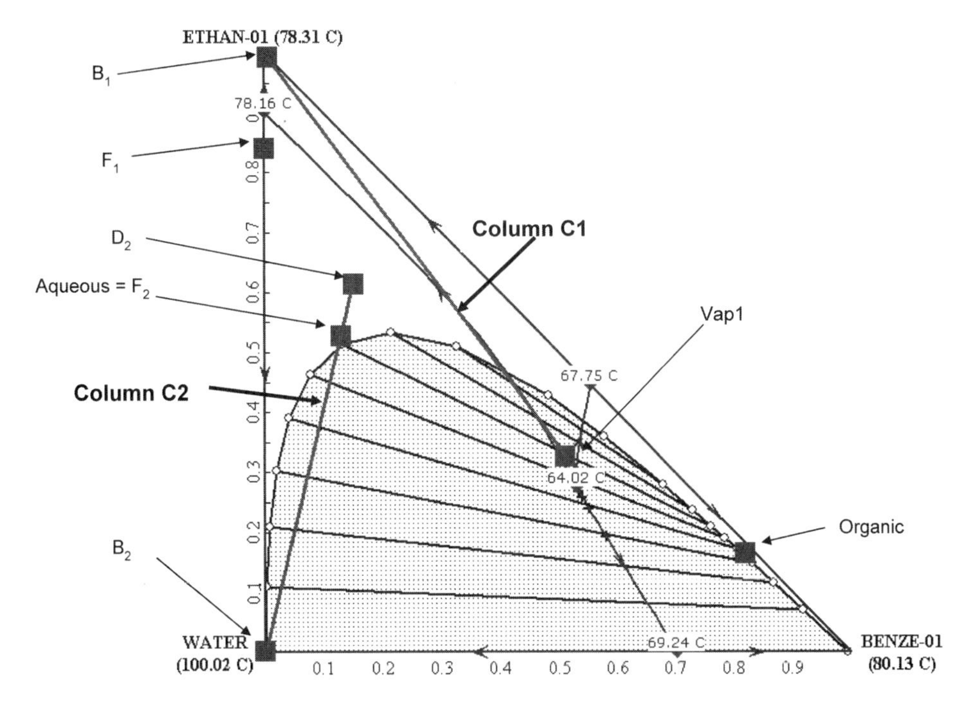

Figure 5.26 Ternary diagram.

condenser for the high-pressure (high-temperature) column can be used as the reboiler in the low-pressure (low-temperature) column.

5.3.1 Flowsheet

Figure 5.27 gives the conceptual flowsheet. The specific system used as an example is methanol/water. Product specifications are 99.9 mol% methanol in the distillate streams (there is one from each column) and 99.9 mol% water in the two bottoms streams. The fresh feed is 1 kmol/s with a composition of 60 mol% methanol and 40 mol% water. The feed is split between the two columns so that the system operates "neat," where the condenser heat removal in the high-pressure column is exactly equal to the reboiler heat input in the low-pressure column. Each column has 32 stages and is fed on the stage that minimizes reboiler heat input.

To achieve the required temperature differential driving force in the condenser/reboiler, the pressures in the two columns must be appropriately selected. The low-pressure column C1 operates at a pressure of 0.6 atm (vacuum conditions, 456 mmHg) that gives a reflux drum temperature of 326 K so that cooling water can be used. The pressure drop per tray is assumed to be 0.0068 atm (0.1 psi). The base temperature of C1 is 367 K.

A reasonable differential temperature driving force is about 20 K. If the ΔT is too small, the heat transfer area of becomes quite large. The pressure in the second column is adjusted to give a reflux drum temperature of $367 + 20 = 387$ K. The pressure in C2 is 5 atm. The base temperature in C2 at this pressure is 428 K, which will determine the pressure of the steam used in this reboiler.

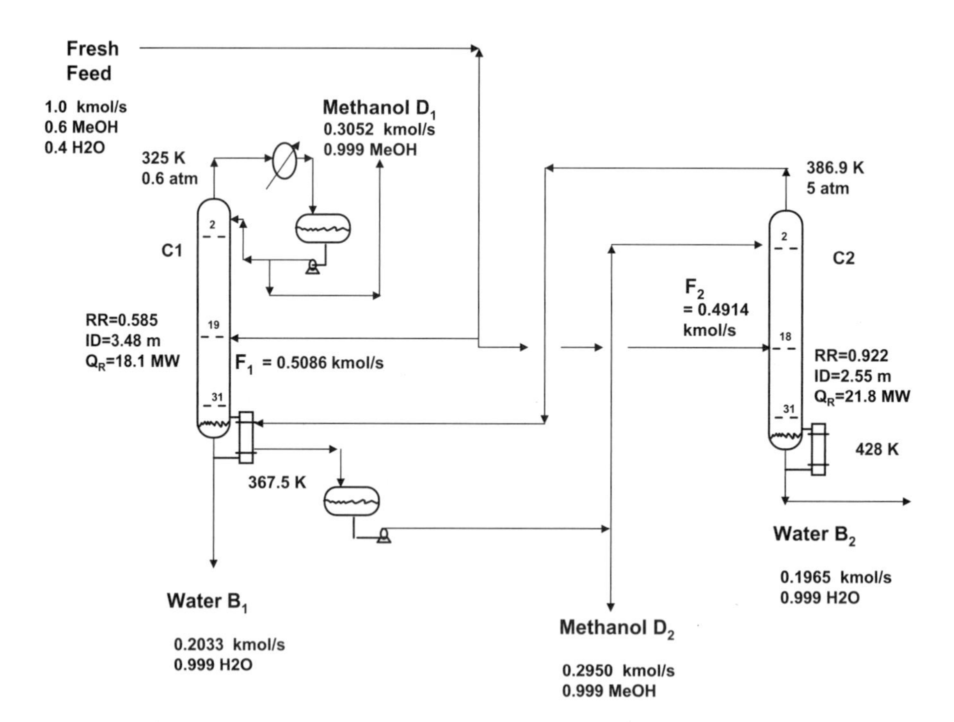

Figure 5.27 Heat-integrated columns.

5.3.2 Converging for Neat Operation

Initially the total feed is split equally between the two columns. This is achieved in the *Splitter* labeled "T1" on the flowsheet shown in Figure 5.28. Two Design Spec/Vary functions are set up in each column to adjust distillate flow and reflux ratio to attain the 99.9 mol% product purities of all four streams. The optimum feed tray location is determined by finding the feed stage that minimizes reboiler heat input. In column C1, it is stage 19; in column C2, it is stage 18.

Under these conditions, the resulting reboiler heat input in the low-pressure column C1 is 17.91 MW. The resulting condenser heat removal in the high-pressure column C2 is 18.62 MW. These are very close, but if the system is to be operated "neat" (with no auxiliary reboilers or condensers), these heat duties must match exactly.

One way to do this is to manually adjust the feed split in "T1" until QR1 is equal to QC2. This can be automated by going to *Flowsheeting Options* on the *Data Browser* window and selecting *Design Spec*. This is similar to the Design Spec/Vary in the column blocks, but now variables from any block can be used. Figure 5.29a shows the window that opens after *New* and *OK* are clicked. On the first page tab *Define* we enter two variables QR1 and QC2, and click on *Edit* to define what they are. Figure 5.29b shows how the QR1 is defined.

On the *Spec* page tab (Fig. 5.29c) a parameter "DELTAQ" is specified with the desired value *Target* and *Tolerance*. Clicking the *Vary* page tab opens the window shown in Figure 5.30. The "T1" block is selected and the variable is *Flow/Frac*. The *ID1* is set at "2" since the flowrate of the stream "2" leaving the splitter is the first variable and is the one specified. The final item is to define DELTAQ on the *Fortran* page tab (Fig. 5.31). Remember that the convention in Aspen Plus is that heat addition is a positive number and heat removal is a negative number. Therefore we want the sum of QR1 and QC2 (in watts) to be small. Running the program yields a feed split with 0.5086 kmol/s fed to the low-pressure column C1 and 0.4914 fed to the high-pressure column C2. The heat duty in the condenser/reboiler is 18.10 MW as shown in Figure 5.32, which is obtained by selecting *Results* under the *DS-1* design spec. The final flowsheet conditions are given in Figure 5.27.

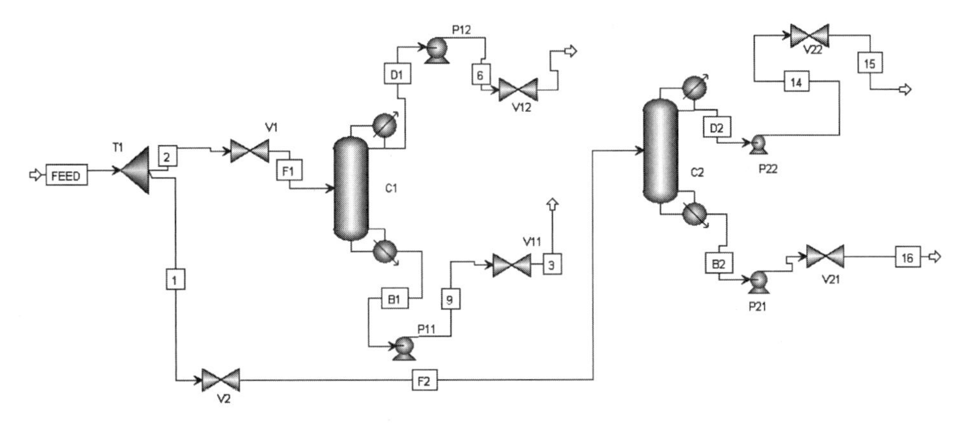

Figure 5.28 Aspen Plus flowsheet.

(a)

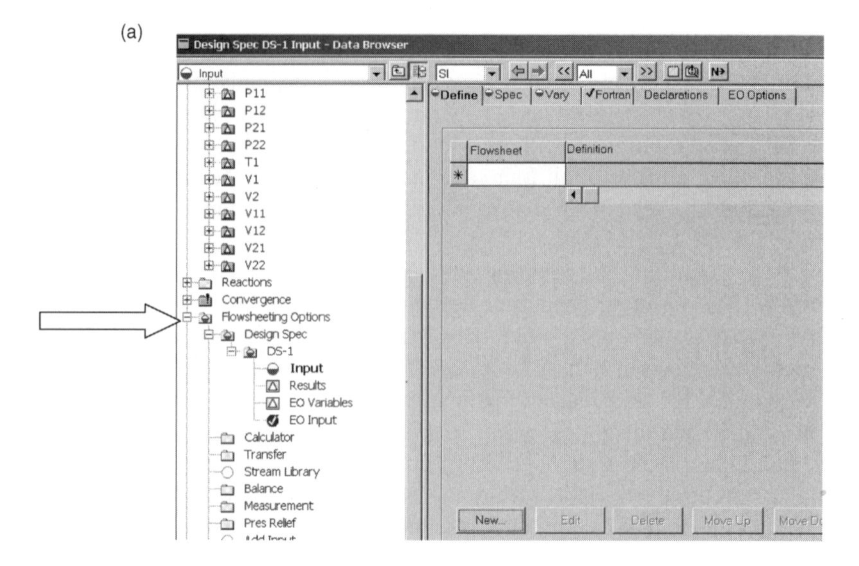

(b)

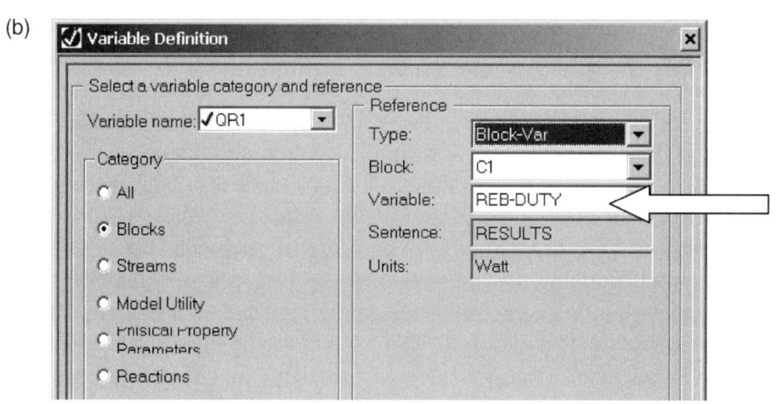

(c)

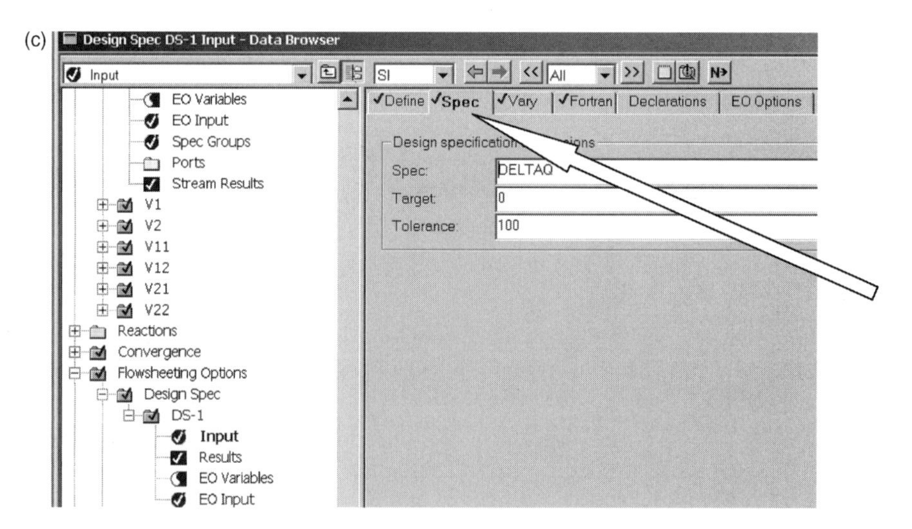

Figure 5.29 (a) Setting up design spec; (b) defining variable; (c) setting specification.

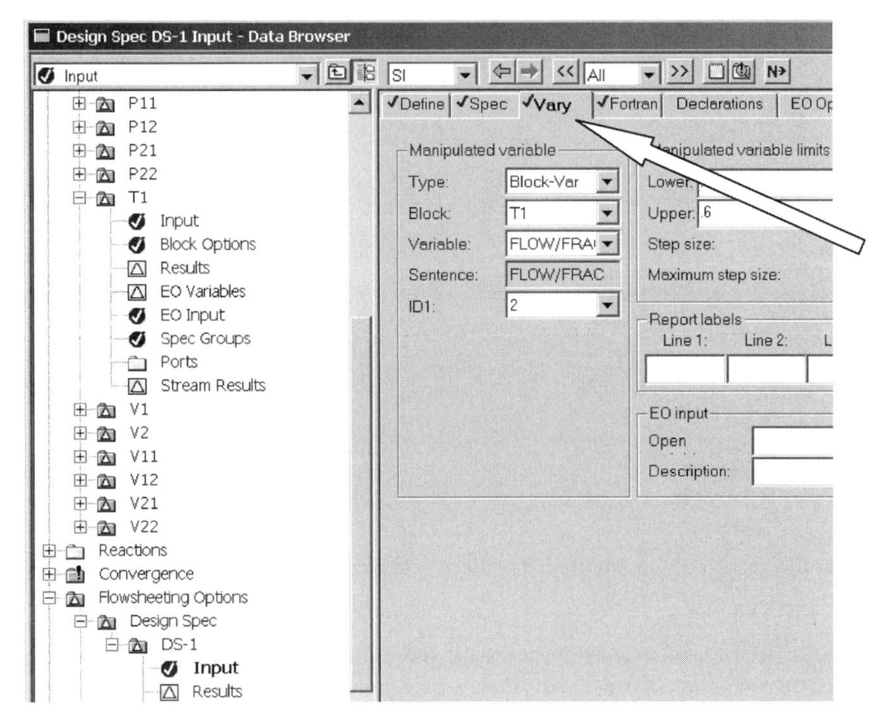

Figure 5.30 Selecting manipulated variable.

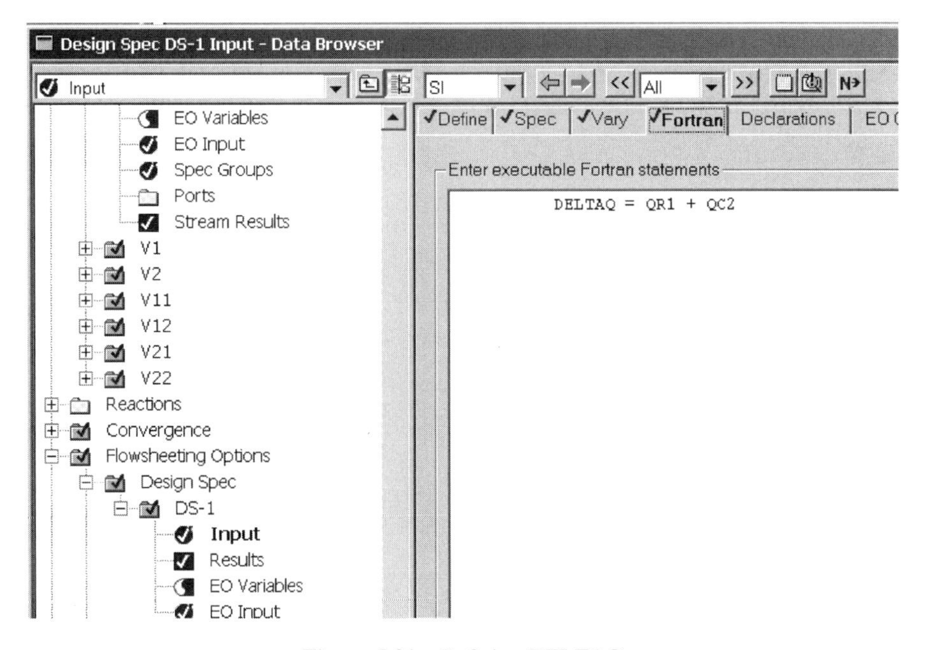

Figure 5.31 Defining DELTAQ.

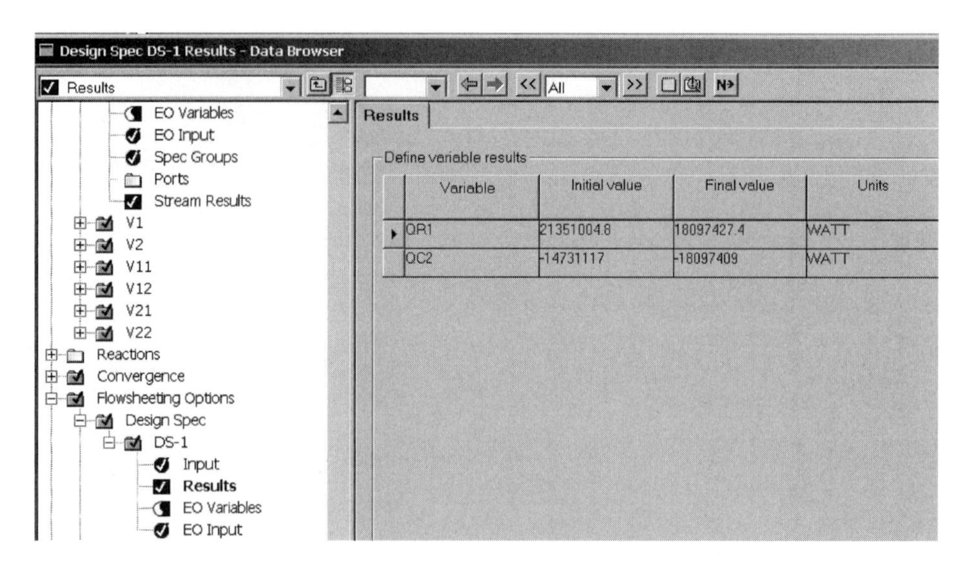

Figure 5.32 Results of design spec.

The last item of interest is to compare the energy and capital costs of this heat-integrated two-column system with those of a single column making exactly the same separation. This single column is the same as the low-pressure column in terms of operating pressure, but its energy consumption, column diameter, and heat exchanger areas will be larger.

The comparison of the two alternative designs is given in Table 5.3. The energy consumption is reduced from 35.58 MW in a single column to 21.81 MW in the high-pressure column of the heat-integrated design. This cuts energy cost from \$2,640,000 per year to \$1,610,000 per year. The same energy value is used in both cases. However, higher-pressure steam is needed in the heat-integrated design because the base temperature is 428 K compared to 367 K in the single column. Using a 34.8 K temperature difference

TABLE 5.3 Comparison of Single and Heat-Integrated Columns

	Single	Low-Pressure	High-Pressure
Stages	32	32	32
N_F	19	19	18
D (m)	4.93	3.48	2.55
Q_C (MW)	34.29	17.44	—
RR	0.585	0.585	0.922
Q_R (MW)	35.58	—	21.81
A_C (m^2)	2890	1433	251
A_R (m^2)	1797	—	161
Shell (10^6 \$)	1.15	0.794	0.255
HX (10^6 \$)	2.25	0.822	0.692
Energy (10^6 \$/year)	2.64	—	1.62
Capital (10^6 \$)	3.40	1.62	1.26
TAC (10^6 \$/year)	3.77		2.58

between the column base temperatures and the condensing steam temperature in the reboiler and a 4 atm pressure drop over the steam valve, give supply steam pressures of 6.8 and 17 atm, respectively, for the two processes. The difference in the cost of these two steam supplies would reduce the energy savings.

Total capital investment is also reduced. This is somewhat counterintuitive because one column with two heat exchangers would be expected to be less expensive than two columns with three exchangers. However, the column diameters and the heat exchanger areas are smaller in the heat-integrated design.

An additional aspect of this heat integration simulation is the calculation of the heat transfer rate in the condenser/reboiler heat exchanger. In this steady-state simulation we have specified that the condenser heat removal in the first column is equal to the reboiler heat input in the second column. This satisfies the first law of thermodynamics. The area of the condenser/reboiler is then calculated on the basis of the heat duty, the differential temperature driving force (the temperature in the reflux drum of the first column minus the temperature in the base of the second column), and an overall heat transfer coefficient. In a dynamic simulation this area is fixed. The heat transfer rate will change dynamically as the two temperatures change. So, in the dynamic simulation the heat transfer rates in the two column must be calculated from $Q = UA(T_{D1} - T_{B2})$ so that the second law of thermodynamics is satisfied. This can be achieved by using *Flowsheet Equations* in Aspen Dynamics, which will be discussed in later chapters.

5.4 CONCLUSION

The complex nonideal distillation columns considered in this chapter provide good examples of the difficulties and capabilities of using simulation in distillation systems for steady-state design. Now we are ready to move to an equally important phase of design in which the dynamics and control of the column or systems of columns and other units are considered. "Simultaneous design" involves both steady-state and dynamic aspects of the process.

CHAPTER 6

STEADY-STATE CALCULATIONS FOR CONTROL STRUCTURE SELECTION

Before we get into the details of converting a steady-state simulation into a dynamic one, it might make sense to discuss some important steady-state calculations that are frequently performed to aid in the selection of a practical, effective control structure for a distillation column.

The majority of distillation columns are designed to attain a specified separation between the two key components. The two design degrees of freedom are usually specified to be the impurity of the heavy-key component in the distillate and the impurity of the light-key component in the bottoms. Therefore, in the operation and control of a distillation column, the "ideal" control structure would measure the compositions of the two products and manipulate two input variables (e.g., reflux flowrate and reboiler heat input) to maintain the desired amounts of the key component impurities in the two product streams.

However, very few distillation columns use this ideal control structure. There are a number of practical reasons for this. Composition analyzers are often expensive to purchase and have high maintenance costs. Their reliability is sometimes inadequate for online continuous control. They also introduce deadtime into the control loop if chromatographic methods are used. In addition, it is often possible to achieve very effective control without using direct composition measurements.

Temperatures are widely used to provide inferential control of compositions. Temperature sensors are inexpensive and reliable and introduce only small measurement lags in the control loop. In a binary system with constant pressure, temperature is uniquely related to composition. This is not true in multicomponent systems, but temperatures at appropriate locations in a distillation column can often provide fairly accurate information about the concentrations of the key components.

In "single end" control structures, the temperature on one tray is controlled. The remaining control degree of freedom is selected to provide the least amount of product quality variability. For example, a constant reflux ratio *RR* can be maintained or the reflux to feed ratio R/F can be fixed. Sometimes the control of two temperatures is required (dual-temperature control). We discuss these alternatives in this chapter.

If tray temperatures are to be used, the issue is selecting the best tray or trays on which temperature is held constant. This problem has been discussed in the distillation literature for over a half-century, and several alternative methods have been proposed. We will review these alternative methods and illustrate their effectiveness for several systems.

It is important to note that all of these methods use only steady-state information, so steady-state process simulators such as Aspen Plus can be easily used to perform the calculations. The methods all require that various variables be held constant while other variables are changed. For example, two product compositions or a tray temperature and reflux flowrate may be held constant while feed composition is changed. The "Design Spec/Vary" feature in Aspen Plus is used to fix the values of the desired independent variables and to calculate all the remaining dependent variables.

In several of the methods, the variable that is changed is the feed composition. The feed flowrate is not considered in any of the methods. This is because feed rate disturbances can be handled by ratioing the flowrates of the manipulated variable directly to the feed flowrate. Maintaining a constant reflux ratio or a constant reflux to feed ratio should produce the same compositions and temperatures on all trays at any feed flowrate. Of course, this assumes that there are no changes in tray efficiencies with throughput. It also assumes that there are no changes in pressure on all the trays, which is seldom the case because of the changes in tray pressure drop and height of liquid on the trays as liquid and vapor rates change. But these effects are usually small enough that they do not adversely impact control to a great extent.

6.1 SUMMARY OF METHODS

6.1.1 Slope Criterion

Satisfaction of the *slope criterion* consists in selecting the tray where there are large changes in temperature from tray to tray.

The temperature profile at design conditions is plotted, and the "slope" of the profile is examined to find the tray where this slope is greatest. Large changes in temperature from tray to tray indicate a region where compositions of important components are changing. Maintaining a tray temperature at this location should hold the composition profile in the column and prevent light components from dropping out the bottom and heavy components from escaping out the top.

6.1.2 Sensitivity Criterion

Satisfaction of the *sensitivity criterion* consists in finding the tray where there is the largest change in temperature for a change in the manipulated variable.

A very small change (0.1% of the design value) is made in one of the manipulated variables (e.g., reflux flowrate). The resulting changes in the tray temperatures are examined to see which tray has the largest change in temperature. The procedure is repeated for thc other manipulated variable (e.g., reboiler heat input). Dividing the change in the tray temperature by the change in the manipulated variable gives the openloop steady-state gain between temperature on that tray and each manipulated variable. The tray with the largest temperature change is the most "sensitive" and is selected to be controlled. A large gain indicates that the temperature on that tray can be effectively controlled by the corresponding manipulated variable. A small gain indicates that valve saturation can easily occur and the operability region could be limited.

6.1.3 Singular Value Decomposition Criterion

Singular value decomposition (SVD) of the steady-state gain matrix is thoroughly treated by Moore.[1]

The steady-state gains between all the tray temperatures and the two manipulated variables are calculated as described in the previous section. A gain matrix K is formed, which has N_T rows (the number of trays) and two columns (the number of manipulated variables). This matrix is decomposed using standard SVD programs (e.g., the *svd* function in Matlab) into three matrices: $K = U\sigma V^T$. The two U vectors are plotted against the tray number. The tray or trays with the largest magnitudes of U indicate locations in the column that can be most effectively controlled. The σ matrix is a 2×2 diagonal matrix whose elements are the singular values. The ratio of the larger to the smaller is the condition number, which can be used to assess the feasibility of dual-temperature control. A large condition number (or small minimum singular value) indicates a system that is difficult to control. The controller is the inverse of the plant gain matrix, and a singular value of zero means that the matrix is "singular" and cannot be inverted.

6.1.4 Invariant Temperature Criterion

With both the distillate and bottoms purities fixed, we change the feed composition over the expected range of values. We select the tray where the temperature does not change as feed composition changes.

The difficulty with this method is that there may be no constant-temperature tray for all feed compositions changes. This is particularly true in multicomponent systems, where the amounts of the nonkey components can vary and significantly affect tray temperatures, especially near the two ends of the column.

6.1.5 Minimum Product Variability Criterion

Satisfaction of the *minimum product variability criterion* consists in selecting the tray that produces the smallest changes in product purities when its temperature is held constant in the face of feed composition disturbances.

Several candidate tray locations are selected. The temperature on one specific tray is fixed, and a second control degree of freedom is fixed such as reflux ratio or reflux flowrate. Then the feed composition is changed over the expected range of values, and the resulting product compositions are calculated. The procedure is repeated for several

[1]C. F. Moore, Selection of controlled and manipulated variables, in *Practical Distillation Control*, Van Nostrand-Reinhold, 1992, Chapter 8.

control tray locations. The tray is selected that produces the smallest changes in product purities when its temperature is held constant in the face of feed composition disturbances.

6.1.6 Summary

In the sections above, we have described the five most frequently used criteria. Sometimes these criteria recommend the same control tray location. In other cases, they recommend different control tray locations. In the next sections we apply these criteria to several typical industrial distillation systems to assess their relative effectiveness.

6.2 BINARY PROPANE/ISOBUTANE SYSTEM

The first separation system examined is a binary mixture of propane and isobutane. The feed flowrate is 1 kmol/s, and the design feed composition is 40 mol% propane. We use the conventional notation that the composition of the feed is z, the composition of the distillate is x_D, and the composition of the bottoms is x_B (all in mole fraction propane). Column pressure is set at 13.5 atm so that cooling water can be used in the condenser (reflux drum temperature is 315 K). The column has 37 stages and is fed on stage 16, using Aspen notation of numbering stages from the reflux drum on down the column.

Distillate purity is specified to be 98 mol% propane. Bottoms impurity is specified to be 2 mol% propane. The reflux ratio required to achieve these purities is 1.08.

6.2.1 Slope Criterion

The upper graph in Figure 6.1 gives the temperature profile at design conditions. The lower graph shows the differences between the temperatures on adjacent trays. The location of the tray with the largest slope is stage 8. There is another tray (stage 29) that has a slope that is almost as large. We will compare the use of both of these later in this section.

6.2.2 Sensitivity Criterion

The upper graph in Figure 6.2 gives the openloop gains between tray temperatures and the two manipulated variables reflux R and reboiler heat input Q_R. The solid lines show reflux flowrate changes, and the dashed lines represent reboiler heat input changes. Very small increases from the steady-state values (+0.1%) of the two inputs are used. As expected, the gains between the tray temperatures and reflux are negative, while they are positive for heat input.

These curves show that stage 8 is sensitive to changes in reflux and both stages 8 and 29 are sensitive to changes in heat input. Therefore stage 8 can be controlled using either reflux or heat input, while stage 29 can be controlled by only heat input.

It should be remembered that these are steady-state results and tell us nothing about dynamics. Temperatures on all trays in the column are quickly affected by changes in heat input, so pairing heat input with any tray temperature is dynamically feasible.

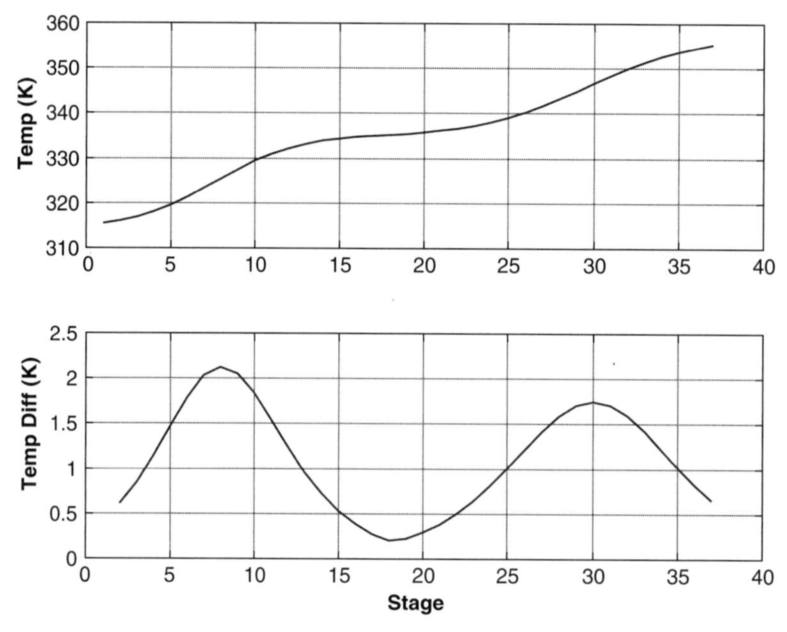

Figure 6.1 Propane/isobutane temperature profile and slope.

However, a change in reflux flowrate takes a significant time to affect temperatures on trays near the bottom of the column because of liquid hydraulic lags (3–6 s per tray). Therefore poor control can be expected when reflux is paired with a tray temperature significantly down from the top of the column.

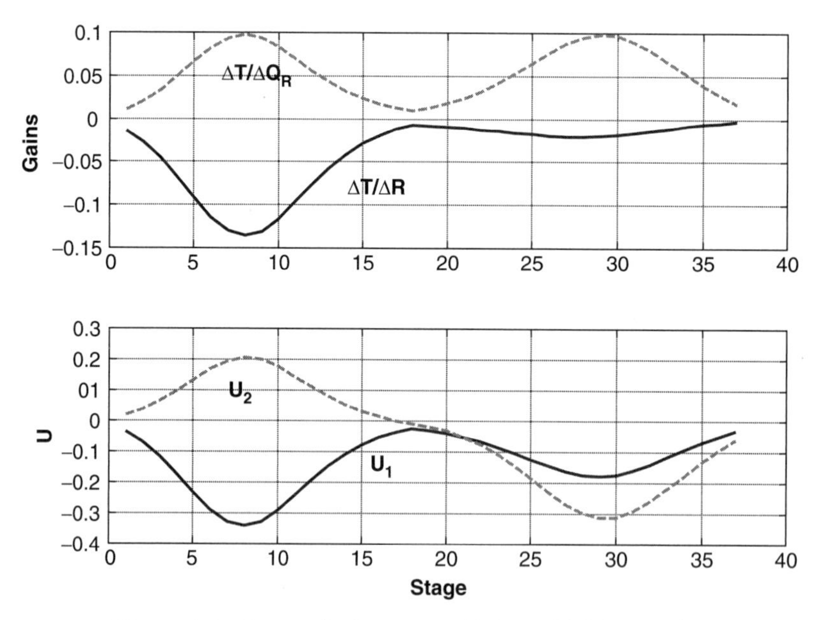

Figure 6.2 Propane/isobutane sensitivity and SVD analysis.

6.2.3 SVD Criterion

The lower graph in Figure 6.2 gives the U_1 and U_2 values from singular value decomposition (SVD) analysis. The first is the solid line and is associated with reflux. The second is the dashed line and is associated with heat input.

The SVD results are similar to the sensitivity results. They suggest that stage 8 can be controlled by reflux and stage 29, by heat input. The singular values of the steady-state gain matrix are $\sigma_1 = 0.479$ and $\sigma_2 = 0.166$, giving a condition number $CN = \sigma_1/\sigma_2 = 2.88$. This indicates that the two temperatures are fairly independent, so a dual-temperature control scheme should be feasible, at least from a steady-state perspective.

6.2.4 Invariant Temperature Criterion

Figure 6.3 gives the changes in the temperature profiles for two feed compositions on either side of the steady-state value (40 mol% propane). The solid lines represent 35 mol% propane and the dashed lines, 45 mol% propane in the feed. The product distillate and bottoms compositions are fixed at 98 and 2 mol%, respectively, for both feed compositions.

As expected in a binary constant-pressure system, fixing the composition fixes the temperature. So the temperatures at the top and at the bottom do not change. In theory, these end temperature could be controlled to achieve constant product purities. In practice, however, small amounts of other components or changes in pressure can render the use of temperatures at the very ends of the column ineffective. This will be demonstrated later when multicomponent systems are considered.

6.2.5 Minimum Product Variability Criterion

Figure 6.4 shows how product purities change when the temperature on a specific tray is held constant and feed composition changes. The second control degree of freedom that is fixed in this figure is the reflux flowrate.

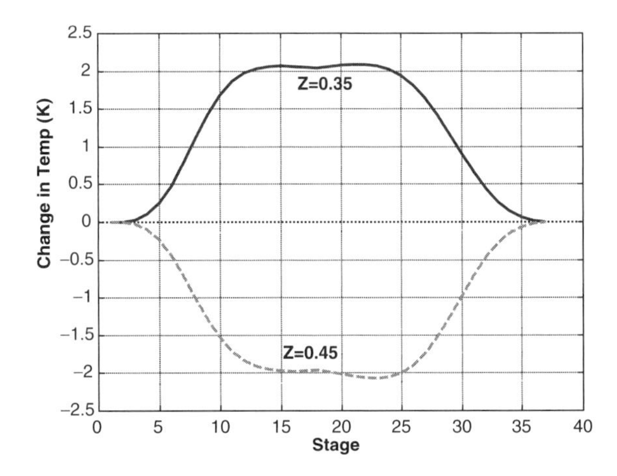

Figure 6.3 Propane/isobutane; changes in temperature profile for feed composition changes with product purities fixed.

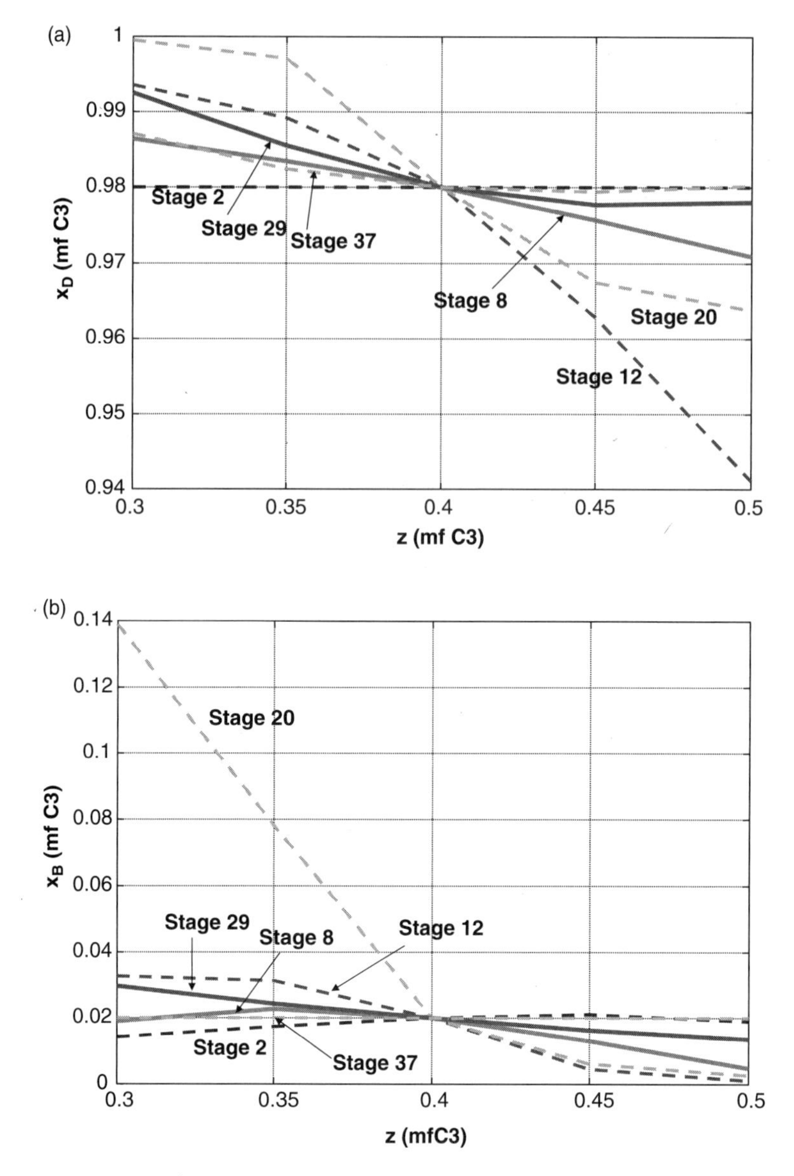

Figure 6.4 (a) Propane/isobutane distillate purity for feed composition changes with fixed reflux and stage temperature; (b) propane/isobutane bottoms impurity for feed composition changes with fixed reflux and stage temperature.

The justification for choosing constant reflux as opposed to constant reflux ratio is given in Table 6.1. The information obtained in the previous section is used. Product purities are fixed, and feed composition is varied. Instead of looking at the temperature profile, we note the required changes in reflux flowrate and reflux ratio. As the results in Table 6.1 clearly show, the required changes in reflux flowrate are much smaller than the required changes in the reflux ratio.

TABLE 6.1 Effect of Feed Composition on Required Reflux Flowrate and Reflux Ratio for Constant Product Purities

System	Product Purities or Impurities (mol %)	z (mol %)	Reflux Flowrate (kmol/s)	Reflux Ratio
D/P	98/2	35 C3	1.0736	3.1233
		40 C3	1.0797	2.7276
		45 C3	1.0798	2.4106
BTX	0.1/0.1	25 B	0.5607	2.247
		30 B	0.5715	1.908
		35 B	0.5830	1.667
Multicomponent	98/2	35 C3/34 IC4	0.8861	2.500
		40 C3/29 IC4	0.8816	2.168
		45 C3/24 IC4	0.8678	1.891
MeAc	0.1/0.1	25 MeAc	0.3091	0.6730
		30 MeAc	0.3642	0.7323
		33 MeAc	0.4436	0.8618

So the reflux flowrate is fixed at 1.0797 kmol/s in Figure 6.4, and the temperature on one stage is held constant (stage 2, 8, 12, 20, 29, or 37). The abscissas in the plots show the mole fraction of propane in the feed; the ordinates, the purity of the distillate x_D and the impurity of the bottoms x_B.

The results shown in Figure 6.4 display some counterintuitive results. Controlling the temperature on stage 8 near the top of the column does a better job in maintaining bottoms purity than does controlling stage 29 near the bottom. The bottoms impurity is held quite close to or under the desired 2 mol% propane. On the other hand, controlling the temperature on stage 29 near the bottom of the column does a better job of maintaining the purity of the distillate at or above the desired 98 mol% propane. Conventional wisdom dictates that a tray located nearer the product stream will hold its purity more constant.

These results indicate that either stage 8 or stage 29 does a fairly good job in maintaining product purities in this binary system when single-end temperature control is used. If dual-temperature control were used and the temperatures at the two ends of the column were controlled, product compositions would be held exactly at their desired values under steady-state conditions if pressure changes do not occur.

6.3 TERNARY BTX SYSTEM

The next separation system examined is a ternary mixture of benzene, toluene, and orthoxylene. The feed flowrate is 1 kmol/s, and the design feed composition is 30 mol% benzene, 30 mol% toluene, and 40 mol% orthoxylene. The operating objective is to separate the light-key component benzene from the heavy-key component toluene. Of course, the heavier-than-heavy-key component orthoxylene goes out the bottom with the toluene. Column pressure is set at 1 atm. The column has 32 stages and is fed on stage 16. Distillate impurity is specified to be 0.1 mol% toluene. Bottoms impurity is specified to be 0.1 mol% benzene. The reflux ratio required to achieve these purities is 1.908.

6.3.1 Slope Criterion

The upper graph in Figure 6.5 gives the temperature profile at design conditions. The lower graph shows the difference between the temperatures on adjacent trays. There is large change right at the feed stage. Because of the introduction of the feed, this is not a good location for temperature control. There is also a large change in temperature near the bottom of the column, which is due to the buildup of the heavier-than-heavy-key component orthoxylene. This is also not a good location for temperature control since we are trying to infer the compositions of benzene and toluene. The slope analysis suggests the use of stage 21 for temperature control.

6.3.2 Sensitivity Criterion

The upper graph in Figure 6.6 gives the openloop steady-state gains between tray temperatures and the two manipulated variables. These curves show that stage 21 is sensitive to changes in heat input and stage 22 is sensitive to changes in reflux.

6.3.3 SVD Criterion

The lower graph in Figure 6.6 gives the U_1 and U_2 values from SVD analysis. The first is the solid line and is associated with reflux. The second is the dashed line and is associated with heat input.

The SVD results are the similar to the sensitivity results. They suggest that stage 21 can be controlled by reflux and stage 23, by heat input. The singular values of the steady-state gain matrix are $\sigma_1 = 9.14$ and $\sigma_2 = 0.518$, which gives a condition number

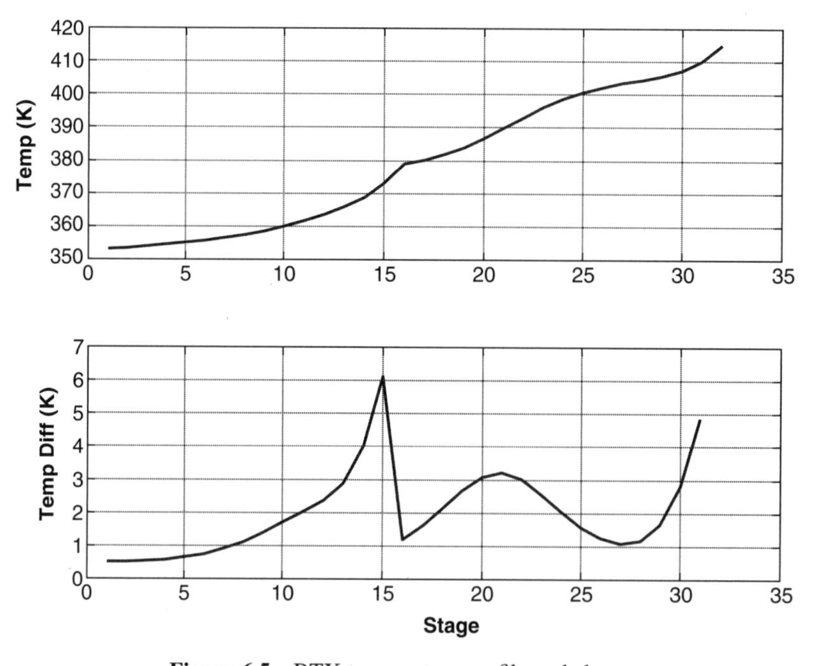

Figure 6.5 BTX temperature profile and slope.

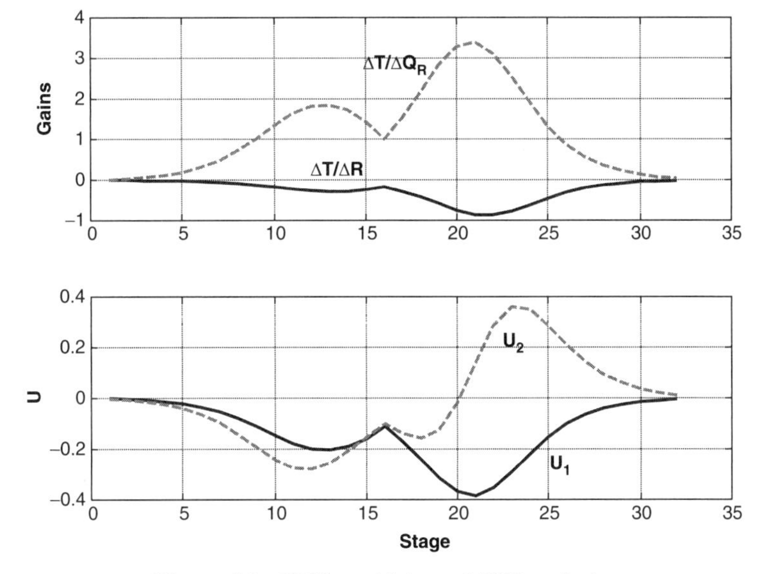

Figure 6.6 BTX sensitivity and SVD analysis.

$CN = \sigma_1/\sigma_2 = 17.6$. This indicates that the two temperatures are not nearly as independent as in the propane/isobutane system, so a dual-temperature control scheme may not be as effective. This makes sense because stages 21 and 23 are too close together to be independent.

6.3.4 Invariant Temperature Criterion

Figure 6.7 gives the changes in the temperature profiles for three different feed compositions in the ternary system. The design feed composition is 30/30/40 mol% benzene/toluene/xylene (BTX). The impurities in the bottoms and in the distillate are kept constant at 0.1 mol% benzene and 0.1 mol% toluene, respectively. The solid lines represent 25/35/40 mol% BTX feed composition; the dashed lines, 35/25/40 mol% BTX feed composition; and the dotted lines, 25/25/50 mol% BTX feed composition.

For changes in the benzene to toluene ratio in the feed, the results show that the temperature on stage 27 does not change for constant product impurities. So, if this is the type of feed composition disturbance expected, controlling stage 27 should provide effective control.

However, for the change in the xylene concentration of the feed, stage 27 changes by almost 3 K.

6.3.5 Minimum Product Variability Criterion

Figure 6.8 shows how product impurities change when the temperature on a specific tray is held constant and feed composition changes. The second control degree of freedom that is fixed in this figure is the reflux flowrate.

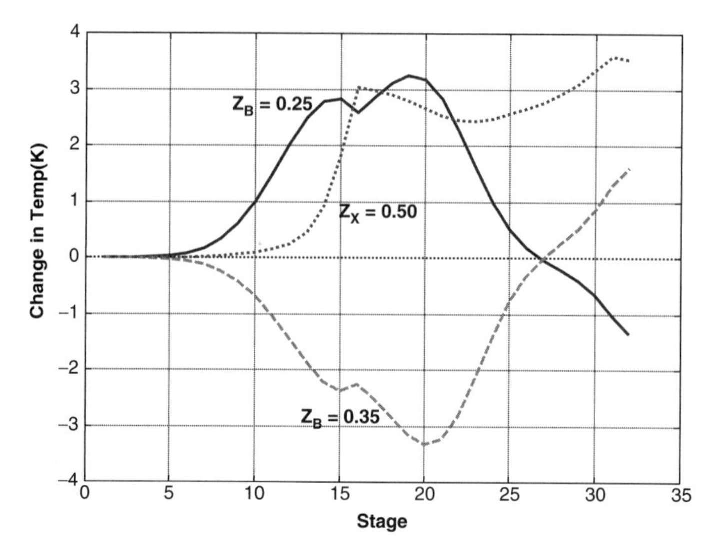

Figure 6.7 BTX changes in temperature profile for benzene and xylene feed composition changes with product purities fixed.

The justification for choosing constant reflux as opposed to constant reflux ratio is given in Table 6.1. The required changes in reflux flowrate are much smaller than the required changes in the reflux ratio in the BTX system. The reflux flowrate is fixed at 0.5715 kmol/s.

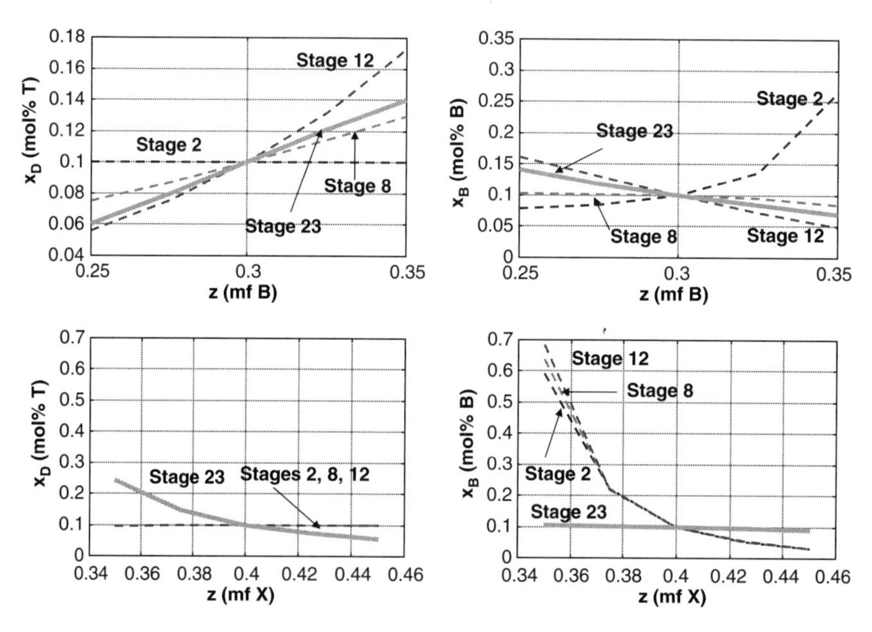

Figure 6.8 BTX product purities for benzene and xylene feed composition changes with fixed reflux and stage temperature.

Figure 6.8 gives results for changes in feed composition when the temperature on stage 2, 8, 12, or 23 is held. The upper graphs indicate changes in the mole fraction of benzene in the feed. The xylene mole fraction is fixed at 0.4, so the toluene composition in the feed changes inversely with the benzene composition. For this type of disturbance, holding stage 23 constant produces less variability in both product purities.

The lower graphs in Figure 6.8 show that holding stage 23 temperature for changes in the xylene composition of the feed is also effective. However, lowering the xylene in the feed strongly affects distillate purity.

These results illustrate the problems with the invariant-temperature criteria. Effectiveness is strongly dependent on which components in the feed change.

6.4 MULTICOMPONENT HYDROCARBON SYSTEM

The next separation system examined is a five-component hydrocarbon mixture. The feed flowrate is 1 kmol/s, and the design feed composition is 1 mol% ethane (C_2), 40 mol% propane (C_3), 29 mol% isobutane (iC_4), 29 mol% normal butane (nC_4) and 1 mol% isopentane (iC_5). The operating objective is to separate the light-key component propane from the heavy-key component isobutane. Of course, the heavier-than-heavy-key components nC_4 and iC_5 go out the bottom with the iC_4. The lighter-than-light key component C_2 goes out the top with the propane. Column pressure is set at 13.5 atm. The column has 37 stages and is fed on stage 18. Distillate impurity is specified to be 2 mol% iC_4. Bottoms impurity is specified to be 2 mol% C_3. The reflux ratio required to achieve these purities is 2.168.

6.4.1 Slope Criterion

The upper graph in Figure 6.9 gives the temperature profile at design conditions. The lower graph shows the differences between the temperatures on adjacent trays. The largest change occurs on stage 31. There is also a large change in temperature at the top of the column due to buildup of the lighter-than-light-key component C_2. There is also a smaller peak at stage 8.

6.4.2 Sensitivity Criterion

The upper graph in Figure 6.10 gives the openloop gains between tray temperatures and the two manipulated variables. These curves show that stage 30 is sensitive to changes in heat input and stage 8 is sensitive to changes in reflux. In this system, the slope criterion and the sensitivity criterion give identical results.

6.4.3 SVD Criterion

The lower graph in Figure 6.10 gives the U_1 and U_2 values from SVD analysis. The SVD results are similar to the sensitivity results. They suggest that stage 8 can be controlled by reflux and stage 30, by heat input. The singular values of the steady-state gain matrix are $\sigma_1 = 0.4066$ and $\sigma_2 = 0.1637$, giving a condition number $CN = \sigma_1/\sigma_2 = 2.48$. This indicates that the two temperatures are quite independent, so a dual-temperature control scheme may be effective if it is required.

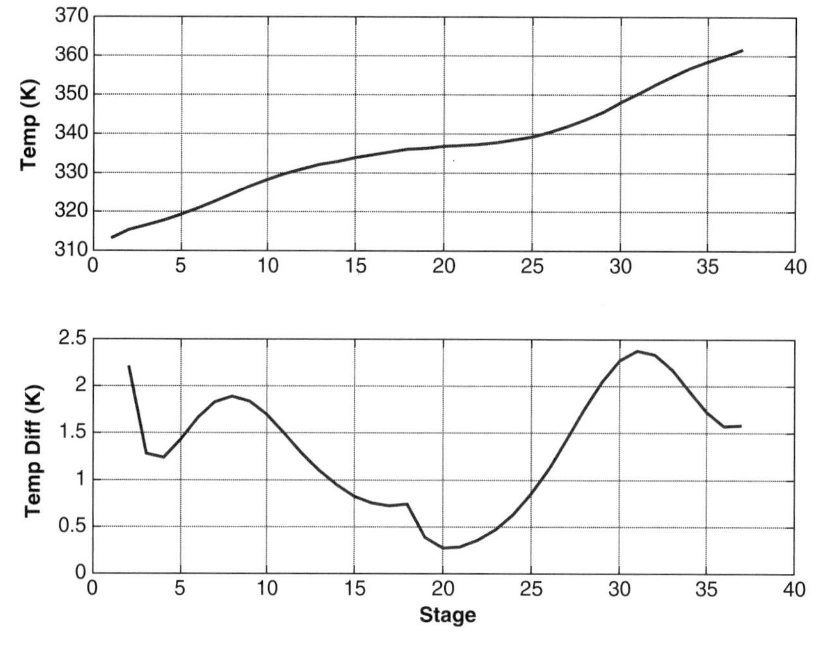

Figure 6.9 Multicomponent temperature profile and slope.

6.4.4 Invariant Temperature Criterion

Figure 6.11 gives the changes in the temperature profiles for four different feed compositions. The design feed composition is 1/40/29/29/1 mol% $C_2/C_3/iC_4/nC_4/iC_5$. The

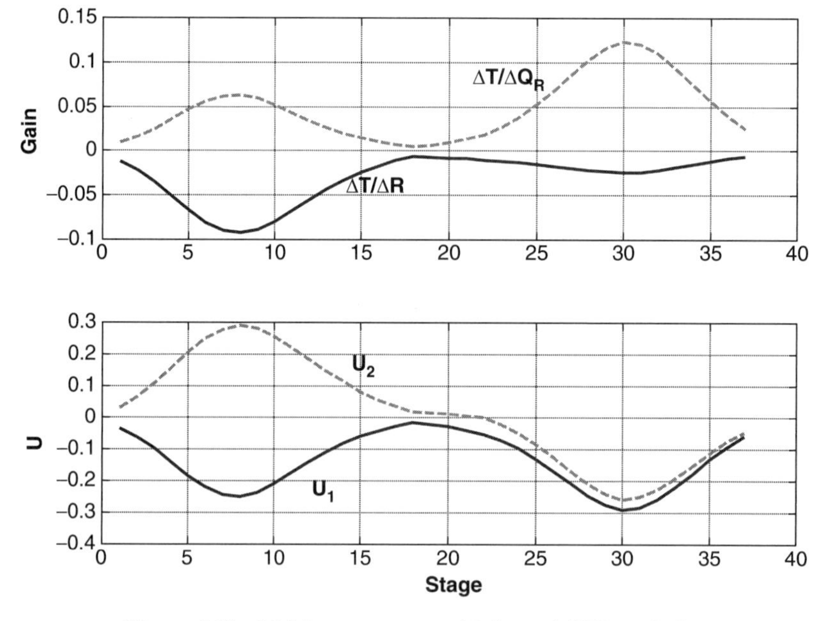

Figure 6.10 Multicomponent sensitivity and SVD analysis.

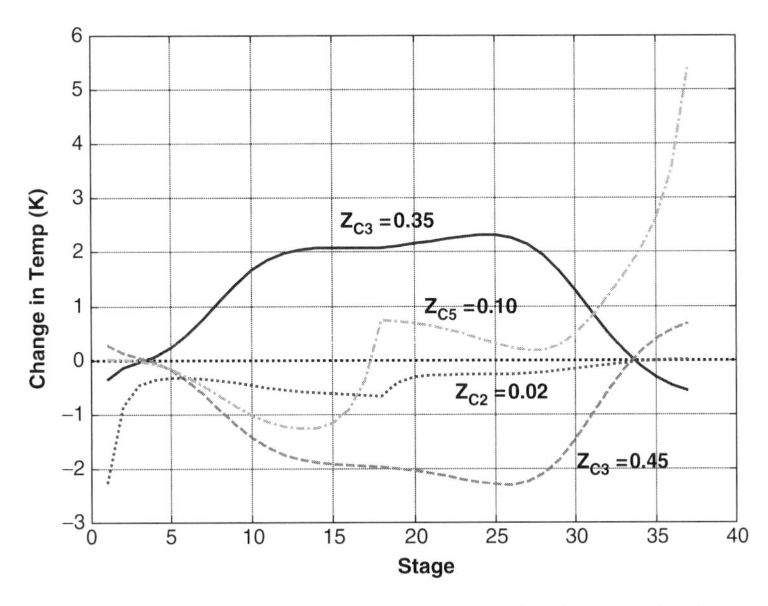

Figure 6.11 Multicomponent changes in temperature profile for C_2 C_3, and iC_5 feed composition changes with product purities fixed.

impurities in the bottoms and in the distillate are kept constant at 2 mol% C_3 and 2 mol% iC_4, respectively. The solid line shows a feed composition in which the propane composition is decreased from 40 to 35 mol% C_3 and the isobutane composition is increased from 29 to 34 mol% iC_4. The dashed line represents a feed composition in which the propane composition is increased from 40 to 45 mol% C_3 and the isobutane composition is decreased from 29 to 24 mol% iC_4. The dotted line shows a feed composition in which the ethane composition is increased from 1 to 2 mol% C_2, while the iC_4 and nC_4 are both reduced from 29 to 28.5 mol%. The dashed–dotted line indicates a feed composition in which the isopentane composition is increased from 1 to 10 mol% iC_5, the isobutane composition decreased from 29 to 25 mol% iC_4 and the n-butane composition decreased from 29 to 24 mol% nC_4.

For changes in the propane to isobutane ratio in the feed or for changes in the C_2 in the feed, the results show that the temperature on stage 34 does not change for constant product impurities. So if these are the types of feed composition disturbance expected, controlling stage 34 should provide effective control.

However, for a change in the iC_5 concentration of the feed, stage 34 changes by $\sim$2 K. Therefore, controlling stage 34 will not handle changes in the composition of the heavier-than-heavy-key component in the feed.

6.4.5 Minimum Product Variability Criterion

Figure 6.12 shows how product impurities change when the temperature on a tray is held constant and feed composition changes. The second control degree of freedom that is fixed in this figure is the reflux flowrate. Table 6.1 shows that the required changes in reflux flowrate are much smaller than the required changes in the reflux ratio in this multicomponent system. The reflux flowrate is fixed at 0.8816 kmol/s.

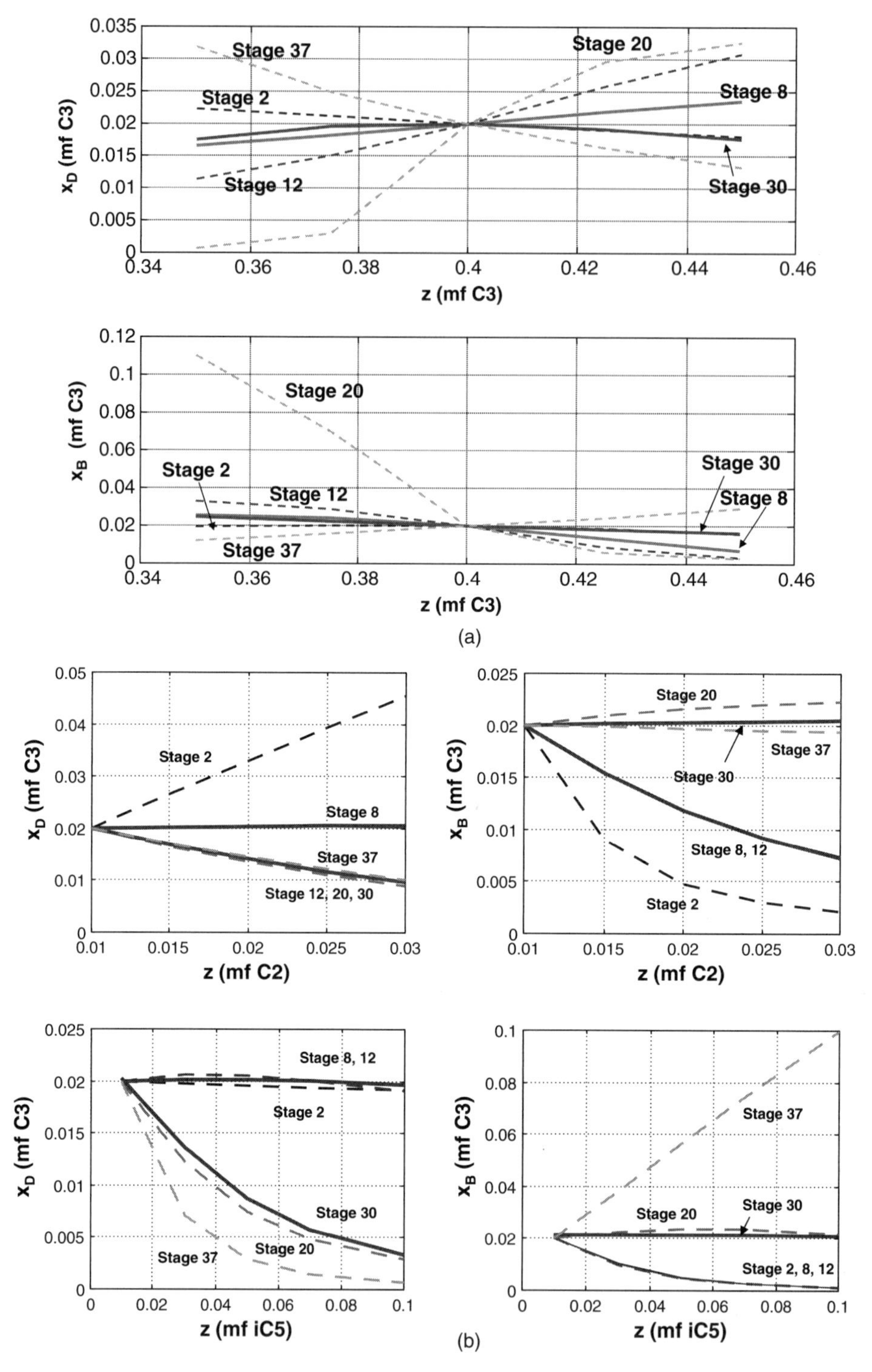

Figure 6.12 (a) Multicomponent product impurities for changes in C_3 feed composition with fixed reflux stage temperature; (b) multicomponent product impurities for changes in C_2 and iC_5 feed composition with reflux and stage temperature.

Figure 6.12a gives results for changes in the propane composition of the feed. Using the SVD-recommended stage 8 or stage 30 holds both the distillate and bottoms impurities close to their specified values of 2 mol%.

The upper graphs in Figure 6.12b give results for changes in the isopentane composition of the feed. The lower graphs in Figure 6.12b give results for changes in the ethane composition of the feed. Using stage 8 or stage 30 keeps both the distillate and the bottoms impurities quite close to or below their specifications.

6.5 TERNARY AZEOTROPIC SYSTEM

Up to this point we have looked at systems with fairly ideal vapor–liquid equilibrium behavior. The last separation system examined is a highly nonideal ternary system of methyl acetate, methanol, and water. Methyl acetate and methanol form a homogeneous minimum-boiling azeotrope at 1.1 atm with a composition of 66.4 mol% methyl acetate and a temperature of 329 K. This means that the overhead product from the distillation column cannot have a composition greater than this azeotropic composition.

The design objectives are to produce a distillate product with 0.1 mol% water and a bottoms product with 0.1 mol% methyl acetate. The feed flowrate is 1 kmol/s, and the design feed composition is 30 mol% methyl acetate (MeAc), 50 mol% methanol (MeOH), and 20 mol% water. Column pressure is set at 1.1 atm. The column has 42 stages and is fed on stage 36 (the stage that minimizes reboiler heat input at design feed composition). The reflux ratio required to achieve the specified purities is 0.7323.

6.5.1 Slope Criterion

The upper graph in Figure 6.13 gives the temperature profile at design conditions. The lower graph shows the differences between the temperatures on adjacent trays. The maximum change occurs on stage 37. There is also a large change in temperature at the very bottom of the column that is due to buildup of the heaviest component water.

6.5.2 Sensitivity Criterion

The upper graph in Figure 6.14 gives the openloop gains between tray temperatures and the two manipulated variables. These curves show that stage 38 is sensitive to changes in heat input and stage 27 is sensitive to changes in reflux.

6.5.3 SVD Criterion

The lower graph in Figure 6.14 gives the U_1 and U_2 values from the SVD analysis. The results suggest that stage 38 can be controlled by either reflux or reboiler heat input. There is a second smaller peak in U_2 at about stage 28 that could be controlled by heat input. The singular values of the steady-state gain matrix are $\sigma_1 = 0.5965$ and $\sigma_2 = 0.0855$, which gives a condition number $CN = \sigma_1/\sigma_2 = 6.98$.

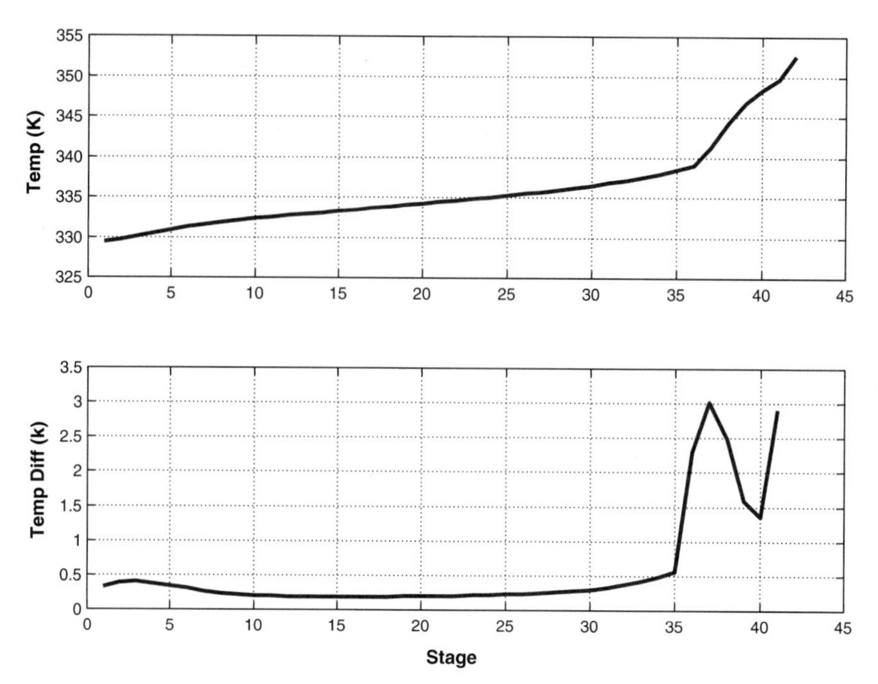

Figure 6.13 MeAc/MeOH/H_2O—temperature profile and slope.

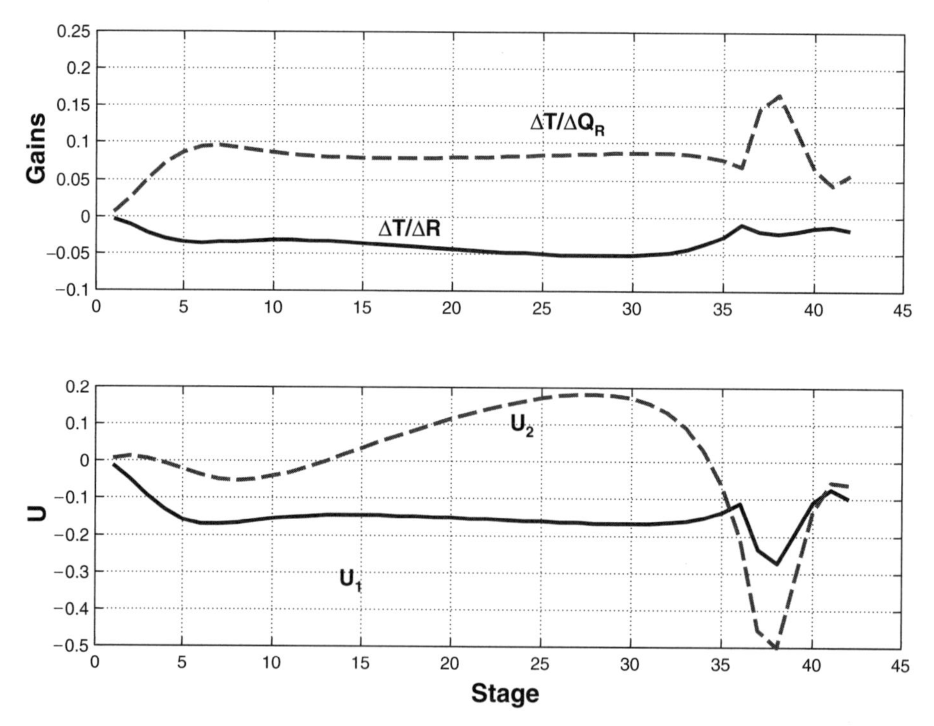

Figure 6.14 MeAc/MeOH/H_2O—sensitivity and SVD analysis.

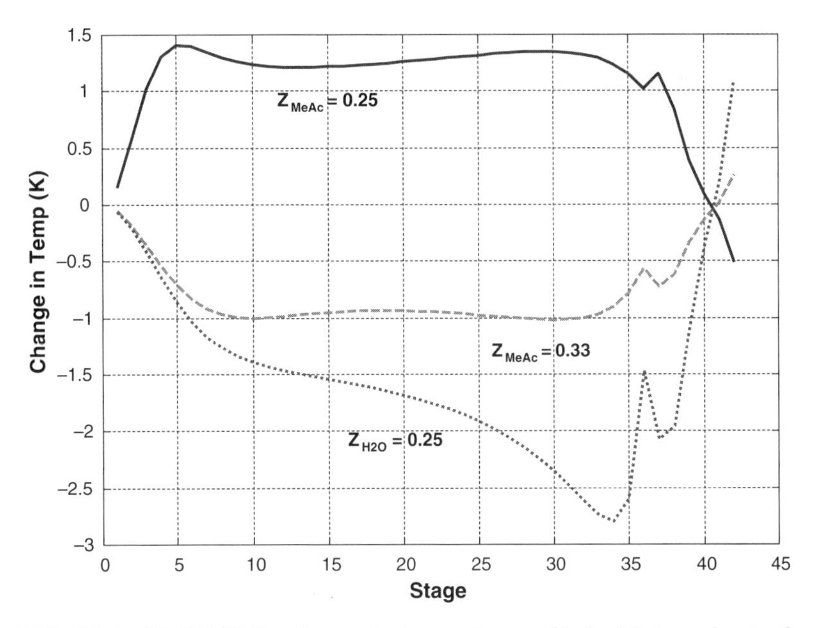

Figure 6.15 MeAc/MeOH/H_2O—changes in temperature profile for MeAc and water feed composition changes with product purities fixed.

6.5.4 Invariant Temperature Criterion

Figure 6.15 gives the changes in the temperature profiles for three different feed compositions. The design feed composition is 30/50/20 mol% MeAc/MeOH/H_2O. The impurities in the bottoms and in the distillate are kept constant at 0.1 mol% MeAc and 0.1 mol% H_2O, respectively. The solid line shows a feed composition in which the MeAc composition is decreased from 30 to 25 mol% and the MeOH composition is increased from 50 to 55 mol%. The dashed line indicates a feed composition in which the MeAc composition is increased from 30 to 33 mol% and the MeOH composition decreased from 50 to 47 mol%. The dotted line indicates a feed composition in which the H_2O composition is increased from 20 to 25 mol% and the MeOH composition is decreased from 50 to 45 mol%.

Results show that the temperature on stage 41 does not change for constant product impurities for all of these feed composition disturbances. So there is a conflict between the SVD and the invariant-temperature criteria. One recommends stage 38, while the other recommends stage 41.

6.5.5 Minimum Product Variability Criterion

Figure 6.16 shows how product impurities change for two different temperature control trays as feed composition changes. In the first case, the temperature of stage 38 is fixed at 344.28 K. In the second case, the temperature of stage 41 is fixed at 349.71 K.

The other control degree of freedom that is fixed in this figure is the *reflux ratio*. Unlike the other systems studied, Table 6.1 shows that the required changes in reflux ratio are smaller than the required changes in the reflux flowrate in this multicomponent, nonideal system. The reflux ratio is fixed at 0.732.

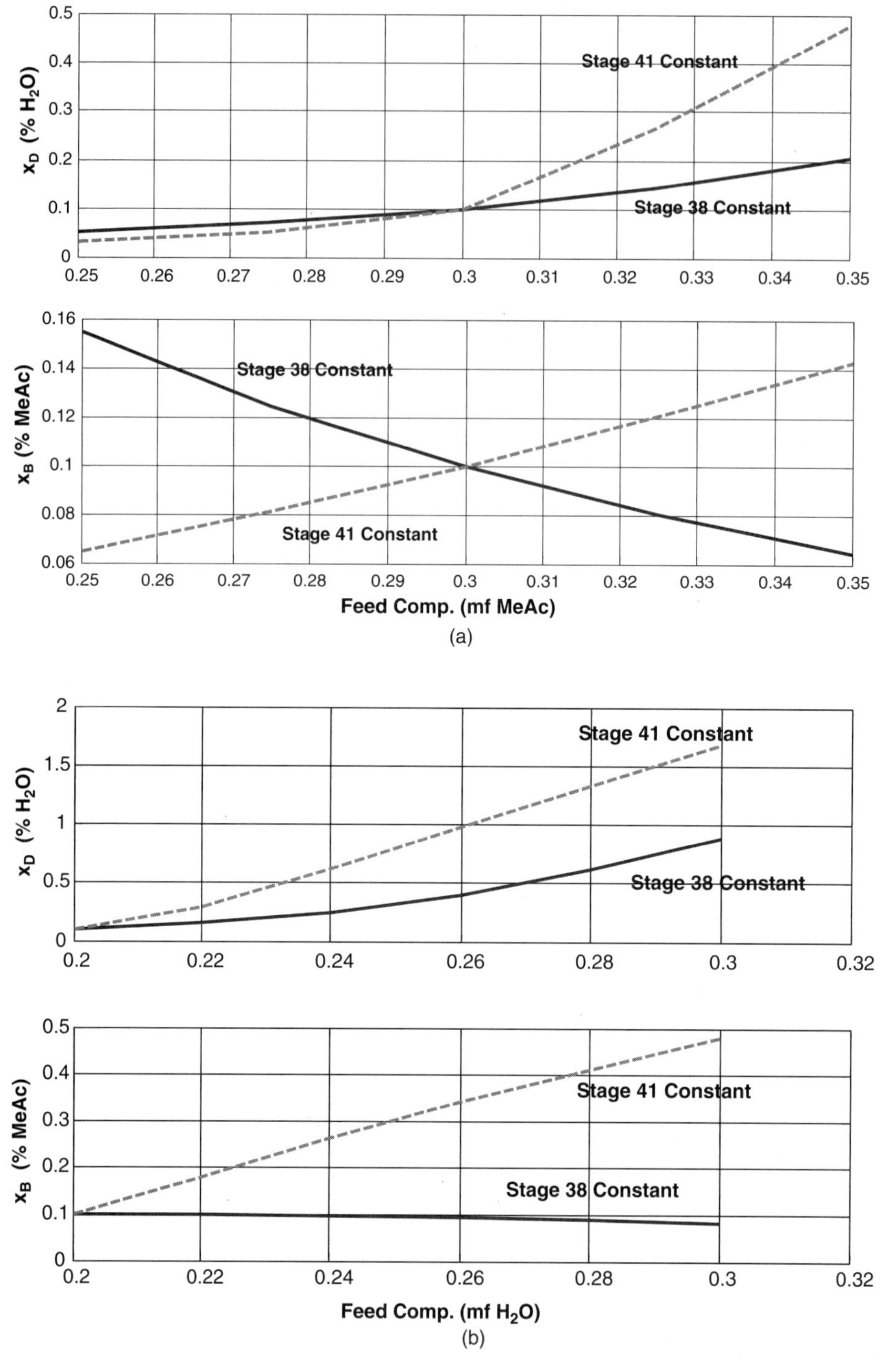

Figure 6.16 (a) MeAc/MeOH/H_2O—changes in product impurities for MeAc feed composition changes with fixed reflux ratio and stage temperature; (b) MeAc/MeOH/H_2O—changes in product impurities for water feed composition changes with fixed reflux ratio and stage temperature.

Figure 6.16a gives results for changes in the MeAc composition of the feed. The control tray recommended by SVD (stage 38) does a better job in holding distillate close to its specified value of 0.1 mol%. The changes in the bottoms composition are about the same for both stages 38 and 41, but in opposite directions.

Figure 6.16b give results for changes in the water composition of the feed. Controlling the temperature on stage 38 keeps both products closer to their specification than when stage 41 is controlled.

6.5.6 Closedloop Multiplicity

One interesting feature of nonlinear systems is the possible appearance of multiple steady states. Most researchers have explored openloop multiplicity. We found that closedloop multiplicity occurs in the methyl acetate/methanol/water system.

The two product impurity levels are fixed at their specified values, and feed composition is varied over a range of values. The required reflux flowrate and reflux ratio are calculated for each case. Figure 6.17 shows what happens when we start with a feed composition of 30/50/20 mol% MeAc/MeOH/H_2O. The reflux ratio is 0.732, and the reflux flowrate is 0.364 kmol/s with this feed composition. As the feed composition is increased, the required reflux and reflux ratio increase. At a feed composition of 33/47/20 mol% MeAc/MeOH/H_2O, the reflux ratio is 0.862 and the reflux flowrate is 0.444 kmol/s. This is labeled "Profile 1" in Figure 6.17.

However, if the feed composition is changed to 34/46/20 mol% MeAc/MeOH/H_2O, there is a huge increase in the required reflux ratio to 1.94 and the reflux flowrate to 0.991 kmol/s. Further increases in MeAc in the feed continue to increase *RR* and *R*.

Now if the feed composition is reduced back to 33/47/20 mol% MeAc/MeOH/H_2O (labeled "Profile 2" in Fig. 6.17), the reflux ratio does not return to 0.862 but changes only to 1.87. Likewise, the reflux flowrate does not return to 0.444 kmol/s but changes

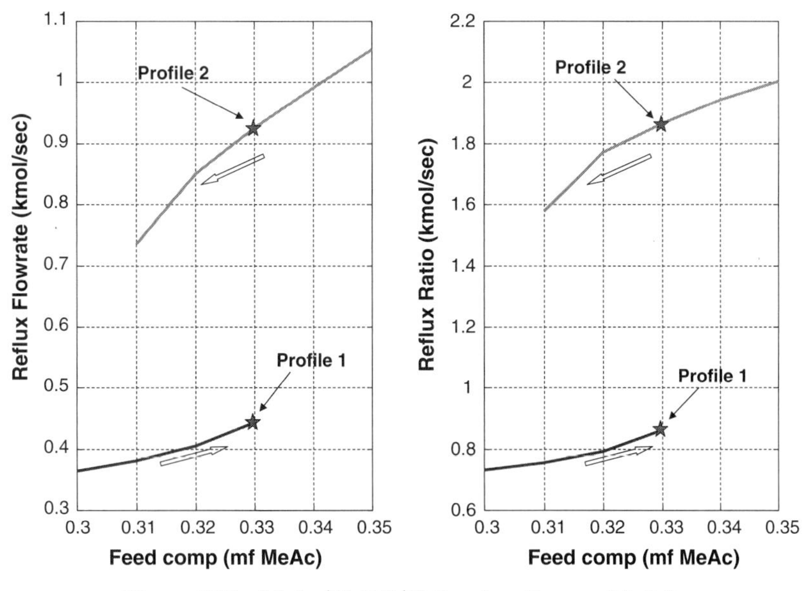

Figure 6.17 MeAc/MeOH/H_2O—closedloop multiplicity.

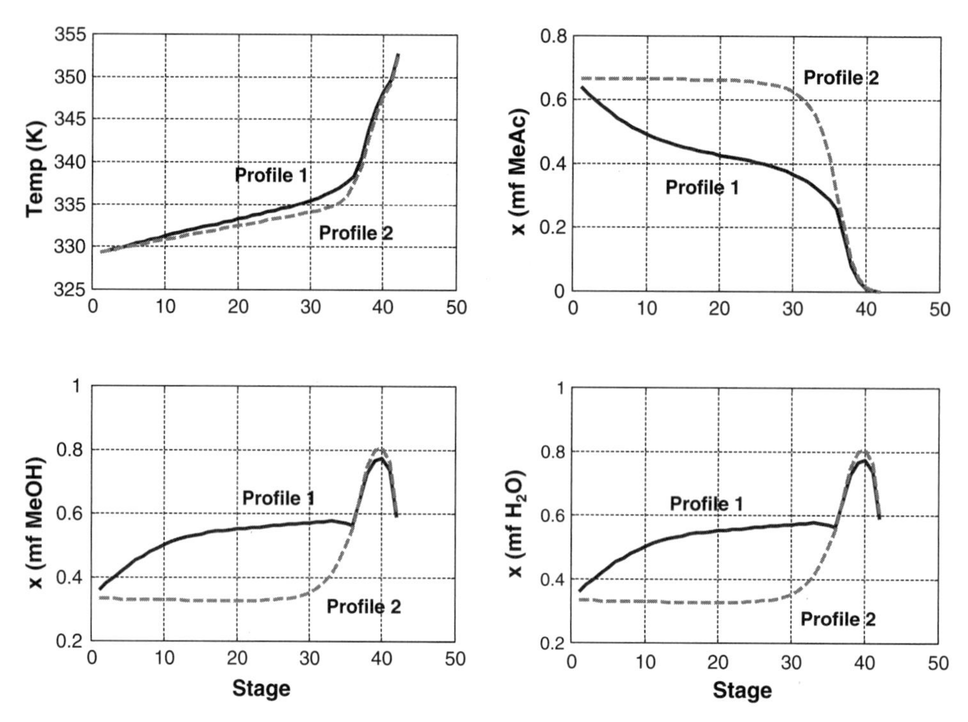

Figure 6.18 MeAc/MeOH/H_2O—profiles at two different steady states.

to 0.925 kmol/s. The reflux ratios and reflux flowrates remain high until the feed composition drops to about 30/50/20 mol%, at which point the required *RR* and *R* drop back to their original levels.

Figure 6.18 gives the composition and temperature profiles of the two steady states. Both have the same feed composition: 33/47/20 mol% MeAc/MeOH/H_2O. Both have the same impurities in the two product streams. But they have different composition and temperature profiles, and they have greatly different reflux ratios and reflux flowrates.

6.6 CONCLUSION

In this chapter we have illustrated how steady-state calculations can be used to provide guidance in the selection of effective control structures for distillation columns.

CHAPTER 7

CONVERTING FROM STEADY STATE TO DYNAMIC SIMULATION

We are now ready to get into the details of converting a steady-state simulation into a dynamic one. Basically the additional information that must be provided is the physical dimensions of the various pieces of equipment.

It is important to remember that pumps and control valves have already been installed in the steady-state simulation. These are not necessary for steady-state simulation, but they are vital for a realistic dynamic simulation. Providing sufficient pressure drop over a control valve at design conditions with the valve at some fraction opening (typically 50%) is crucial for dynamic controllability. If valve pressure drop is too small, changing the valve opening from 50% to 100% will, in piping systems with other equipment taking pressure drops, result in only a fairly small increase in flowrate. If a valve saturates, controllability is lost.

The propane/isobutane column developed in Chapter 3 is used in this chapter as a numerical example. The control valves all have pressure drops of $\sim$3 atm. The column has 32 stages, is fed on stage 14, operates at 16.8 atm, and produces distillate and bottoms products with impurity levels of 2 mol% isobutane and 1 mol% propane, respectively.

7.1 EQUIPMENT SIZING

The dynamic response of a flow system depends on the flowrate and the volume. For a given flowrate, the smaller the volume, the faster the transient response.

The procedure for sizing the distillation column shell has already been discussed (in Chapter 3). The only remaining issues are the sizes of the reflux drum and the column base. A commonly used heuristic is to set these holdups to allow for 5 min of liquid holdup when the vessel is 50% full, based on the total liquid entering or leaving the vessel. For the reflux drum, this is the sum of the liquid distillate and the reflux. For the column base, it is the liquid entering the reboiler from the bottom tray.

Distillation Design and Control Using Aspen™ Simulation, By William L. Luyben

These volumetric liquid flowrates can be found by clicking on *Profiles* under the block C1 and then opening the *Hydraulics* page tab. Figure 7.1 shows the window that opens on which the volumetric liquid flowrate for the reflux drum (stage 1) is given as 0.1782 m^3/s. Scrolling down to the bottom tray (stage 31) gives a volumetric liquid flowrate of 0.3438 m^3/s. The total volume of the reflux drum should be $0.1782(60)(10) = 106.9$ m^3 and that of the column base $0.3438(60)(10) = 206.3$ m^3. Assuming a length to diameter ratio of 2, the diameters and lengths can be calculated:

$$\text{Volume} = \frac{\pi D^2}{4}(2D)$$

The reflux drum is 4.08 m in the diameter and 8.16 m in length. The column base (or reboiler or "sump") has a diameter of 5.08 m and a length of 10.16 m.

The values are entered by clicking the *Dynamics* button on the top toolbar (see Fig. 7.2). If this button is not displayed, click the *View* button, then *Toolbar*, and check the *Dynamics* box. The window that opens has several page tabs. On the *Reflux Drum* page tab, the appropriate diameter and length are entered. The same is done on the *Sump* page tab. Finally, the *Hydraulics* page tab is clicked, and the window shown in Figure 7.3 opens, on which stage numbers (2 through 31) and the column diameter (5.91 m) are entered. The default values of weir height and tray spacing are 0.05 and 0.6096 m, respectively.

At this point all equipment has been sized. There remain two items to take care of. The pressure of the feedstream leaving valve V1 must be exactly equal to the pressure on the stage where it is fed. The pressure on stage 14 is found by looking in *Profiles* (1,709,839.11 N/m^2). The outlet pressure of valve V1 is set equal to this value, and the simulation is run again. The last thing to do is to click the *Pressure Checker* button that is just to the right of the *Dynamics* button on the top toolbar (see Fig. 7.2). If the plumbing has been correctly specified, the window shown in Figure 7.4 appears. We are ready to go into Aspen Dynamics.

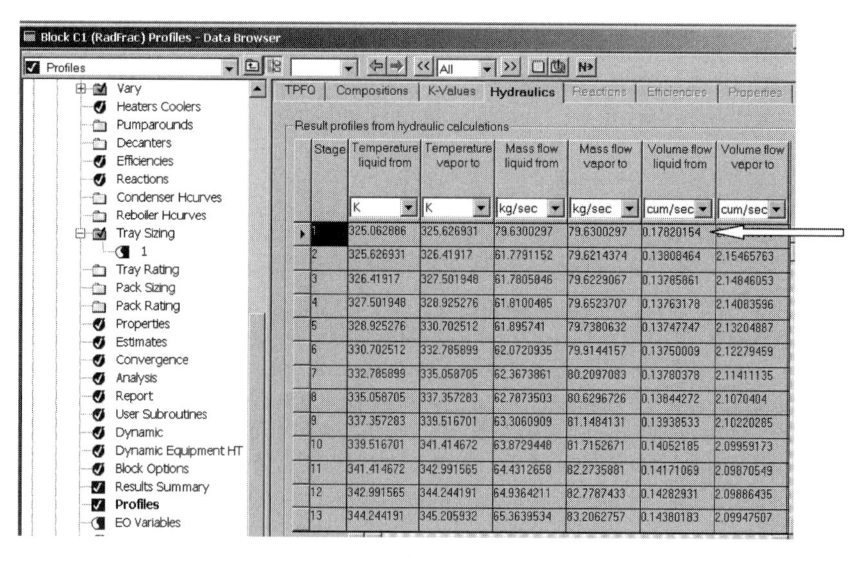

Stage	Temperature liquid from	Temperature vapor to	Mass flow liquid from	Mass flow vapor to	Volume flow liquid from	Volume flow vapor to
	K	K	kg/sec	kg/sec	cum/sec	cum/sec
1	325.062886	325.626931	79.6300297	79.6300297	0.17820154	
2	325.626931	326.41917	61.7791152	79.6214374	0.13808464	2.15465763
3	326.41917	327.501948	61.7805846	79.6229067	0.13785861	2.14846053
4	327.501948	328.925276	61.8100485	79.6523707	0.13763178	2.14083596
5	328.925276	330.702512	61.895741	79.7380632	0.13747747	2.13204887
6	330.702512	332.785899	62.0720935	79.9144157	0.13750009	2.12279459
7	332.785899	335.058705	62.3673861	80.2097083	0.13780378	2.11411135
8	335.058705	337.357283	62.7873503	80.6296726	0.13844272	2.1070404
9	337.357283	339.516701	63.3060909	81.1484131	0.13938533	2.10220285
10	339.516701	341.414672	63.8729448	81.7152671	0.14052185	2.09959173
11	341.414672	342.991565	64.4312658	82.2735881	0.14171069	2.09870549
12	342.991565	344.244191	64.9364211	82.7787433	0.14282931	2.09886435
13	344.244191	345.205932	65.3639534	83.2062757	0.14380183	2.09947507

Figure 7.1 *Hydraulics* page tab.

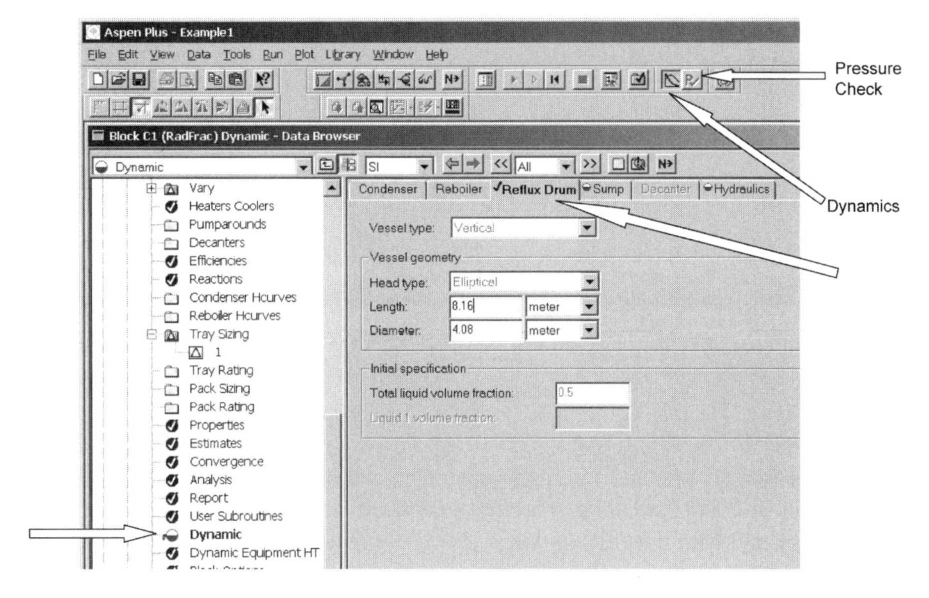

Figure 7.2 *Dynamics* button.

7.2 EXPORTING TO ASPEN DYNAMICS

Aspen Dynamics uses the steady-state information generated in Aspen Plus, but they are two different programs with different files. The Aspen Plus file is *filename.apw*, and there is also a backup file, *filename.bkp*, that is generated. The latter file can be used to upgrade to newer versions of Aspen Plus.

The information from Aspen Plus is "exported" into Aspen Dynamics by generating two additional files. The first is a *filename.dynf* file that is used in Aspen Dynamics and is modified to incorporate controllers, plots, and other features. The second file is a

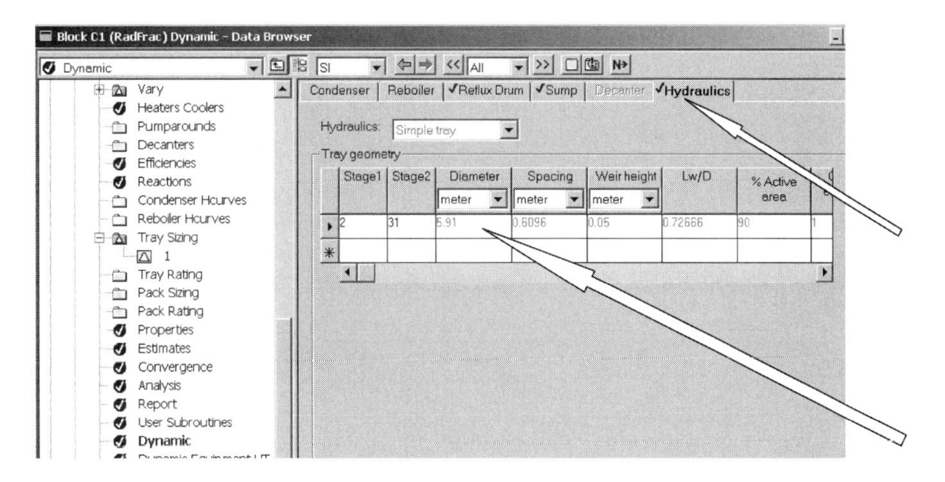

Figure 7.3 Specifying diameter and weir height.

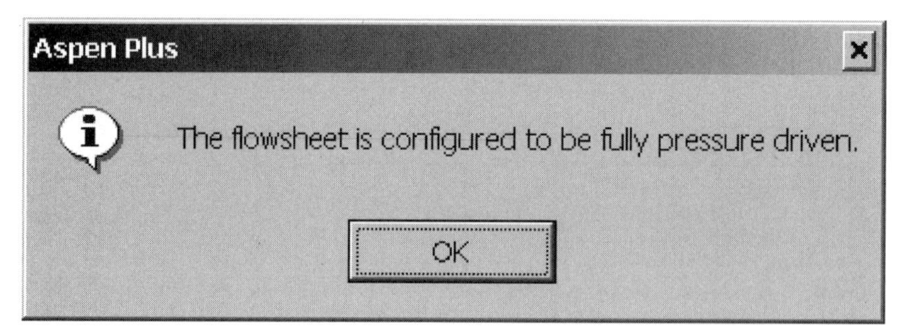

Figure 7.4 Pressure check.

filenamedyn.appdf file that contains all the physical property information to be used in Aspen Dynamics.

In the propane/isobutane column example, the Aspen Plus files are called *Example1.apw* and *Example1.bkp*. The files generated and used in Aspen Dynamics are *Example1.dynf* and *Example1dyn.appdf*.

The procedure for "exporting" is to click on *File* at the top left corner of the Aspen Plus window and select *Export*. The window shown in Figure 7.5 opens, and the dropdown menu is used to select *P driven Dyn Simulation*, which is the tenth item on the list. Then the *Save* button is clicked.

An error appears (Fig. 7.6a), which informs us that the specified pressure drop (0.0068 atm) is too small for the given weir height and vapor rates. Aspen Dynamics calculates tray pressure drops rigorously, and they change with vapor and liquid rates. We have two options: increase the specified pressure drop or decrease the weir height. The latter has no effect on the steady-state solution, so we go back to *Dynamics* under the C1 block, select the *Hydraulics* page tab, and change the weir height from 0.05 to 0.025 m. Running the program again, pressure checking, and "Exporting" again cause the

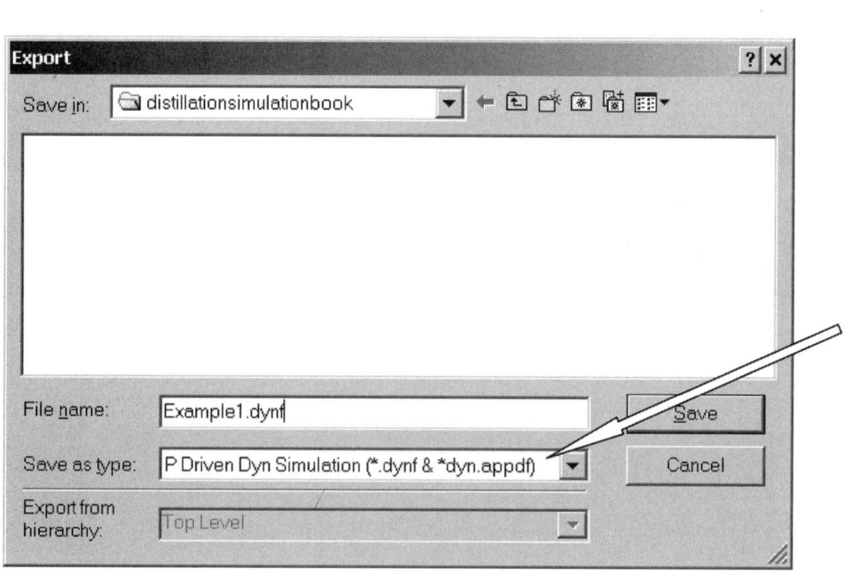

Figure 7.5 Exporting.

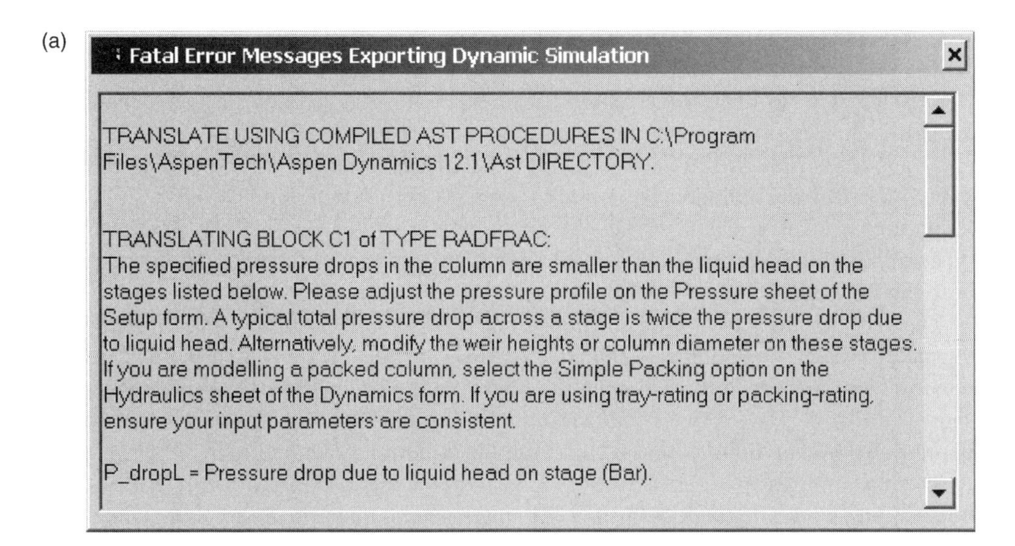

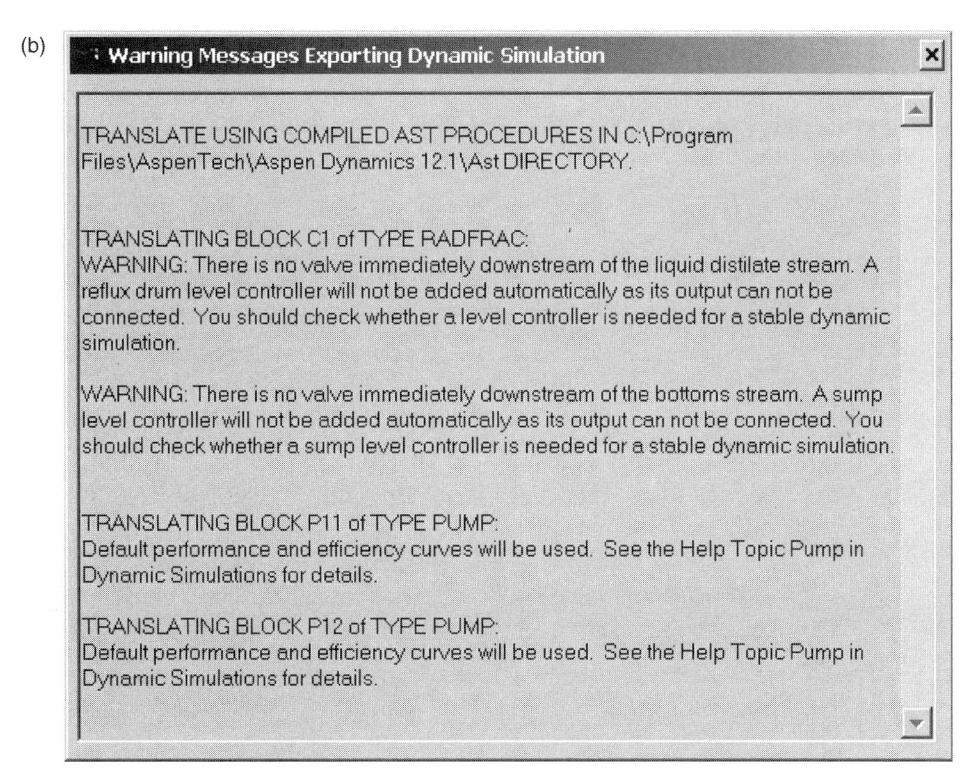

Figure 7.6 (a) Error message; (b) warning message.

window shown in Figure 7.6b to open. These messages just remind us that level controllers should be installed once we get into Aspen Dynamics and the typical pump curves are used in the pumps.

The "export" is successful, and we are ready to go into Aspen Dynamics.

7.3 OPENING THE DYNAMIC SIMULATION IN ASPEN DYNAMICS

The easiest way to open the *Example1.dynf* file is to double-click on the filename in *Windows Explorer*. The screen that opens contains several windows (see Fig. 7.7).

The *Process Flowsheet Window* is where the control structure will be developed. The *Simulation Messages* window is where the progress of the simulation is shown and simulation time is displayed. The window in the upper left corner, *Exploring*, is where various types of controllers, control signals, and other elements can be found to "drop and drag" onto the flowsheet.

The very first thing to do with any newly imported file is to make an "initialization" run to make sure that everything is running. At the very top of the screen there is a little window that says *Dynamic*. Clicking the arrow to the right opens the dropdown menu shown in Figure 7.8a. Select *Initialization* and click the *Run* button, which is just to the upper right. If everything is set up correctly, the window shown in Figure 7.8b opens.

The next thing to do is to make sure that the integrator is working correctly. This is done by changing from *Initialization* back to *Dynamic* and clicking the *Run* button again. The *Simulation Messages* window at the bottom of the screen should start displaying advancing simulation time. An example is shown in Figure 7.8c. Note the green block at the bottom of the screen. If something is wrong, this block will turn red and you will not be able to run. The run is stopped by clicking the *Pause* button, which is the second button to the right of the *Run* button.

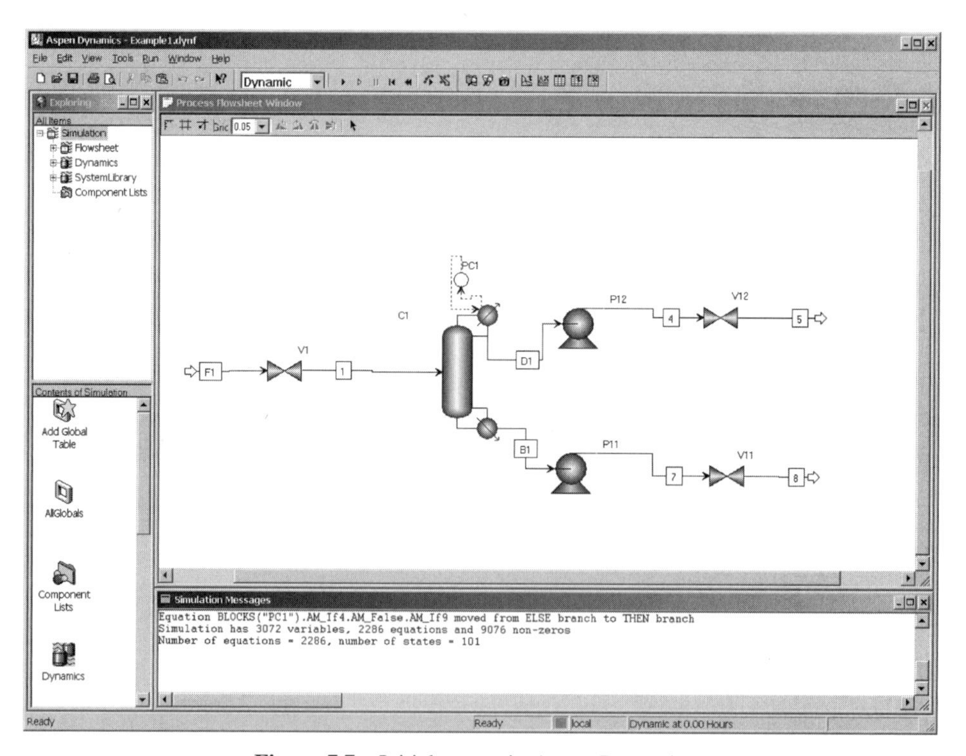

Figure 7.7 Initial screen in Aspen Dynamics.

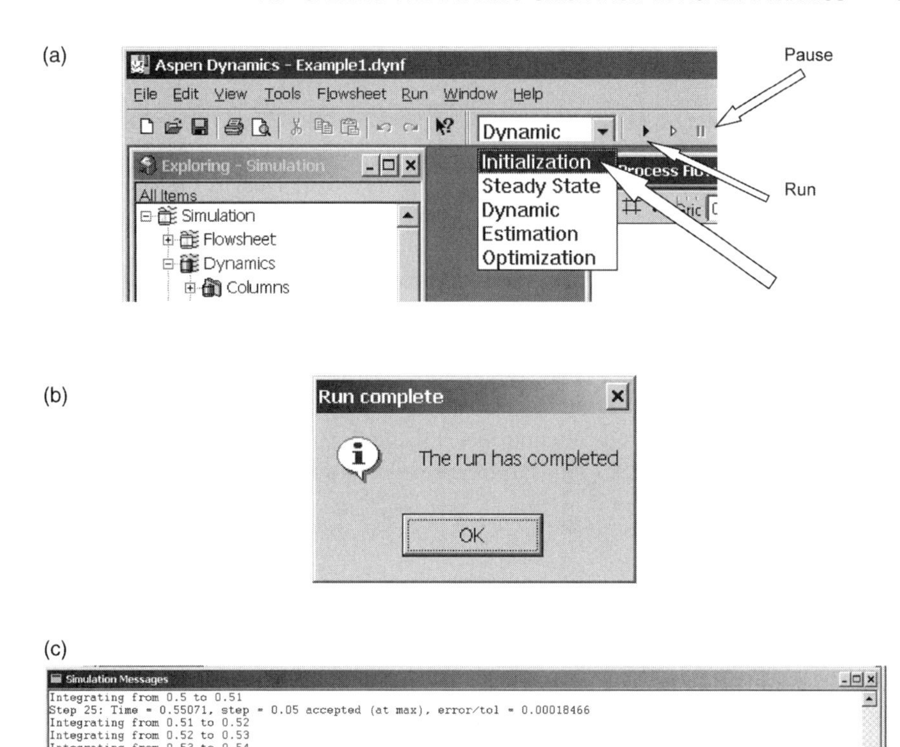

Figure 7.8 (a) Selecting initialization; (b) initialization run successful; (c) making *Dynamic* run.

The initial flowsheet has some default controllers already installed. In this simple, single-column process there is only one default controller, the pressure controller. It is configured to measure condenser pressure and manipulate condenser heat removal. The action of the controller, the range of the pressure transmitter, the maximum heat removal rate, and the controller tuning constants are all set up at some nominal values. We will come back and look at these later. But this indicates the number of items that must be specified for each new controller that we add to the flowsheet. At a minimum, four additional controllers must be added to achieve effective operation of the column:

1. Reflux drum-level controller
2. Base-level controller
3. Feedflow controller
4. Tray temperature controller

Which of the four remaining manipulated variables selected to control each of these four is the issue of control structure selection. A variety of control structures will be discussed here and in subsequent chapters.

We have not included reflux in the list above because the default configuration in Aspen Dynamics is to fix the mass flowrate of the reflux. We will discuss how this setup can be changed later in this chapter.

7.4 INSTALLING BASIC CONTROLLERS

Let us go through the details of installing a level controller on the base of the column. Go to the *Exploring* window, click on *Dynamics* under *Simulation*, and select *Control Models*. A long list of alternative controller types and dynamic elements is displayed, as shown in Figure 7.9. We will use several of these extensively: *Dead_time*, *Lag_1*, *Multipy*, *PID*, and *PIDIncr*.

For simple level controllers, which should be proportional only, the *PID* model is used. Place the cursor on the *PID* icon, **press and hold** the left mouse button, and drag the cursor to the flowsheet window. Releasing the mouse button places a circle on the flowsheet, which is labeled "B2" (see Fig. 7.10). This can be renamed by clicking the circle, right-clicking, selecting *Rename block*, and typing in the desired label "LC11."

Now the control signals must be connected. The process variable signal *PV* is the variable to be controlled. For this level controller, it is the liquid level on the last stage of the column (stage 32). We go back to the *Exploring* window and click on *Stream Types* under *Dynamics*, as shown in Figure 7.11.

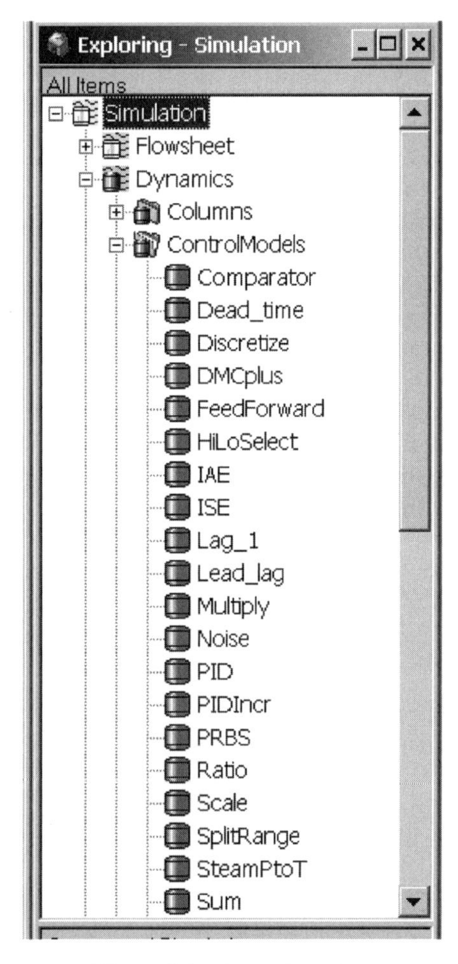

Figure 7.9 Control models.

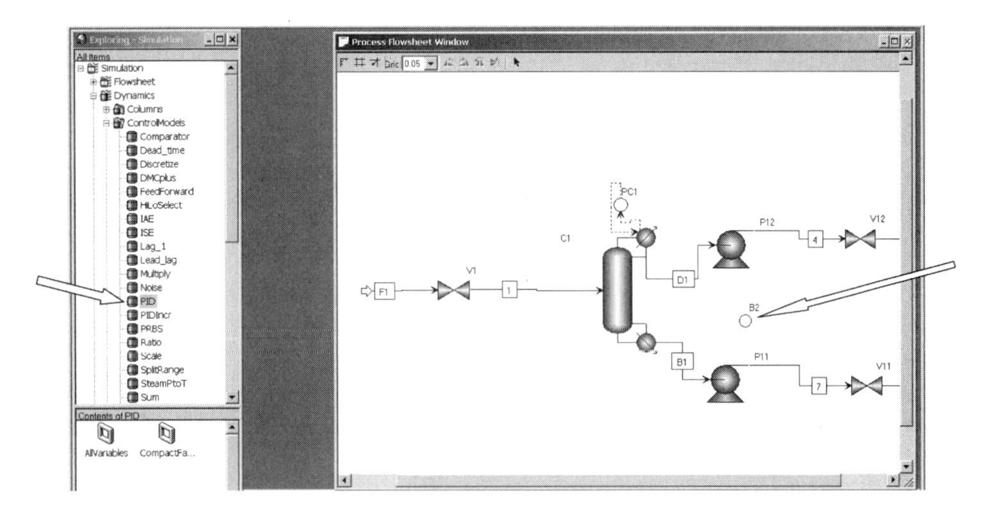

Figure 7.10 Controller placed on flowsheet.

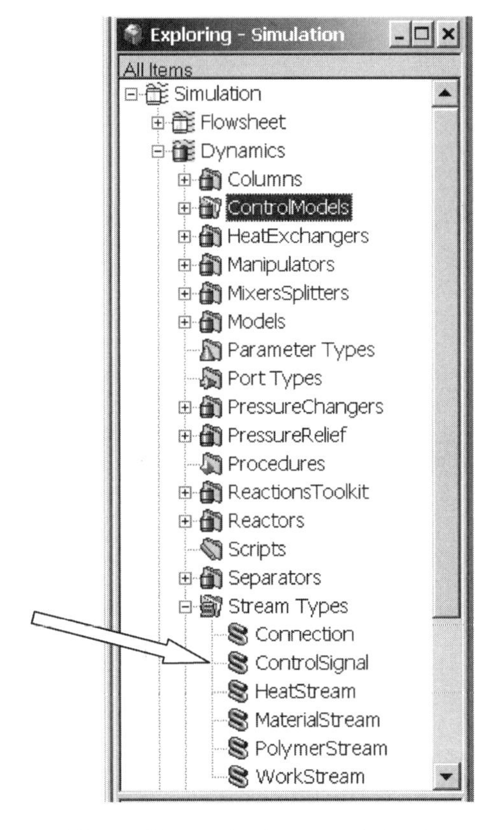

Figure 7.11 Selecting control signal.

To place a control signal on the flowsheet, we place the cursor over the *ControlSignal* icon, **click and hold** the left mouse button, and move the cursor to the flowsheet. A number of blue arrows show up that indicate where a control signal can be placed. Placing the cursor on the arrow coming out of the reboiler and releasing the mouse button opens the window shown in Figure 7.12, on which we can scroll down to stage 32 and select *Level*.

Clicking *OK* connects the control signal line to the reboiler and lets us connect the other end to the controller (see Fig. 7.13a). Clicking on the blue arrow on the left side of the controller opens the window shown in Figure 7.13b, where we select *LC11.PV*. Clicking *OK* completes the control signal connection between the reboiler and the level controller.

The next step is to connect a control signal from the controller to the valve V11 on the bottoms stream. Dragging and dropping another control signal at the arrow exiting the controller opens the window shown in Figure 7.14a. We select the *LC11.OP* and click *OK*. Then the line is connected to the valve. The final LC11 loop is shown in Figure 7.14b.

To set up the controller, double-click on the "LC11" icon on the flowsheet. The controller faceplate shown in Figure 7.15a opens. Controller faceplates are where we keep track of what is going on in the dynamic simulation, set controller parameters, switch from manual to automatic control, change setpoints, and perform other tasks. The very first thing to do is to click the *Configure* button. This opens the window shown in Figure 7.15b. Then we click on the *Initialize Values* button at the bottom of this window. This provides the steady-state values of the base level (6.35 m) and the control valve opening (*Bias* = 50%).

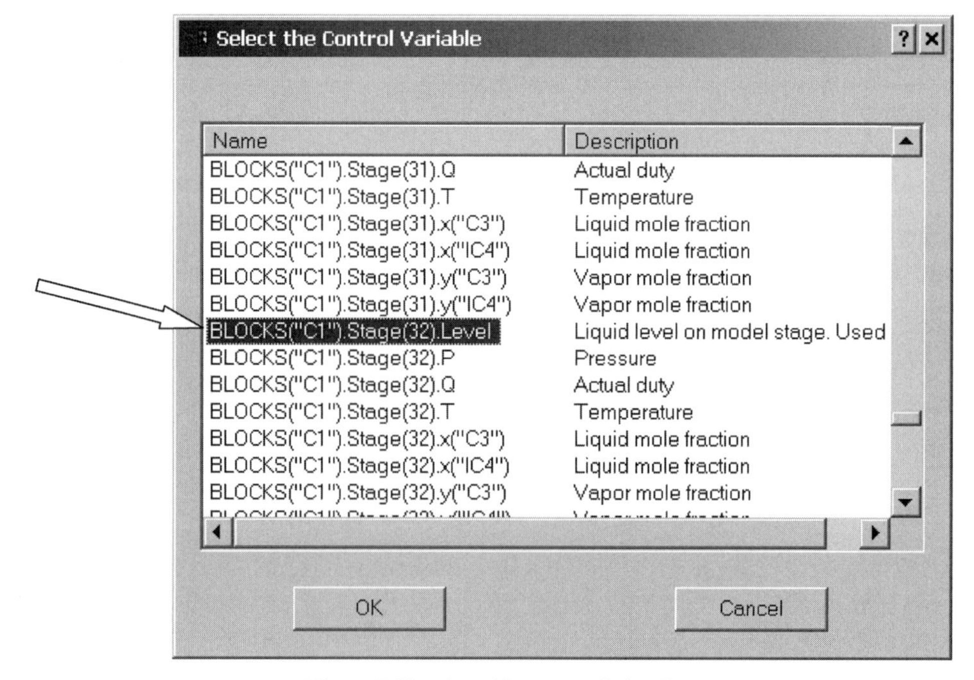

Figure 7.12 Attaching control signal.

(a)

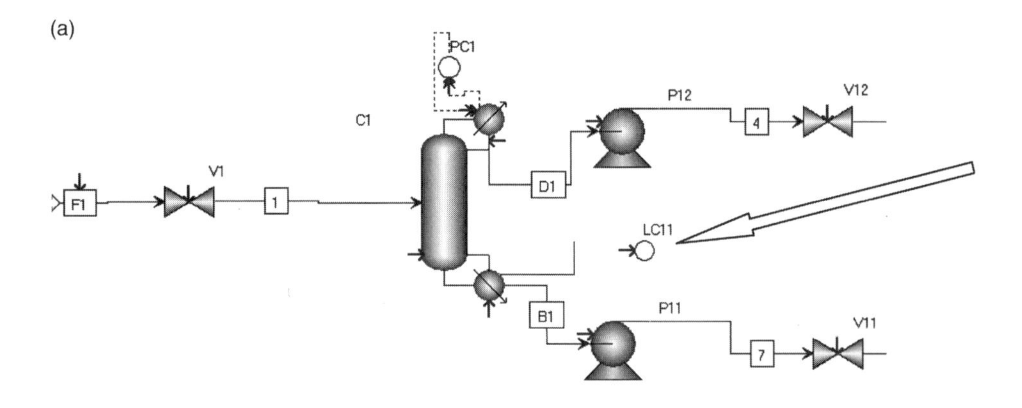

(b)

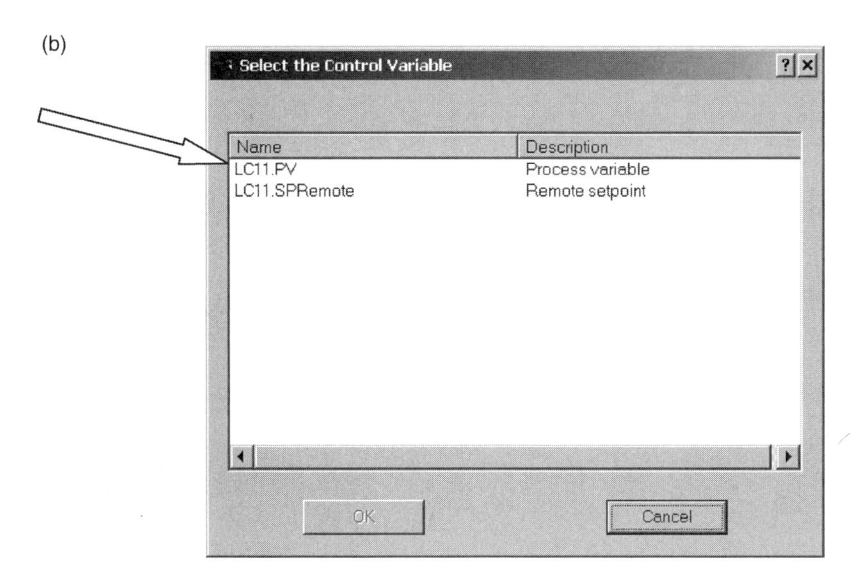

Figure 7.13 (a) Control signal attached to reboiler; (b) attaching to controller.

The action of the controller should be *Direct* (under *Controller action* in Fig. 7.15b) because if the level increases, the signal to the valve should increase ($PV\uparrow$, $OP\uparrow$) to remove more bottoms. In some columns, base level is controlled by manipulating a valve in the feed to the column. In that control structure, the base level controller action should be *Reverse*.

Since we want proportional-only control, the controller gain is set equal to 2 and the integral time is set at a very large number (9999 min, as shown in Fig. 7.15b).

Next the *Ranges* page tab is clicked, which opens the window shown in Figure 7.16. The default value for the level transmitter range is 0–12.7 m. The default value of the controller output range is 0–100%. Both of these are what we want, so they require no changes. The *Configure* window is then closed.

(a)

(b)

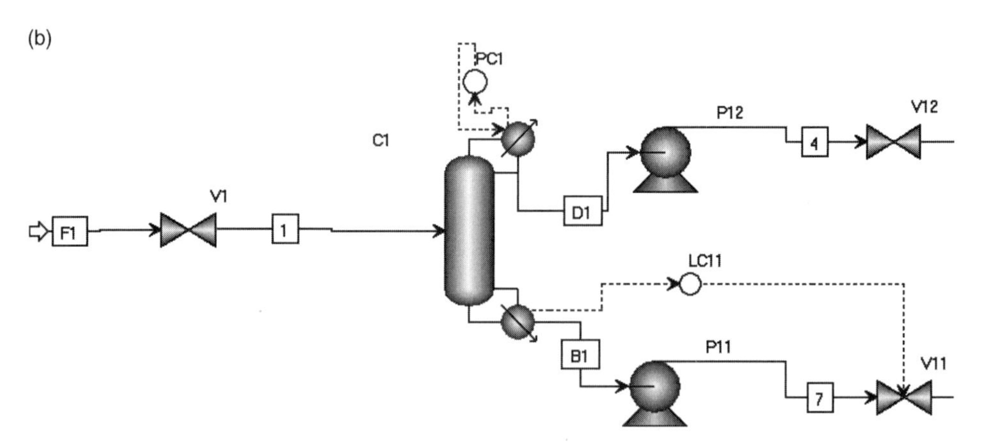

Figure 7.14 (a) Selecting *OP* signal; (b) final loop with signal connected.

(a)

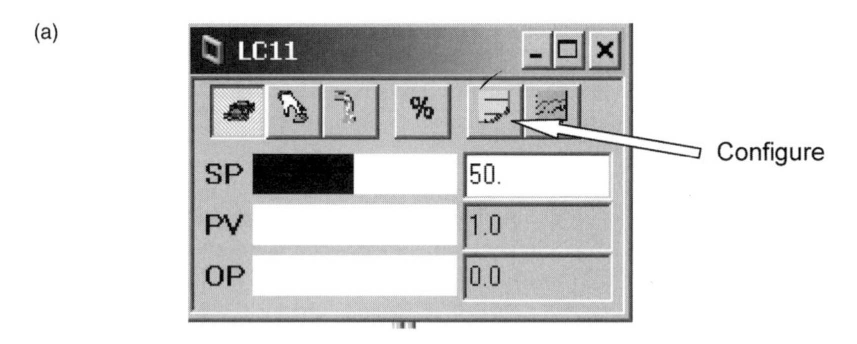

Figure 7.15 (a) Initial controller faceplate; (b) *tuning* page tab.

(b)

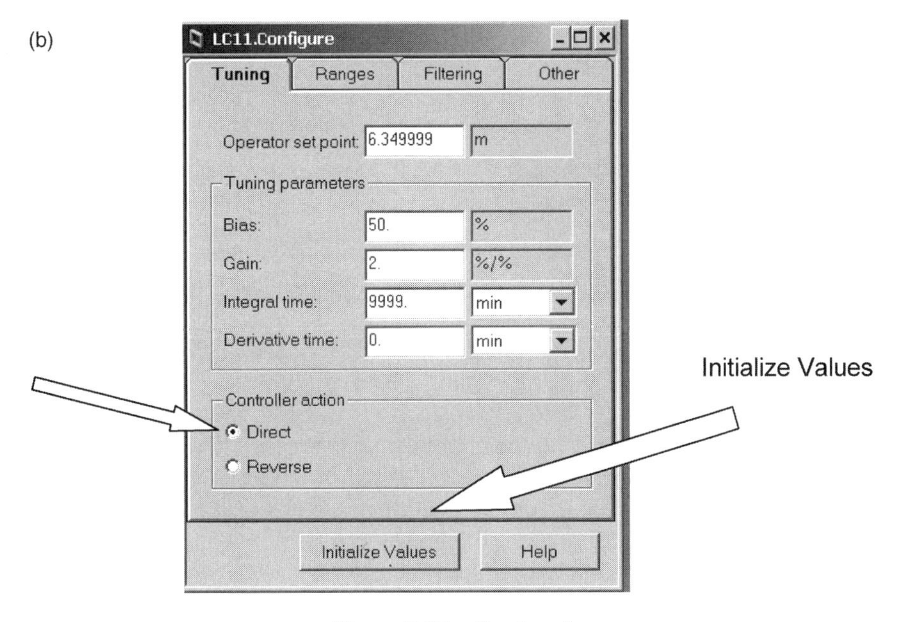

Figure 7.15 *Continued.*

The faceplate is located at some convenient spot in the window where we can keep an eye on what is going on with this level loop. Remember that there will eventually be five controller faceplates.

Let us look in detail at the faceplate. As shown in Figure 7.17, there are six buttons at the top. The first button on the left is the *Auto* button, the second is *Manual*, and the third is *Cascade*.

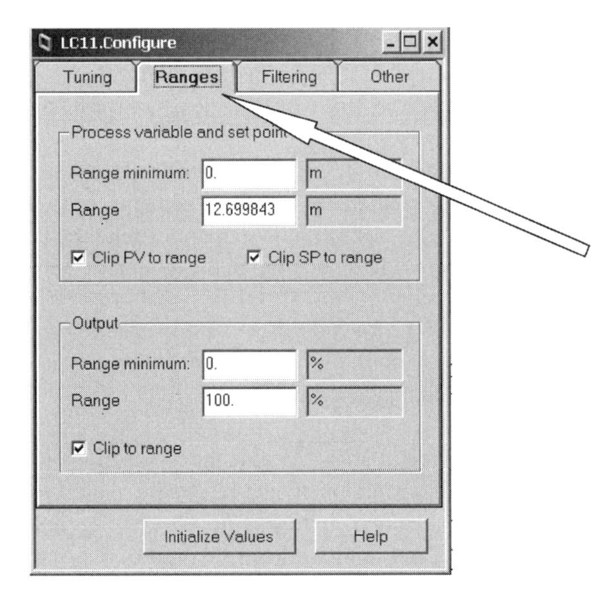

Figure 7.16 *Ranges* page tab.

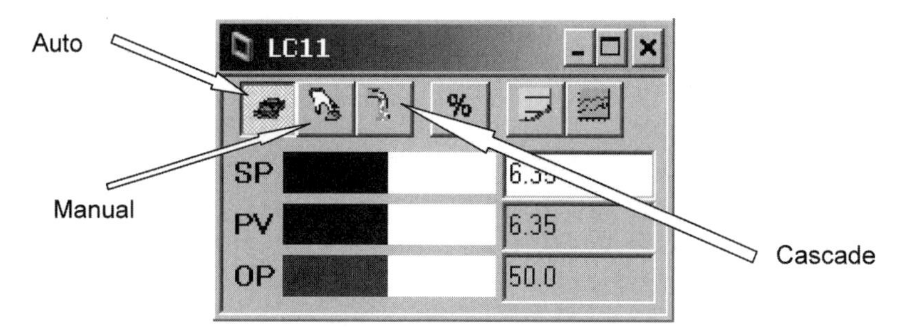

Figure 7.17 Faceplate details.

When the *Auto* button is pushed, the controller changes the *OP* signal automatically according to the current values of the setpoint *SP* and the *PV*. The value of the setpoint can be changed by double-clicking on the number in the box to the right of *SP*, typing in the desired number, and hitting *Enter* on the keyboard.

When the *Manual* button is pushed, you can manually set the *OP* signal. This is done by double-clicking on the number in the box to the right of *OP*, typing in the desired number, and hitting *Enter* on the keyboard. When the *Cascade* button is pushed, the controller receives its setpoint signal from some other control element. We will illustrate this later in this chapter.

Now would be a good time to check out the pressure controller, which was automatically set up when we started Aspen Dynamics. Figure 7.18a shows the faceplate (which appears after double-clicking the icon on the flowsheet) and the *Configure* page tab. The default controller tuning constants are a gain of 20 and an integral time of 12 min. These work pretty well in most column simulations. Note that the controller output is not a "% of scale" signal sent to a valve. It is a heat removal rate in the condenser. As a result, the controller is set up to be reverse-acting: when pressure goes up, the controller output signal goes down. The reason for this action becomes clear when we look at the *Ranges* page tab shown in Figure 7.18b. Note that the controller output ranges from a minimum of −45,347,000 W to a maximum of 0 W. This conforms with the Aspen convention that heat removal is negative.

In my experience the only modification that sometimes needs to be made to the pressure controller is to change to a more convenient pressure transmitter range. For example, in this column the operating pressure is 16.8 atm. We might change the pressure transmitter range to 14–19 atm from the very wide range used in the default setup. Of course, the gain should be correspondingly reduced.

The second-level controller LC12 for the reflux drum is installed and connected in the same way. The *PV* signal comes from the level on stage 1. The *OP* goes to valve V12. A direct-acting proportional-only controller is specified.

The final basic controller that we need to set up is a flow controller on the feed. A *PID* controller is placed on the flowsheet. Its *PV* signal is the molar flowrate of feedstream F1. Its *OP* signal goes to valve V1. After opening the *Tuning* page tab and clicking the *Initialize Values* button, we set the controller to be *Reverse*-acting, and use conventional flow controller tuning ($K_C = 0.5$ and integral time = 0.3 min), as shown in Figure 7.19a. The most common error in setting up the flow controller is to forget to specify *Reverse* action. Since flow control is very fast and essentially algebraic, it seems to help the numerical integrator to use some filtering in a flow controller. The *Filtering* page tab is selected, the *Enable filtering* box is checked, and a small filter time constant (0.1 min) is typed in (see Fig. 7.19b).

The flowsheet now has four controller faceplates displayed, as shown in Figure 7.20. We have one more controller to add, a temperature controller that holds the temperature on a selected tray by adjusting the reboiler heat input.

It is important to clarify what is happening to the reflux flow. The column icon does not show the plumbing details of a reflux drum, pump, and reflux valve. As mentioned earlier, the default condition in Aspen Dynamics is that the **mass** flowrate of reflux is constant, unless otherwise adjusted. For example, if we wanted to control reflux drum level with reflux flowrate, the level controller *OP* signal would be connected to the column and the *Reflux.FmR* would be selected. The second common application would be if we wanted to ratio the reflux flowrate to the feed flowrate. We will illustrate these by examples in this and later chapters.

A word of caution is appropriate at this point. During the initialization of controllers, quirky things sometimes occur. There are some bugs in Aspen Dynamics that sometimes set the *OP* signal at the wrong initial value (e.g., at 100 instead of 50%) or the *PV* at a value

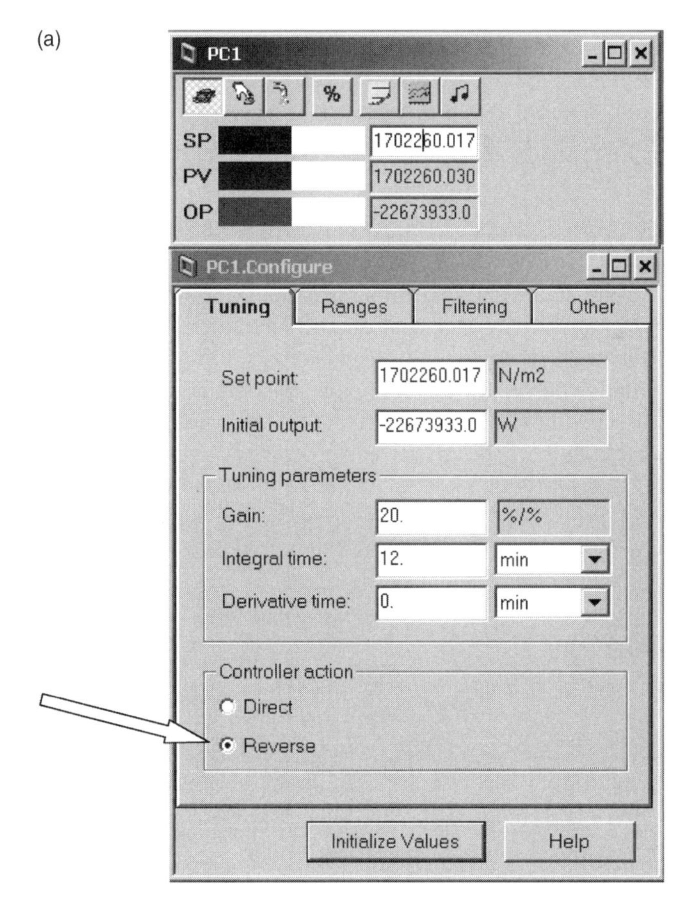

Figure 7.18 (a) Pressure controller; (b) *Ranges* page tab.

(b)

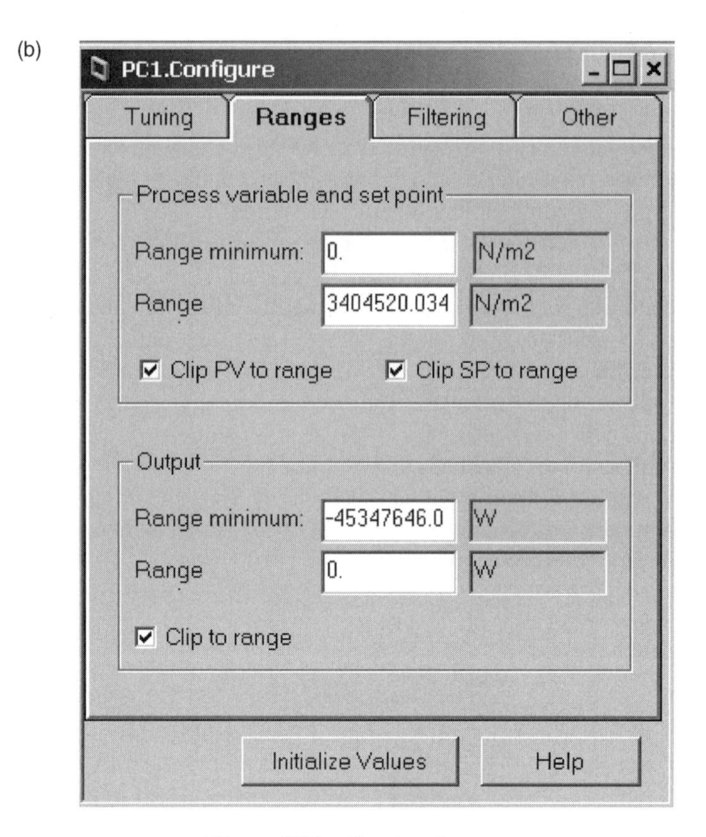

Figure 7.18 *Continued.*

not equal to the steady-state value. To circumvent these problems, switch the controller to manual, type in the correct *OP* value, and run the simulation for a while. Then switch the controller to automatic and run out to a steady state.

While we are on the subject of running, once the simulation runs out in time and converges to a steady state, the file should be saved. It is convenient to save the file with time set equal to zero so that it can be used to establish initial conditions for new runs. To do this, make an *Initialization* run and then switch to *Dynamic* (but do not perform a run). Click the *Rewind* button, which is the fifth one from the right on the upper toolbar (see Fig. 7.20). A window opens (Fig. 7.21) on which you can select the *Initialization Run* as the *Select rewind snapshot* and then save the simulation file. Note that the "Sim Time" is 0. You can then "rewind" to these conditions whenever you want to start again at this steady state.

7.5 INSTALLING TEMPERATURE AND COMPOSITION CONTROLLERS

Installing temperature and composition controllers is somewhat more involved than installing level and flow controllers because of three issues:

1. We need to include additional dynamic elements in the loop. Temperature and composition measurements have significant inherent dynamic lags and deadtimes.

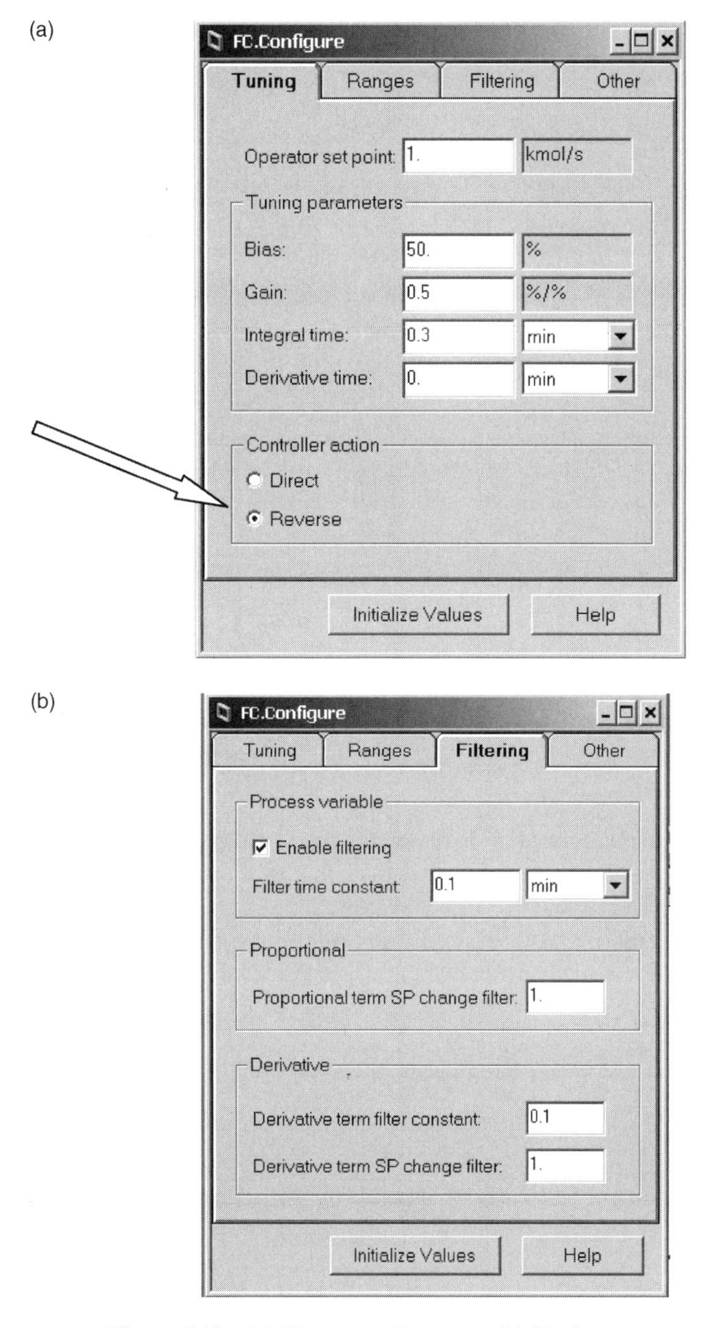

Figure 7.19 (a) Flow controller setup; (b) filtering.

These should be incorporated in the control loop. This is necessary so that we use realistic controller tuning constants and do not predict dynamic performance that is better than reality.

2. The tuning of temperature and composition controllers is more involved than simply using heuristics as is done for flow and level controllers. Some convenient and

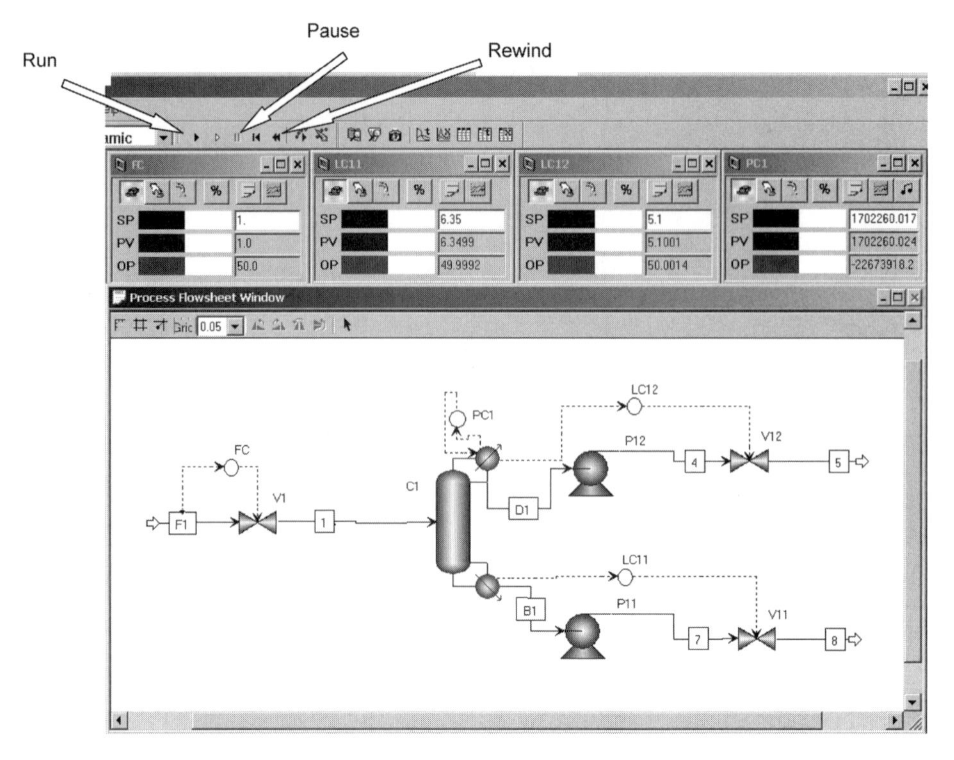

Figure 7.20 Controller faceplates.

effective tuning procedure is required. One of the best methods is to run a relay–feedback test to find the ultimate gain and ultimate frequency, and then use the conservative Tyreus–Luyben tuning settings.

3. The appropriate location for the temperature or composition sensor may not be obvious. Some method for making a good selection must be used. As discussed

Rewind

Select rewind snapshot:

Description	Sim Time	Date & Time	Converged	Run Number
Initialization Run	0.000000	3/18/2005 9:07:14 AM	Yes	2
File saved	17.492000	3/16/2005 5:41:19 PM	Modified	1
File saved	6.605000	3/16/2005 5:05:29 PM	Yes	1
Dynamic Initialization	0.000000	3/16/2005 4:42:48 PM	Yes	1
Initialization Run	0.000000	3/16/2005 4:41:52 PM	Yes	1
Initial Specification	2.110000	3/16/2005 4:41:51 PM	No	0

Rewind   Cancel

Figure 7.21 Rewinding to *Snapshot*.

in Chapter 6, there are several ways to approach this problem. These include looking at the shape of the temperature profile in the column, calculating steady-state gains, and using SVD analysis.

7.5.1 Tray Temperature Control

Let us first discuss using a temperature controller to maintain a tray temperature in the column. Looking at the temperature profile in Aspen Plus, we see that stage 9 displays a fairly steep slope. Its temperature is 337.36 K.

A controller is installed on the flowsheet in the normal way, except that instead of using a *PID* controller, we select a *PIDIncr* controller. The important difference between these two is that the *PIDIncr* controller has a built-in relay–feedback test capability, which make this dynamic test a breeze.

The *PV* is selected to be the temperature on stage 9. The *OP* is selected to be the reboiler heat input *QRebR*. Figure 7.22a shows the controller faceplate and the *Tuning* page tab after the *Initialize Values* button has been clicked. The normal controller output is 27,166,000 W. The controller action should be set at *Reverse* because if the tray temperature is going up, the reboiler heat input should be decreased. It is convenient to change the range of the temperature transmitter from the default 273–401 K to a more convenient and narrower range of 320–370 K, as shown on the *Ranges* page tab in Figure 7.22b.

The program is run to make sure that everything works okay without a lag or a deadtime in the loop. Now we back up and insert a deadtime element on the flowsheet between the column and the TC9 temperature controller. The reason for installing the controller initially without the deadtime element is to avoid initialization problems that sometimes crop up if you attempt to install the deadtime and the controller all in one shot.

Before we proceed, it might be wise to save some of the work. Since a fair amount of time has been spent in setting up the faceplates and arranging them on the screen, we can avoid having to do this again by clicking on *Tools* in the toolbar at the top of the screen and selecting *Capture screen layout*. The window shown in Figure 7.23 opens, on which we enter an appropriate name. When the program is restarted, the screen layout can be reinstalled by going to the *Exploring* window, clicking *Flowsheet*, and double-clicking on the icon in the lower *Flowsheet Contents* window with the name you provided.

Now let us install the deadtime element. The line from stage 9 temperature is selected. Right-clicking, selecting *Reconnect Destination*, and placing the icon on the arrow pointing to the deadtime icon, we connect the input to the deadtime. A new control signal is inserted between the deadtime and the controller. The deadtime icon then is selected. Right-clicking, selecting *Forms* from the dropdown list, and selecting *All Variables*, we open the window shown in Figure 7.24a. The *DeadTime* value is initially 0 min. Note that the *Input* and *Output* values are set at a default number, not the actual 337.36 K value. A deadtime of 1 min is entered, and performing an *Initialization* run fills in the correct values, as shown in Figure 7.24b. The final flowsheet and controller faceplates are shown in Figure 7.25.

Everything is ready for the relay–feedback test. Clicking the *Tune* button opens the window shown in Figure 7.26a. We specify a *Closed loop ATV* as the *Test method*. The default value of the *Relay output amplitude* is 5%, which is usually good. For a very nonlinear column, the amplitude may have to be reduced.

(a)

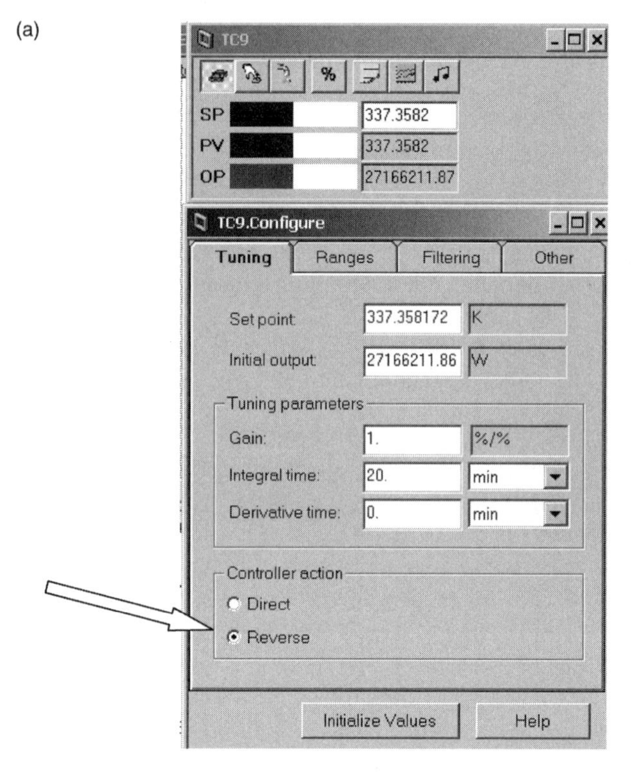

(b)

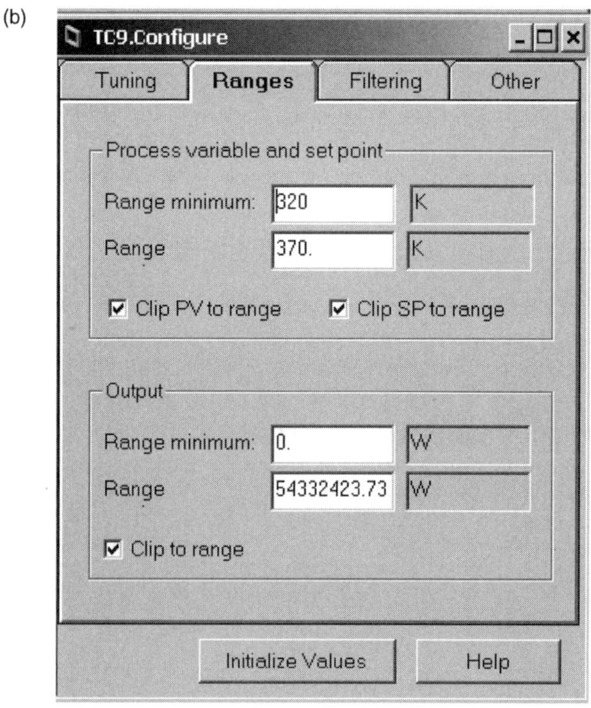

Figure 7.22 (a) Initial installation of temperature controller; (b) *Ranges* page tab.

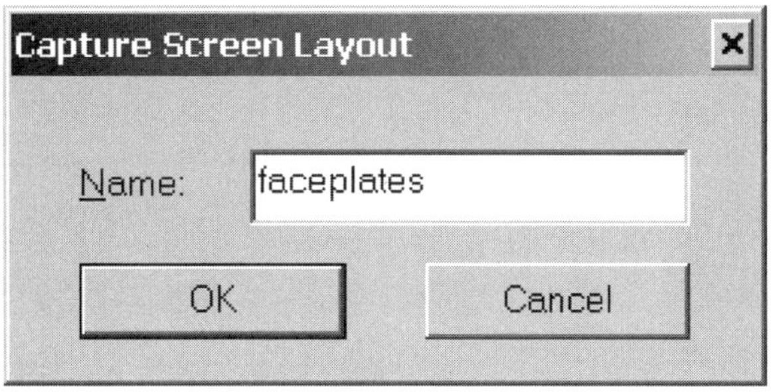

Capture Screen Layout

Name: faceplates

OK Cancel

Figure 7.23 Saving screen layout.

To start the test, click the *Run* button at the top of the screen and click the *Start test* button on the *Tune* window. To be able to see the dynamic responses, click the *Plot* button at the top of the controller faceplate. After several (4–6) cycles have occurred, click the *Finish test* button. Figure 7.26b gives the results. The predicted ultimate gain is 5.9, and the ultimate period is 4.5 min.

The timescale in Figure 7.26b is fairly small. To get good-looking plots, the plot time interval must be reduced from the default value of 0.01 h. Aspen Dynamics calls this parameter "communication time," and it can be accessed by going to the top toolbar in the Aspen Dynamics window and selecting *Run* and *Run Options*. A window opens on which *Communication Time* can be set. A value of 0.0005 h was used to get the plots shown in Figure 7.26b. This parameter does not affect the results of a dynamic simulation, except for slowing it down somewhat. It only affects the appearance of plots.

Finally the *Tuning parameters* page tab is clicked, the *Tyreus–Luyben Tuning rule* (Fig. 7.26c) is selected, and the *Calculate* button is pushed. The resulting controller settings are gain $K_C = 1.84$ and integral time $\tau_I = 9.9$ min. The Tyreus–Luyben tuning

(a)

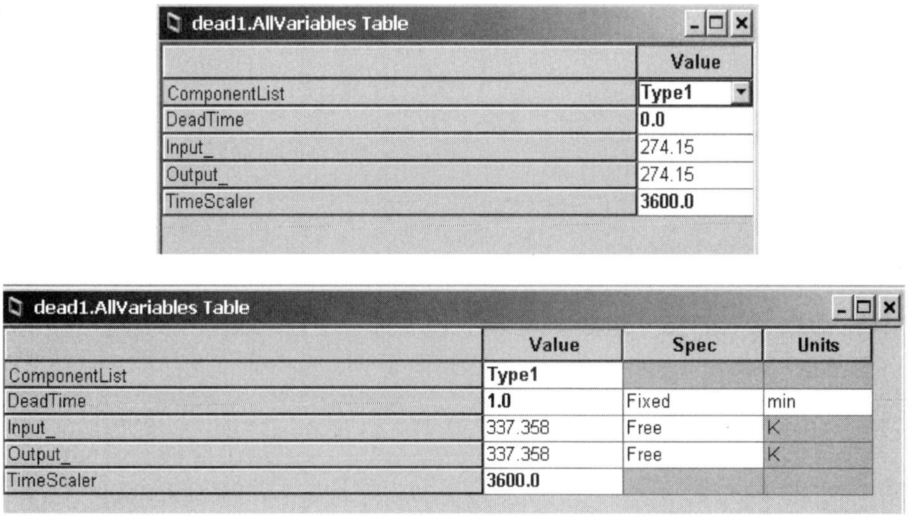

dead1.AllVariables Table

	Value
ComponentList	Type1
DeadTime	0.0
Input_	274.15
Output_	274.15
TimeScaler	3600.0

(b)

dead1.AllVariables Table

	Value	Spec	Units
ComponentList	Type1		
DeadTime	1.0	Fixed	min
Input_	337.358	Free	K
Output_	337.358	Free	K
TimeScaler	3600.0		

Figure 7.24 (a) Deadtime *All Variables* table; (b) after initialization run.

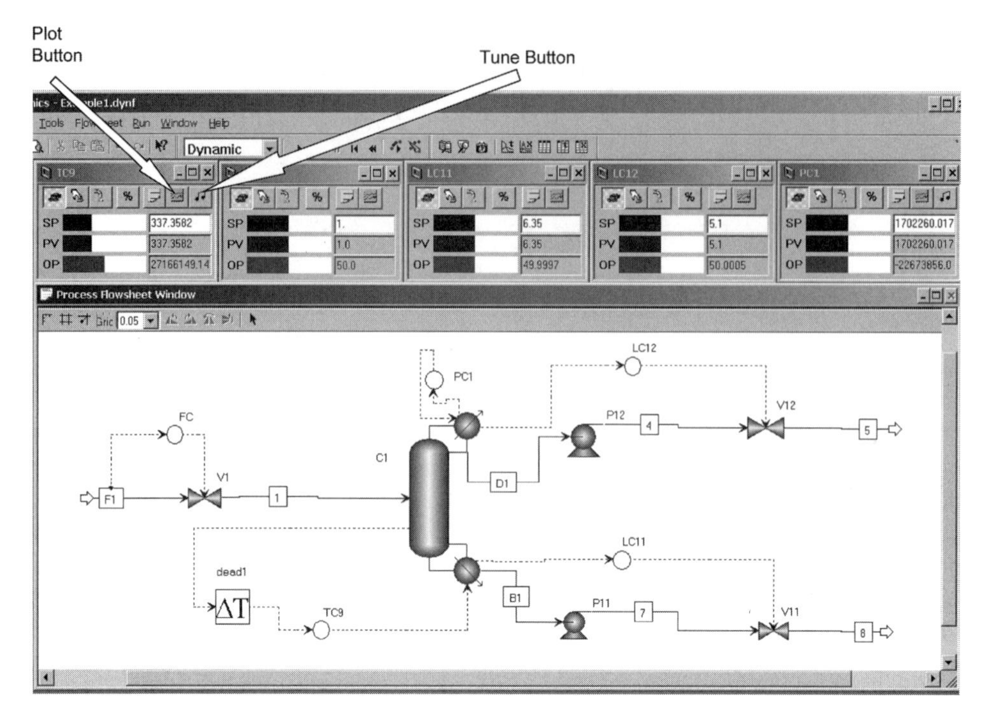

Figure 7.25 Flowsheet with controller faceplate.

formulas are

$$K_C = \frac{K_U}{3.2}$$

$$\tau_I = 2.2P_U$$

These are loaded into the controller by clicking the *Update controller* button. Run the simulation out in time for a while to see how well these settings work in terms of bringing the column to steady state. In the next section we will subject the column to disturbances and evaluate the performance of several control structures.

7.5.2 Reflux to Feed Ratio

Before we illustrate the use of a composition controller, it might be instructive to show how a reflux to feed ratio structure is set up. In Chapter 5, steady-state calculations indicated that a R/F ratio scheme should do a pretty good job of maintaining product purities in the propane/isobutane system in the face of feed composition disturbances and, of course, feed flowrate changes.

The R/F structure is implemented by using a multiplier block. The input of this block is the mass flowrate of the feed, which is 52.513 kg/s. The properties of any stream can be found by clicking on the steam name, right-clicking, selecting *Forms*, and then selecting *Results*. Figure 7.27 gives the stream information for the feed F1.

(a)

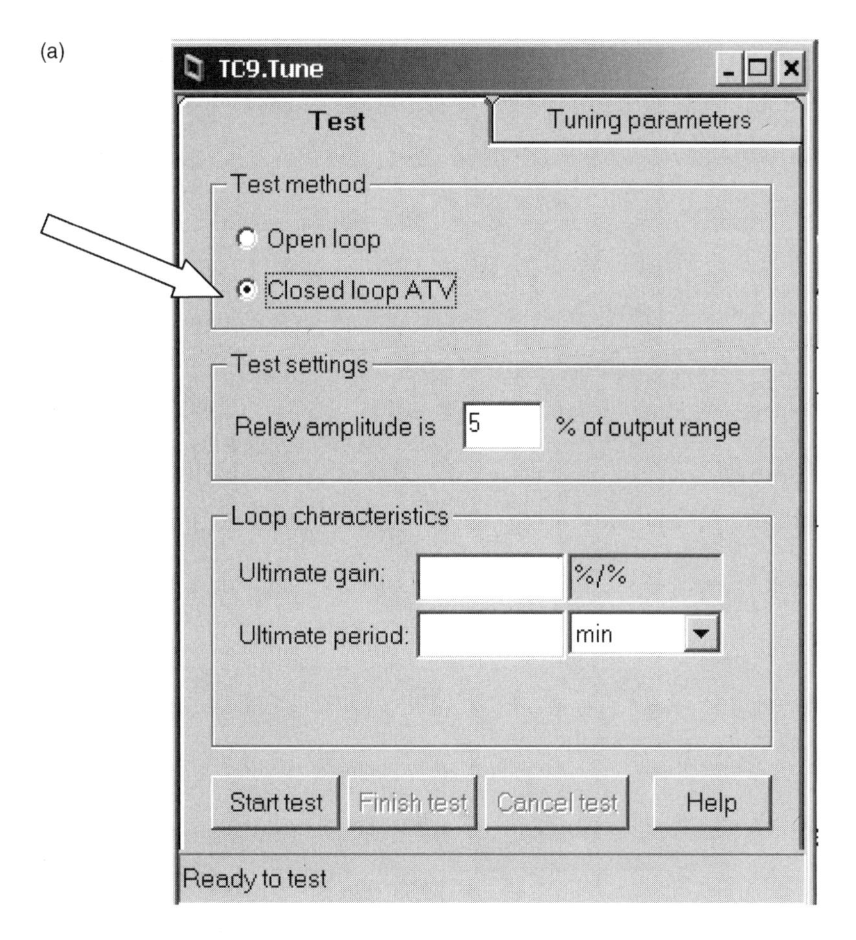

(b)

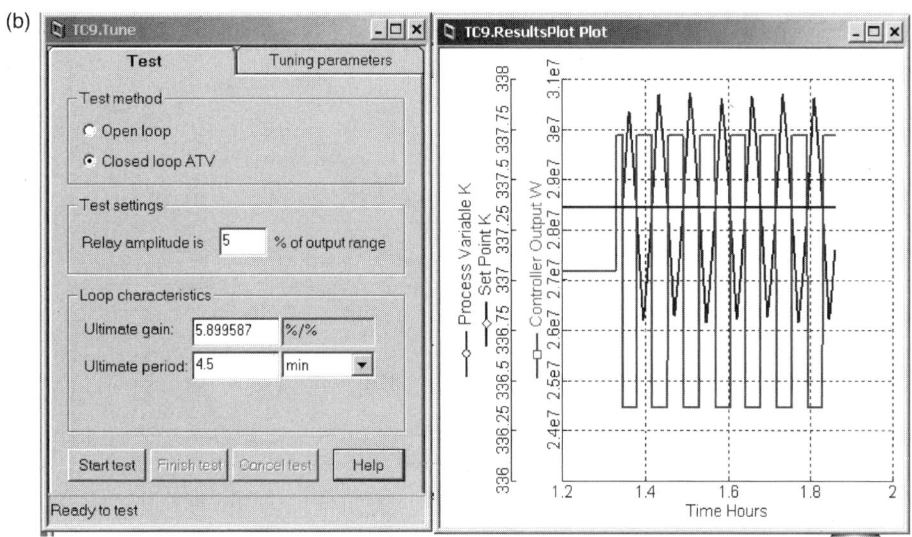

Figure 7.26 (a) Setting up the relay–feedback test; (b) relay–feedback test results; (c) calculated controller settings.

(c)

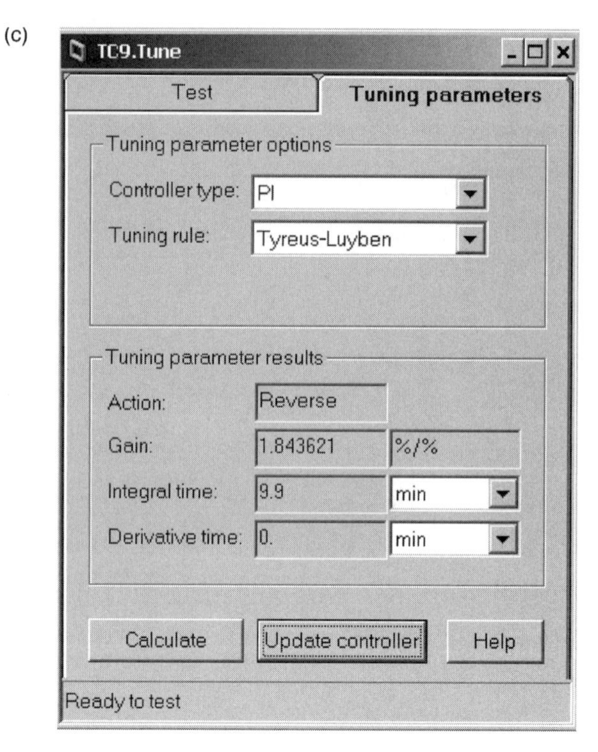

Figure 7.26 *Continued.*

F1.Results Table

	Description	Value	Units
F	Total mole flow	1.0	kmol/s
Fm	Total mass flow	52.5126	kg/s
Fv	Total volume flow	0.105737	m3/s
T	Temperature	**322.0**	K
P	Pressure	**2.0265e+006**	N/m2
vf	Molar vapor fraction	0.0	
h	Molar enthalpy	-1.3807e+008	J/kmol
Rho	Molar density	9.45744	kmol/m3
Rhom	Mass density	496.635	kg/m3
MW	Molar weight	52.5126	kg/kmol
zn(*)			
Zn("C3")	Mole fraction	0.4	kmol/kmol
Zn("IC4")	Mole fraction	0.6	kmol/kmol
zmn(*)			
Zmn("C3")	Mass fraction	0.335893	kg/kg
Zmn("IC4")	Mass fraction	0.664107	kg/kg
Fcn(*)			
Fcn("C3")	Component mole flow	0.4	kmol/s
Fcn("IC4")	Component mole flow	0.6	kmol/s
Fmcn(*)			
Fmcn("C3")	Component mass flow	17.6386	kg/s
Fmcn("IC4")	Component mass flow	34.874	kg/s

Figure 7.27 Results table for stream F1.

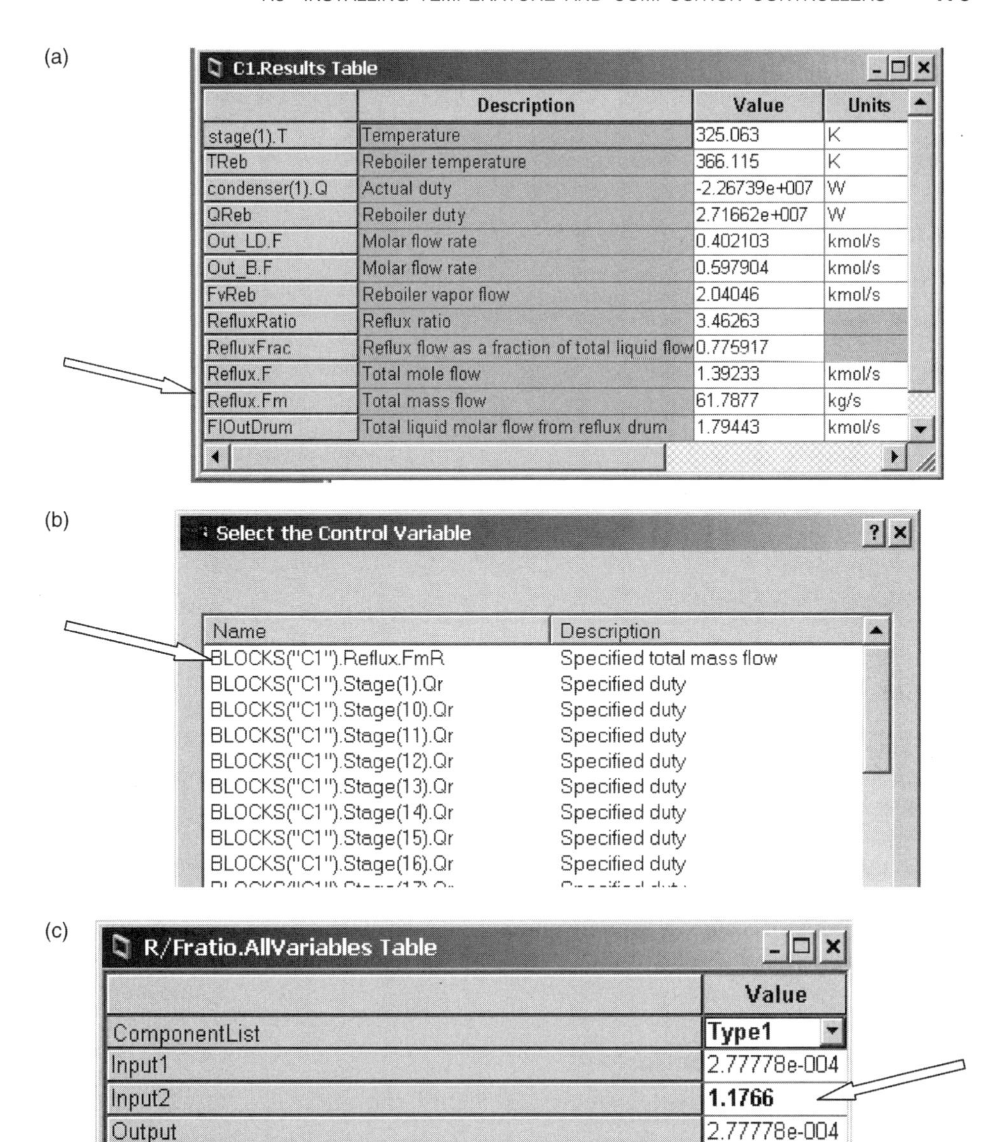

Figure 7.28 (a) Results table for column; (b) choices; (c) entering ratio constant.

The output of the multiplier block will be the mass flowrate of the reflux. To determine the design value of this variable, we can click the column icon, right-click, select *Forms*, and select *Results*. This opens the table shown in Figure 7.28a. The second line from the bottom gives the reflux mass flowrate of 61.7877 kg/s. So the multiplier block should multiply the feed mass flowrate by the number $(61.7877/52.513) = 1.1766$.

A *Multiplier* control model is placed on the flowsheet. A control signal is connected from the feed F1 (mass flowrate) to the multiplier (labeled "R/F ratio" in Figure 7.29). This is "Input1." A second control signal is connected from the multiplier output to the column and connected to the blue arrow pointing to the line below the condenser. A list

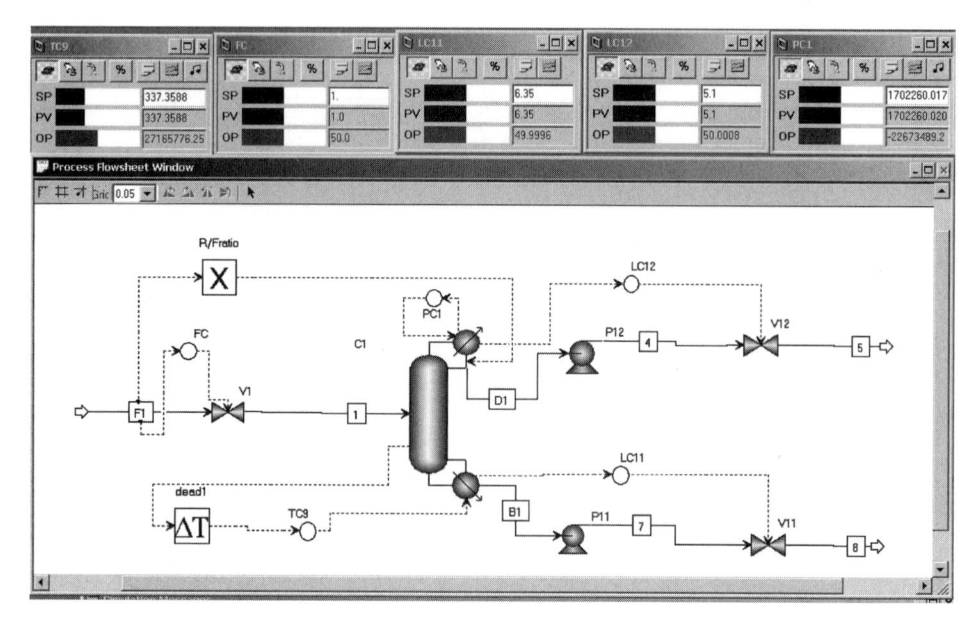

Figure 7.29 Flowsheet with R/F structure.

of alternatives opens, and the top one, *Reflux.FmR*, is selected (see Fig. 7.28b). To set the constant in the multiplier, the icon is clicked and right-clicked. *Forms* and *All Variables* are selected, and the window shown in Figure 7.28c opens, on which we enter the number "1.1766" for "Input2." In this example, one of the inputs is a constant. In other examples, both of the inputs can vary with time.

The final flowsheet and controller faceplates are shown in Figure 7.29. We will compare the performance of this control structure with some alternatives later in this chapter. First we want to illustrate the use of a composition controller.

7.5.3 Composition Control

We want to compare tray temperature control with two types of composition control. In both, the composition of the propane product is measured directly and controlled at 2 mol% isobutane. The first type is "direct composition control," in which a single PI controller is used with reboiler heat input manipulated. The second type uses a cascade composition-to-temperature control structure.

Composition measurement typically has larger deadtime and lags than does temperature control. We assume a 3-min deadtime in the composition measurement.

First, we add a *PIDIncr* controller to the flowsheet and make the appropriate connections and do not use a deadtime, which will be added later. The controller should be set to *Reverse*. The *PV* is the mole fraction of isobutane in the distillate stream. The *OP* is reboiler heat input. A composition transmitter range of 0–0.05 mF isobutane is used, as shown in Figure 7.30.

After the simulation is run, a 3-min deadtime is inserted. *Initialization* and *Dynamic* runs are made to converge to steady-state conditions. Then a relay-feedback test is run. Results are shown in Figure 7.31. Note that the timescale on the plot is very different

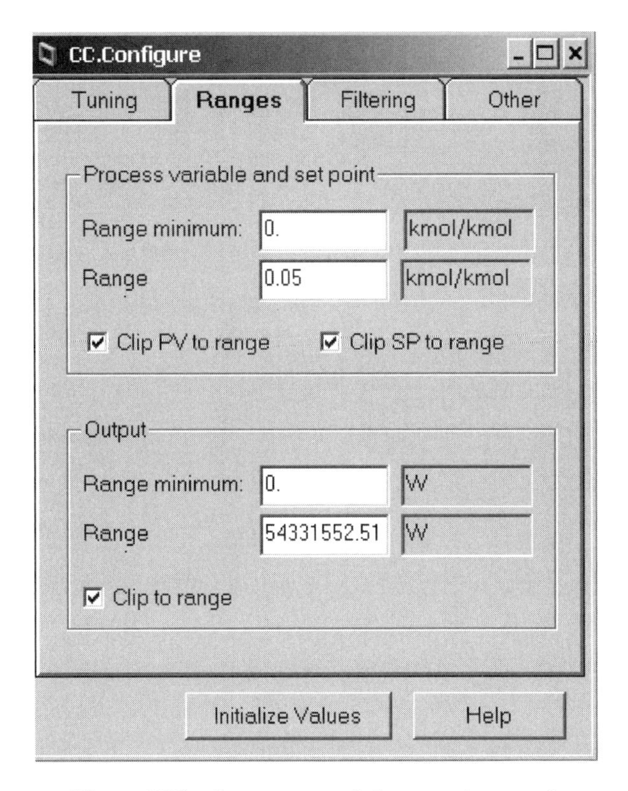

Figure 7.30 *Ranges* page tab for cascade control.

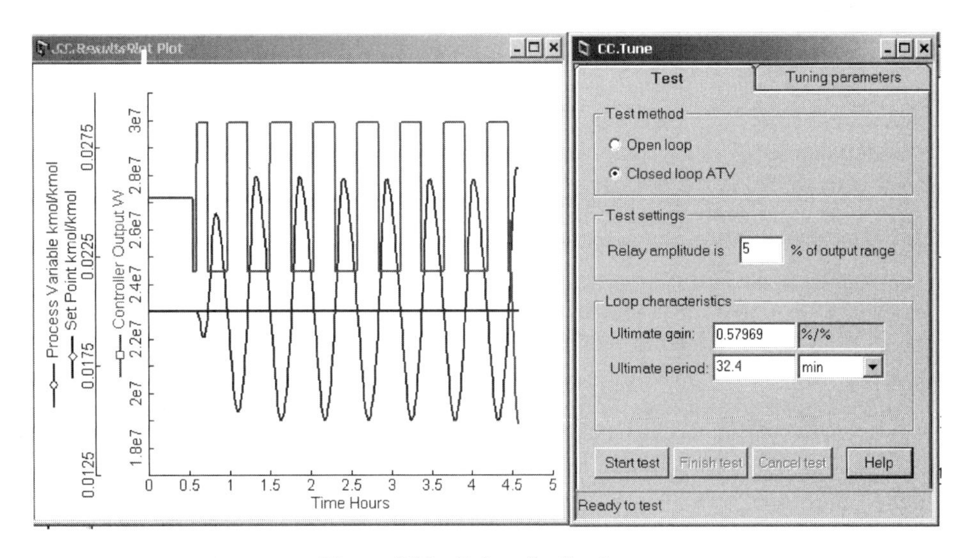

Figure 7.31 Relay–feedback test.

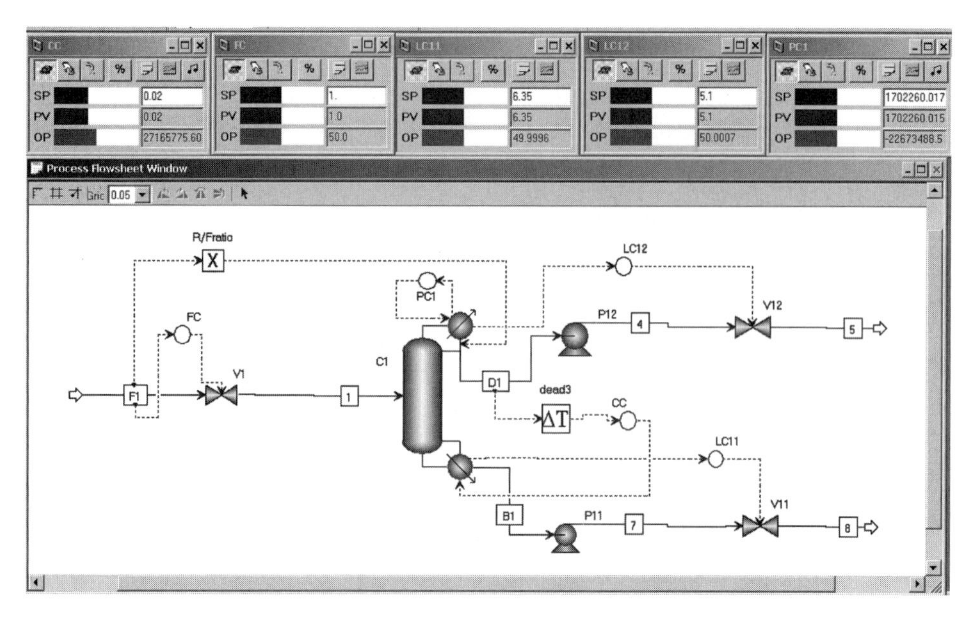

Figure 7.32 Final flowsheet with cascade control.

from that for the temperature controller. The ultimate gain is 0.58, and the ultimate period is 32.4 min. The Tyreus–Luyben settings are calculated and inserted in the composition controller. The final flowsheet is given in Figure 7.32.

7.5.4 Composition/Temperature Cascade Control

Temperature control has the advantage of being fast, but it may not hold the product purity constant. Composition control is slow, but it will drive product purity to the desired value. The final control structure studied in this chapter is a cascade combination of composition and temperature control that achieves both fast control and the maintenance of product purity.

The tray temperature controller is the secondary controller. It is set up in exactly the same way as we did in the previous section. It looks at tray temperature and manipulates reboiler heat input. However, its setpoint is not fixed. The setpoint signal is the output signal of the composition controller, which is the primary controller.

The tuning of the secondary temperature controller remains unchanged. The primary composition controller must be retuned since its ouput signal is now a temperature setpoint. With the temperature controller set on automatic, the relay–feedback test is run on the composition controller. Figure 7.33 shows the relay feedback test results. The ultimate gain and ultimate period are 0.98 and 15.9 min, respectively, compared to the direct composition results of 0.58 and 32.4 min. We can see immediately that a higher gain and smaller integral time result, which indicates tighter control with the cascade control structure. Figure 7.34 shows the cascade control structure and controller faceplates. Note that the TC7 temperature controller is "on cascade" (meaning that its set-point signal is the output signal of the composition controller).

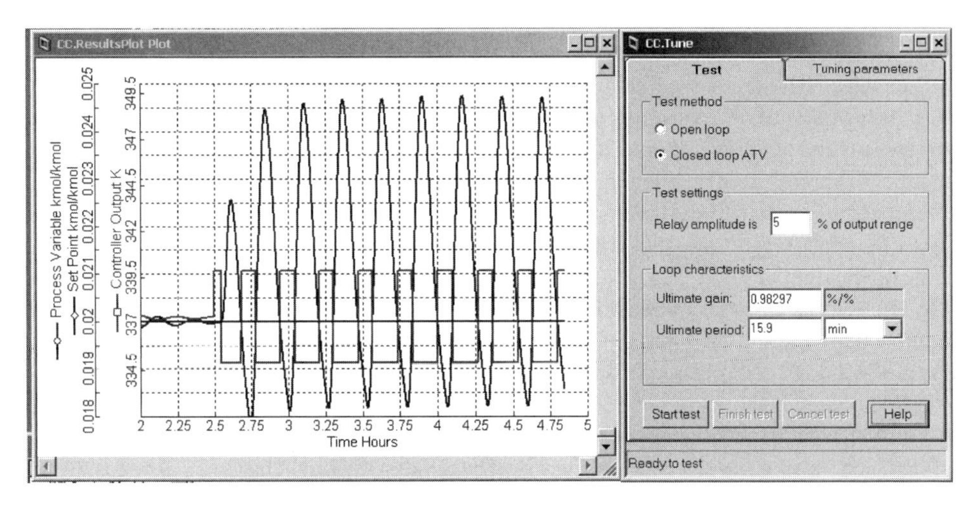

Figure 7.33 Cascade control relay–feedback test.

7.6 PERFORMANCE EVALUATION

We want to see how well the three alternative control structures developed above perform in the face of disturbances, specifically, how close to the desired values of temperature and composition these variables are maintained, both at steady state and dynamically. A disturbance is made and the transient responses are plotted.

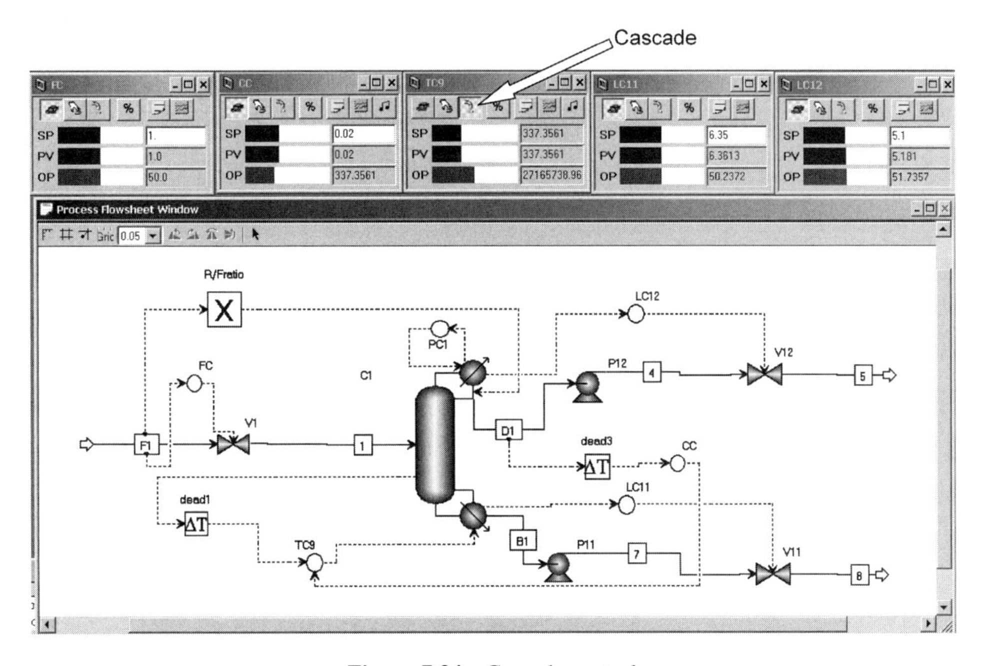

Figure 7.34 Cascade control.

7.6.1 Installing a Plot

To see what is going on, the first thing to do is to set up a plot or a stripchart. This will show how the variables of interest change dynamically with time. To open a plot, go to *Tools* on the top toolbar and click *New Plot*. A plot window opens, as shown on the left in Figure 7.35a. A number of variables will be plotted: flowrates of the feed, distillate, bottoms, and reflux; stage 7 temperature; reboiler heat input; and distillate impurity (mf isobutane) and bottoms impurity (mf propane). We "drop and drag" each one of these variables onto the plot window. The table on the right of Figure 7.35a is opened by clicking the icon for the F1 stream, right-clicking, and selecting *Forms* and then *Results*. Place the cursor just to the right of *F* (total molar flow) in this table and click once. Then click and hold the left mouse button, drag the cursor to the plot window, and release. This procedure is repeated for each variable. The plot with all variables added is shown in Figure 7.35b. Now we need to define the scales of each variable, simplify the labels, and make other "beautification" changes to the plot. This is done by right-clicking on the plot and selecting *Properties*. The window shown in Figure 7.36 opens, which contains a number of page tabs. On the *Axis Map* page tab, click the box *One for Each* to get different scales, which are then set on the *Axis* page tab. If you want fixed scales, uncheck the box to the left of *Reset axis range to data* for each of the variables and define minimum and maximum values for each.

Figure 7.37 illustrates the type of plot that results from making a dynamic run. At time equal 0.1 h, the setpoint of the feedflow controller is changed from 1 to 1.2 kmol/s. To

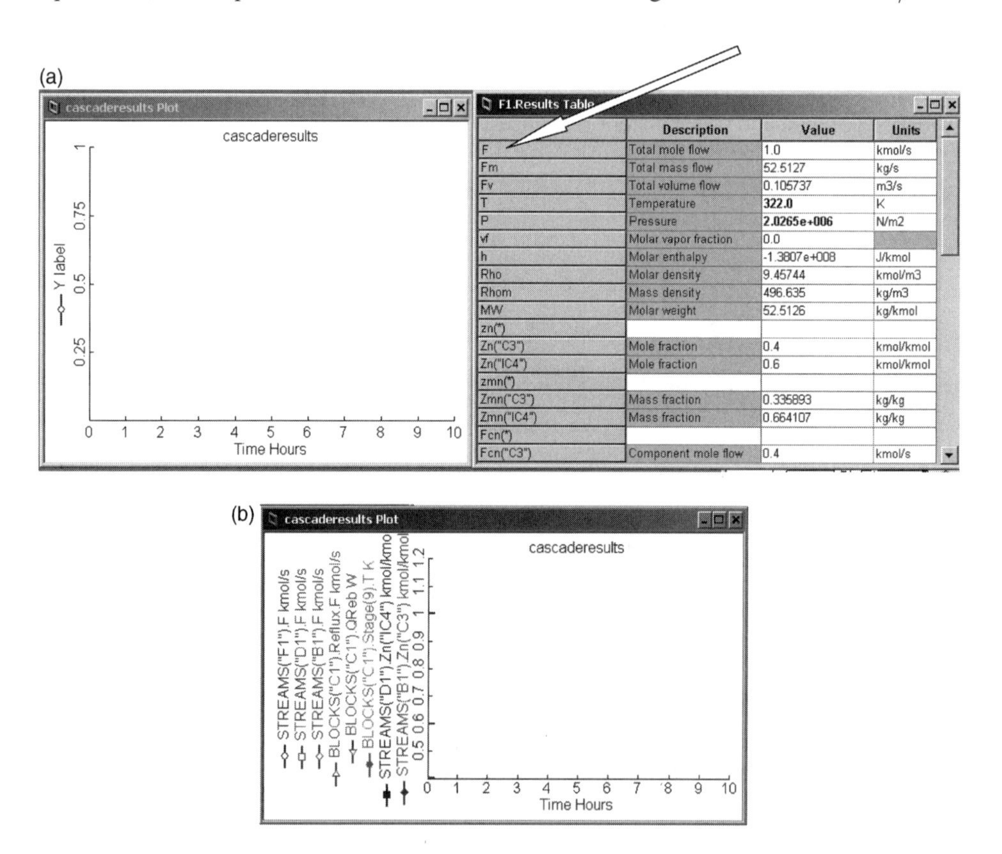

Figure 7.35 (a) New plot and F1 results table; (b) plot with variables.

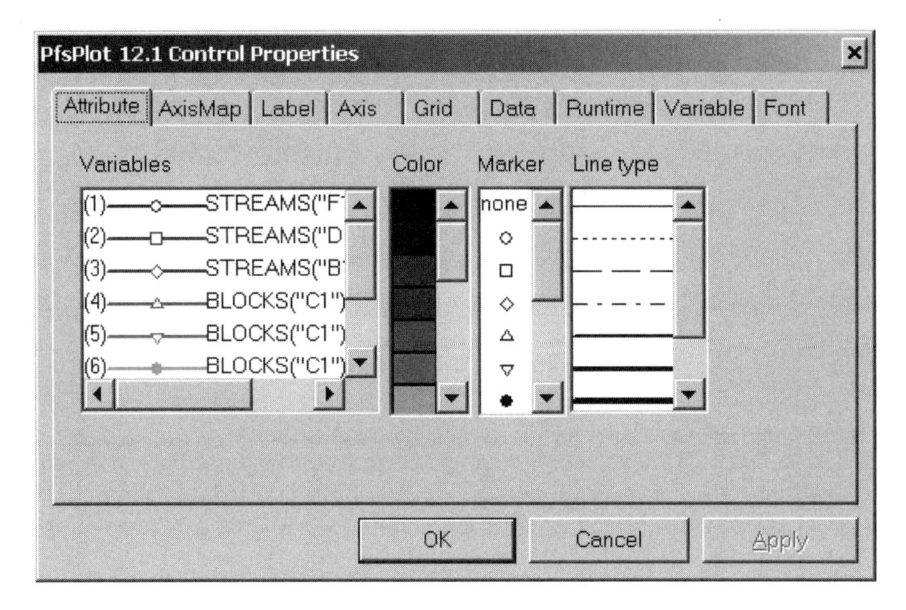

Figure 7.36 Properties of plot.

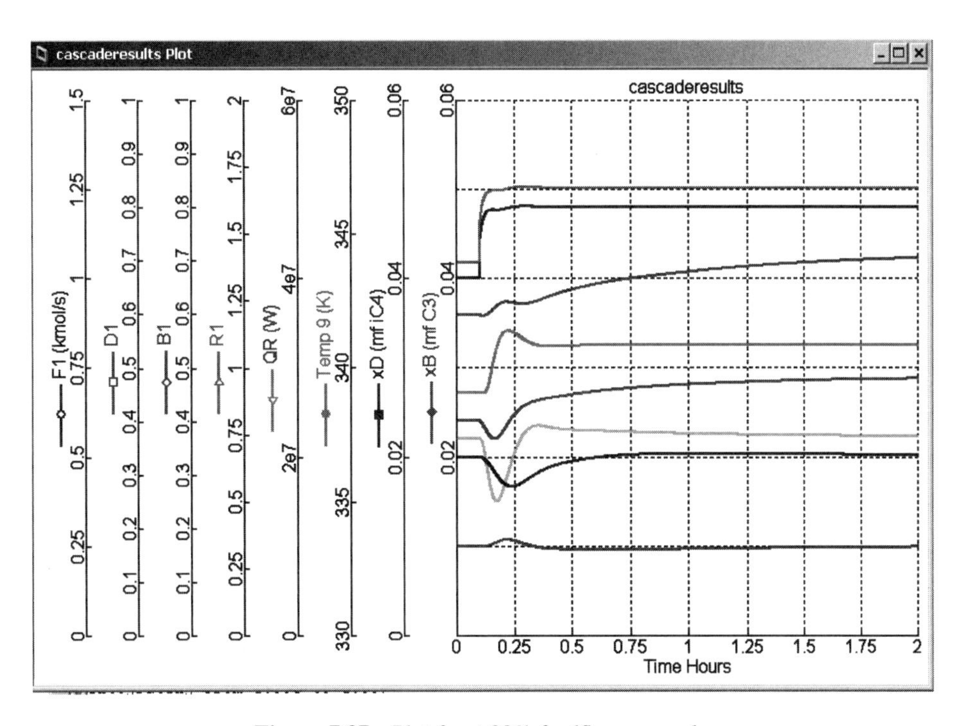

Figure 7.37 Plot for +20% feedflow; cascade.

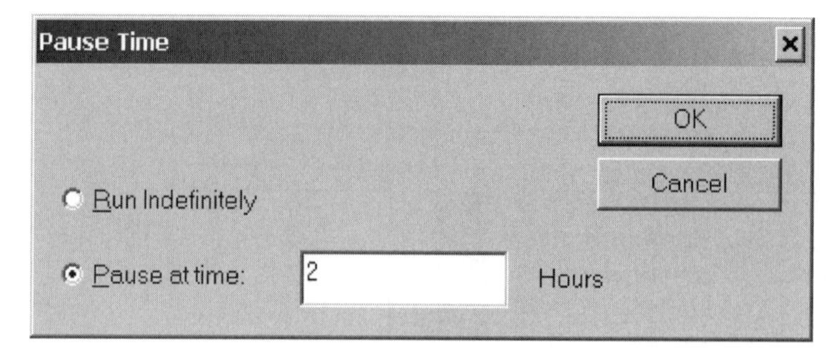

Figure 7.38 Pause window.

pause the simulation at some point in time, go to the top toolbar and click *Run* and then *Pause At*, which opens the window shown in Figure 7.38.

Although these Aspen Dynamics plots are useful to see what is going on during each run, they have a fixed structure (multiple variables per plot) and do not permit comparisons of results from different runs. Fortunately, this is easily handled by storing the data in a file and then producing your own plots using your favorite plotting software. To store the data, right-click the plot and select *Show as History*. Figure 7.39 shows the resulting table, which can be copied and pasted into a file.

Runs are made for all three control structures with two disturbances. Feed flowrate is changed by changing the feedflow controller setpoint. Feed composition is changed by clicking on the F1 stream icon, right-clicking, and selecting *Forms* and then *Manipulate*.

cascaderesults History Table

Time	BLOCKS("C1").QReb	BLOCKS("C1").Stage(9).T	STREAMS("D1").Zn("IC4")	STREAMS("B1").Zn("C3")
Hours	W	K	kmol/kmol	kmol/kmol
0.0	2.71657e+007	337.356	0.0199971	0.0100044
1.e-003	2.71657e+007	337.356	0.0199971	0.0100044
0.002	2.71657e+007	337.356	0.0199971	0.0100044
0.003	2.71657e+007	337.356	0.0199971	0.0100044
0.004	2.71657e+007	337.356	0.0199971	0.0100044
0.005	2.71657e+007	337.356	0.0199971	0.0100044
0.006	2.71657e+007	337.356	0.0199971	0.0100044
0.007	2.71657e+007	337.356	0.0199971	0.0100044
0.008	2.71657e+007	337.356	0.0199971	0.0100044
0.009	2.71657e+007	337.356	0.0199971	0.0100044
0.01	2.71657e+007	337.356	0.0199971	0.0100044
0.011	2.71657e+007	337.356	0.0199971	0.0100044
0.012	2.71657e+007	337.356	0.0199971	0.0100044
0.013	2.71657e+007	337.356	0.0199971	0.0100044
0.014	2.71657e+007	337.356	0.0199971	0.0100044
0.015	2.71657e+007	337.356	0.0199971	0.0100044
0.016	2.71657e+007	337.356	0.0199971	0.0100044
0.017	2.71657e+007	337.356	0.0199971	0.0100044
0.018	2.71657e+007	337.356	0.0199971	0.0100044
0.019	2.71657e+007	337.356	0.0199971	0.0100044
0.02	2.71657e+007	337.356	0.0199971	0.0100044
0.021	2.71657e+007	337.356	0.0199971	0.0100044
0.022	2.71657e+007	337.356	0.0199971	0.0100044
0.023	2.71657e+007	337.356	0.0199971	0.0100044
0.024	2.71657e+007	337.356	0.0199971	0.0100044
0.025	2.71657e+007	337.356	0.0199971	0.0100044

Figure 7.39 *Show as History* table.

F1.Manipulate Table

	Description	Value	Units	Spec
FR	Specified total molar flow	1.0	kmol/s	Free
T	Temperature	322.0	K	Fixed
P	Pressure	2.0265e+006	N/m2	Fixed
vfR	Specified molar vapor fraction	-0.375766		Free
ZR("*")				
ZR("C3")	Specified mole fraction	0.5		Fixed
ZR("IC4")	Specified mole fraction	0.5		Fixed

Figure 7.40 Manipulate window.

The window shown in Figure 7.40 opens, on which the mole fractions of propane and isobutane are changed from 0.40/0.60 to 0.50/0.50.

The data are loaded into Matlab for plotting. Figure 7.41 compares the responses for a 20% increase in feedflow. The solid lines indicate when only a tray temperature is controlled. The dashed lines show when the composition/temperature cascade control structure is used. The direct composition control structure results are not shown because the responses are very poor. This is as we would expect because of the very low controller gain and the very large integral time.

The cascade control structure gives tighter control with smaller peak transient disturbances and a shorter settling time. Both control structures return the product purities to their desired values because of the R/F ratio used in both.

Figure 7.42 gives results for feed composition disturbance. At time equal 0.1 h, the feed composition is changed from 40 mol% propane to 50 mol% propane. The temperature control structure holds the temperature on stage 9 constant. Both distillate and bottoms

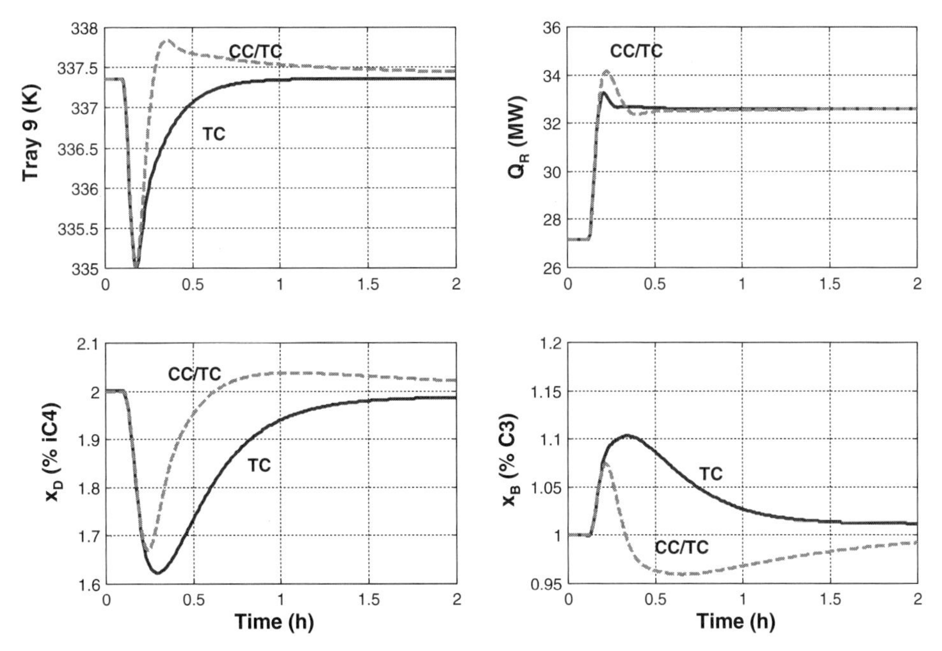

Figure 7.41 TC and cascade; feedflow disturbance.

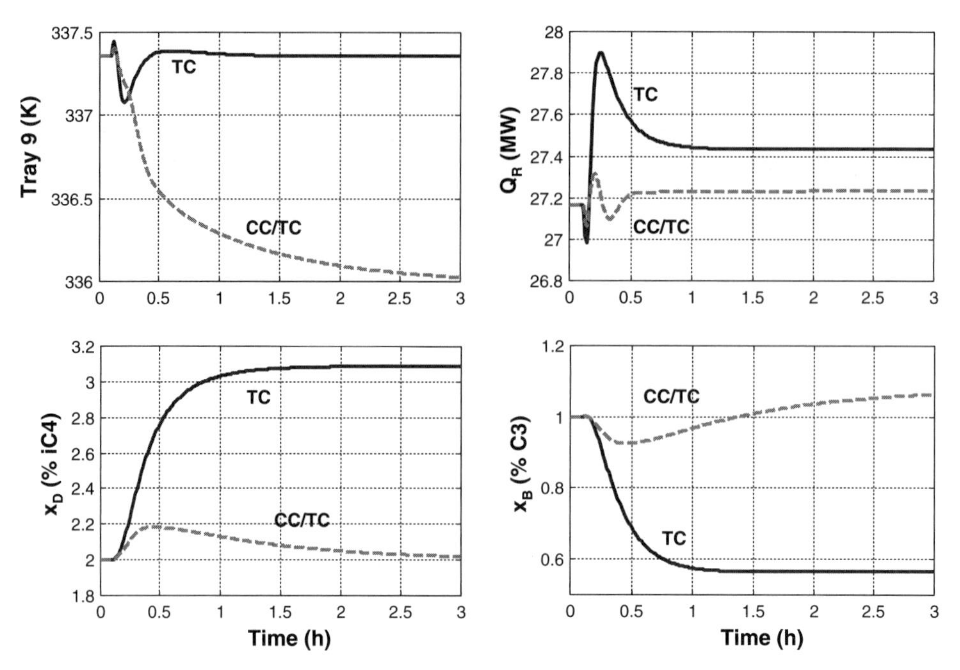

Figure 7.42 TC and cascade; feed composition 40–50% C3.

impurities change significantly. The distillate impurity climbs from the desired 2 mol% isobutane level to over 3 mol%. The bottoms stream becomes overly pure. Using the CC/TC cascade control structure, the distillate purity is maintained. The setpoint of the temperature controller is decreased by the composition controller to drive the distillate impurity to 2 mol% isobutane.

These results illustrate that single-end temperature control provides stable base-level regulatory control of the column. However, product purities change somewhat when feed composition changes.

7.7 COMPARISON WITH ECONOMIC OPTIMUM DESIGN

Now that all the basic dynamic simulation tools have been developed and their application illustrated, the last issue to be covered in this chapter is the effect of column design on control performance. In Chapter 4 the economic optimum design for this system was shown to be a 44-stage column operating with a reflux ratio of 3. So far in this chapter a 32-stage column with a 3.46 reflux has been simulated and controllers developed. This section provides a comparison of these two alternative designs. Is the economic optimum design also the best design from the standpoint of dynamic controllability?

A dynamic simulation of the 44-stage column is developed. The smaller column diameter (5.22 vs. 5.91 m) and lower reboiler heat input (24.8 vs. 27.2 MW) gives a slightly smaller reflux drum ($D = 4$ m) and base ($D = 5$ m). Stage 9 is selected for temperature control (336 K at steady state).

A relay–feedback test gives an ultimate gain of 5.31 and an ultimate period of 4.8 min. The corresponding values for the 32-stage column are 5.9 and 4.5 min. These results

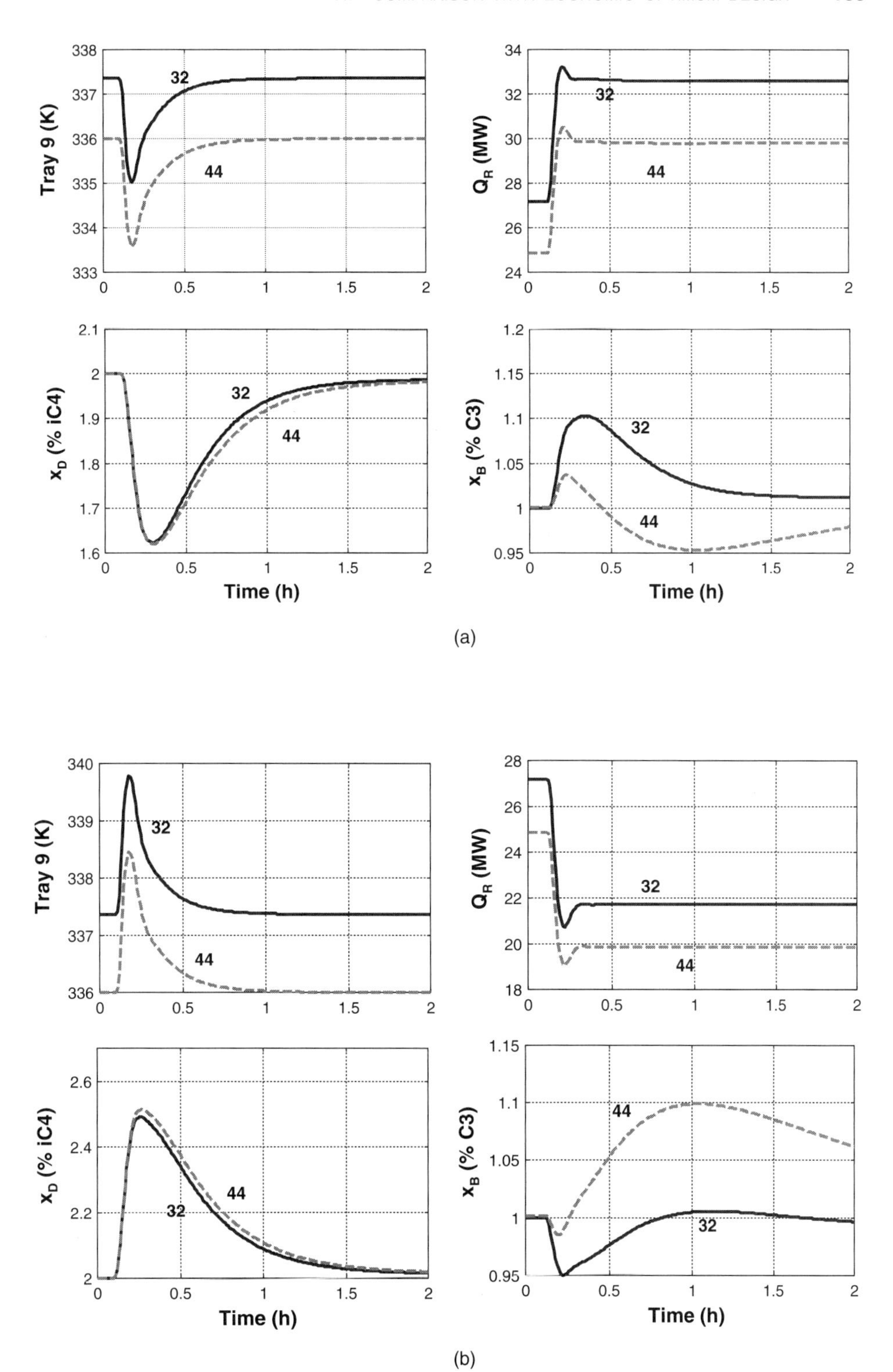

Figure 7.43 Columns with 32 and 44 stages: (a) +20% feedflow; (b) −20% feedflow.

indicate that somewhat tighter control can be achieved in the 32-stage column because of the larger controller gain and smaller integral time.

These predictions are **not** borne out by dynamic simulations. Figure 7.43a gives results for the two columns with a 20% step increase in feedflow. For this disturbance, the biggest difference between the two columns is in the bottoms impurity. The 44-stage column displays a much smaller transient peak departure from the desired value of 1 mol% propane. Although the response of the 32-stage column is faster, the dynamic deviation is larger.

The explanation for this difference is that there are more trays between the feedpoint and the base (feed stage is 19 and base is 44, giving 25 trays) in the 44-stage column. In the 32-stage column the feed stage is 14 and the base is 32, giving only 18 trays. Therefore the disturbance takes longer to reach the base and the temperature controller has more time to take corrective action.

We could try to advance the general conclusion that having more trays in the column helps the dynamic response. This certainly is the conventional wisdom in distillation control.

However, the response of the system to other disturbances brings this conclusion into question. Figure 7.43b gives results for the two columns with a 20% step *decrease* in feedflow. Now the deviation in bottoms composition is larger in the 44-stage column and the response is quite slow.

It should be remembered that a reflux to feed ratio is adjusting the reflux flow to the top of the column as soon as the feed flowrate changes. The change in reflux affects the control tray quickly because it is only eight stages down in both columns. For an increase in feed flowrate, the increase in reflux starts to reduce the temperature on the control tray, and the temperature controller increases the heat input, which affects the bottoms composition quickly. So the bottoms composition is being affected by three

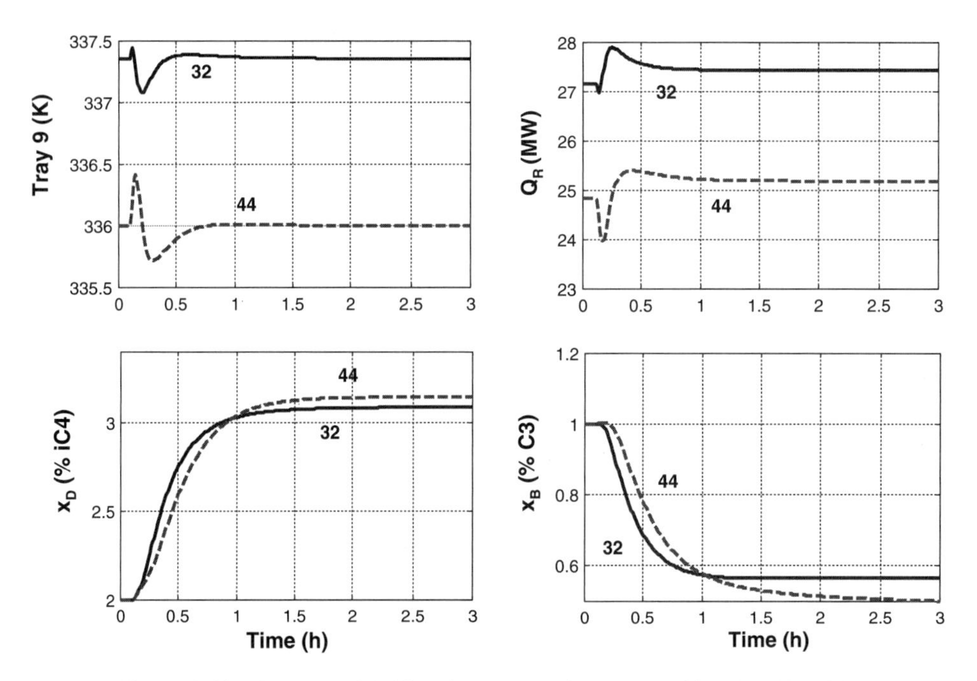

Figure 7.44 Columns with 32 and 44 stages; feed composition 40–50% C3.

inputs: the feed flowrate, the reflux flowrate, and the reboiler heat input. These inputs all have different timescales in terms of the time when they affect the bottoms composition. The net effect is difficult to predict. This demonstrates the usefulness of a rigorous dynamic simulation.

Figure 7.44 gives results for a feed composition change. The difference between the two columns is quite small. There are no changes in reflux or feed flowrates. So the variables affecting bottoms composition are the feed composition and the vapor boilup changes that result from the feed composition affecting the control tray temperature. Note that the energy consumption in the 32-stage column is higher than that in the 44-stage column.

These results demonstrate that distillation column dynamics are not as simple as one might expect, even in this ideal VLE system. The dynamic interplay between the effects of disturbances, measurements, and manipulated variables can lead to some unexpected and counterintuitive results.

7.8 CONCLUSION

A very full bag of distillation dynamic simulation techniques has been developed and demonstrated in this chapter. The example considered is a simple binary ideal VLE column. As the remaining chapters in this book demonstrate, these techniques can be readily extended to much more complex flowsheets and phase equilibria.

CHAPTER 8

CONTROL OF MORE COMPLEX COLUMNS

In this chapter we apply the techniques learned in Chapter 7 for the simple binary column to more complex phase equilibria and more complex distillation flowsheets.

8.1 METHYL ACETATE COLUMN

The first example is the ternary mixture of methyl acetate, methanol, and water. In Chapter 5 we found that the economic optimum column has 27 stages, is fed on stage 21, has a reflux ratio of 1.002, and operates at 1.1 atm. The feed is 0.1 kmol/s of a 30/50/20 mol% mixture of MeAc/MeOH/H_2O.

The diameter is 1.38 m. The diameters of the reflux drum and column base that give 10-min holdups are 1.34 and 1.27 m, respectively. The temperature gradient is the largest at stage 23, so this will be controlled at 342 K by manipulating reboiler heat input. The constant reflux ratio policy was found in Chapter 5 to be better than the constant reflux to feed policy, so this control structure will be implemented.

The file is exported to Aspen Dynamics, and the additional controllers are installed in the normal way. Figure 8.1 gives the control structure and the controller faceplates.

Note that the distillate mass flowrate is measured and fed to the multiplier "RR" as input1. Input2 is the constant 1.002. The output of the multiplier is the mass flowrate of the reflux stream.

A 1-min deadtime is used in the temperature loop. Relay–feedback testing gives controller tuning constants $K_C = 2.0$ and $\tau_I = 7.9$ min. The temperature transmitter span is 300–400 K.

Figure 8.2 shows the response of the system to positive and negative 20% step changes in feed flowrate. The control structure provides stable regulatory level control. Note that the reflux flow changes directly with the distillate.

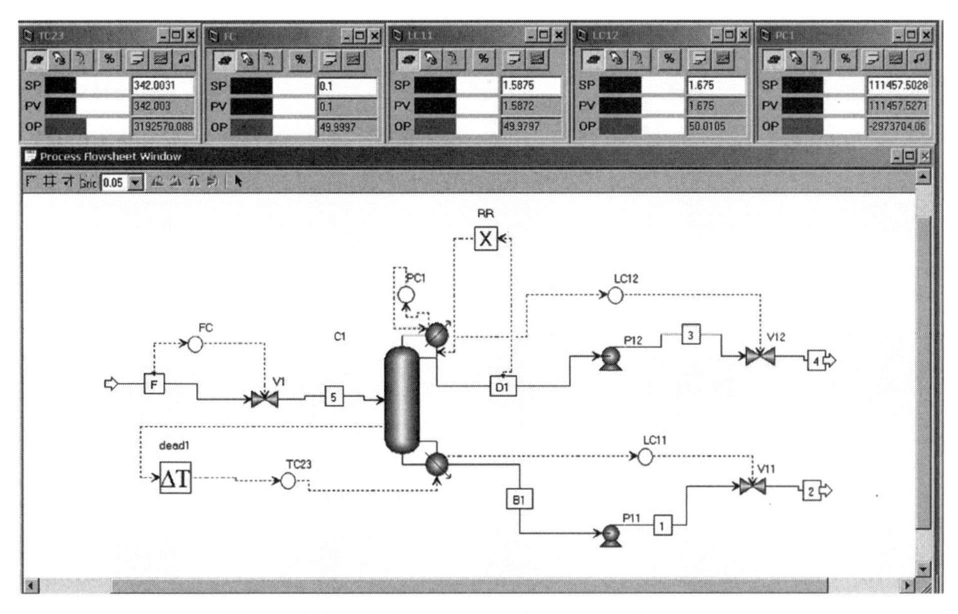

Figure 8.1 Methyl acetate column control structure.

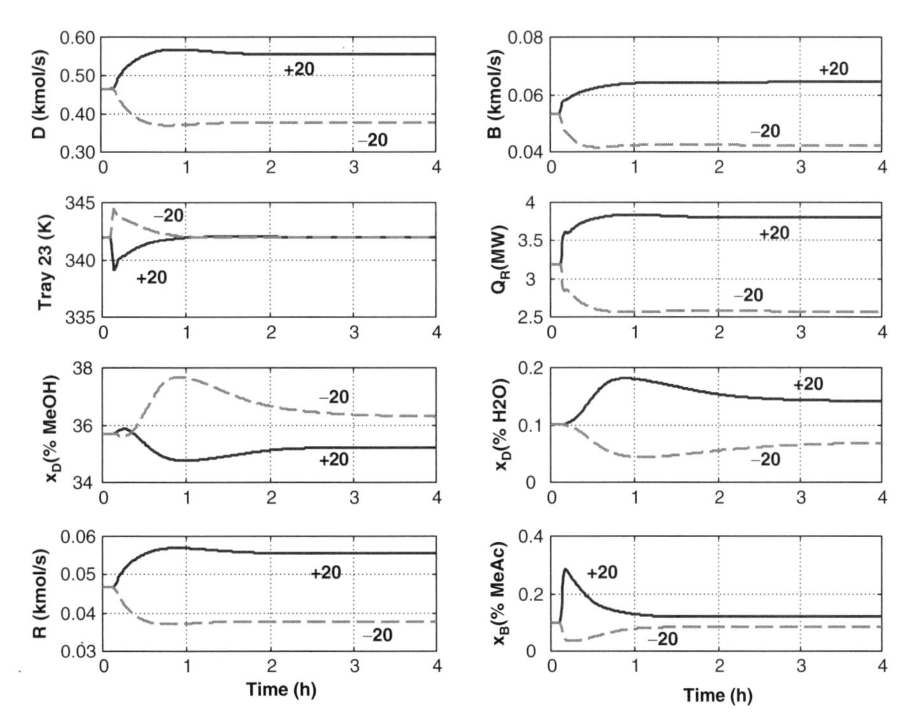

Figure 8.2 MeAc feed rate changes.

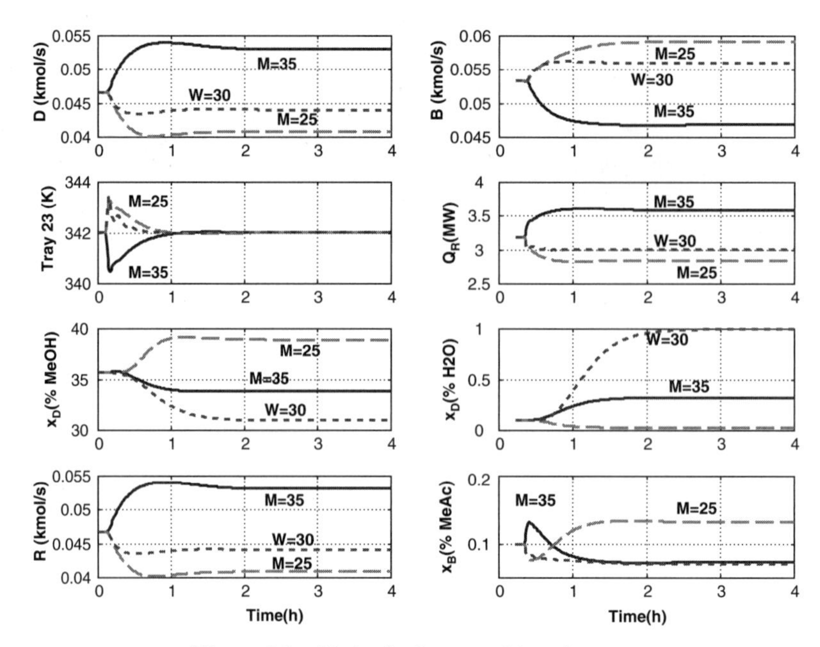

Figure 8.3 MeAc feed composition changes.

One interesting result is that the product compositions do not return exactly to their design values for these feed flowrate changes, despite the fact that the reflux ratio is held constant. In theory, if the flowrates are all held as ratios and the temperature on any tray is held constant, all temperatures and compositions throughout the column should return to their original values.

However, the pressures throughout the column are **not** constant in an Aspen Dynamics simulation. Therefore, holding stage 23 at a constant temperature does not keep all compositions constant because the pressure on stage 23 is not constant. For example, the initial pressure is 126,621 Pa. For the 20% increase in feed flowrate, the higher liquid and vapor rates in the column give a higher final pressure on stage 23 (128,686 Pa). For the 20% decrease in feed flowrate, the lower liquid and vapor rates in the column give a lower final pressure on stage 23 (124,807 Pa). The water impurity in the distillate seems to be affected the most.

Figure 8.3 gives results for three feed composition disturbances. In the first (solid lines), the feed composition is changed from 30/40/20 to 35/45/20 mol% MeAc/MeOH/H_2O. In the second (dashed lines), the feed composition is changed to 25/55/20 mol%. In the third (dotted lines), the feed composition is changed to 30/40/30 mol%.

The increase in the water content of the feed has the most effect on product purities. The distillate water impurity increases to almost 1 mol%.

8.2 COLUMNS WITH PARTIAL CONDENSERS

All the distillation columns considered up to this point in the book have used total condensers, where the distillate product is a liquid. However, many industrial columns have partial condensers in which the distillate product is removed as a vapor stream. This is commonly employed when there are very light components in the feed to the column that would require a high column pressure or a low condenser temperature to completely condense these very volatile components. The use of a partial condenser can avoid the use of costly refrigeration in the condenser.

The control of partial condenser columns is more complex because of the interaction among the pressure, reflux drum level, and tray temperature control loops. Both pressure and level in the reflux drum need to be controlled, and there are several manipulated variables available. The obvious are reflux flow, distillate flow, and condenser heat removal, but even reboiler heat input can be used. In this section, we explore three alternative control structures for this type of system, under two different design conditions: (1) a high vapor distillate flowrate (moderate reflux ratio) and (2) a very low vapor distillate flowrate (high reflux ratio).

As the dynamic simulation results will show, the preferred control structure depends on the control objectives of the entire process. For example, when the distillate goes to a downstream unit and wide variability in its flowrate is undesirable, the control structure should control pressure with condenser heat removal, control level with reflux, and maintain a constant reflux ratio.

The material in the section is taken from a paper[1] in which English engineering units are used, so these are retained.

8.2.1 Process Studied

The numerical example used to study partial condensers is a depropanizer with a feed that contains a small amount of ethane but mostly propane, isobutane, and *n*-butane. Two cases are considered. The first has a feed composition that is 2 mol% ethane and 40 mol% propane, so the distillate flowrate is large and the reflux ratio is moderate ($RR = 2.6$). In the second case, the propane in the feed is only 4 mol% (with 0.02 mol% ethane), which gives a small vapor distillate flowrate and a large reflux ratio ($RR = 20$). Table 8.1 gives design parameters for the two cases. The Chao–Seader physical properties are used.

Design specifications are 1 mol% isobutane impurity in the distillate and 0.5 mol% propane impurity in the bottoms. The column contains 30 trays (32 stages) and is fed in the middle.

If the column is designed with a vapor distillate product, the column operates with a reflux drum pressure of 210 psia, which gives a reflux drum temperature of 110°F and permits the use of cooling water in the condenser. If a total condenser were used, the column pressure would have to be 230 psia to give a reflux drum temperature of 110°F. Of course, higher ethane concentrations in the feed would increase the difference between the operating pressures of total and partial condenser columns.

The major difference between the two cases is the distillate flowrate: 42.15 lb·mol/h in the first case and only 3.76 lb·mol/h in the second. The small vapor flowrate in the latter case corresponds to a volumetric flowrate of only 1.44 ft^3/min. Considering the total volume of the 30-tray column (41.8 ft^3) and the volume of vapor in the half-full reflux drum (9.5 ft^3), the overall pressure time constant of the process is 36 min. This indicates that control of pressure using the small vapor distillate flow will be difficult and slow. The dynamic results given in a later section confirm this expected performance.

The diameters of the columns for the two cases are 1.6 and 1.5 ft, respectively. The reflux drum and base dimensions were sized to give 10-min holdups.

[1] W. L. Luyben, *Ind. Chem. Research* **43**, 6416 (2004).

TABLE 8.1 Design Parameters for Two Cases

		High Distillate, Low *RR*	Low Distillate, High *RR*
Reflux ratio		2.63	20.3
Flows (lb·mol/h)	D	42.15	3.76
	B	57.85	96.24
	F	100	100
	R	111	76.1
Compositions			
(mf C_2/C_3)	z	0.02/0.40	0.0002/0.04
(mf iC_4)	x_D	0.01	0.01
(mf C_3)	x_B	0.005	0.005
Pressure (psia[a])		210	210
Temperatures (°F)	Reflux drum	110	110
	Base	196	197
Heat duty (10^6 Btu/h)	Condenser	0.641	0.440
	Reboiler	1.04	0.742
Diameter (ft)		1.6	1.5
Total stages		32	32

[a]Pounds per square inch absolute.

8.2.2 Alternative Control Structures

Three alternative control structures are studied. In the first two, pressure is controlled by manipulating the flowrate of the vapor distillate stream from the reflux drum. This is the conventional configuration that is recommended in most papers and books. In the third control structure, pressure is controlled by condenser heat removal.

Fundamentals Conventional single-end control is used in all configurations since dual-composition control is rarely used with this propane/butane separation. The temperature on a tray in the rectifying section of the column, where the temperature changes from tray to tray are large, is controlled by manipulating reboiler heat input. The steady-state design shows a temperature on stage 8 of 128°F in the large-distillate case and 155°F in the small-distillate case. Figure 8.4 gives the temperature profiles for the two design cases.

The tray temperature controllers are tuned by inserting a 1-min deadtime in the loop and using the relay–feedback test to determine the ultimate gain and ultimate frequency. Then the Tyreus–Luyben settings are used. Table 8.2 gives the tuning constants.

Note that the tuning constants listed in Table 8.2 are different for the two design cases because of the different vapor distillate flowrates. Note also that the tuning constants are different for some of the different control structures because the effect of reboiler heat input (or the equivalent vapor flowrate to the condenser) on pressure varies depending on what manipulated variable is used to control pressure. Proportional level controllers are used with gains of 2. The default pressure controller tuning constants from Aspen Dynamics are used.

Control Structure CS1 Figure 8.5a shows the control structure that is probably most commonly used for distillation columns with partial condensers. The main features of this

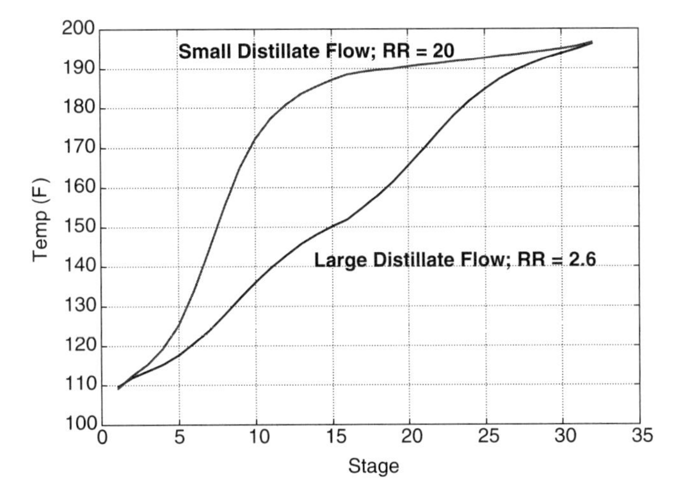

Figure 8.4 Temperature profiles.

structure are pressure controlled by manipulating vapor distillate flowrate and reflux drum level controlled by manipulating condenser heat removal. Reflux flowrate is fixed or ratioed to feed.

The interaction between the level and pressure loops is present because any disturbance that affects either loop will propagate to the other loop. For example, suppose that the feed composition changes and more ethane enters the column. The temperature in the reflux drum will drop and the rate of heat transfer in the condenser will decrease for a fixed flowrate of cooling water. The rate of condensation will decrease. Pressure will *increase*, so the pressure controller will increase the distillate flowrate. The drop in condensation will also *decrease* the reflux drum level. When the level controller increases cooling water flowrate to increase the level, pressure will decrease. This interaction can cause the pressure controller and the level controller to start fighting each other.

One fundamental concept in distillation control is to prevent rapid changes in pressure. If pressure *increases* too quickly, vapor rates through the trays decrease, which can cause weeping and dumping. If pressure *decreases* too rapidly, vapor rates increase, which can

TABLE 8.2 Temperature Controller Tuning Constants

Control Structure		Large Distillate, Moderate *RR*	Small Distillate, Large *RR*
CS1/CS2	K_u	4.8	8.1
	P_u (min)	5.1	4.2
	K_C	1.6	2.5
	τ_I (min)	11	9.2
CS3	K_u	3.6	2.8
	P_u (min)	4.2	5.4
	K_C	1.1	0.86
	τ_I (min)	9.2	12

(a)

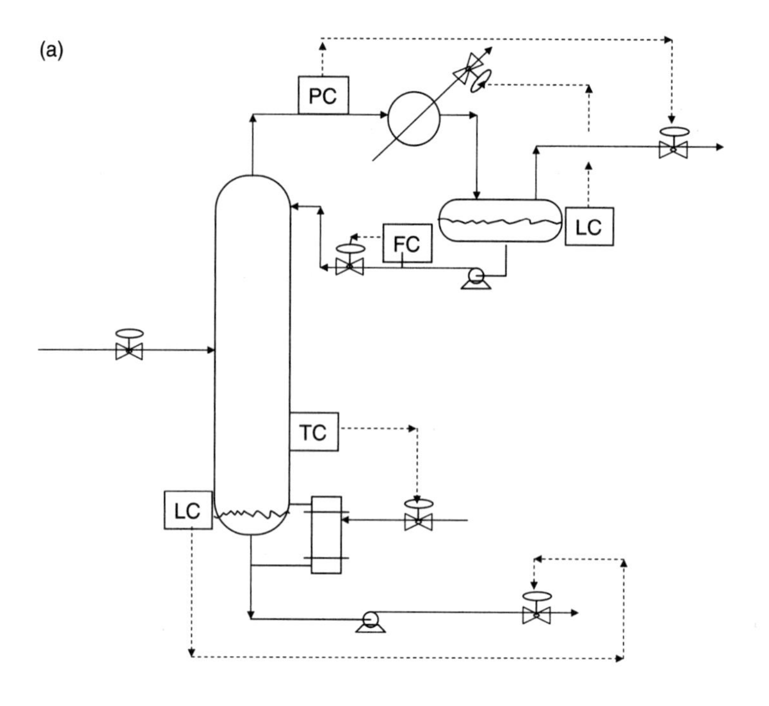

(b)

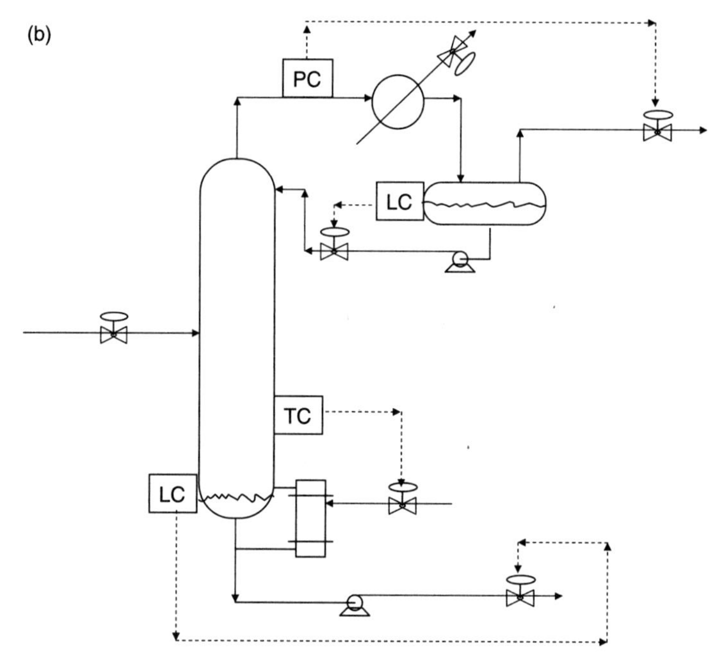

Figure 8.5 (a) CS1 (R fixed); (b) CS2 (Q_c fixed); (c) CS3 (RR fixed).

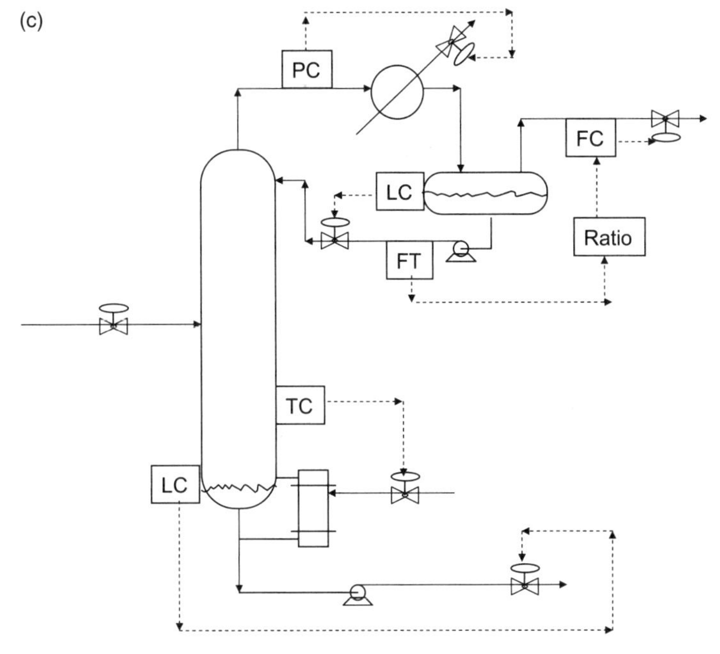

Figure 8.5 *Continued.*

cause flooding. Therefore fairly tight pressure control is required. This leads to large and rapid changes in the vapor distillate flowrate when it is used to control pressure. If the distillate is fed to a downstream unit, these large flowrate changes represent severe disturbances.

One of the main advantages of the CS1 structure is that the constant reflux flowrate establishes steady liquid flowrates down through the trays of the column. Changing vapor rates can be achieved fairly rapidly (in 20–30 s). Changing liquid rates takes much longer because of the hydraulic lags introduced by weirs and baffles. The rule of thumb is about 3–6 seconds per tray. So in a 30-tray column, it will take 2–3 minutes for a change in reflux flow to work its way down to the base of the column.

Control Structure CS2 Figure 8.5b presents an alternative control structure in which the condenser heat removal is fixed instead of fixing the reflux flowrate. Since pressure is controlled by manipulating the flowrate of vapor distillate, this structure has the same problem of distillate flowrate variability to a downstream unit.

This structure does not keep reflux flowrate constant, so internal liquid rates can fluctuate, which can lead to poor hydraulic performance.

Some distillation columns with partial condensers are constructed with the condenser installed at the top of the column inside the shell. There is usually no reflux drum. Vapor flows upward through the tubes of the condenser. The condensate liquid flows downward and drops into a liquid distributer above the top tray. These "dephlegmator" systems are frequently used when very toxic or dangerous chemicals are involved because they eliminate some pumps, extra vessels, and fittings and thus reduce potential leak problems.

In this type of system there are only two manipulated variables: vapor distillate leaving the top of the condenser and condenser heat removal. The absence of a reflux drum means that there is no surge capacity to attenuate disturbances. As a result, these systems have very poor dynamic disturbance behavior and should be avoided if possible.

A viable alternative is to place a large total trapout tray below the condenser that can serve as an internal reflux drum. Liquid reflux can be taken from this trapout tray and fed to the top tray through a control valve. This modified system requires additional column height, which means higher capital investment. But its dynamic controllability is much better.

Control Structure CS3 Figure 8.5c shows the third alternative control structure studied. Now condenser heat removal is used to control pressure, and reflux is used to control level.

Since the distillate flowrate cannot be held constant, the control scheme must permit it to change. This is achieved in this control structure by ratioing the distillate flowrate to the reflux flowrate.

Of course, maintaining a constant reflux ratio may or may not be the best structure to handle feed composition changes when single-end control is used in a distillation column. To address this issue, the curves shown in Figure 8.6 are generated by varying the feed composition (in terms of the light- and heavy-key components) while maintaining the purities at both ends of the column. The required changes in the reflux and reflux ratio are plotted in terms of ratios to their design values at the design feed composition (40 and 4 mol% propane, respectively, for the two cases).

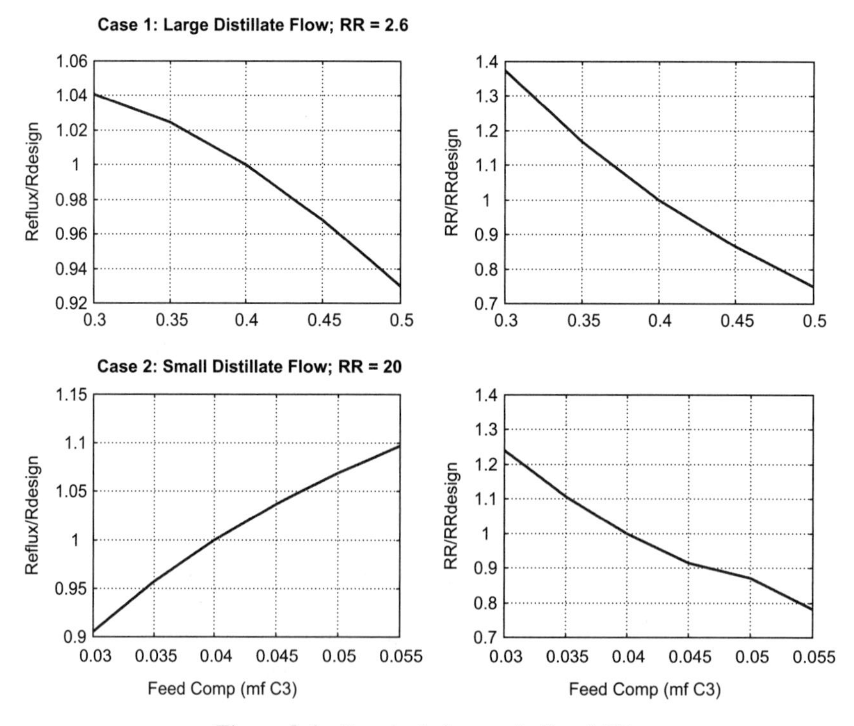

Figure 8.6 Required changes in *R* and *RR*.

The required changes in the reflux flowrate are less than those required in the reflux ratio. Therefore a fixed reflux-to-feed structure should do a better job in maintaining, at steady state, the desired product purities than a fixed reflux-ratio structure for this column.

Control structure CS3 may have a steady-state disadvantage, but it may provide dynamic advantages because of lower variability in the vapor distillate flowrate. The dynamic simulation results presented in the next section illustrate these effects.

8.2.3 Dynamic Performance

The three control structures are simulated in Aspen Dynamics, controllers are tuned, and feed flowrate disturbances are imposed on the system. At time equal to 0.2 h, the feed flowrate is increased from 100 to 120 lb·mol/h. At time equal to 4 h, the feed is dropped to 80 lb·mol/h. Finally, at time equal to 7 h, the feed is increased to 120 lb·mol/h. These very large disturbances are handled with different degrees of effectiveness by the three control structures.

Control Structure CS1 Figure 8.7 gives results using control structure CS1. Reflux is ratioed to the feed flowrate, pressure is controlled by distillate flowrate, and level is controlled by condenser heat removal.

Figure 8.7a gives dynamic responses for the large-distillate, low-reflux-ratio case. This control structure produces very large changes in the distillate flowrate as well as fairly large deviations in pressure. When feed is increased and reflux is increased, the level starts to drop, which increases condenser heat removal. Note that heat removal is

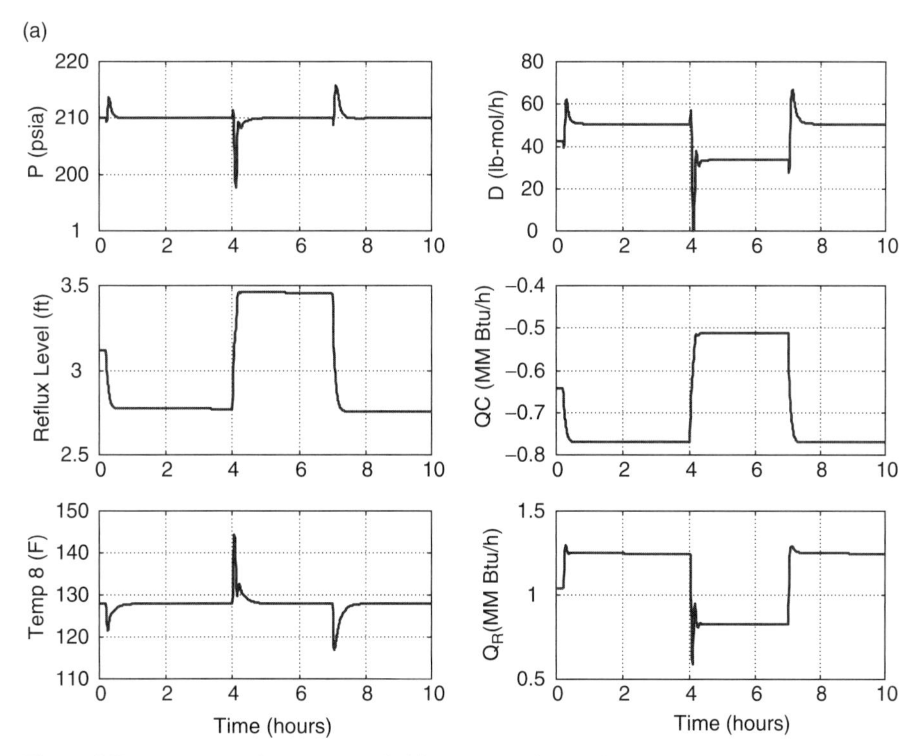

Figure 8.7 Responses for (a) large-distillate case (CS1) and (b) small-distillate case (CS1).

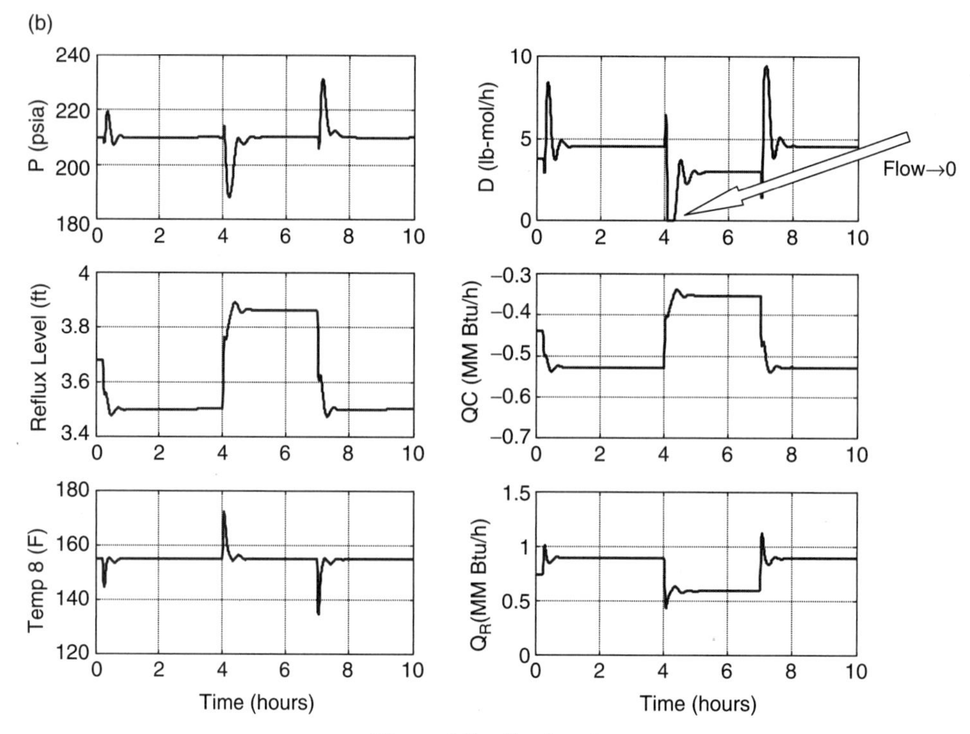

Figure 8.7 *Continued.*

shown as a negative number, using Aspen notation, so the lower the Q_C curve, the more heat removal. This tends to reduce pressure. However, at the same time, the temperature controller sees a drop in stage 8 temperature caused by the increase in feed flowrate, so it increases reboiler heat input. The net effect is a small initial drop in pressure followed by a large increase.

The disturbance that produces the largest deviations from normal is the 50% step decrease from 120 to 80 lb·mol/h in feed flowrate at time equal to 4 h. This produces a drop in pressure of over 10 psi and an increase in temperature of 15°F. The distillate flowrate goes all the way to zero, which would make life very difficult for a downstream unit.

Figure 8.7b gives dynamic responses for the small-distillate, high-reflux-ratio case. The changes in the distillate flowrate are even larger (on a percentage basis). In fact, the distillate is completely shut off for about 20 min following the large drop in feed flowrate at time equal to 4 h. The deviations in pressure are much larger than those seen in the previous case.

These results show that this control structure has poor dynamic performance, particularly when the distillate is fed to a downstream unit.

Control Structure CS2 Figure 8.8 gives results using control structure CS2. Condenser heat removal is fixed, pressure is controlled by distillate flowrate, and level is controlled by reflux. Figure 8.8a illustrates the large-distillate case and Figure 8.8b, the small-distillate case.

The pressure deviations are much less than with the previous control structure because condenser heat removal is constant and does not contribute to the changes in pressure. The variability in the flowrate of the vapor distillate is also significantly reduced. This is true for both the large- and small-distillate flowrate cases. Therefore, from the perspective of plantwide control, this structure is dynamically better than the previous one.

Control Structure CS3 Figure 8.9 gives results using control structure CS3. Pressure is controlled by condenser heat removal, level is controlled by reflux, and distillate flowrate is ratioed to the reflux flowrate. Figure 8.9a gives results for the large-distillate case and Figure 8.9b, the small-distillate case.

The deviations in pressure are even smaller than in either of the previous control structures. In addition, the changes in the vapor distillate flowrate are more gradual.

However, note that stage 8 temperatures take longer to return to the setpoint value and that the changes in reboiler heat input are larger. This occurs because both the distillate and the reflux flowrates change for a change in feed flowrate, which necessitates larger changes in vapor boilup.

Thus this control structure is better from a plantwide control perspective, but it may not be as good from an individual column control perspective.

Comparisons Figure 8.10 provides direct comparisons among the three alternative control structures for the two cases. The solid lines represent the small-distillate case; the dashed lines, the large-distillate case.

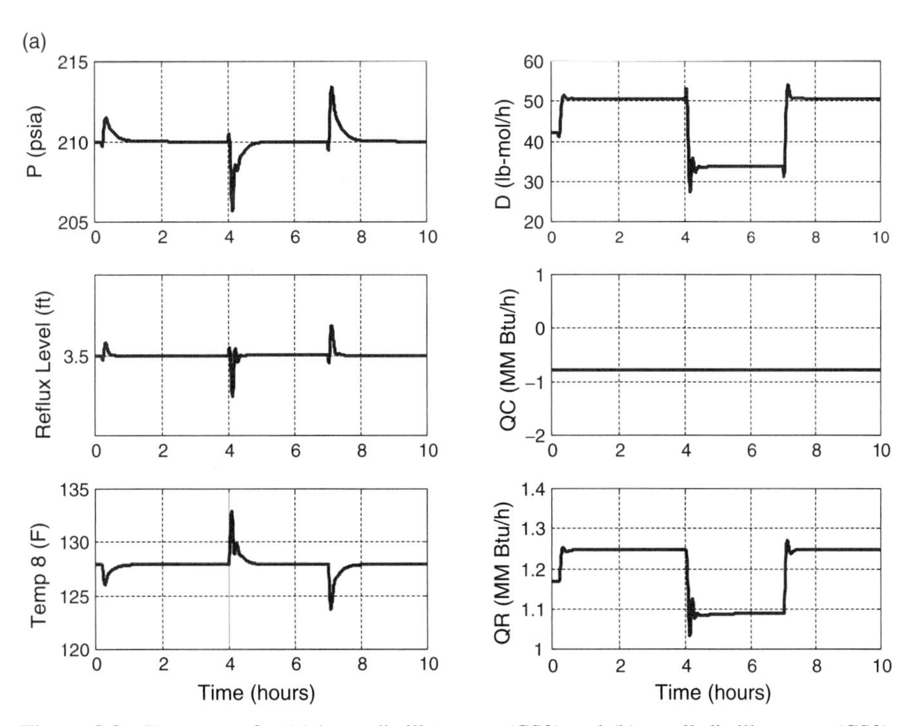

Figure 8.8 Responses for (a) large-distillate case (CS2) and (b) small-distillate case (CS2).

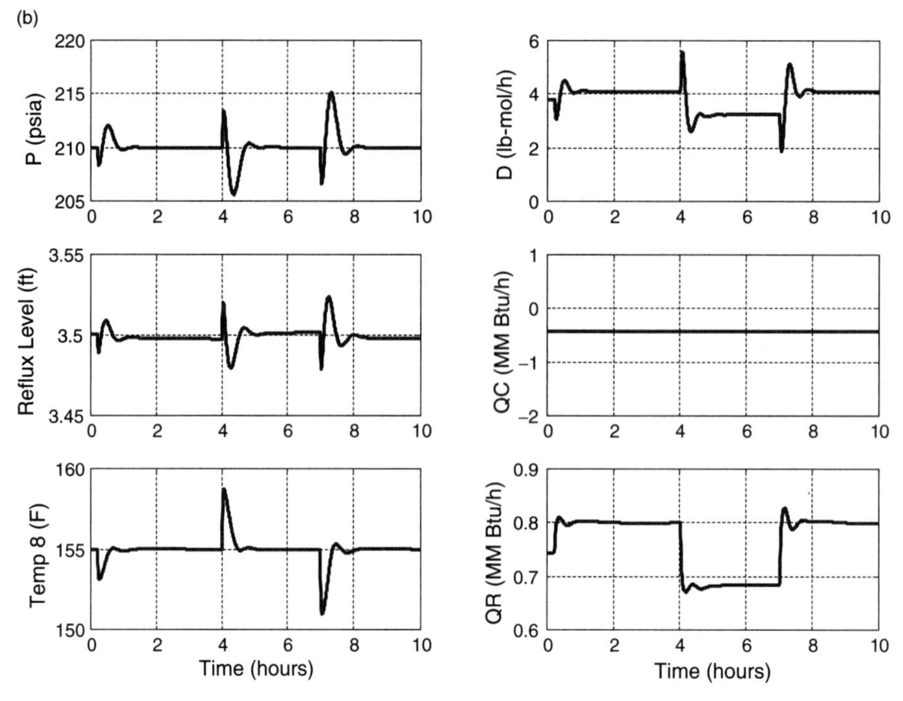

Figure 8.8 *Continued.*

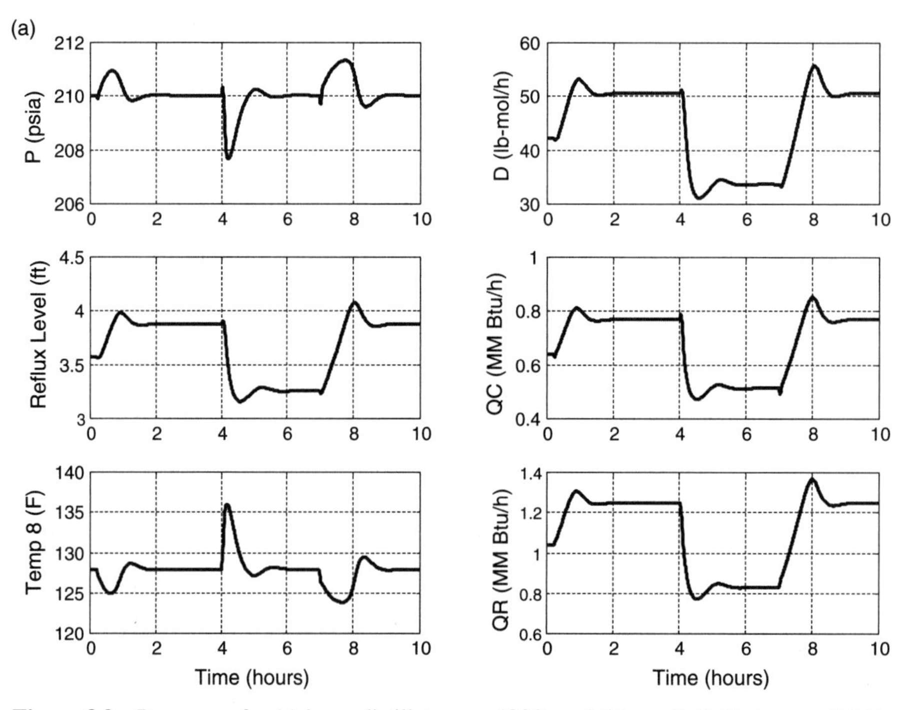

Figure 8.9 Responses for (a) large-distillate case (CS3) and (b) small-distillate case (CS3).

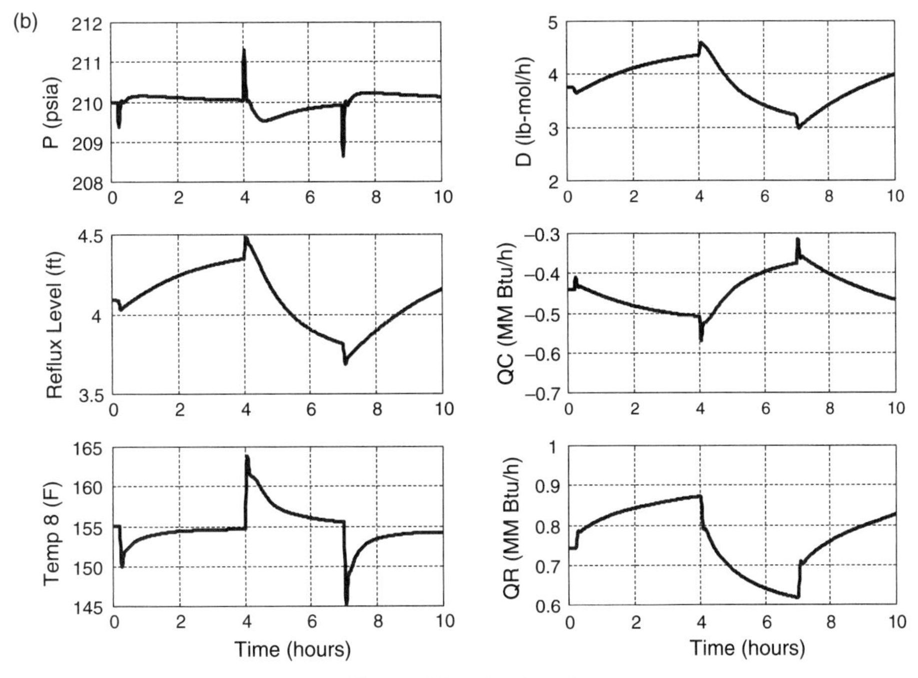

Figure 8.9 *Continued.*

Figure 8.10a shows how pressure responds to the three disturbances with the three control structures. Figure 8.10b shows the responses in the distillate flowrate. Figure 8.10c gives the responses in stage 8 temperatures.

These results illustrate the important conclusion that control structure CS3 provides smaller and more gradual changes in the vapor distillate flowrate, which would be desirable from a plantwide control perspective if this stream were fed to a downstream process.

8.2.4 Product Composition Performance

The results presented up to this point have only looked at flows, levels, pressures, and tray temperatures. The effects of control structure and steady-state distillate flowrate have been examined for disturbances in feed flowrate. We have not looked at feed composition disturbances.

The job of the column is to achieve the desired product purities in the face of disturbances. Recall that the design specifications are 1 mol% iC_4 in the distillate and 0.5 mol% C_3 in the bottoms. We are using single-end control with a temperature on a tray in the column controlled by manipulating reboiler heat input. The other degree of freedom is held constant and depends on the control structure. In CS1 it is the reflux to feed ratio, and in CS3 it is the reflux ratio. In control structure CS2, the other degree of freedom is a fixed condenser heat removal.

Feed Flowrate Changes Both CS1 and CS3 structures should handle feed rate changes, at least from a steady-state standpoint; that is, product compositions should return to their design values for feed rate changes. This is expected because we are

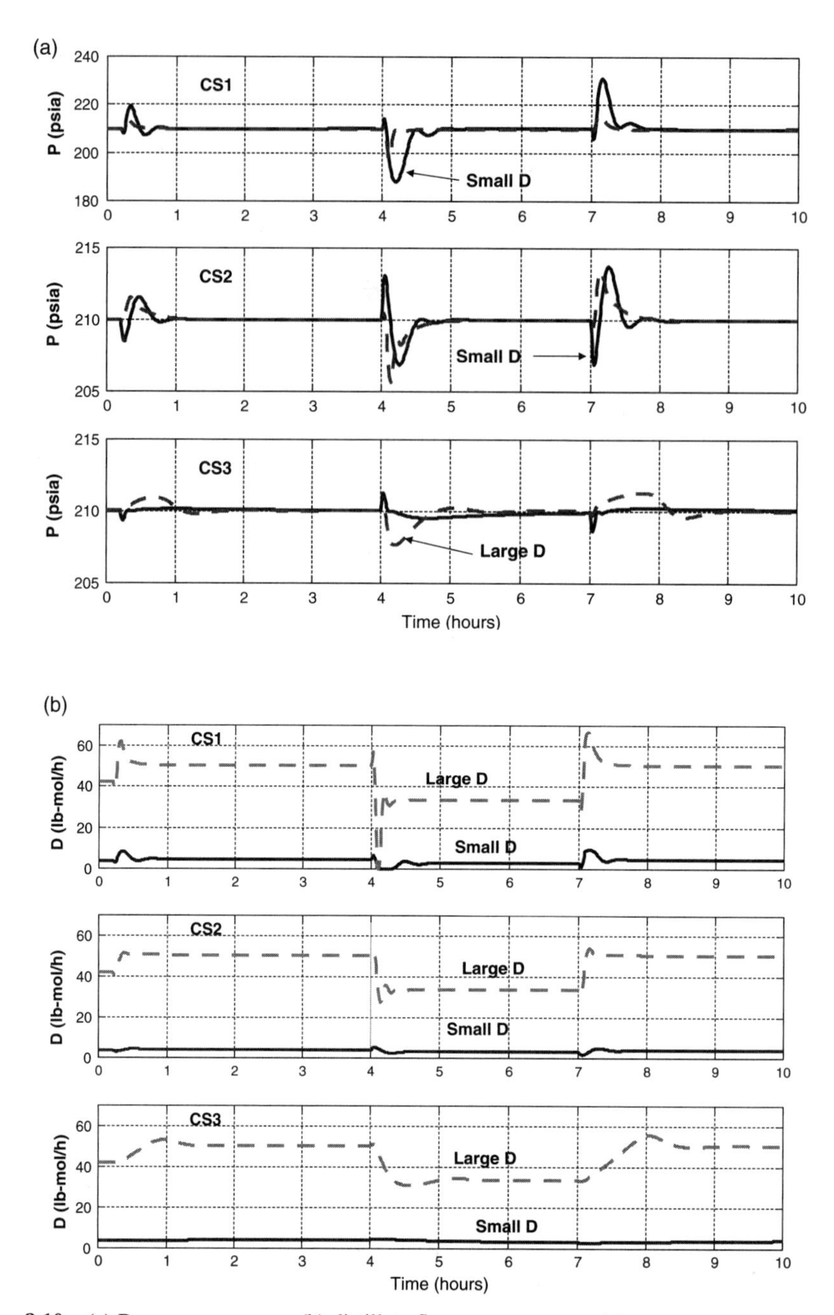

Figure 8.10 (a) Pressure responses; (b) distillate flowrate responses; (c) tray temperature responses.

maintaining flow ratios. Dynamically the product compositions will not be constant, as the results in Figure 8.11 illustrate. Figures 8.11a and 8.11c show how the impurities in the bottoms ($x_{B,C3}$) and in the distillate ($x_{D,iC4}$) vary during the feed flowrate disturbances described earlier. Results are given for the large distillate design case for all three structures.

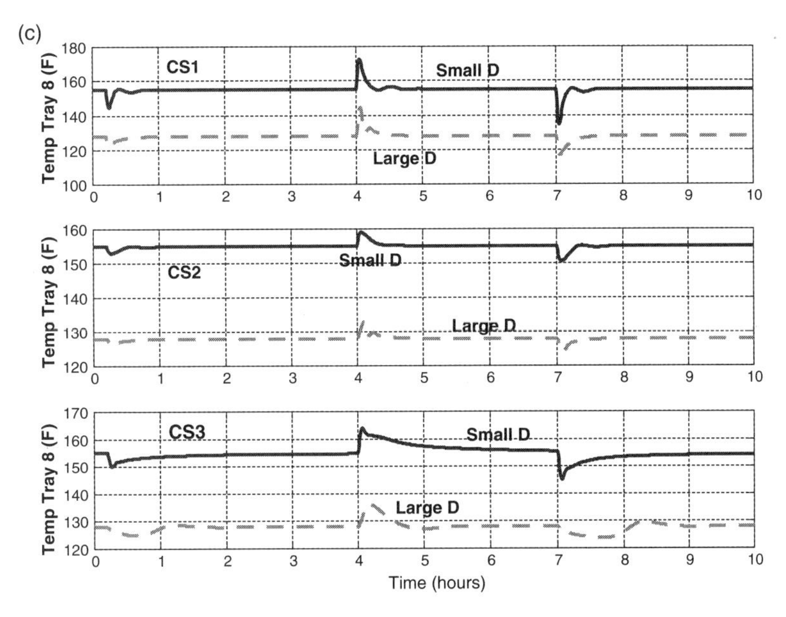

Figure 8.10 *Continued.*

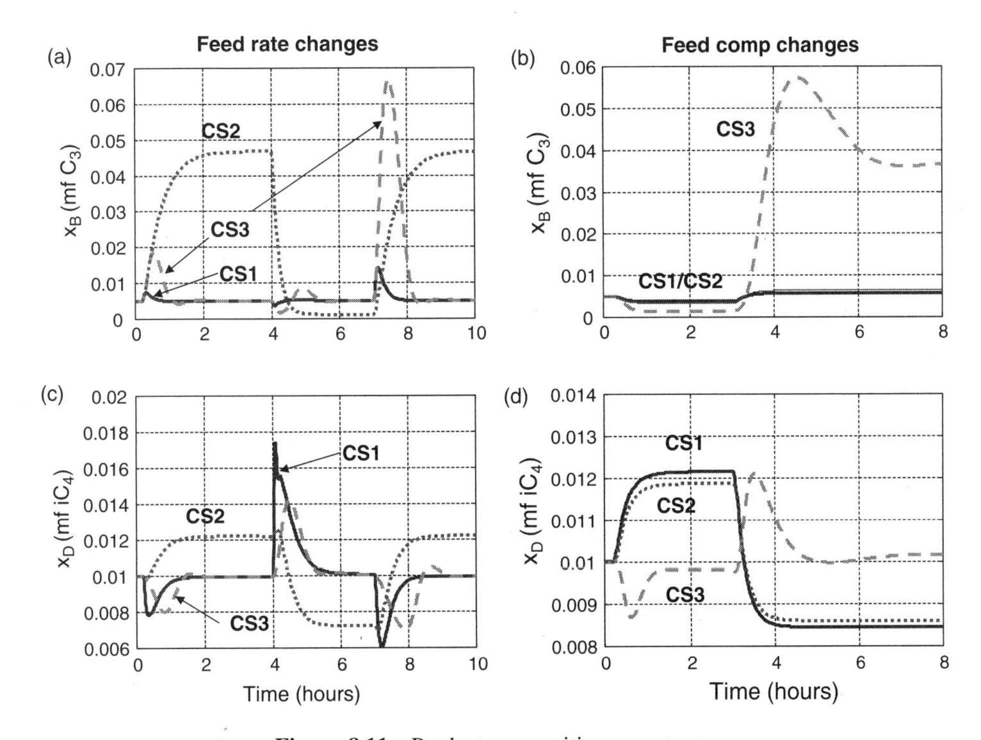

Figure 8.11 Product composition responses.

Control structures CS1 and CS3 drive both product compositions back close to the desired values at the new steady state. There are significant dynamic departures, particularly in the bottoms composition when using CS3. Consider the effect of the large increase in feed flowrate at time equal to 7 hours. There is a dynamic tenfold increase in the impurity of propane in the bottoms at ~7.5 h. This occurs because the increase in feed flowrate brings more light material into the column, which affects bottoms composition quickly before the corresponding drop in stage 8 temperature can increase reboiler heat input to compensate for the increase in feed flowrate. Remember that the feed is liquid, so it affects the bottoms much more quickly and drastically than the distillate. In addition, there is a one-minute deadtime in the temperature loop. Also keep in mind that the tray selected is in the rectifying section above the feed tray.

This same problem exists in the other control structures, but in CS1 the increase in feed flowrate is accompanied by an immediate increase in reflux flowrate. This quickly affects stage 8 temperature, and reboiler heat input increases in time to limit the peak in the propane impurity in the bottoms to about 1.5 mol%.

The CS3 control structure with its fixed condenser heat removal does not return the product purities to their desired values for feed flowrate changes.

Obviously all three of these control structures could be improved by using feedforward control. In CS1 and CS3 the reboiler heat input could be ratioed to the feed flowrate (with the ratio reset by the temperature controller), and in CS3 the condenser heat removal could be ratioed to the feed flowrate. Figure 8.12 illustrates the improvement that the Q_R/F ratio provides for CS3 with the large-distillate case.

The solid lines represent CS3 feed rate disturbances with the ratio and the dashed lines, without. The worst-case peak in the propane impurity in the bottoms ($x_{B,\ C3}$) is reduced

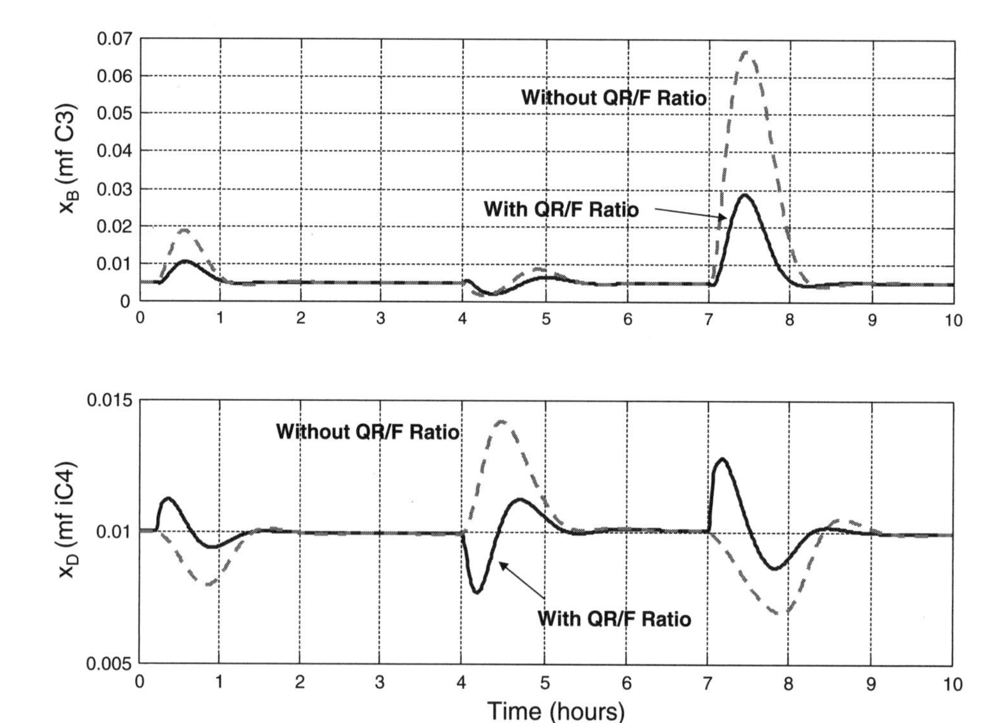

Figure 8.12 CS3 feed rate disturbances with and without Q_R/F ratio.

from about 6.6 mol% to less than 3 mol%. This is still very large compared to the specification of 0.5 mol%, but keep in mind that the disturbance is very large (a 50% step increase in feed flowrate), which is probably much larger than that to which an industrial column would typically be subjected.

Feed Composition Changes Perhaps more important are the performances of the three alternative control structures to disturbances in feed compositions, since feedforward is seldom an option to handle these disturbances. Figures 8.11b and 8.11d compare the three alternative control structures for two step changes in feed composition. At time equal to 0.2 h, the feed composition is changed from 40 to 45 mol% propane and from 30 to 25 mol% isobutane. Then at time equal to 3 h, the feed composition is changed from 45 to 36 mol% propane and from 25 to 34 mol% isobutane.

The performances of structures CS1 and CS2 are quite similar because in both the reflux flowrate is essentially constant since only feed composition is changing. There is some steady-state deviation from the desired product purities. Increasing propane content in the feed with a fixed stage 8 temperature produces (somewhat counterintuitively) a higher level of isobutane impurity in the distillate and a lower level of propane impurity in the bottoms.

The response of structure CS3 shows some important results, both steady-state and dynamic. From the analysis discussed earlier and presented in Figure 8.6, we would expect the constant reflux ratio strategy used in CS3 would not handle feed composition changes as well as the constant reflux flowrate strategies of the other two structures, at least from a steady-state perspective. Figure 8.6 shows that a flowrate slightly higher than design reflux (4% higher for the high-distillate case and 10% for the low-distillate case) would give on-specification product purities over the range of feed composition. To achieve the same product purities with a constant reflux ratio strategy would require operating with 35% and 24% higher reflux ratios than the values at the design feed composition for the two cases. Thus it is no surprise that CS3, with its reflux ratio fixed at the design value, does not do as good a job in maintaining product purities for feed composition disturbances.

The results in Figure 8.11 show that an increase in propane concentration in the feed produces only a small steady-state shift in distillate purity when CS3 is used, which is less than that produced by the other two control structures. However, the change in the bottoms purity is larger than that produced by the other structures. The same occurs for a decrease in feed propane composition.

The results shown in Figure 8.6 are for the case when both distillate and bottoms purities are maintained. The results in Figure 8.11 reveal an asymmetric behavior in which distillate purity changes little but bottoms purity changes drastically. One fundamental reason for this is our selection of a rectifying section tray to control. Had we selected a tray in the stripping section, bottoms purity would be better maintained at the expense of larger changes in distillate purity. These results suggest that the method for selecting the control tray temperature depends on which product is more important.

Of course, this problem of steady-state shift in product purity for feed composition changes could be solved by using a cascade composition/temperature control (CC/TC) structure. Keep in mind, however, that the reflux ratio would have to be fixed at the highest value needed to handle the range of feed compositions.

The results shown above are for the high-distillate case. Similar results were obtained for the small-distillate case. However, the dynamics for the feed composition disturbances are much slower because the changes made in the propane concentration of the feed are much smaller (changed from 4 to 5 mol% C_3).

8.2.5 Dual-Composition Control

The control structures studied up to this point all use single-end inferential control. A single tray temperature is controlled with the objective of maintaining a temperature profile in the column that we hope will hold product purities close to their specifications. This goal was achieved with varying degrees of success, depending on the control structure used and the disturbance. If tight product composition control is required, a dual-composition control structure can be used. However, this would require two online composition analyzers, which are expensive and require high maintenance.

To illustrate the improvement in control that is achieved by using dual-composition control, the CS3 control structure is augmented by two composition controllers, one controlling propane impurity in the bottoms (CCxB) and the second controlling isobutane impurity in the distillate (CCxD). Figure 8.13a shows the Aspen Dynamics flowsheet. Three-minute deadtimes are used in the composition loops.

The controller output signal from the CCxD controller changes the ratio of the distillate to reflux. It is tuned using the relay–feedback test and Tyreus–Luyben tuning ($K_C = 1.8$ and $\tau_I = 54$ min with a composition transmitter span of 5 mol% isobutane). The controller output signal from the CCxB controller changes the setpoint of the stage 8 temperature controller, which in turn changes the ratio of the reboiler heat input to feed flowrate. It is tuned using the relay–feedback test and Tyreus–Luyben tuning ($K_C = 1.0$ and $\tau_I = 37$ min with a composition transmitter span of 2 mol% propane).

Setting up the multiplier for the reboiler heat input to feed ratio is a little tricky. The multiplier uses metric units: flows in kmol/h and heat in GJ/h. So the steady-state value of the second input to the multiplier, which is the temperature controller output signal, is calculated:

$$\frac{Q_R(\mathrm{GJ/h})}{F(\mathrm{kmol/h})} = \frac{(1.038 \times 10^6\,\mathrm{Btu/h})(1.055\,\mathrm{GJ}/10^6\,\mathrm{Btu/h})}{(100\,\mathrm{lb \cdot mol/h})(0.454\,\mathrm{kmol/h})} = 0.0241$$

Figure 8.13b shows the controller faceplates. Note that the flow controller on the vapor distillate (FCD) has a remote setpoint (on "cascade") coming from a multiplier ("ratio") whose two inputs are the reflux flowrate and the CCxD controller output signal. The temperature controller is also on "cascade" with its setpoint coming from the CCxB controller.

Figure 8.14 demonstrates the effectiveness of this dual-composition control structure. Figure 8.14a shows how product purities vary in the face of the same scenario of feed flowrate disturbances and feed composition disturbances used previously. A comparison of these results with the CS3 results given in Figure 8.11 reveals a very significant reduction in product quality variability, both dynamically and at steady state. Both products are returned to the specifications, even for feed composition disturbances.

Figure 8.14b shows the changes in other key variables. The solid lines represent feed flowrate disturbances; the dashed lines, feed composition disturbances. Note that the CCxB composition controller changes the setpoint temperature for the feed composition

(a)

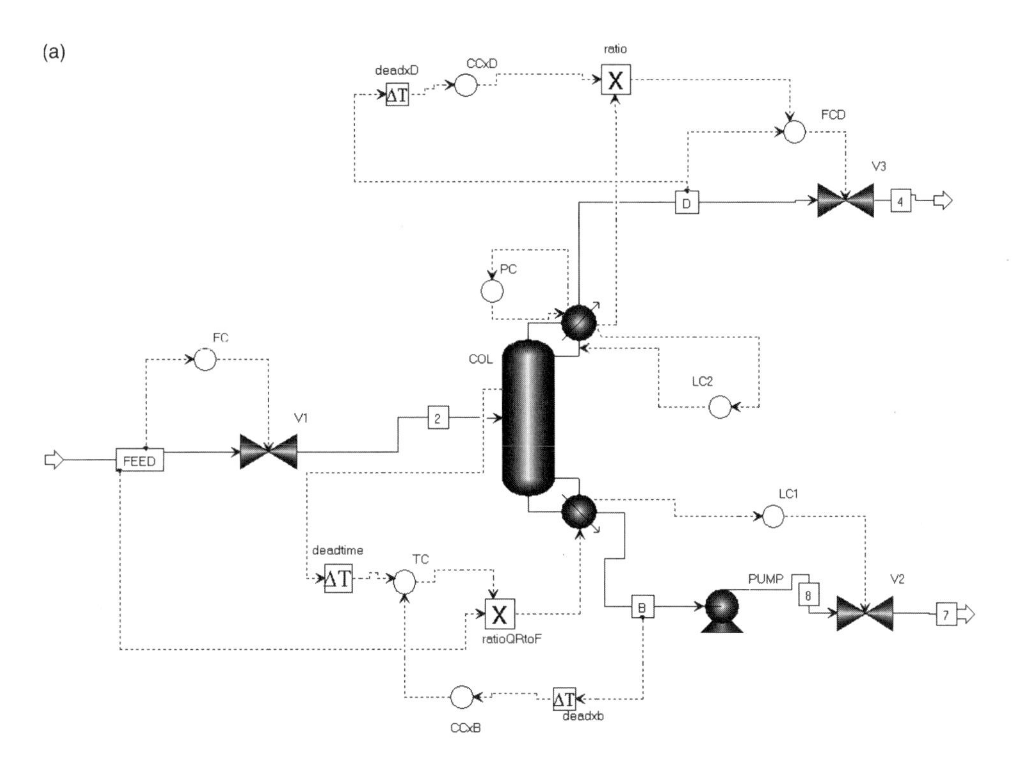

(b)

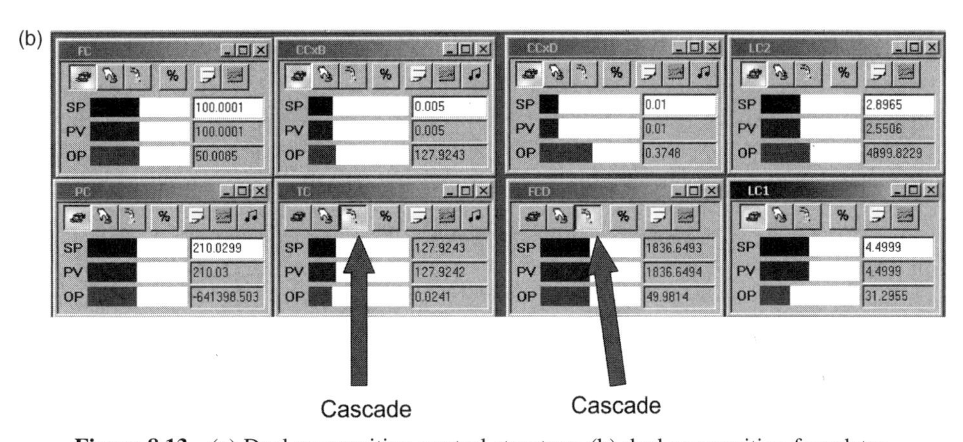

Figure 8.13 (a) Dual-composition control structure; (b) dual-composition faceplates.

disturbances, shifting it lower than the design 128°F for higher propane compositions in the feed and higher for the lower propane compositions in the feed. Likewise the reflux ratio is adjusted by the CCxD controller from its design value of 2.63 so that the distillate purity is maintained for feed composition disturbances.

The variability in the vapor distillate flowrate is still much less than with the other control structures, so even with dual-composition control the downstream unit is not subjected to large and rapid disturbances.

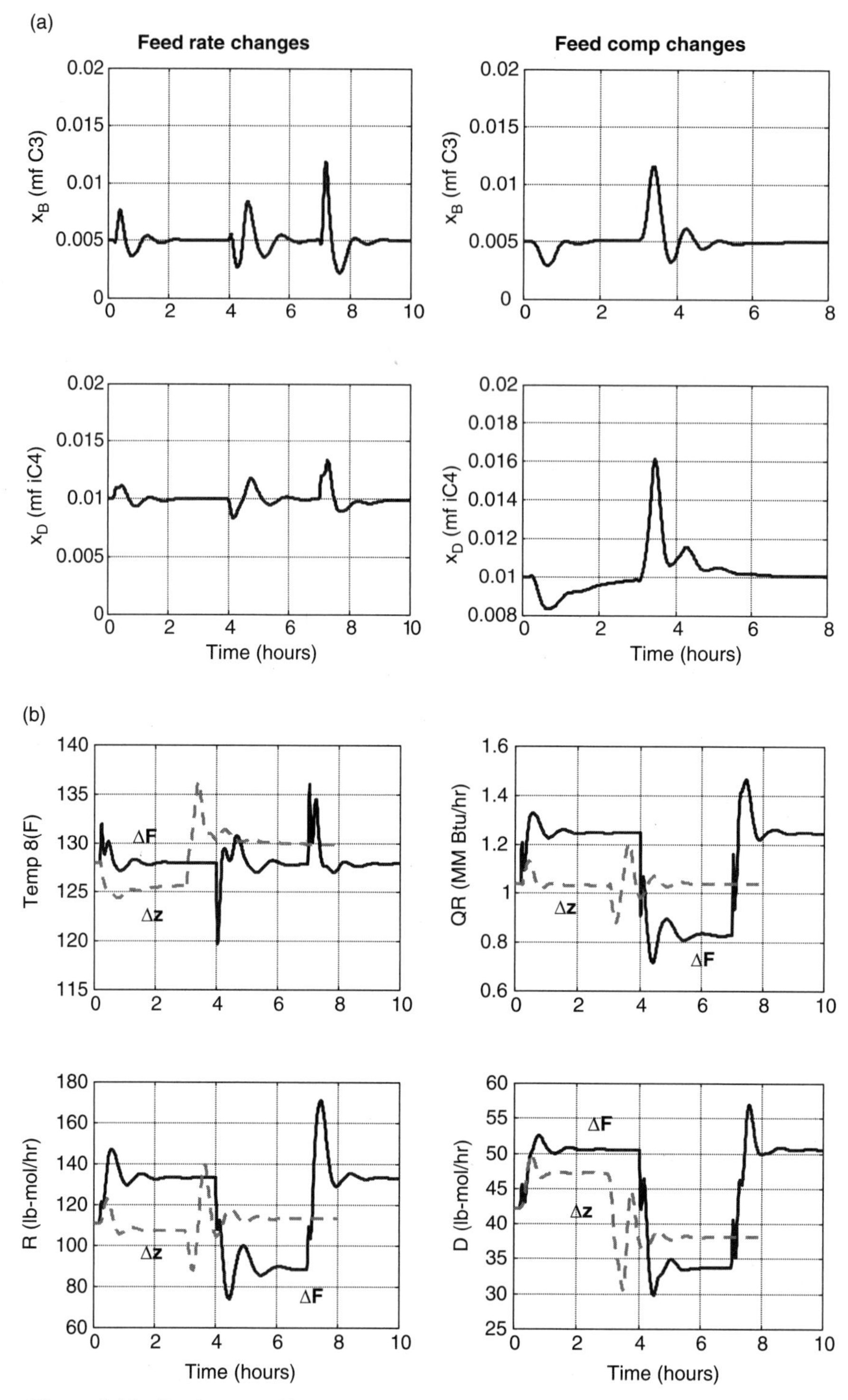

Figure 8.14 Dual-composition control: (a) product purities; (b) flows and temperature.

These results illustrate the improvement in dynamic and steady-state performance that is achievable with conventional PI control structures through the use of ratio and cascade control schemes. Of course, online composition measurements are required for dual-composition control.

8.2.6 Conclusion

This process provides an interesting and important example of the conflicts between individual unit performance and plantwide performance. Considering only the column in isolation, control structure CS1 gives the best results in terms of product purities for disturbances in both feed flowrate and feed composition. However, this control structure produces large and rapid changes in the vapor distillate flowrate, which could seriously degrade the performance of downstream units. Control structure CS3 provides more gradual changes in the distillate flowrate, but it does not hold product purities as close to their specifications as control structure CS1 does in the face of feed composition disturbances.

Modifying the CS3 structure by the addition of dual-composition control provides effective product quality control with essentially the same low variability in distillate flowrate.

A final comment might be useful in situations where the coolant is a boiling liquid (e.g., a refrigerant such as propane). The heat transfer rate in this situation depends on the pressure and composition of the coolant and on the coolant liquid level (since this affects the heat transfer area). Normally the level is maintained to keep all the tubes covered. The coolant pressure is adjusted to control the heat removal rate, which can be conveniently determined by simply measuring the flowrate of the coolant vapor leaving the condenser. Thus this more complex system with cooling-side dynamics can be handled in a straightforward way, and the performance of the alternative control structures should be similar to that found in this study in which the heat load is assumed to be manipulated directly.

8.3 CONTROL OF HEAT-INTEGRATED DISTILLATION COLUMNS

The steady-state simulation of a two-column heat-integrated distillation system was developed in Chapter 5. The two columns run "neat," so all the energy released in the condenser of the high-pressure column is used in the reboiler of the low-pressure column. In the steady-state design, this balance was achieved by using a *Flowsheeting Options Design Spec* that split the total feed between the two columns such that the two heat duties balanced.

Now we want to study the dynamics of this complex flowsheet. The problem of balancing the heat duties becomes one of the major aspects of designing an effective control structure. The other issue is what to manipulate to control a tray temperature in each column. In the high-pressure column, reboiler heat input can be manipulated. But in the low-pressure column, reboiler heat input is not adjustable since it is fixed by the condenser heat duty in the high-pressure column. Table 8.3 summarizes the major design features of the two columns.

TABLE 8.3 Parameters for Two Columns

		C1—Low Pressure	C2—High Pressure
Pressure (atm)		0.6	5
Stages		32	32
Feed stage		19	18
Reflux ratio		0.585	0.922
Shell diameter (m)		3.48	2.55
Flows (kmol/s)	F	0.5085	0.4914
	B	0.2033	0.1964
	D	0.3052	0.2950
Compositions			
(mf MeOH)	z	0.60	0.60
(mf MeOH)	x_D	0.999	0.999
(mf MeOH)	x_B	0.001	0.001
Temperatures (K)	Reflux drum	325.2	**386.88**
	Base	**367.5**	428.1
Heat duty (MW)	Condenser	17.44	**18.10**
	Reboiler	**18.10**	21.81
Diameter (m)	Reflux drum	2.0	2.2
	Base	1.7	1.8
Control stage		26	24
Control stage temperature (K)		356.52	417.08

8.3.1 Development of Control Structure

The column shell, reflux drum, and column base are sized for each column, and the file is exported into Aspen Dynamics. The flowsheet opens with two default pressure controllers manipulating condenser duties. Four level controllers are installed. Reflux drum levels are controlled by manipulating the distillate flowrates. Base levels are controlled by manipulating the bottoms flowrates. A temperature controller "TC2" is installed on the high-pressure column C2, which controls stage 24 at 417.08 K by manipulating reboiler heat input Q_{R2}.

The major issue is to make the reboiler heat input Q_{R1} to the low-pressure column C1 equal to the heat removal rate Q_{C2} in the condenser of the high-pressure column C2. The latter is being calculated by the pressure controller in the high-pressure column.

Aspen Dynamics has the capability of using "flowsheet equations" to solve this "neat" operation problem. In the *Exploring* window, select *Flowsheet.* The window below *Exploring* is labeled *Contents of Flowsheet* as shown in Figure 8.15.

In this window there are two parallel blue bars. Double-clicking opens a window called *Text Editor—Editing Flowsheet,* on which equations can be written. As shown in Figure 8.16, the heat duty in the low-pressure column C1 is equated to the negative of the heat duty in the high-pressure column C2. The syntax must be precise.

Reboiler heat input in C1: `Blocks(''C1'').QReb`

Condenser heat removal in C2: `Blocks(''C2'').Condenser(1).QR`

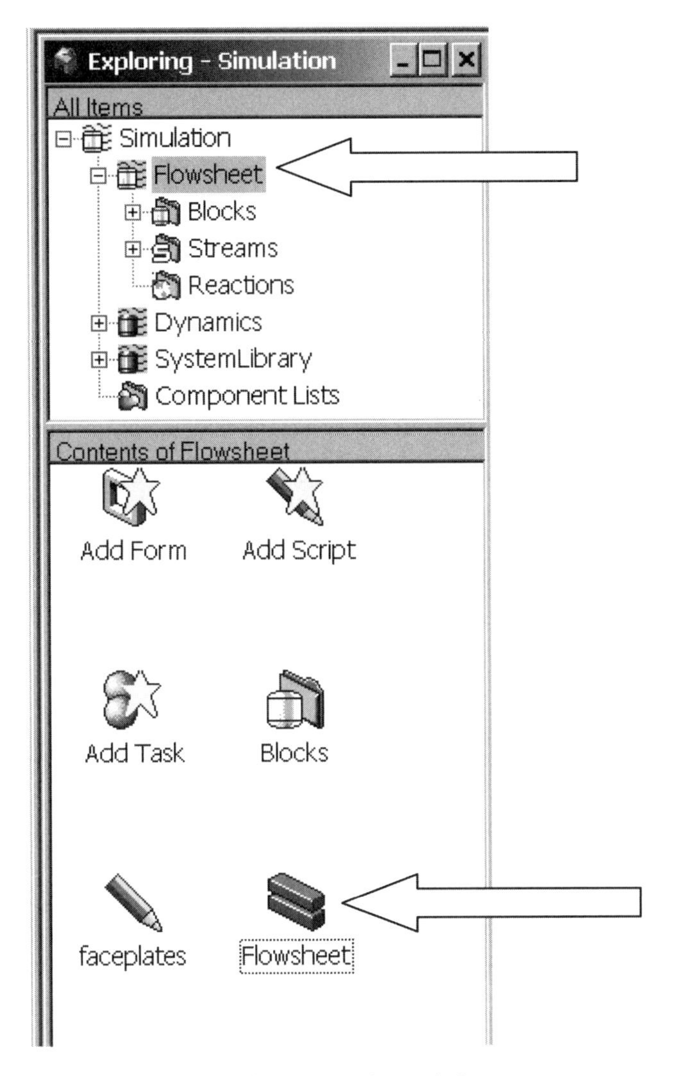

Figure 8.15 Flowsheet window.

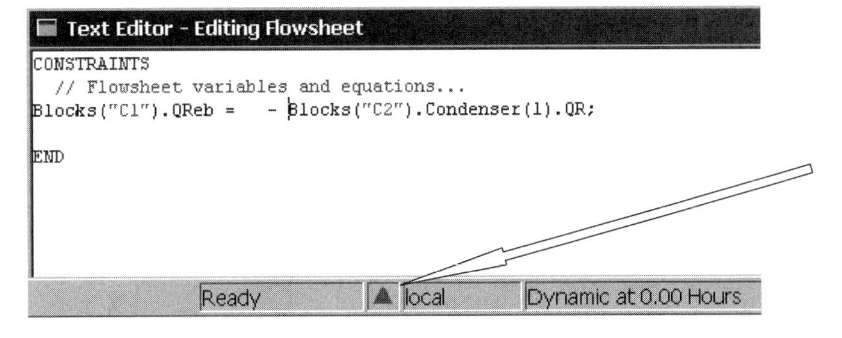

Figure 8.16 Flowsheet equations.

Note also the minus sign. The equation must end with a semicolon. Right-click the *Text Editor* window and select *Compile.* If your equations are correct, a message appears on the *Simulation Messages* window stating that no compilation errors occurred.

Note the little red block at the bottom of Figure 8.16. The simulation is overspecified by one variable, so the reboiler heat input in column C1 must be changed from *Fixed* to *Free*. This is done by clicking on the C1 icon, right-clicking, selecting *Forms*, and clicking *All Variables.* As shown in Figure 8.17, scroll down until *QReb* is found. Click the arrow and select *Free*. The red block turns green, and the simulation is ready to run.

It is important to clearly understand what the setup is. As the simulation runs along in time, the pressure controller PC2 in the high-pressure column C2 is controlling pressure by manipulating Q_{C2}. The flowsheet equations set the reboiler input Q_{R1} in the low-pressure column C1 equal to the negative of Q_{C2}.

The temperature controller TC2 in the high-pressure column is tuned by inserting a one-minute deadtime and relay–feedback testing. The tuning constants are $K_C = 1.15$ and $\tau_I = 10.6$ min. A temperature transmitter range of 350–450 K is used. The action of this temperature controller is *Reverse.*

But we need some way to control the temperature in the low-pressure column C1. One way to do this is to adjust the feed F1 to the low-pressure column. A temperature controller is installed that maintains the temperature on stage 26 at 356.5 K by manipulating the control valve V1. This controller is set up to be *Direct*-acting. If the temperature is increasing, there is too little feed for the given reboiler heat input, so we should feed more to the column. Remember, we cannot adjust the reboiler heat input in this column. This is whatever is dictated by column C2.

The final element of the control scheme involves setting the total feedflow to the unit. This is done by installing a flow controller that looks at the total fresh feed and manipulates the valve in the feedline to the high-pressure column. Figure 8.18 shows this control structure.

Obviously these two temperature controllers are interacting. An increase in temperature in the high-pressure column will reduce reboiler heat input in column C2. This will reduce the condenser duty in column C2, which means a reduction in the heat input to column C1. This will decrease the temperatures in column C1, and the temperature controller TC1 will reduce the feed to column C1. However, now the total flow controller will increase the feed to column C2, which will tend to drop the temperature in column C2.

One way to take this interaction into account is to first tune TC2 with TC1 on manual. Then TC2 is placed on automatic and a relay–feedback test is run on TC1. With a

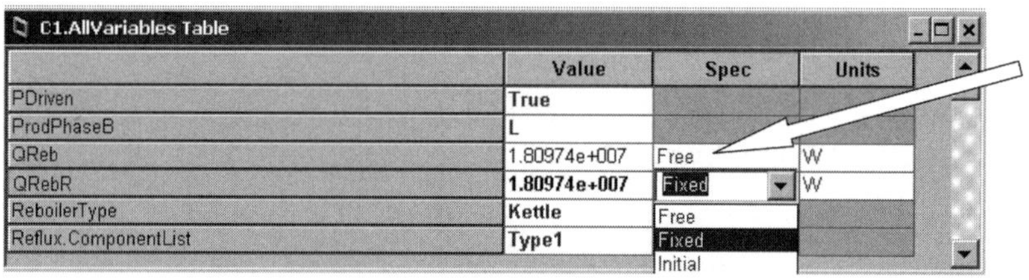

C1.AllVariables Table

	Value	Spec	Units
PDriven	True		
ProdPhaseB	L		
QReb	1.80974e+007	Free	W
QRebR	1.80974e+007	Fixed	W
ReboilerType	Kettle	Free	
Reflux.ComponentList	Type1	Fixed	
		Initial	

Figure 8.17 Changing Q_{Reb} to *Free.*

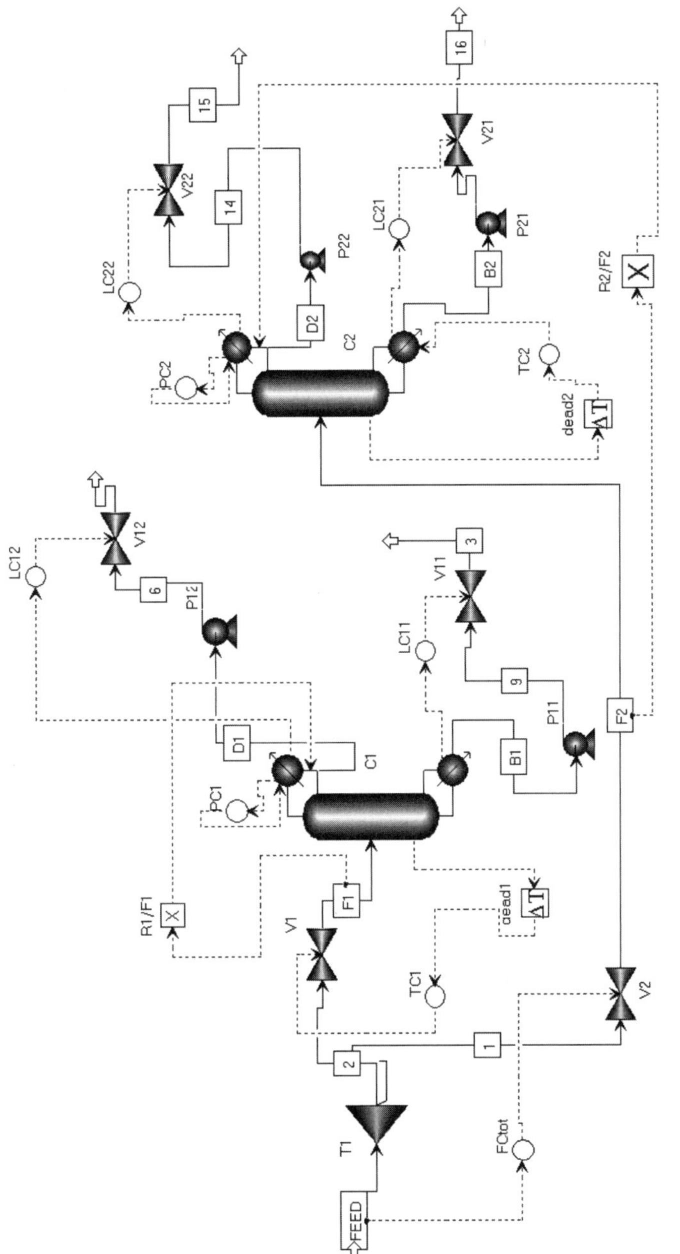

Figure 8.18 Flowsheet with control structure.

1-min deadtime in this loop and a temperature transmitter range of 300–400 K, the resulting controller tuning constants are $K_C = 0.535$ and $\tau_I = 33$ min. These numbers indicate that control will be fairly slow, which is to be expected in this highly interactive system.

The only remaining issue is how to handle reflux flow in both columns. Steady-state runs in Aspen Plus with product purities fixed were made over a range of feed compositions. Results showed that the constant reflux-to-feed structure is better than constant reflux ratio. Therefore, these R/F ratios are installed on both columns. The R/F mass ratio in the first column is 0.42557 and in the second is 0.67045.

8.3.2 Response of Single-End Temperature Control Structure

The responses of this highly interactive system to 20% disturbances in total feed flowrate are shown in Figure 8.19. The control structure provides stable base-level regulatory control. Figure 8.19a shows that flow and temperature responses take about 2 h to settle out.

The effect of the step increase in the setpoint of the total feedflow controller is an immediate increase in F2. This causes a decrease in temperature in column C2, and the temperature controller increases reboiler heat input. This increases the vapor boilup in column C1, which increases stage 26 temperature. The TC1 temperature controller then brings in more feed F1.

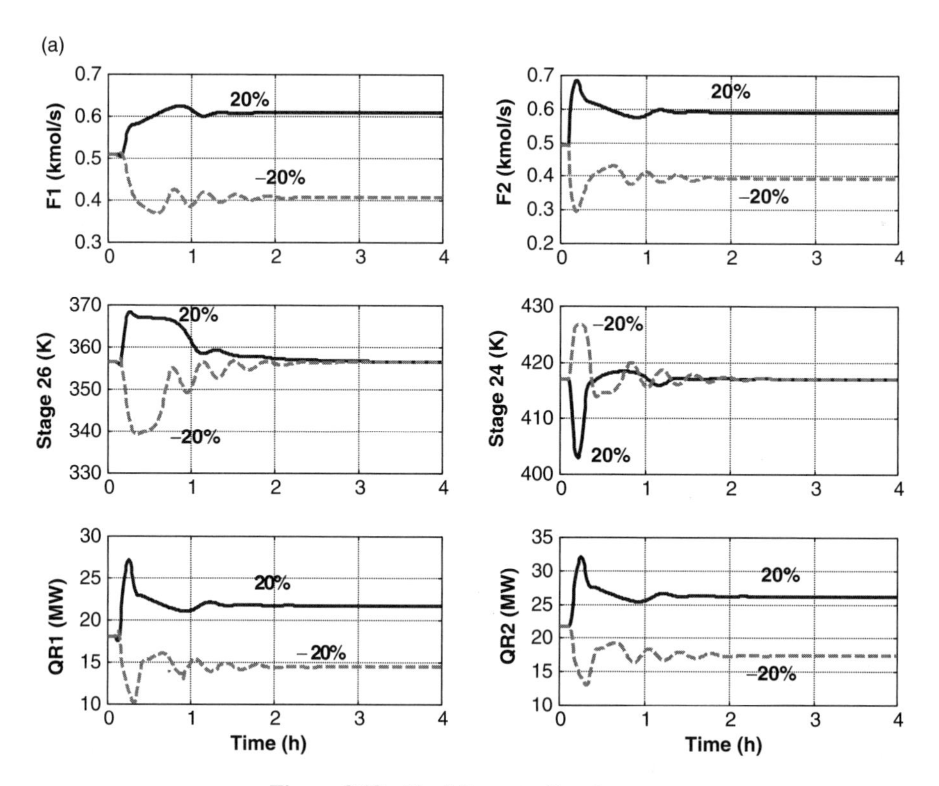

Figure 8.19 Feed flowrate disturbances.

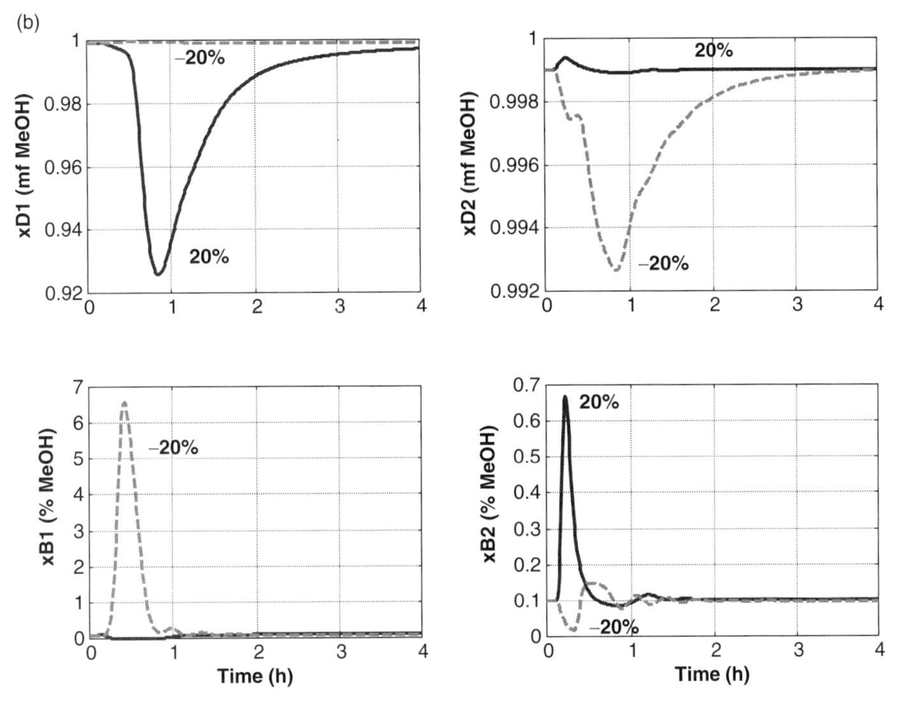

Figure 8.19 *Continued.*

As shown in Figure 8.19b, the distillate and bottoms product purities return to values close to their specification levels. However, there are very large dynamic drops in product purities. An increase in feed flowrate causes large drops in distillate purities of the distillate in column C1 and the bottoms in column C2. The reverse effect occurs for a decrease in feed flowrate.

We would expect that the deviations in the C2 bottoms purity could be improved by using a steam to feed ratio in this column. However, this might cause a more rapid increase in heat input in column C1, which could make the deviations in the distillate purity of column C1 even worse. This control structure is shown in Figure 8.20.

Setting up the ratio requires converting the two signals to metric units. The steady-state value of F2 is 0.49137 kmol/s (1768.9 kmol/h), and this is *Input1* to the multiplier. The steady-state value of reboiler heat input is 21.80 MW (78.48 GJ/h), so this should be the output of the multiplier. The value of *Input2* to the multiplier is calculated to be 0.04437. Change the initial value of the temperature controller output to 0.04437, and specify the output range of the temperature controller to be 0–0.1.

The TC2 controller is retuned with TC1 on manual, giving $K_C = 1.06$ and $\tau_I = 9.2$ min. Then TC1 is retuned with TC2 on automatic, giving $K_C = 1.16$ and $\tau_I = 14.5$ min, which is quite different from the previous values of $K_C = 0.535$ and $\tau_I = 33$ min. These results indicate that much tighter control is possible with the steam to feed ratio.

The performance of this control structure for 20% step changes in feed flowrate are given in Figures 8.21a and 8.21b. The improvement is very striking. The peak transient deviations in product purities are reduced by an order of magnitude. However, the performance of the system for feed composition disturbances is not very good.

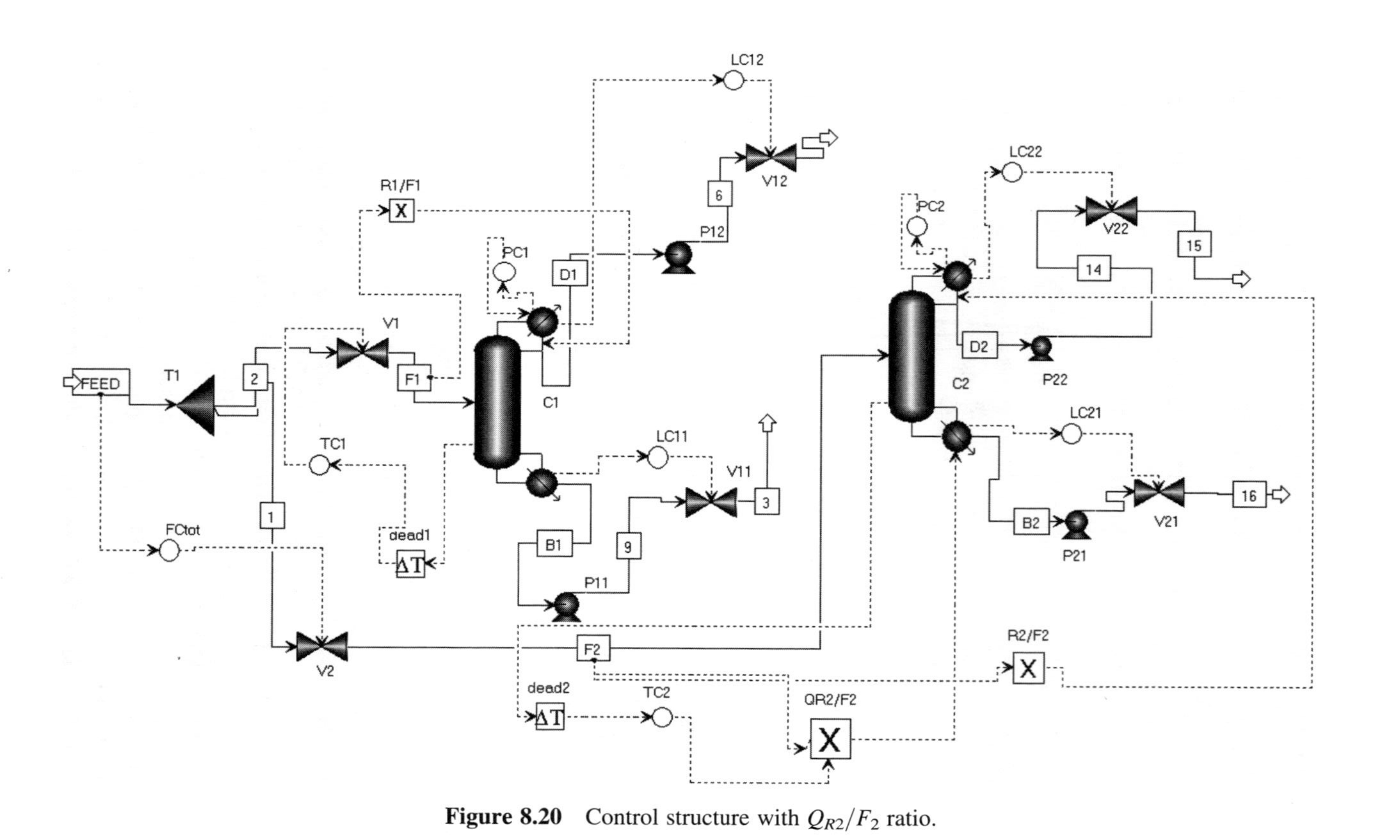

Figure 8.20 Control structure with Q_{R2}/F_2 ratio.

Figure 8.21c shows what happens to product purities for two types of feed composition changes. In the first, feed composition is changed from 60 mol% methanol to 70 mol%. In the second, the feed composition is changed to 50 mol%. This second disturbance causes the impurities of the distillate and bottoms streams from both columns to climb significantly higher than the desired 0.1 mol%. The problem is a steady-state one, not a dynamic one. The fixed R/F structure with single-end temperature control does not maintain product purities in the face of large feed composition disturbances.

If the process is subject to large and frequent feed composition disturbances and if tight control of product quality is required, it may be necessary to go to a dual-temperature-control structure.

8.3.3 Dual Temperature Control Structure

If dual-temperature control is to be used in both columns, we need to select two appropriate trays in each. The SVD method is used to guide in this selection. Figure 8.22 gives the steady-state gains and SVD results for both columns. In the high-pressure column C2, the manipulated variables are the conventional ones: reboiler heat input and reflux flowrate. However, in the low-pressure column, the manipulated variables are reflux flowrate and feed flowrate. The steady-state gains are obtained by making 0.1% changes in the inputs and calculating the changes of the temperatures on all trays.

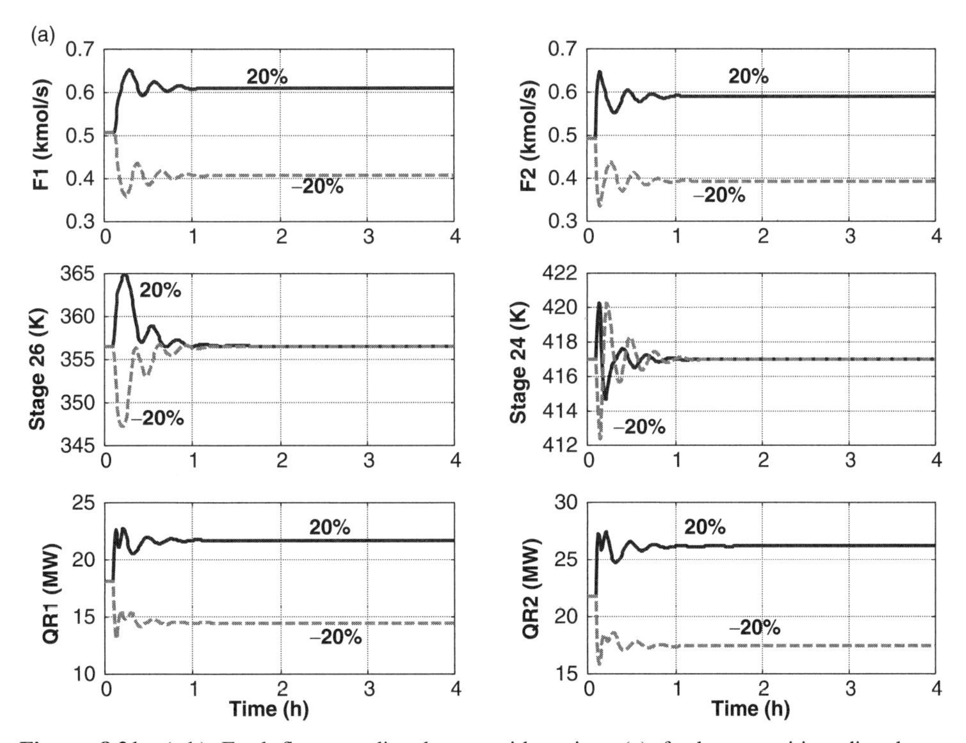

Figure 8.21 (a,b) Feed flowrate disturbance with ratios; (c) feed composition disturbance with ratio.

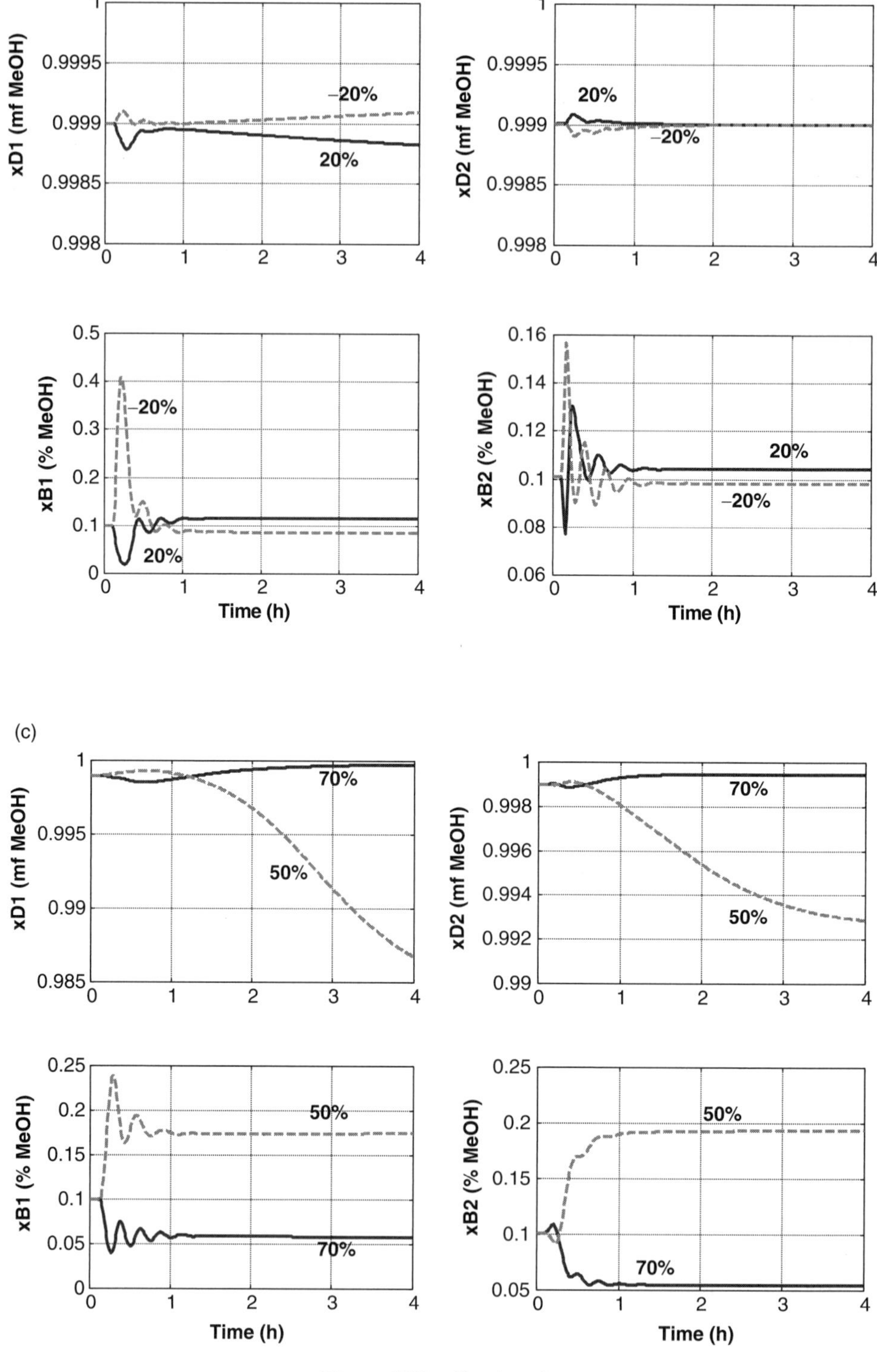

Figure 8.21 *Continued.*

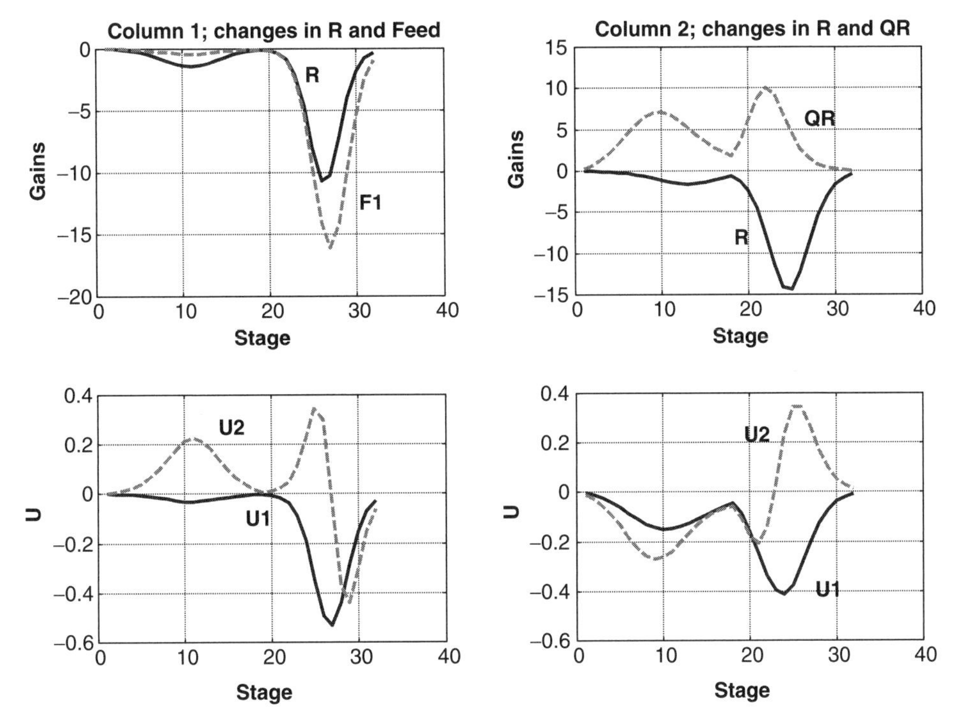

Figure 8.22 Gains and SVD; heat-integrated columns C1 and C2.

For the low-pressure column C1, the SVD results suggest the use of stages 11 and 26. The temperature on stage 26 is controlled by manipulating the feedstream F1 using a temperature controller labeled "TC126." This controller has already been tuned in the single-end control structure. The temperature on stage 11 is controlled by manipulating the reflux to feed ratio using a temperature controller labeled "TC111." This controller was tuned with the TC126 controller on automatic, using a relay–feedback test. The resulting controller tuning constants were $K_C = 6.1$ and $\tau_I = 28$ min. The resulting response was found to be too underdamped, so the gain was cut to 3.

For the high-pressure column C2, the SVD results suggest the use of stages 10 and 24. The temperature on stage 24 is controlled by manipulating the reboiler heat input using a temperature controller labeled "TC224." This controller has already been tuned in the single-end control structure. The temperature on stage 10 is controlled by manipulating the reflux to feed ratio using a temperature controller labeled "TC210." This controller is tuned with the TC224 controller on automatic, using a relay–feedback test. The resulting controller tuning constants are $K_C = 5.1$ and $\tau_I = 15$ min.

Figure 8.23 gives the complete dual-temperature control structure. The two-column heat-integrated process has a fairly complex control structure. There are four temperature controllers and three ratios, in addition to the conventional four level controllers, two pressure controllers, and a total feedflow controller.

The performances of this structure for feed composition disturbances ranging from 60 to 70 mol% methanol and from 60 to 50 mol% methanol are shown in Figure 8.24. A comparison with the single-temperature control results given in Figure 8.21c shows a very significant improvement. Product purities are maintained much closer to their specification levels.

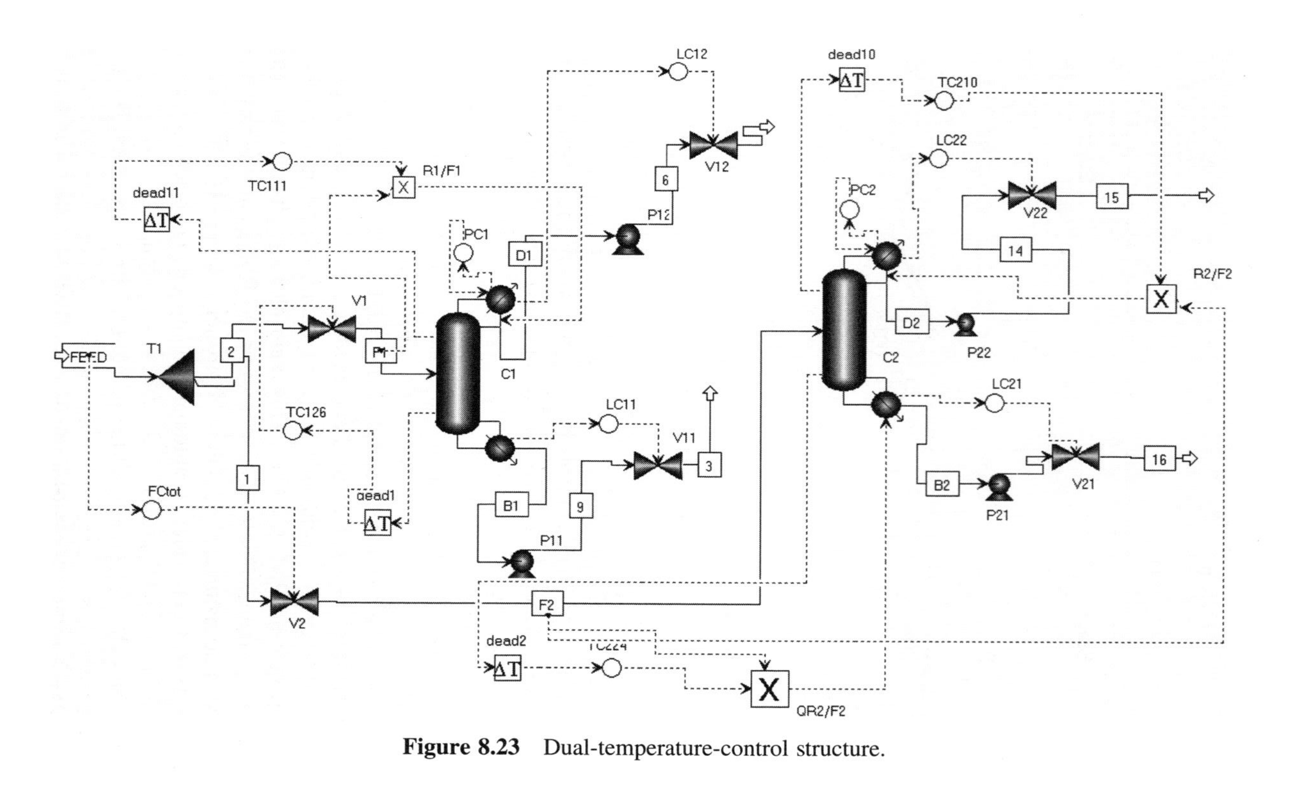

Figure 8.23 Dual-temperature-control structure.

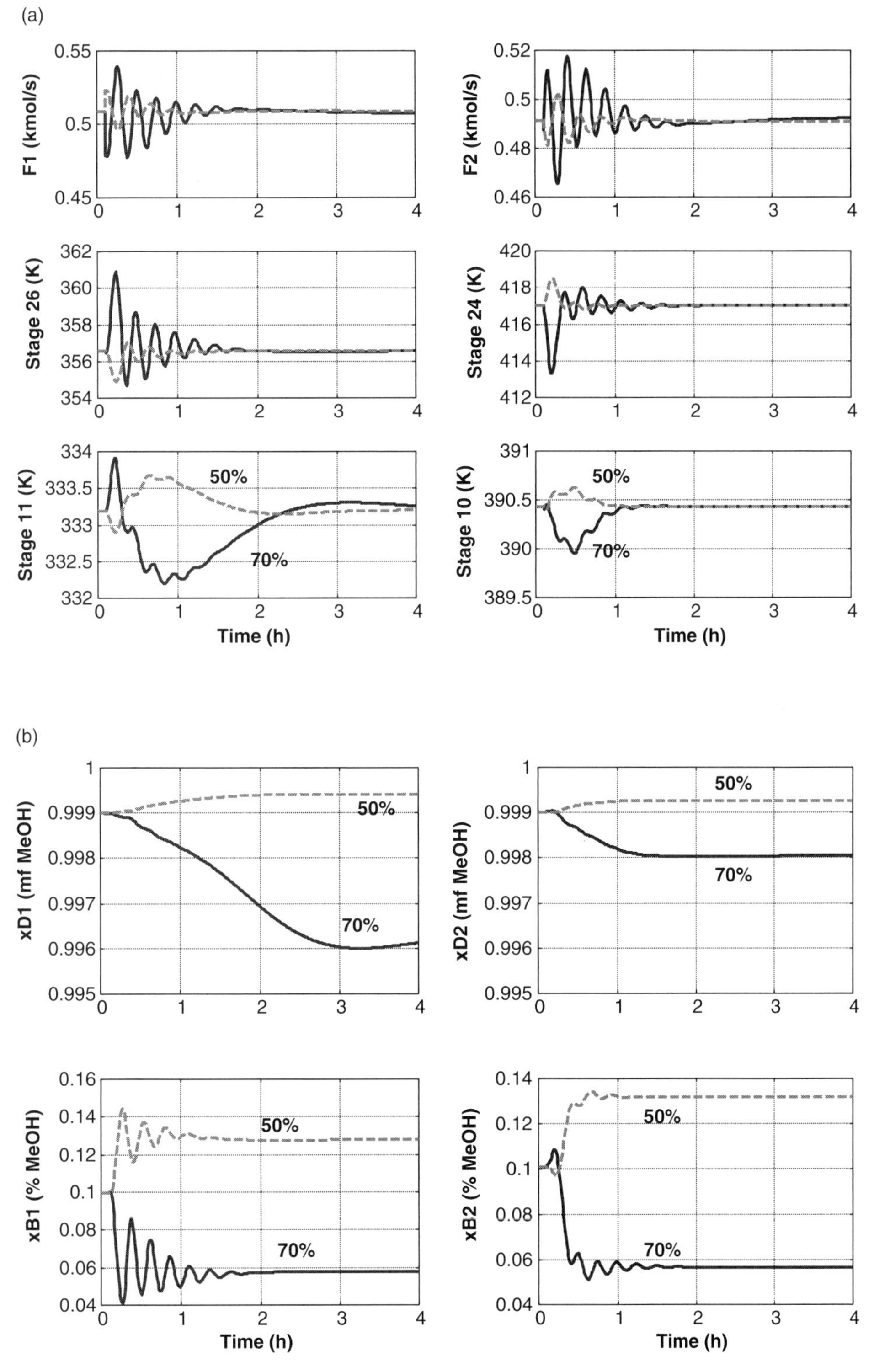

Figure 8.24 Feed composition disturbance with dual temperature.

8.3.4 More Rigorous Simulation

The modeling of the two heat-integrated column discussed above assumes that the pressure controller in the high-pressure column can freely manipulate the heat removal rate in the condenser/reboiler heat exchanger. This is rigorously not true because the heat transfer area is fixed, and the heat transfer rate really depends on the temperature difference between the reflux drum of the high-pressure column and the base of the low-pressure column.

A more rigorous dynamic simulation of this system could be fairly easily put together by using *Flowsheet Equations* in Aspen Dynamics. The heat-transfer rate could be calculated using the area A_{HX}, overall heat transfer coefficient U, and the differential temperature (temperature on stage 1 in the high-pressure column minus the temperature on stage 32 in the low-pressure column). This would fix Q_{C2} and Q_{R1}, which must be equal but of opposite sign. No pressure controller would be used on the high-pressure column. Its pressure would float. The low-pressure column would have the normal pressure controller manipulating condenser heat removal.

The heat-integrated process provides an excellent example of the power and usefulness of dynamic simulation of distillation column systems. Alternative control structures can be easily and quickly evaluated.

8.4 CONTROL OF AZEOTROPIC COLUMNS/DECANTER SYSTEM

The dehydration of ethanol using benzene as a light entrainer was studied in Chapter 5. The process consisted of two distillation columns, one decanter, and two recycle streams. One of the recycle streams was successfully closed, but the second would not converge using steady-state Aspen Plus.

In this section we demonstrate how this second recycle loop can be successfully converged in Aspen Dynamics. A plantwide control structure is developed and its effectiveness evaluated. A very counterintuitive level control loop is shown to be required for stable operation.

8.4.1 Converting to Dynamics and Closing Recycle Loop

The usual base and reflux drum sizing procedure gives column base diameters of 1.76 and 0.855 m, respectively, in columns C1 and C2. Assuming as aspect ratio (L/D) of 2 gives the lengths. There is no reflux drum in the first column. The decanter is sized to provide 20 min of holdup based on the total liquid entering in the two liquid phases (aqueous and organic). The resulting decanter has a diameter of 2 m and a length of 4 m. Horizontal orientation is specified to aid in the phase separation. With the equipment sized and the flowsheet pressure checked, it is exported into Aspen Dynamics. Figure 8.25a shows the initial flowsheet that opens. Pressure controllers are installed on both columns. In column C1, pressure is held by manipulating the position of valve V12 in the overhead vapor line upstream of the condenser. Note that the Reflux recycle loop is not closed.

As discussed in Chapter 5, the column C1 is very sensitive to the amount of organic reflux. If too much is fed, the benzene comes out the bottom with the ethanol product.

If too little is fed, the water is not entrained overhead and comes out the bottom with the ethanol. Therefore, the organic reflux is selected to control the temperature on stage 28 in column C1, where there is a sharp change in the temperature profile. We know that the temperature control will not be very tight because of the liquid hydraulic lag between a change in reflux and a tray temperature 27 trays down the column. This controller is added before the reflux recycle loop is closed, as shown in Figure 8.25b. An *Initialization* run is made to make sure that everything is running okay.

Now we are ready to close the recycle loop. The block "V1" and the streams "REFLUX2" and "ORGREC" are deleted from the flowsheet. The "REFLUX" stream is selected and right-clicked. Selecting *Reconnect Source*, this stream is attached to the mixer "M1." The little red light appears at the bottom of the window. Double-clicking opens the window shown in Figure 8.26a, telling us that the simulation is overspecified by two variables. Double-clicking the *Analyze* button opens Figure 8.26b, where changing temperature of the reflux stream from *Fixed* to *Free* is suggested. This makes sense because the reflux stream is now coming from the organic liquid phase in the decanter. The other suggestion is to make a change in the temperature controller, which is not reasonable. What is reasonable is to change the pressure of the reflux from *Fixed* to *Free*. This is done selecting the stream "REFLUX," right-clicking, selecting *Forms* and then *All Variables*. Scroll down to *P* and change to *Free*, as shown in Figure 8.26c. The green light appears at the bottom, indicating that the simulation is now "square" (as many variables as equations, i.e., 0 degrees of freedom). An *Initialization* run and a *Dynamics* run are made to check that the integrator is working okay with both recycle streams connected.

8.4.2 Installing the Control Structure

The other controllers are now added in the usual way. Base levels are held by bottoms flowrate. Reflux drum level in column C2 is held by the distillate flowrate (the RECYCLE stream back to column C1). The tray with the sharpest temperature change in column C2 is stage 20 (365.14 K at steady state). The temperature controller TC2 manipulates reboiler heat input. A third temperature controller is added on the heat exchanger before the decanter to control the temperature of the stream entering the decanter.

The control of the two liquid inventories in the decanter is critical. Since only a very small amount of benzene is lost, the organic level basically floats up and down as changes occur in the reflux flowrate. An organic phase level controller adjusts the benzene makeup stream, but it is so small that the level changes are significant. This does not hurt anything as long as the decanter does not overfill or the organic level is lost.

The control of the aqueous level would appear to be straightforward. A level controller would manipulate the valve "VD1." Conventional wisdom says that this controller is *Direct*-acting. If the level goes up, more is fed to column C2. Very surprisingly, this setup was found to **not** work. The system shut down. However, making the controller *Reverse*-acting produced a stable control structure. We demonstrate this in the next section.

Figure 8.27 shows the final control structure. Features not mentioned include ratioing both the reboiler heat and the reflux flowrate input in column C1 to the feed flowrate. The temperature controller TC1 in column C1 changes the reflux to feed ratio.

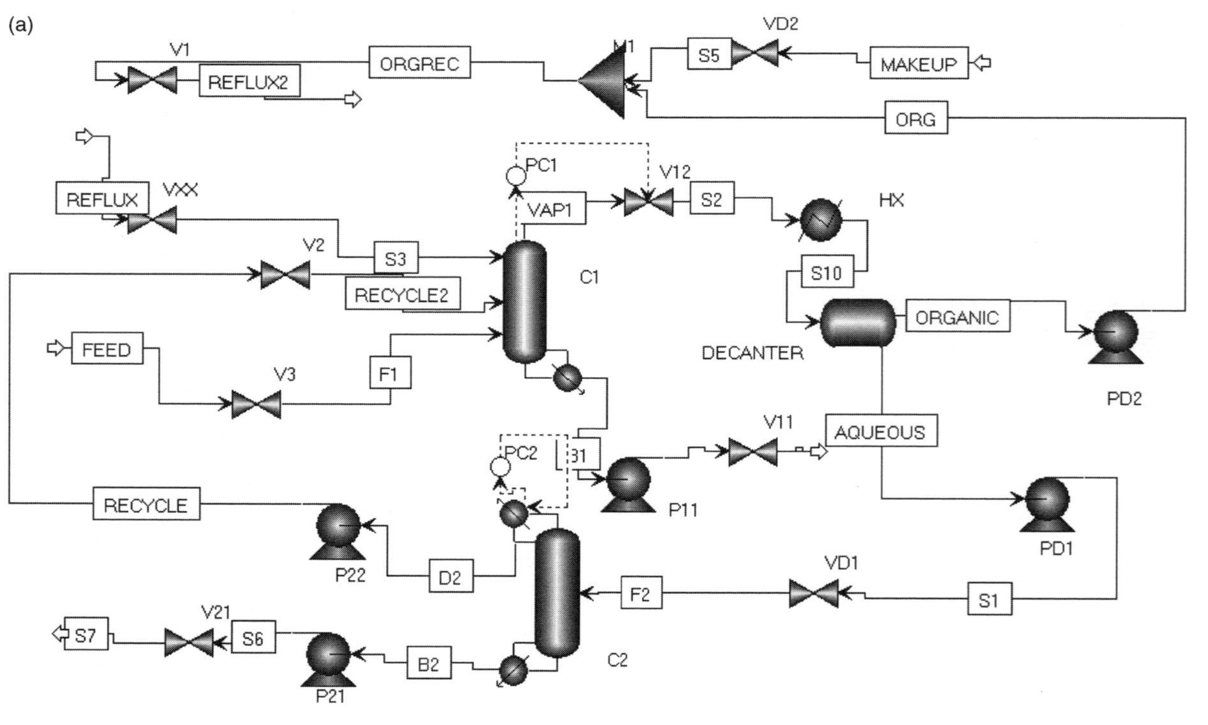

Figure 8.25 (a) Initial flowsheet in Aspen Dynamics; (b) temperature T_{C1} added.

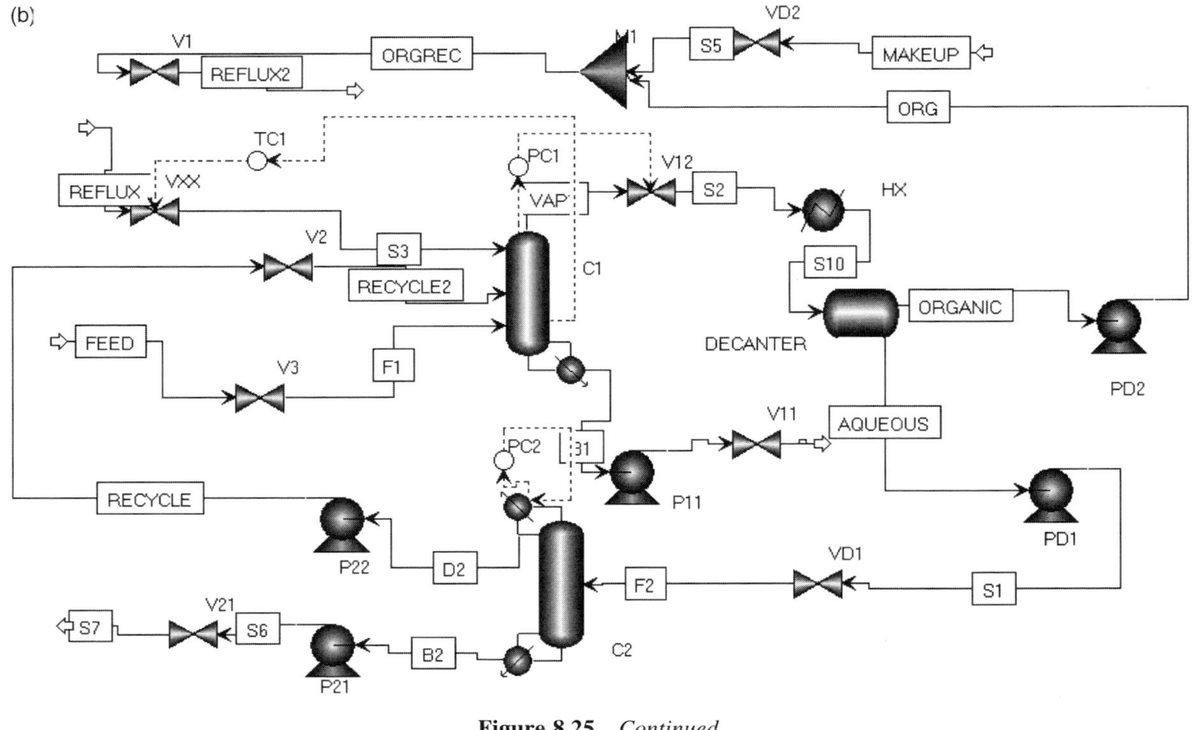

Figure 8.25 *Continued.*

(a)

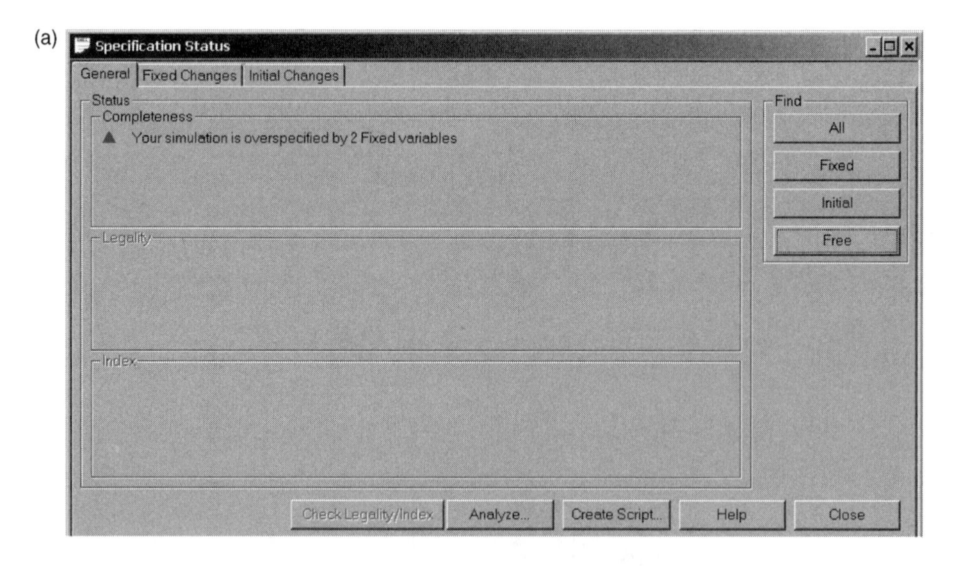

(b)

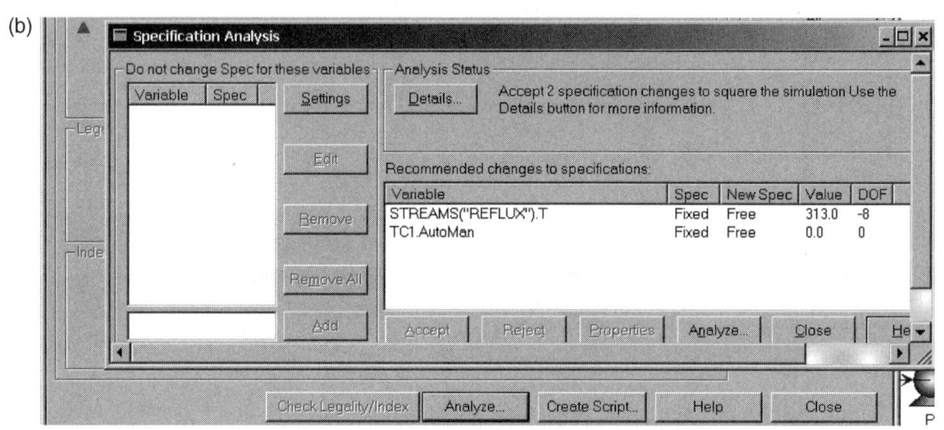

(c)

REFLUX.AllVariables Table

	Value	Spec	Units
Out_P.V	0.084419	Free	m3/kmol
Out_P.z("BENZE-01")	0.823	Free	kmol/kmol
Out_P.z("ETHAN-01")	0.159	Free	kmol/kmol
Out_P.z("WATER")	0.018	Free	kmol/kmol
Out_P.Zrev("BENZE-01")	0.333333	Free	kmol/kmol
Out_P.Zrev("ETHAN-01")	0.333333	Free	kmol/kmol
Out_P.Zrev("WATER")	0.333333	Free	kmol/kmol
P	506625.0	Free	
PassThrough	**Normal**		
PDriven	**True**		
Rho	11.853	Free	kmol/m3
Rhom	852.666	Free	kg/m3
SensorActive	**No**		
Solvent			

Paused | local | Dynamic at 1.26 Hours

Figure 8.26 (a) Simulation overspecified; (b) suggested changes; (c) changing reflux pressure to *Free*.

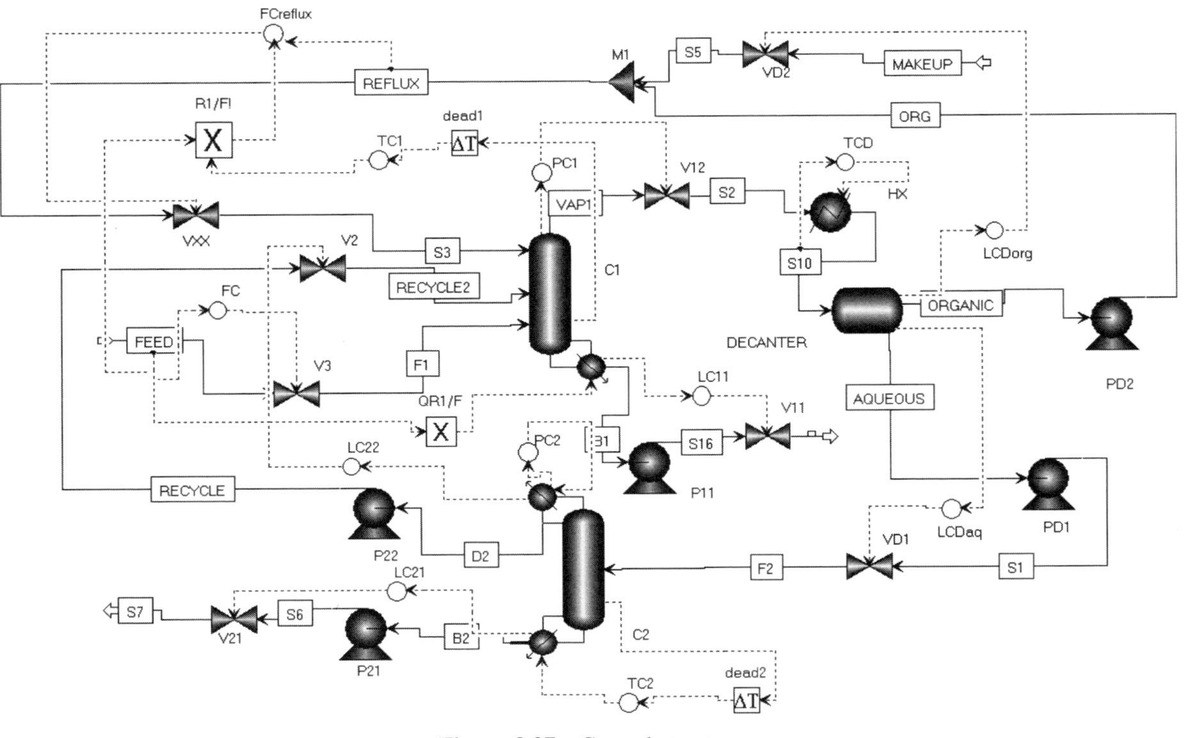

Figure 8.27 Control structure.

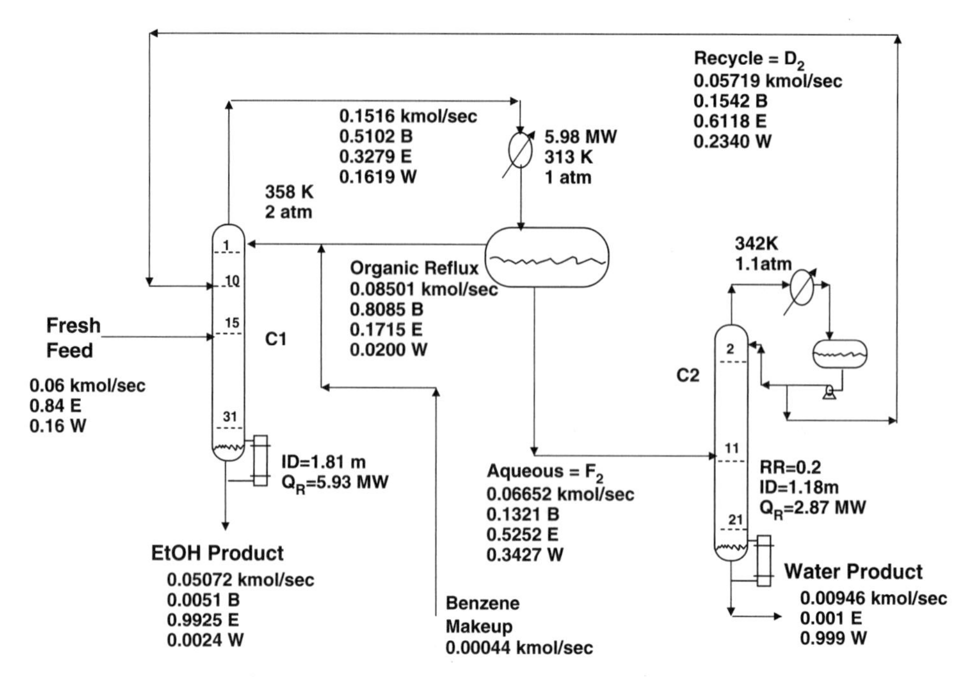

Figure 8.28 Steady state from dynamic simulation.

Deadtimes of 1 min are included in both column temperature controllers, and relay–feedback tests are run to find controller settings. In TC1 they are $K_C = 1.15$ and $\tau_I = 44$ min (using a temperature range of 300–400 K and an output range of 0–2). As expected, this control is fairly slow because of the liquid hydraulic lags. In TC2 the controller settings are $K_C = 0.707$ and $\tau_I = 9.2$ min (using a temperature range of 300–400 K and an output range of 0–5.84 MW).

With this control structure in place, the simulation is run out to a steady state. Figure 8.28 gives the conditions, which are quite close to those found in Chapter 5.

8.4.3 Performance

The first thing to show is that a direct-acting aqueous level controller does not work. Figure 8.29a shows what happens. There is no disturbance. The controller is simply switched from *Reverse* to *Direct.* The system shuts down in about 2 h.

In contrast, Figure 8.30a shows the responses of the system when a reverse-acting aqueous level controller is used. The disturbances are positive and negative 20% step changes in the setpoint of the feed flow controller. Stable regulatory control of the highly nonideal distillation system is achieved. Both the ethanol and the water products are kept close to their desired purity levels.

Further evidence of the counterintuitive response of the decanter is given in Figure 8.29b. The aqueous level controller is placed on manual, and its controller output is increased from 53.49% to 55%. The flowrate of the aqueous stream increases. But the aqueous level **increases**!

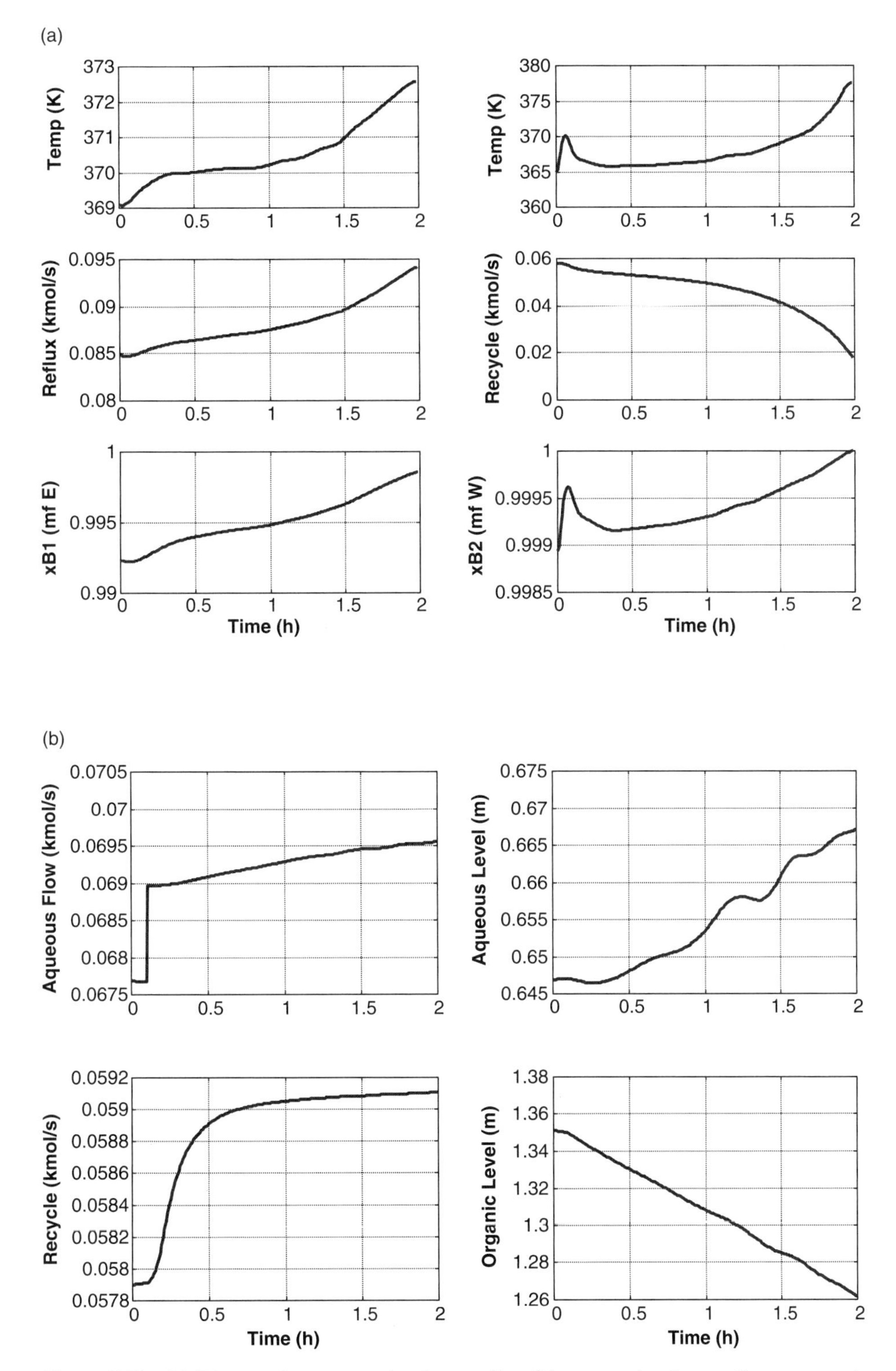

Figure 8.29 (a) Direct-acting aqueous level controller; (b) aqueous level controller on manual.

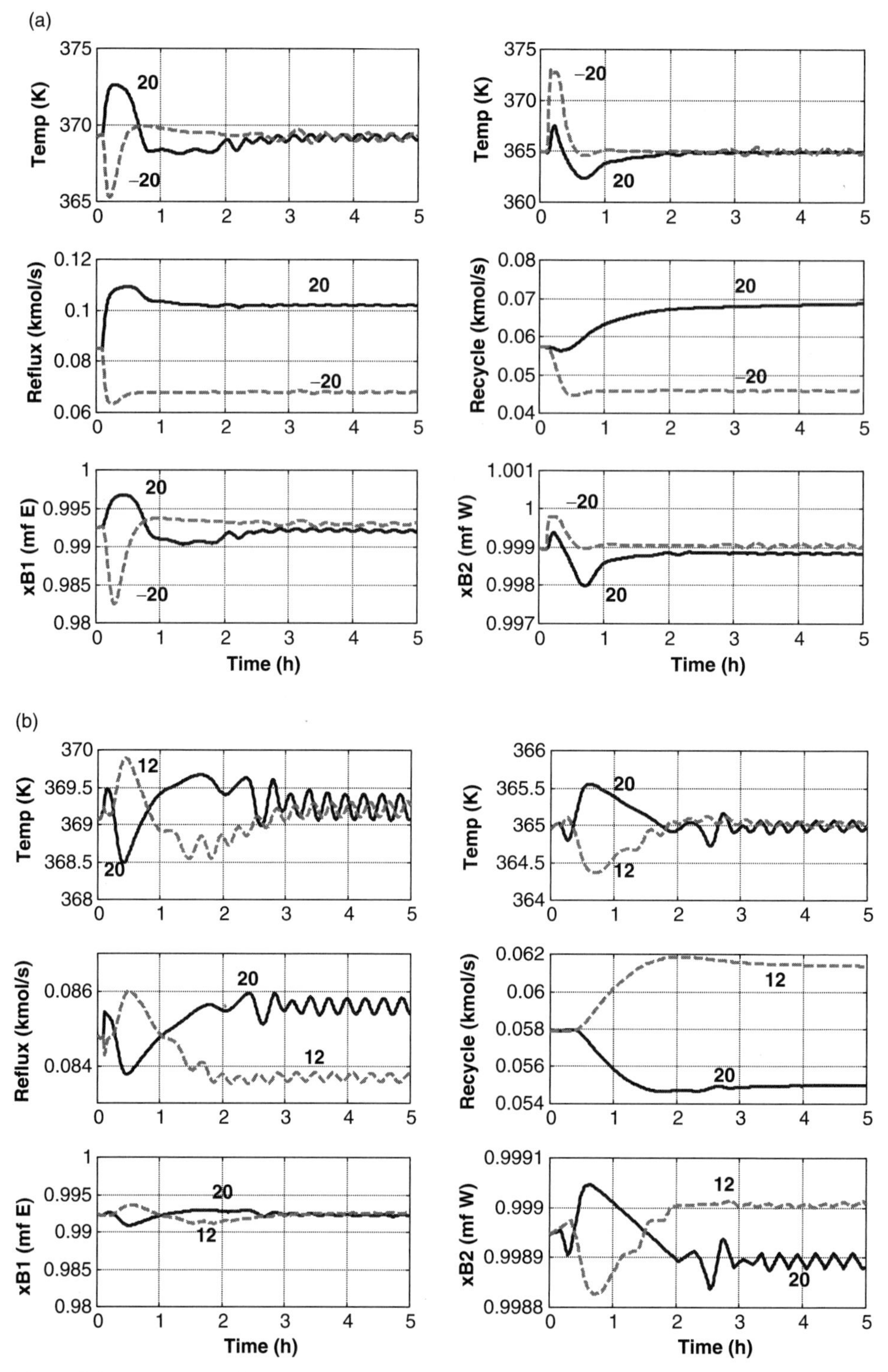

Figure 8.30 Reverse-acting aqueous level controller: (a) feed rate changes; (b) feed composition changes.

This counterintuitive behavior and the unconventional action required in the aqueous level controller has been observed in real plant operation.

Figure 8.30b shows responses to changes in feed composition from 16 to 20 mol% water and from 16 to 12 mol% water, with the corresponding changes in ethanol. Both the ethanol and the water products are kept close to their desired purity levels. More water in the feed requires more organic reflux and produces less recycle.

This study provides an excellent example of the utility of distillation simulation in both designing and controlling a very complex nonideal system.

8.5 CONCLUSION

Several complex distillation systems have been examined in this chapter. The complexity can arise in either the vapor–liquid equilibrium or the configuration of multiple interconnected units.

CHAPTER 9

REACTIVE DISTILLATION

9.1 INTRODUCTION

Reactive distillation columns incorporate both phase separation and chemical reaction. They have economic advantages in some systems over conventional reactor, separation, and/or recycle flowsheets, particularly for reversible reactions in which chemical equilibrium constraints limit conversion in a conventional reactor. Because both reaction and separation occur in a single vessel operating at some pressure, the temperatures of reaction and separation are not independent. Therefore, reactive distillation is limited to systems in which the temperatures conducive to reaction are compatible with temperatures conducive for vapor–liquid separation.

Pressure in conventional distillation design is usually set by a minimum temperature in the reflux drum (so that cooling water can be used) or a maximum temperature in the reboiler (to prevent fouling or thermal decomposition). Establishing the optimum pressure in a reactive distillation column is more complex because of the interplay between reaction and phase separation. Most VLE relationships show an increase in volatility with decreasing temperature. On the other hand, reaction rates decrease with decreasing temperature. If the reaction is exothermic, the chemical equilibrium constant increases with decreasing temperature. So low operating pressure or temperature, which facilitates the phase separation, may require lots of catalyst or liquid tray holdup to compensate for the low reaction rates.

In conventional distillation design, tray holdup has no effect on steady-state composition. In reactive distillation, tray holdup (or amount of catalyst) has a profound effect on conversion, product composition, and column composition profiles. So, in addition to the normal design parameters of reflux ratio, number of trays, feed tray location, and

Distillation Design and Control Using Aspen™ Simulation, By William L. Luyben

pressure, reactive distillation columns have the additional design parameter of tray holdup. If there are two reactant feedstreams, an additional design parameter is the location of the second feed.

Reactive distillation is usually applied to systems in which the relative volatilities of the reactants and products are such that the products can be fairly easily removed from the reaction mixture while keeping the reactants inside the column. For example, consider the classical reactive distillation system with reactants A and B reacting to form products C and D in a reversible reaction.

$$\mathrm{A} + \mathrm{B} \Longleftrightarrow \mathrm{C} + \mathrm{D}$$

For reactive distillation to be effective, the volatilities of the products C and D should be greater or less than the volatilities of the reactants A and B. Suppose the volatilities are

$$\alpha_{\mathrm{C}} > \alpha_{\mathrm{A}} > \alpha_{\mathrm{B}} > \alpha_{\mathrm{D}}$$

Reactant A would be fed into the lower section of a reactive column and rise upward. Reactant B would be fed into the upper section and flow downward. As the components react, product C would be distilled out the top of the column, and product D would be withdrawn out the bottom. The reactants can be retained inside the column by vapor boilup and reflux while the products are removed. Figure 9.1 illustrates this ideal case.

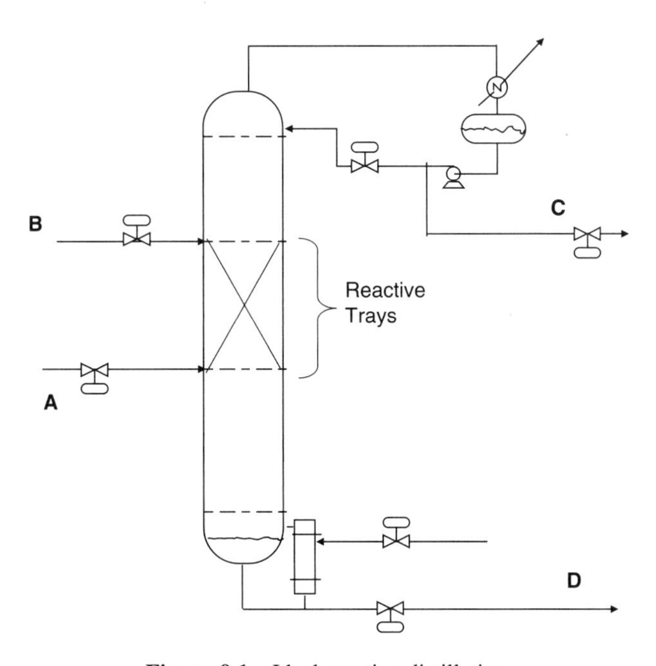

Figure 9.1 Ideal reactive distillation.

9.2 TYPES OF REACTIVE DISTILLATION SYSTEMS

There are many types of reactive distillation systems because several types of reactions are carried out in reactive columns. There are also several types of process structures that are used, some with recycle of an excess reactant and others without any reactant recycle.

9.2.1 Single-Feed Reactions

Reactions with a single reactant producing two products are easy to design and control because there is no need to balance the stoichiometry:

$$A \Longleftrightarrow B + C$$

Only one reactant is fed to the column. The two products are removed out the two ends of the column. Olefin metathesis is an example of this type of reactive distillation column. Figure 9.2 illustrates this system and gives an effective control scheme. A C_5 olefin reacts to form a light C_4 olefin, which is removed in the distillate, and a heavy C_6 olefin, which is removed in the bottoms. The two temperature controllers are used to maintain conversion and product quality. The production rate is set by a feed flow controller.

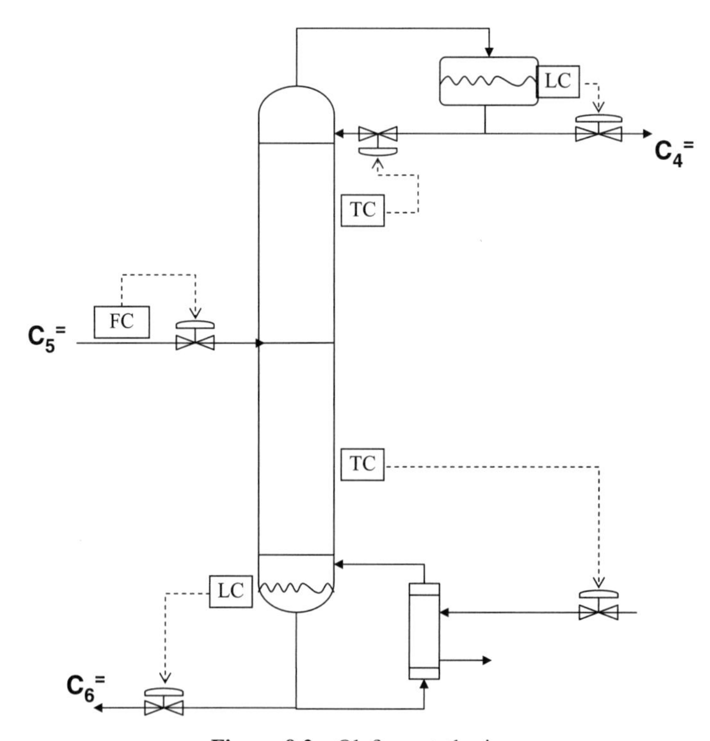

Figure 9.2 Olefin metathesis.

9.2.2 Irreversible Reaction with Heavy Product

The ethylene glycol reactive distillation system is an example of a reactive distillation system with two reactants that are consumed in a fast and irreversible reaction.

$$\text{Ethylene oxide} + \text{water} \longrightarrow \text{ethylene glycol}$$

Figure 9.3 shows the system and an effective control structure. Ethylene oxide is very volatile, and ethylene glycol is very heavy, so the product is removed from the bottom of the column. The ethylene oxide concentrates in the top of the column. **No** distillate product is removed. The water feed is introduced to hold the liquid level in the reflux drum. This level loop achieves the necessary balancing of the reaction stoichiometry by adjusting the makeup water flowrate to exactly match the water consumption by reaction with ethylene oxide. Production rate is set by flow-controlling the ethylene oxide.

9.2.3 Neat Operation versus Use of Excess Reactant

If the reaction involves two reactant feedstreams, two basic flowsheets that are used. Consider the reaction $A + B \Longleftrightarrow C + D$. One way to design the process is to feed an excess of one of the reactants into the reactive distillation column along with the other reactant. Figure 9.4 shows a system in which an excess of reactant B is fed. In most cases this excess must be recovered. A second distillation column is used in Figure 9.4 to achieve this recovery. The fresh feed of reactant B is mixed with the recycle of B coming from the recovery column.

The control of this system is fairly easy. The total flow of B to the reactive column is controlled by manipulating the fresh feed of reactant B. The fresh feed of reactant A sets

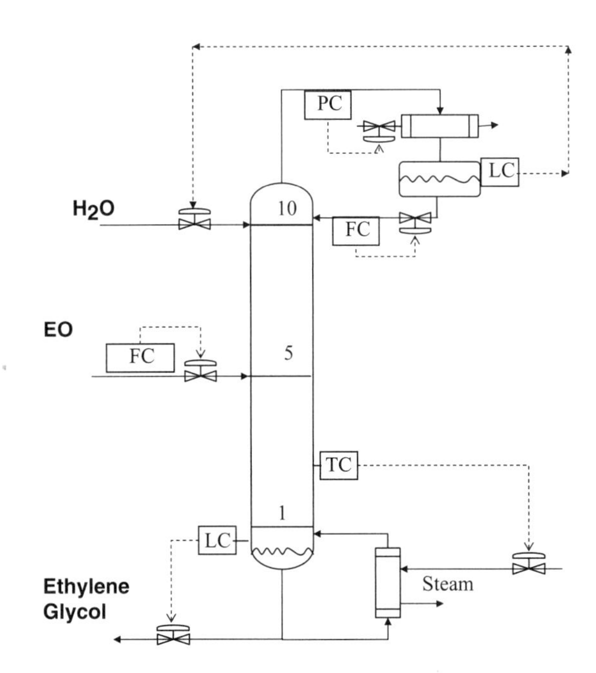

Figure 9.3 Ethylene glycol.

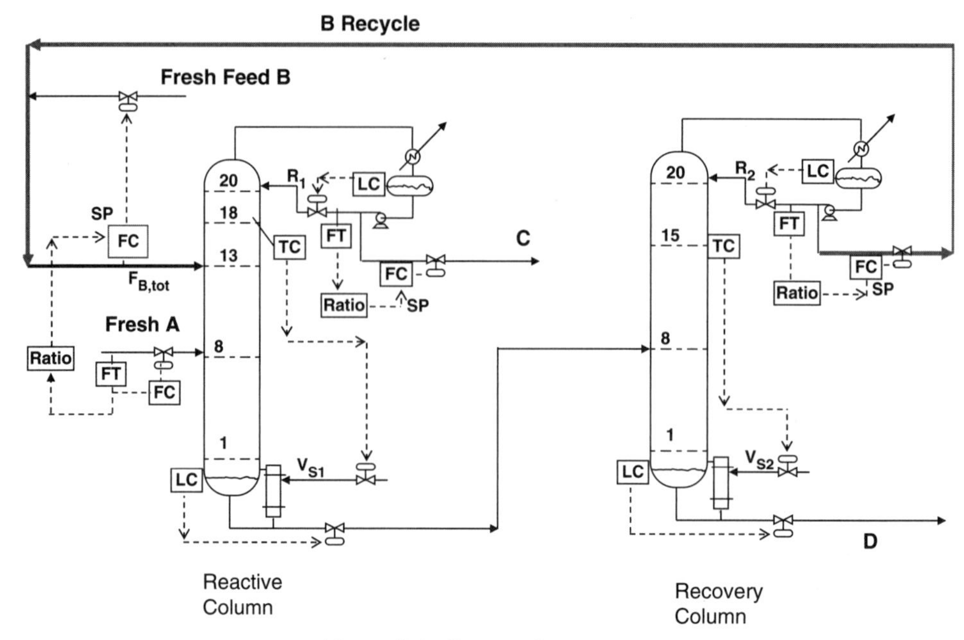

Figure 9.4 Excess of reactant B.

the production rate, and the setpoint of the total B flow controller is ratioed to the flowrate of A. The control scheme features reflux ratio control and temperature controllers in both columns.

The alternative flowsheet used just one column and is more economical, but it presents a much more difficult control problem. The operation is "neat," meaning that the amounts of the two reactants fed are exactly balanced so as to satisfy the reaction stoichiometry. Figure 9.5 illustrates the system for the case when the reaction produces two products ($A + B \Longleftrightarrow C + D$), which go out the two ends of the column. The two temperature controllers achieve the balancing of the reactants. With two products, the column temperature information can be used to detect whether more or less of each reactant should be fed. The production of methyl acetate and water from methanol and acetic acid is an example of this type of reactive distillation system.

A commonly committed error in these two-reactant feed systems is to assume that a control structure with one feed ratioed to the other will provide effective control. This scheme does not work because of inaccuracies in flow measurements and changes in feed composition. Remember in neat operation that the reactants must be balanced down to the last molecule. This can be achieved only by using some sort of feedback information from the process that indicates a buildup or depletion of reactant.

However, consider the case when there is only one product: the reaction $A + B \Longleftrightarrow C$. Now the column temperature information is not rich enough to use to balance the stoichiometry. This means that the measurement and control of an internal column composition must be used in this neat operation. An example of this type of system is shown in Figure 9.6. The production of ETBE from ethanol and isobutene produces a heavy product, which goes out the bottom of the column. The C_4 feedstream contains inert

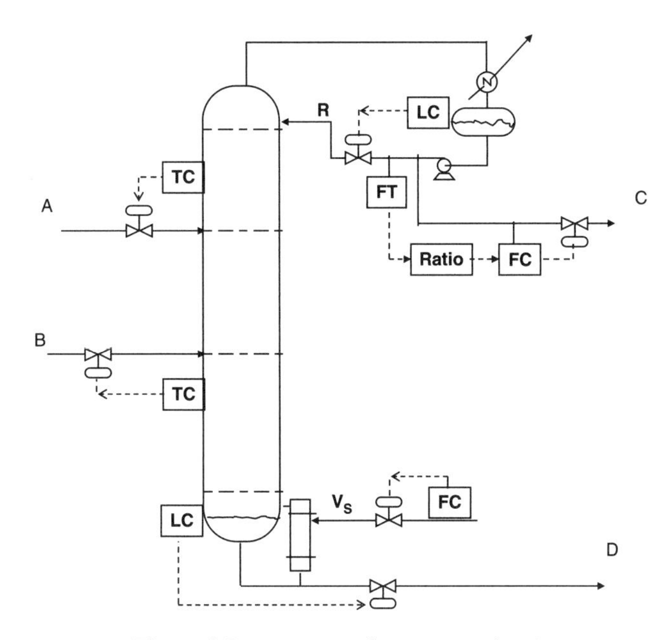

Figure 9.5 Neat operation (two products).

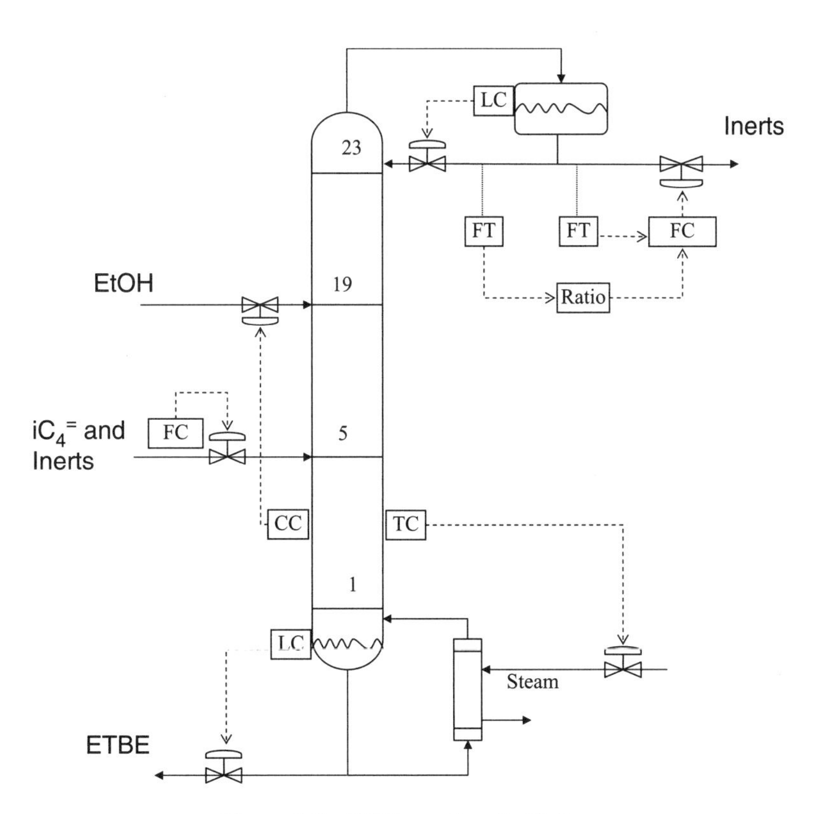

Figure 9.6 ETBE reactive distillation.

components in addition to the isobutene. These inerts go out the top of the column. The production rate is set by the flow controller on the isobutene feedstream. The ethanol concentration on a suitable tray in the column is maintained by manipulating the ethanol fresh feed. Reboiler heat input controls a tray temperature in the stripping section to maintain ETBE product quality.

The *tert*-amyl methyl ether (TAME) reactive distillation system considered in the next section has similar chemistry (two reactants and only one product), and an internal composition controller is required to balance the reaction stoichiometry. There is a recycle stream of one of the reactants in the TAME system, but this occurs because of the existence of azeotropes that carry some of this reactant out of the column with the inerts present in the feed.

9.3 TAME PROCESS BASICS

In this section we study the simulation and control of the *tert*-amyl methyl ether (TAME) process as a typical example of a reactive distillation system. There are two feedstreams: (1) methanol and (2) a mixture of reactants and inert components.

The C_5 feedstream to the TAME process contains about 24 mol% reactive isoamylenes: 2-methyl-1-butene (2M1B) and 2-methyl-2-butene (2M2B). The remaining components are pentanes and pentenes (largely isopentane, iC_5), which are inert in the TAME reaction. TAME is the highest-boiling component, so it leaves in the bottoms stream from the reactive distillation column. The lighter C_5s leave in the distillate stream along with a significant amount of methanol.

Methanol forms minimum boiling azeotropes with many of the C_5s. The reactive column operates at 4 bar, which is the optimum pressure that balances the temperature requirements for reaction with those for vapor–liquid separation. At this pressure, isopentane and methanol form an azeotrope at 339 K that contains 26 mol% methanol. Therefore the distillate from the reactive column contains a significant amount of methanol, which must be recovered.

Since the iC_5/methanol azeotrope is pressure-sensitive (79 mol% iC_5 at 10 bar and 67 mol% iC_5 at 4 bar), it is possible to use a pressure-swing process with two distillation columns, operating at two different pressures, to separate methanol from the C_5 components. An alternative separation process for this system is extractive distillation, which is studied in this chapter.

Figure 9.7 gives the flowsheet of the process. There is a prereactor upstream of the reactive distillation column C1. The flowsheet contains three distillation columns (one reactive), and there are two recycle streams (methanol and water). The design of the prereactor and reactive column are based on a study by Subawalla. and Fair.[1]

9.3.1 Prereactor

The prereactor is a cooled liquid-phase tubular reactor containing 9544 kg of catalyst. The C5 fresh feed (flowrate 1040 kmol/h) and 313 kmol/h of methanol are fed to the reactor.

[1]H. Subawalla and J. R. Fair, Design guidelines for solid-catalyzed reactive distillation systems, *Ind. Eng. Chem. Research* **38**, 3693 (1999).

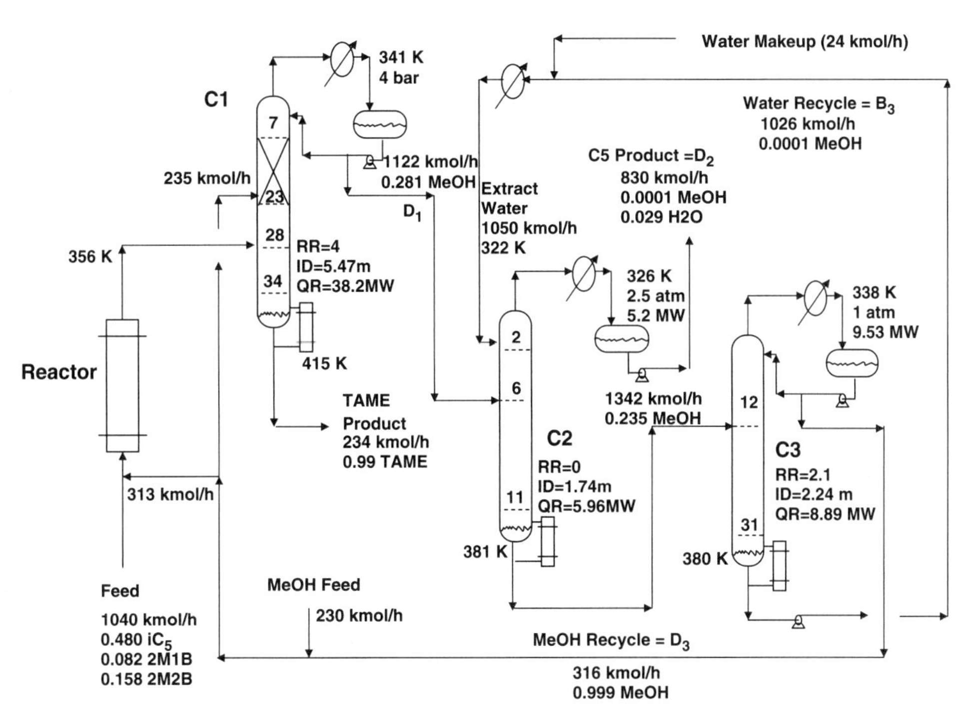

Figure 9.7 TAME process with extractive distillation methanol recovery.

9.3.2 Reactive Column C1

The reactor effluent is fed into a 35-stage reactive distillation column (C1) on stage 28. Catalyst is present on stages 7–23. The reactor effluent is fed five trays below the reactive zone. A methanol stream is fed at the bottom of the reactive zone (stage 23). The flowrate of the methanol fed to the reactive column is 235 kmol/h.

Figure 9.8 gives composition and temperature profiles in the reactive column C1. The reflux ratio is 4, which gives a bottoms purity of 99.2 mol% TAME and a distillate impurity of 0.1 ppm TAME. Reboiler heat input and condenser heat removal are 38.2 and 39 MW, respectively. The operating pressure is 4 bar. The column diameter is 5.5 m. The overall conversion of 2M1B and 2M2B in the C_5 fresh feed is 92.4%. Table 9.1 gives stream information for the prereactor and column C1.

The distillate D_1 has a methanol composition (28 mol% methanol) that is near the azeotrope at 4 bar. It is fed at a rate of 1122 kmol/h to stage 6 of a 12-stage extraction column. Water is fed on the top tray at a rate of 1050 kmol/h and a temperature of 322 K, which is achieved by using a cooler (heat removal 1.24 MW). The column is a simple stripper with no reflux. The column operates at 2.5 atm so that cooling water can be used in the condenser (reflux drum temperature is 326 K). Reboiler heat input is 5.96 MW. The overhead vapor is condensed and constitutes the C_5 product stream.

This column is designed by specifying a very small loss of methanol in the overhead vapor (0.01% of methanol fed to the column) and finding the minimum flowrate of extraction water that achieves this specification. Using more than 10 trays or using reflux did not affect the recovery of methanol.

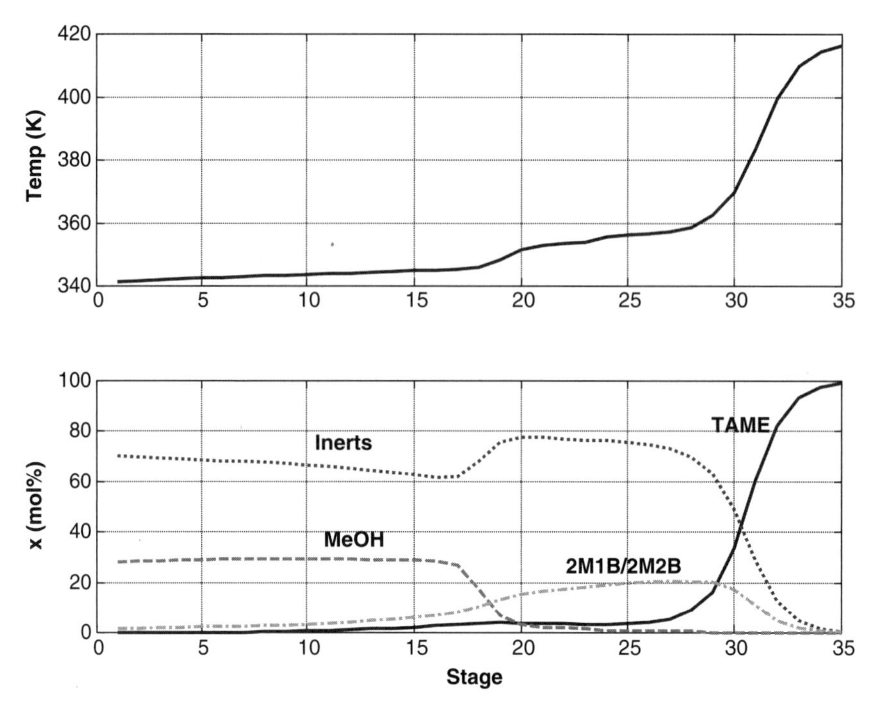

Figure 9.8 C1 temperature and composition profiles.

The bottoms is essentially a binary methanol/water mixture (23.5 mol% methanol), which is fed to a 32-stage column operating at atmospheric pressure. The number of trays in the second column is optimized by determining the total annual cost of the column over a range of tray numbers. Reboiler heat input and condenser heat removal are 8.89 and 9.53 MW, respectively. The column diameter is 2.24 m.

TABLE 9.1 Stream Information for Prereactor and Column C1

	Fresh Methanol (kmol/h)	Feed (kmol/h)	MeOH Reactor (kmol/h)	Reactor Effluent (kmol/h)	MeOH C1 (kmol/h)	B1 (kmol/h)	D1 (kmol/h)
MeOH	230	—	313	188	235	—	316
2M1B	—	85.6	—	13.7	—	0.01	3.31
2M2B	—	165	—	112	—	0.60	14.1
TAME	—	—	—	125	—	232	—
nC_5	—	88.4	—	88.4	—	0.22	88.2
iC_5	—	501	—	501	—	0.17	501
1-Pentene	—	38.1	—	38.1	—	0.18	37.9
2-Pentene	—	162	—	162	—	0.76	161
Total	230	1040	313	1228	235	234	1122
T (K)	325	343	351	355	351	415	341
P (atm)	17	10	7	6	4	4.29	3.95

TABLE 9.2 Stream Information for Columns C2 and C3

	B2 (kmol/h)	D2 (kmol/h)	B3 (kmol/h)	D3 (kmol/h)	Water Makeup (kmol/h)	Extract Water to C2 (kmol/h)
MeOH	316	0.03	1.03	315	—	—
2M1B	—	3.31	—	—	—	—
2M2B	—	14.1	—	—	—	—
TAME	—	—	—	—	—	—
nC_5	—	88.4	—	—	—	—
iC_5	—	501	—	—	—	—
1-Pentene	—	38.0	—	—	—	—
2-Pentene	—	161	—	—	—	—
Water	1025	—	1026	0.32	24	1050
Total	230	830	1027	315	24	1050
T (K)	325	326	379	338	325	322
P (atm)	17	2.5	1.2	1.0	7	2.5

A reflux ratio of 2.1 produces 316 kmol/h of high-purity methanol in the distillate (99.9 mol% MeOH) and 1026 kmol/h of high-purity water in the bottoms (99.9 mol% H_2O). The methanol is combined with 230 kmol/h of fresh methanol feed, and the total is split between the methanol feed streams to the prereactor and to the reactive column. The water is combined with a small water makeup stream, cooled, and recycled back to the extractive column C2.

Some makeup water is needed because a small amount of water goes overhead in the vapor from column C2 (2.9 mol% water). The solubility of water in pentanes is quite small, so the reflux drum of column C2 would form two liquid phases (not shown in Fig. 9.1). The aqueous phase would be 19.9 kmol/h with 99.9 mol% water. The organic phase would be 809 kmol/h with 0.5 mol% water. Table 9.2 gives stream information around columns C2 and C3.

The convergence of the steady-state flowsheet was unsuccessful, so it was converged in the dynamic simulation, using the methods discussed in Chapter 8.

9.4 TAME REACTION KINETICS AND VLE

The liquid-phase reversible reactions considered are

$$\begin{aligned} &2M1B + MeOH \Longleftrightarrow TAME \\ &2M2B + MeOH \Longleftrightarrow TAME \\ &2M1B \Longleftrightarrow 2M2B \end{aligned}$$

The kinetics for the forward and reverse reactions are given in Table 9.3. These reaction rates are given in units of kmol s^{-1} kg_{cat}^{-1} and are converted to the Aspen-required units of kmol s^{-1} m^{-3} by using a catalyst bulk density of 900 kg/m^3. The concentration units in the reaction rates are in mole fractions. The reactive stages in the column each contain

TABLE 9.3 Reaction Kinetics

Reaction	A_F (kmol s^{-1} kg^{-1})	E_F (kJ/mol)	A_R (kmol s^{-1} kg^{-1})	E_R (kJ/mol)	ΔH_{RX} (kJ/mol)
R1	1.3263×10^8	76.103737	2.3535×10^{11}	110.540899	−34.44
R2	1.3718×10^{11}	98.2302176	1.5414×10^{14}	124.993965	−26.76
R3	2.7187×10^{10}	96.5226384	4.2933×10^{10}	104.196053	−7.67

1100 kg of catalyst. This corresponds to 1.22 m^3 on each tray, which gives a weir height of 0.055 m for a reactive column with a diameter of 5.5 m.

The reactions and all the kinetic parameter must be set up in Aspen Plus. In the *Exploring* window, click on *Reactions* and then the second *Reactions.* Right-click and select *New.* This opens the window shown in Figure 9.9a, in which the type of reaction is selected

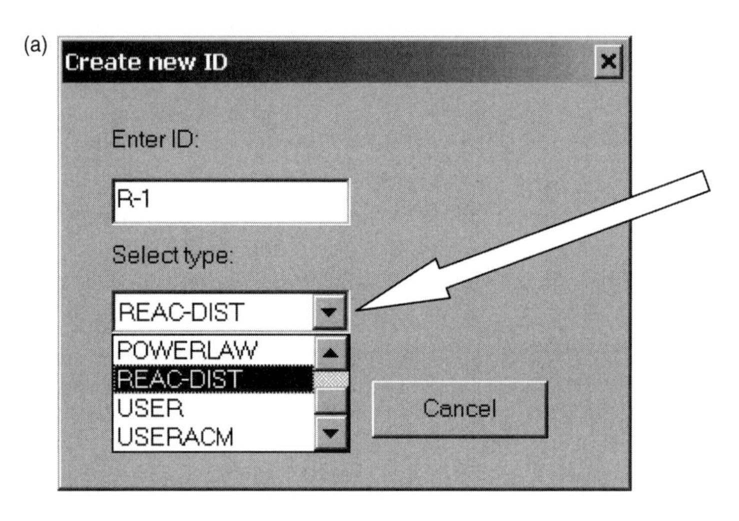

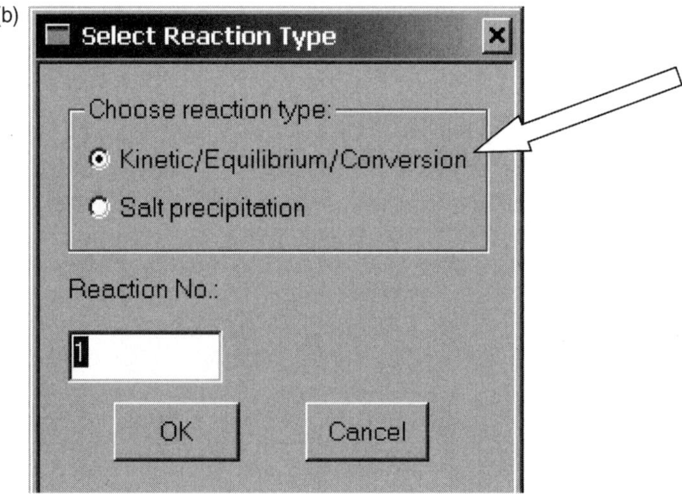

Figure 9.9 (a) Create new reaction; (b) select reaction type; (c) input reactants and products.

(c)

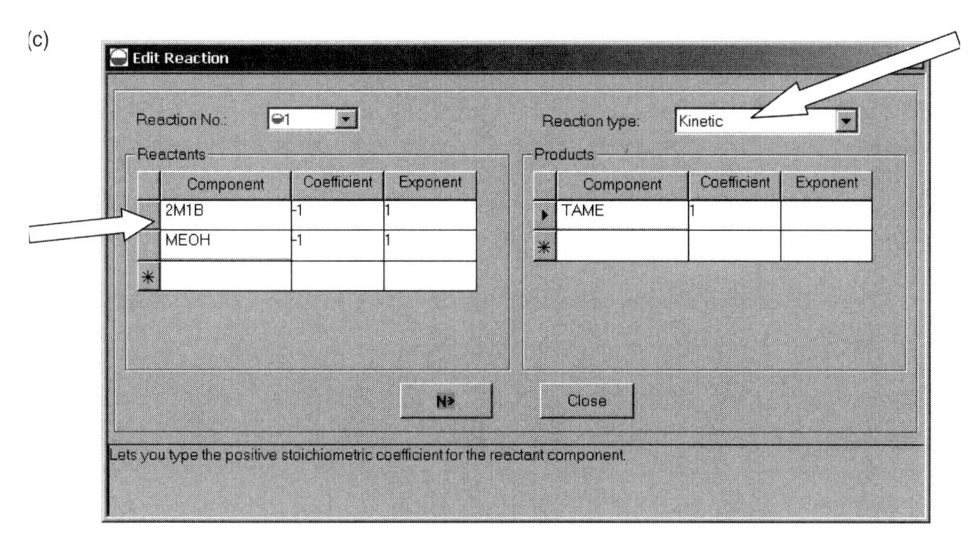

Figure 9.9 *Continued.*

to be *REAC-DIST*. Then click on the new reaction *R-1*, which opens the window shown in Figure 9.9b, on which *Kinetic* is selected.

Clicking *OK* opens the window shown in Figure 9.9c, on which reactant and product components are selected. The reactant coefficients are negative, and the product coefficients are positive. The *Exponent* is the power-law exponent used in the reaction rate expression. Make sure to select *Kinetic* in the upper right corner.

The procedure is repeated for the three forward reactions and for the three reverse reactions. These are shown in Figure 9.10. All of this input is done on the *Stoichiometry* page tab. Clicking the *Kinetic* page tab and selecting one of the six reactions opens the window shown in Figure 9.11a, on which all the kinetic parameters are entered for that reaction. Remember to select *Liquid* for reacting phase and *mole fraction* for the concentration basis (see Fig. 9.11b).

It is important to point out that for reactive distillation Aspen Plus does not list *activities* as a standard option for the concentration basis. This is a distinct limitation because many of the reactions that occur in reactive distillation systems use activities. Special-purpose programs can be written, but these are beyond the scope of this book. Later versions of Aspen Plus should remove this limitation.

Now the reactions have been set up. Go to the C1 block and click *Reactions*. Enter the starting and ending stages on which reaction occurs on the *Specifications* page tab in Figure 9.12a and select the reaction *R-1*. Note that *R-1* is a set of six reactions.

Clicking the *Holdups* page tab opens the window shown in Figure 9.12b, in which the molar or volumetric holdups on each of the reactive trays are entered. The reactive liquid volume on each tray is set at 1.22 m^3, which corresponds to a liquid height of 0.055 m for a reactive column with a diameter of 5.5 m.

It is important to note that the diameter of the column is not known initially because this depends on vapor velocities that are unknown until the column is converged to the desired specifications. So column sizing in a reactive distillation column is an iterative procedure. A diameter is estimated, tray holdups calculated, and the column is converged. Then the

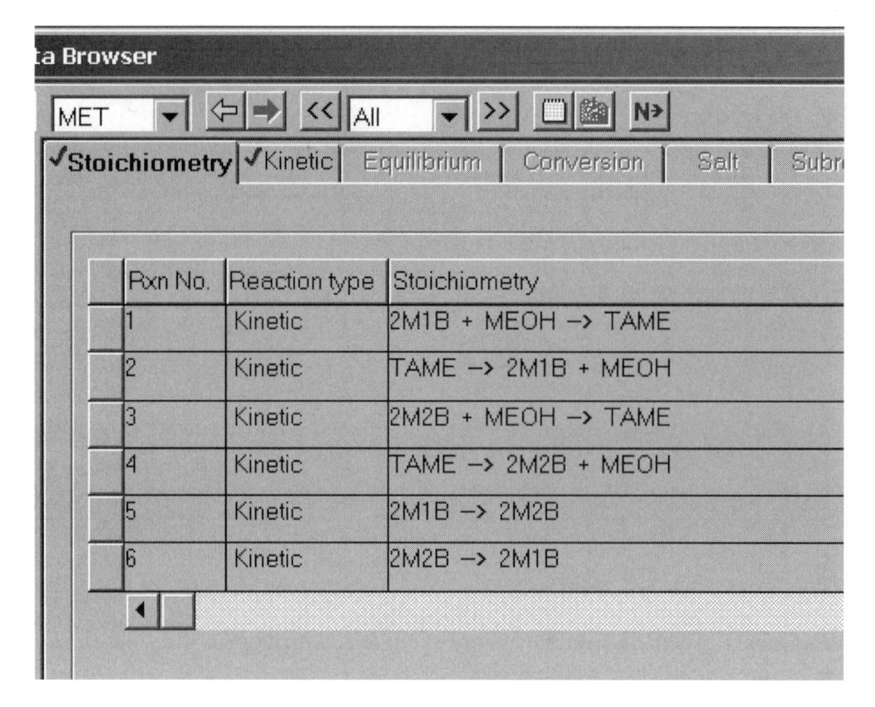

Figure 9.10 Six reactions specified.

(a)

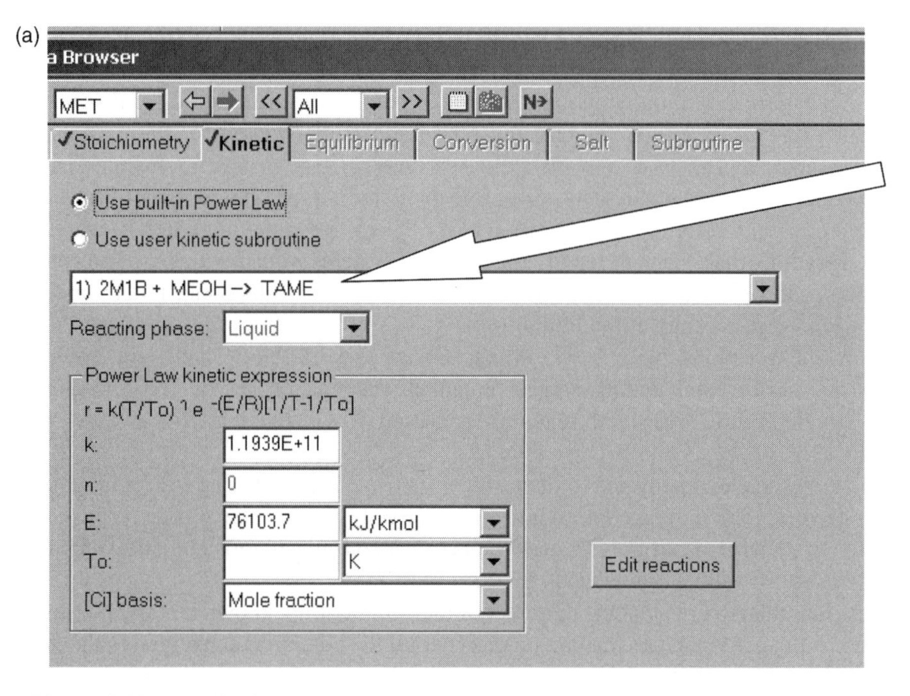

Figure 9.11 (a) Kinetic parameters; (b) specify concentration basis and reactive phase.

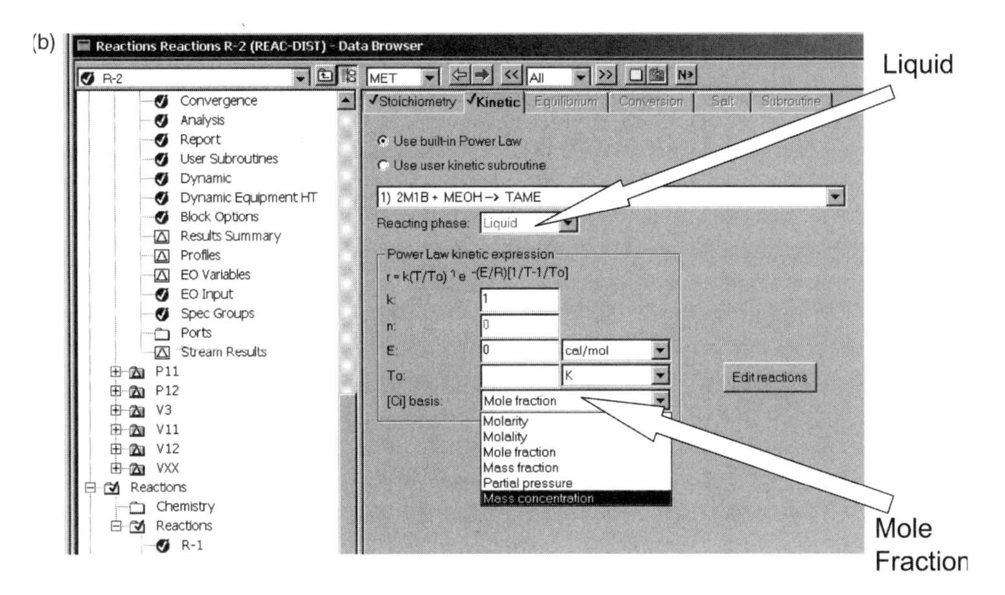

Figure 9.11 *Continued.*

diameter calculated in *Tray Sizing* is compared with the estimated diameter and the calculations repeated.

The other important issue is the height of liquid on a tray. Hydraulic limitations prevent excessive liquid depth because this would cause large pressure drops. Liquid depths are limited to about 0.1 m. If more liquid holdup is needed, the column diameter can be increased beyond the minimum calculated from sizing calculations. Of course, more reactive trays could also be added to the column. However, reactive distillation columns have the interesting feature that there is an optimum number of reactive stages. Having too few reactive stage results in high energy consumption because the reactant concentrations in the reactive zone must be large, and this requires large vapor rates to keep the reactants from leaving in the bottoms or the distillate. Adding more reactive stages reduces the

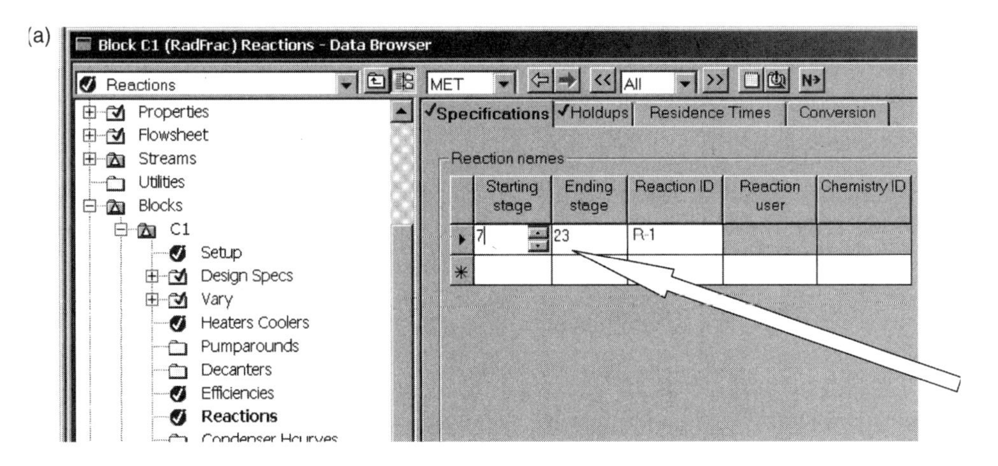

Figure 9.12 (a) Specifying reactive trays and holdups; (b) tray holdups.

(b)

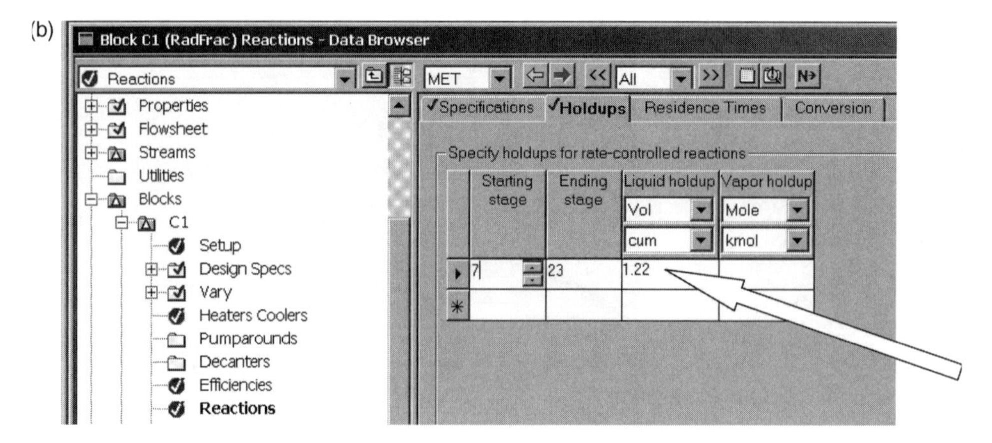

Figure 9.12 *Continued.*

vapor boilup requirements because reactant concentrations in the reactive zone decrease. However, beyond some point, adding more reactive stages begins to increase energy consumption because the reactant concentrations in the reactive zone start to increase due to a larger contribution from the reverse reactions.

Figure 9.13 gives a ternary diagram for the isopentane/methanol/TAME system at 4 bar. The phase equilibrium of this system is complex because of the existence of azeotropes. The Unifac physical property package in Aspen Plus is used to model the vapor–liquid equilibrium in all units except the methanol/water column, where the van Laar equations are used because of their ability to accurately match experimental data.

9.5 PLANTWIDE CONTROL STRUCTURE

In preparation for exporting the steady-state flowsheet into Aspen Dynamics, all equipment is sized. Column diameters are calculated by Aspen tray sizing. Reflux drums and column bases are sized to provide 5 min of holdup when 50% full, based on the total liquid entering the surge capacity. Pumps and control valves are specified to give adequate dynamic rangeability. Typical valve pressure drops are 2 atm.

When the flowsheet with a tubular reactor was exported into Aspen Dynamics, the program would not run. A liquid-filled plug-flow reactor will not run in version 12 of Aspen Dynamics. To work around this limitation, the tubular reactor was replaced by two continuous stirred tank reactors (CSTRs) in series. Operating temperatures in both reactors were set at 355 K and volume at 10 m^3. This design gave the same reactor effluent as the tubular reactor.

The plantwide control structure is shown in Figure 9.14. The tray temperature is controlled in each column by manipulating reboiler heat input. The trays are selected by finding the location where the temperature profile is steep: stage 31 in column C1 (see Fig. 9.8), stage 7 in column C2, and stage 7 in column C3. In addition, an internal composition in column C1 is controlled by manipulating the flowrate of methanol to the column. Stage 18 is selected (see Fig. 9.8). The flowrate of methanol to the reactor is ratioed to the feed flowrate.

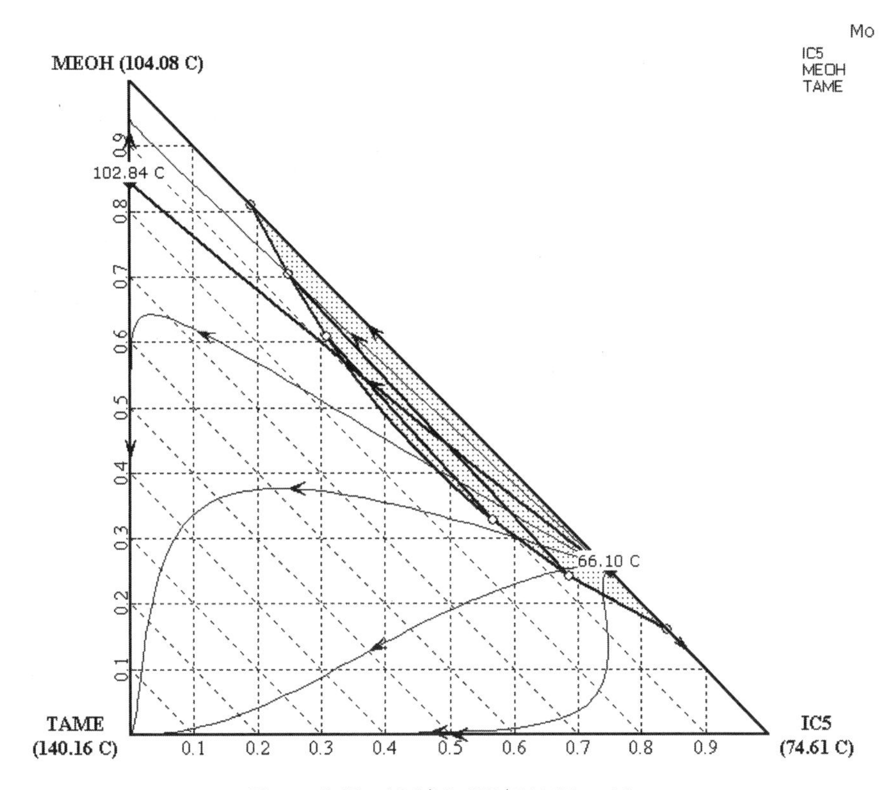

Figure 9.13 iC5/MeOH/TAME at 4 bar.

The flowrate of extraction water fed to the top of column C2 is ratioed to the feed to this column D_1 by using a multiplier and a remotely set flow controller. The temperature of the extraction water is controlled by manipulating cooling water to the cooler. Base level is controlled by manipulating bottoms, and reflux drum level is controlled by manipulating distillate. The binary methanol/water mixture from the bottom of column C2 is fed to column C3. A constant reflux ratio is maintained in this column by adjusting reflux flowrate.

There are two key plantwide material balance loops associated with column C3. The level in the reflux drum provides a good indication of the inventory of methanol in the system. If this level is going down, more methanol is being consumed in the reaction than is being fed into the process. Therefore the control structure maintains the reflux drum level in C3 by manipulating the methanol fresh feed.

Note that the flowrate of the total methanol (D_3 plus fresh methanol feed) is fixed by the two downstream flow controllers setting the flowrates to the reactor and to column C1. This means that there is an immediate effect of fresh feed flowrate on reflux drum level. The distillate flow D_3 changes inversely with fresh feed flow because the downstream flowrate is fixed. Thus the reflux drum level sees the change in the methanol fresh feed instantaneously.

At the other end of the column, the base level provides a good indication of the inventory of water in the system. Ideally there should be no loss of water since it just circulates

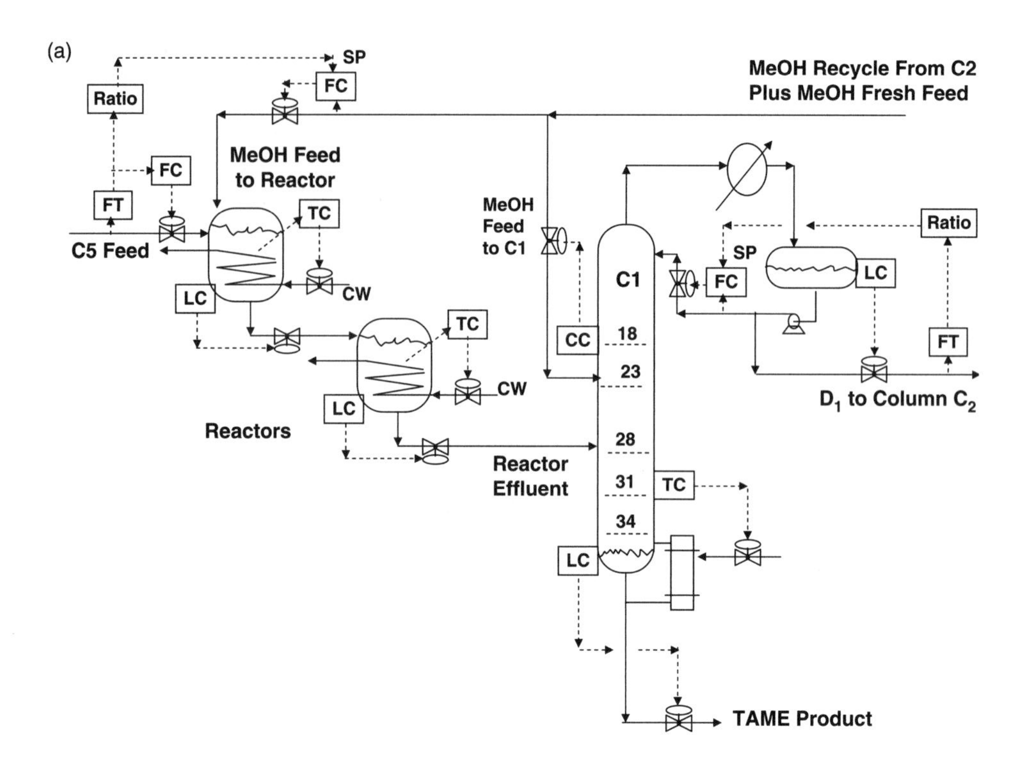

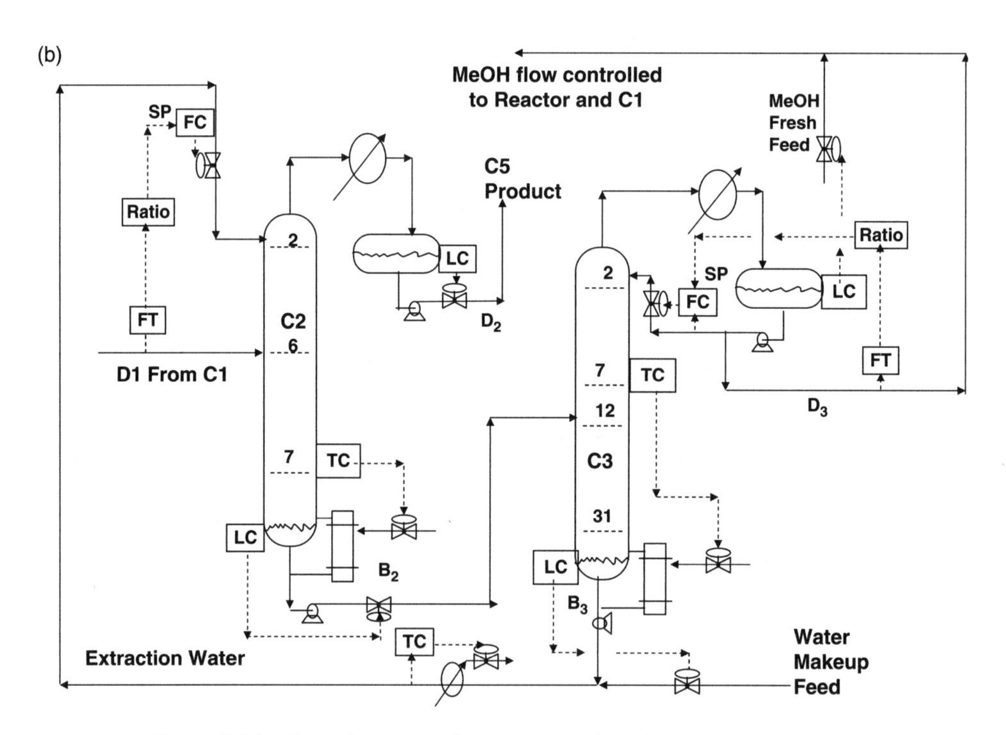

Figure 9.14 Control structure for reactors and (a) C1 and (b) C2 and C3.

around between the extractive column and the recovery column. However, there is a small amount of water lost in the overhead from column C2. A water makeup stream is used to control the liquid level in the base of column C3. This makeup flow is very small compared to the water circulation, so the base of column C3 must be sized to provide enough surge capacity to ride through disturbances.

All temperature and composition controllers have one-minute deadtimes. The PI controllers are tuned by running a relay–feedback test and using the Tyreus–Luyben settings. All liquid levels are controlled by proportional controllers with gains of 2 for all level loops except the two reactors, which have gains of 10. Liquid levels in reflux drums are controlled by manipulating distillate flowrates. The reflux ratios in all columns are controlled by manipulating reflux. Column pressure controllers use default controller settings and manipulate condenser heat removal.

Figure 9.15a gives the responses of the process to 20% changes in feed flowrate. Figure 9.15b gives responses to changes in feed composition. Effective plantwide control is achieved. The control structure provides stable base-level regulatory control for large disturbances. The purity of the TAME product is held quite close to its specification.

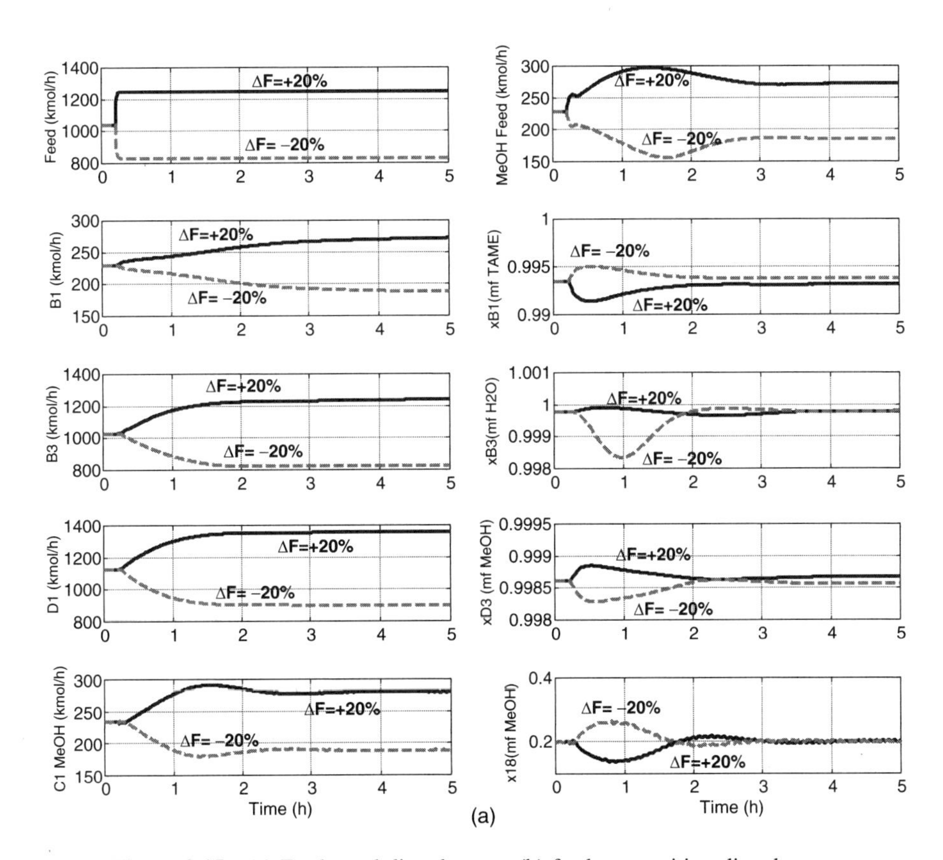

Figure 9.15 (a) Feed rated disturbances; (b) feed composition disturbances.

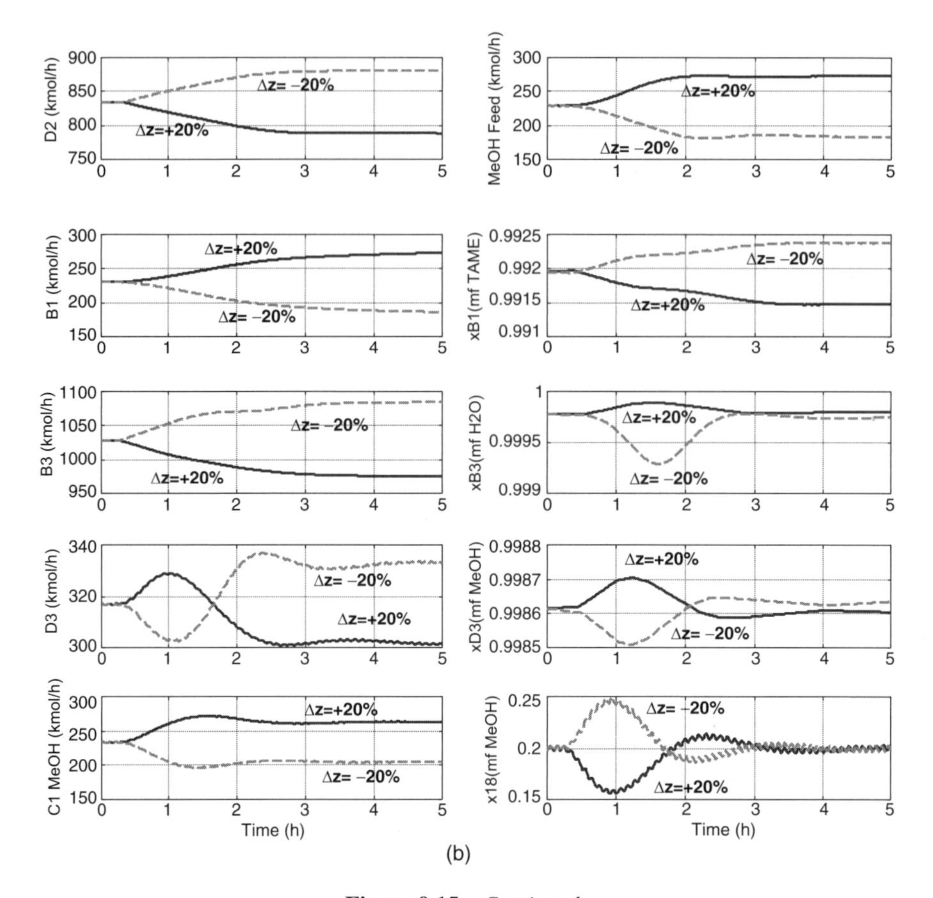

Figure 9.15 *Continued.*

9.6 CONCLUSION

The design and simulation of reactive distillation systems have been discussed in this chapter. The reactive distillation column is more complex than a plain distillation column because the effects of both phase equilibrium and chemical reaction must be considered simultaneously.

CHAPTER 10

CONTROL OF SIDESTREAM COLUMNS

In this chapter we study distillation columns that have more than the normal two product streams. These more complex configurations provide savings in energy costs and capital investment in some systems. Sidestream columns are used in many ternary separations, and the examples in this chapter illustrate this application. However, a sidestream column can also be used in a binary separation if different purity levels are desired. For example, two grades of propylene products are sometimes produced from a single column. The bottoms stream is propane, the sidestream is medium-purity propylene, and the distillate is high-purity polymer-grade propylene.

The most widespread use of sidestream columns occurs in petroleum fractionators, which have multiple sidestream products that are complex mixtures of many components. The sidestreams have progressively higher boiling point ranges as we move down the column. The top sidestream may be kerosene. The next may be a diesel cut or jet fuel. The next is a light gasoil. The final may be a heavy gasoil. The liquid streams withdrawn from the main column are each fed to a small stripping column. Open steam is fed at the bottom of each sidestream stripper to strip out light components from the liquids withdrawn from the main column. We discuss these types of sidestream columns in Chapter 11.

Sidestream columns come in several flavors. Both liquid and vapor sidestreams are used. Sometimes the sidestream is a final product. Because the purity attainable in a sidestream is limited, the sidestream from the main tower is sometimes fed to a second column (usually a stripper or a rectifier) for further purification with a recycle stream back to the main column. Several examples are studied in this chapter.

Distillation Design and Control Using Aspen™ Simulation, By William L. Luyben

10.1 LIQUID SIDESTREAM COLUMN

Liquid sidestream columns are frequently used when the feed stream is a ternary mixture in which the concentration of the lightest component is small. This lightest component is removed in the distillate product. The intermediate component is removed in a sidestream that is withdrawn as a liquid from a tray higher up in the column above the feed tray. A liquid sidestream is used because the concentrations of the lightest component in the *liquid* phase on all the trays above the feed tray are smaller than the concentrations of the lightest component in the *vapor* phase. This indicates a limitation in the application of a liquid sidestream column; the volatility between the lightest and intermediate components must be fairly large. If this volatility is too small, a high-purity sidestream cannot be produced.

10.1.1 Steady-State Design

The specific numerical case used is a ternary mixture of dimethyl ether (DME), methanol (MeOH), and water. The feed composition is 5 mol% DME, 50 mol% MeOH, and 45 mol% water. The feed flowrate is 100 kmol/h, and the feed is fed on stage 32 of a 52-stage column. The liquid sidestream is withdrawn from stage 12. The column pressure is set at 11 atm so that cooling water can be used in the condenser (reflux drum temperature is 323 K with a distillate composition of 98 mol% DME and 2 mol% MeOH). The NRTL physical property package is used.

The presence of a sidestream provides an additional degree of freedom. Three purities can be set. We use three Design Spec/Vary functions to achieve the following specifications:

1. Distillate impurity is set at 2 mol% MeOH by varying the distillate flowrate.
2. Sidestream impurity is set at 2 mol% water by varying the sidestream flowrate.
3. Bottoms purity is set at 2 mol% MeOH by varying the reflux flowrate.

Note that the other impurity in the sidestream (DME) cannot be specified with a fixed column configuration because there are no remaining degrees of freedom. However, the number of stages and the locations of the feed and the sidestream can be changed to alter the sidestream DME composition.

The key separation in this liquid sidestream column is between DME and methanol in the section above the feed tray. Since all the DME in the feed must flow up the column past the sidestream drawoff tray, the concentration of DME in the vapor phase is significant. The liquid-phase concentration, however, is smaller if the relative volatility between DME and methanol is large. The normal boiling points of these two components (DME = 248.4 K and MeOH = 337.7 K) are quite different. This gives a relative volatility at the sidestream drawoff tray of ~24. Thus the vapor composition of 4.04 mol% DME has a liquid in equilibrium with it that is only 0.16 mol% DME. The column diameter is 0.61 m. The reboiler heat input is 1.346 MW.

Figure 10.1 gives the flowsheet and steady-state parameters for the process. Note that the distillate flowrate is quite small because of the low concentration of DME in the feed. This gives a very high reflux ratio ($RR = 41$), which means that the reflux drum level will have to be controlled by manipulating the reflux flowrate.

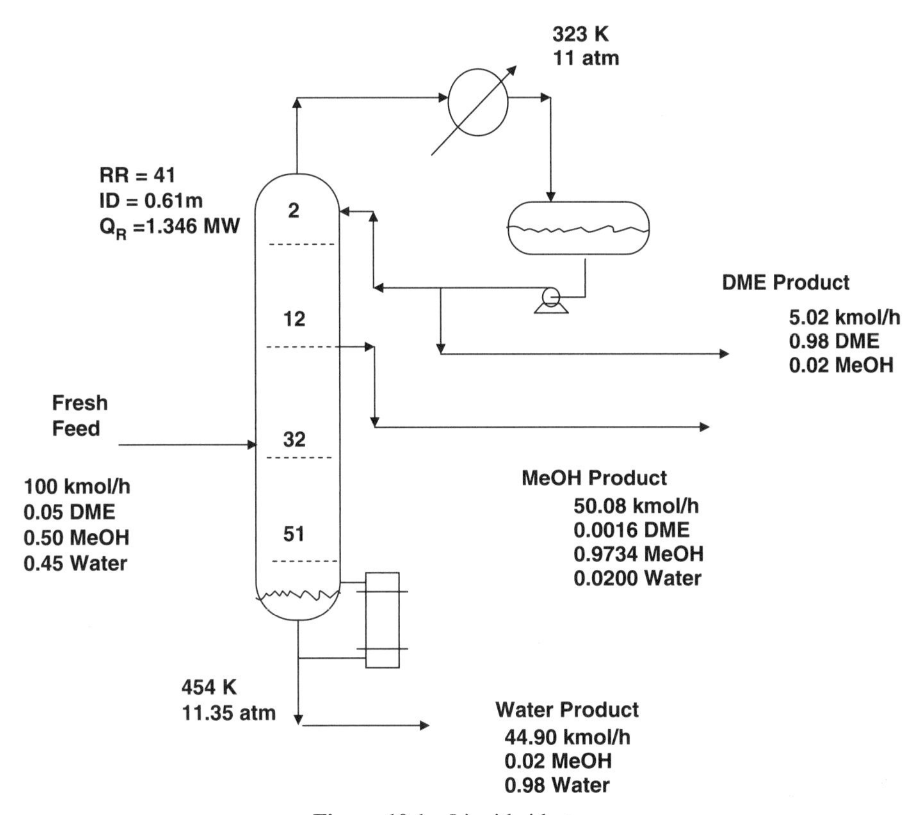

Figure 10.1 Liquid sidestream.

10.1.2 Dynamic Control

The usual sizing calculations are performed, the flowsheet is pressure-checked, and the file is exported to Aspen Dynamics. Flow controllers are installed on the feed, distillate, and sidestream. Base level is controlled by manipulating bottoms flowrate. Reflux drum level is controlled by manipulating reflux flowrate.

The major control structure issues of this sidestream column are how to manipulate the sidestream flowrate and how to manipulate the distillate flowrate. They cannot be fixed or just ratioed to the feed flowrate because changes in feed composition required that the flowrate of the distillate and the sidestream must change to achieve the desired purities. Figure 10.2 shows a control structure that provides effective control of this complex column. Figure 10.3 gives the controller faceplates. Note that two of the flow controllers are on "cascade" (remote setpoints).

The control scheme shown in Figure 10.2 controls the temperature on stage 3 by manipulating the distillate flowrate and controls the temperature on stage 51 by manipulating the reboiler heat input. These stages are selected by studying the temperature and composition profiles shown in Figure 10.4. The two temperature controllers (TC3 and TC51) have one-minute deadtimes and 100 K temperature transmitter spans. They are tuned by running a relay–feedback test and using the Tyreus–Luyben settings.

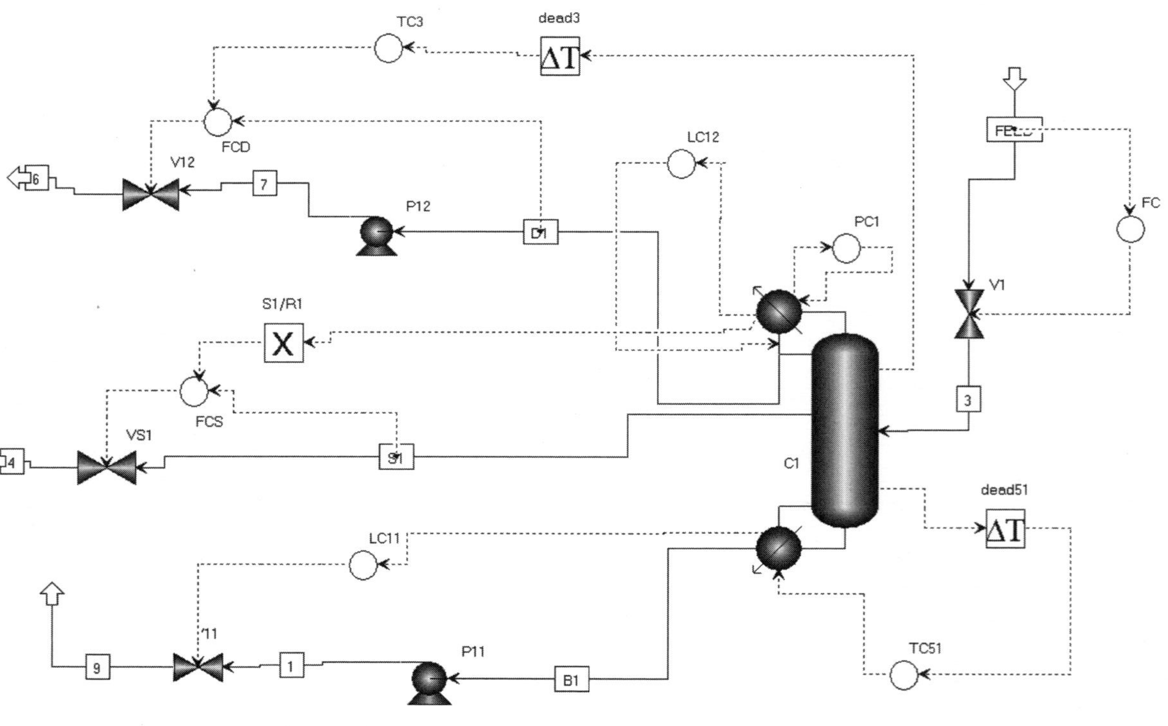

Figure 10.2 Control structure for liquid sidestream column.

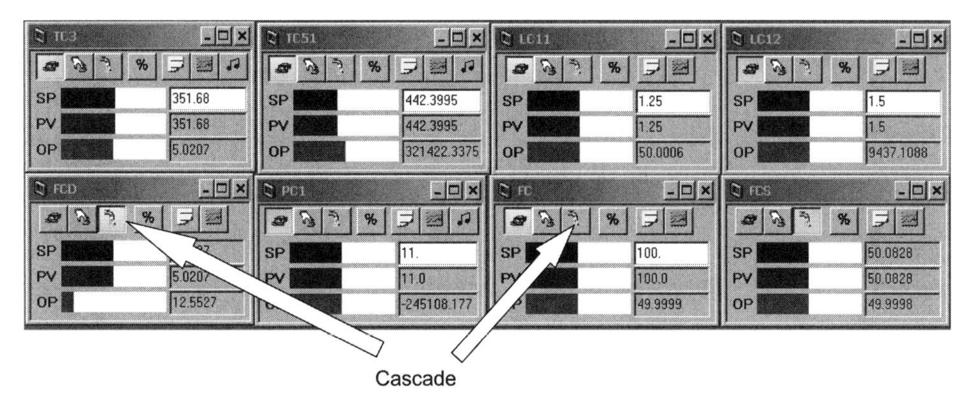

Figure 10.3 Controller faceplates.

Now the remaining issue is how to manipulate the sidestream flowrate. One effective method is to ratio the sidestream flowrate to the reflux flowrate. This lets the sidestream flowrate move as disturbances occur. For example, suppose that too much methanol starts to move down the column. This will decrease the temperature on stage 51, and the TC51 temperature controller will increase the reboiler heat input. This sends more vapor up the column, which increases the reflux drum level. The level controller increases reflux flow, and the S_1/R_1 ratio causes the sidestream flowrate to increase, which pulls more methanol out of the column.

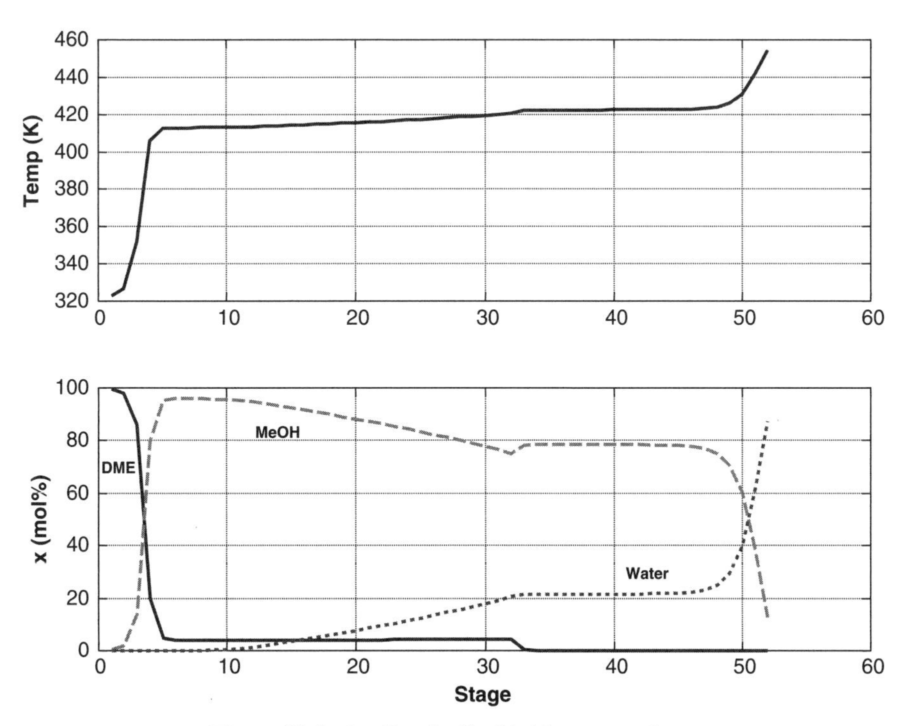

Figure 10.4 Profiles for liquid sidestream column.

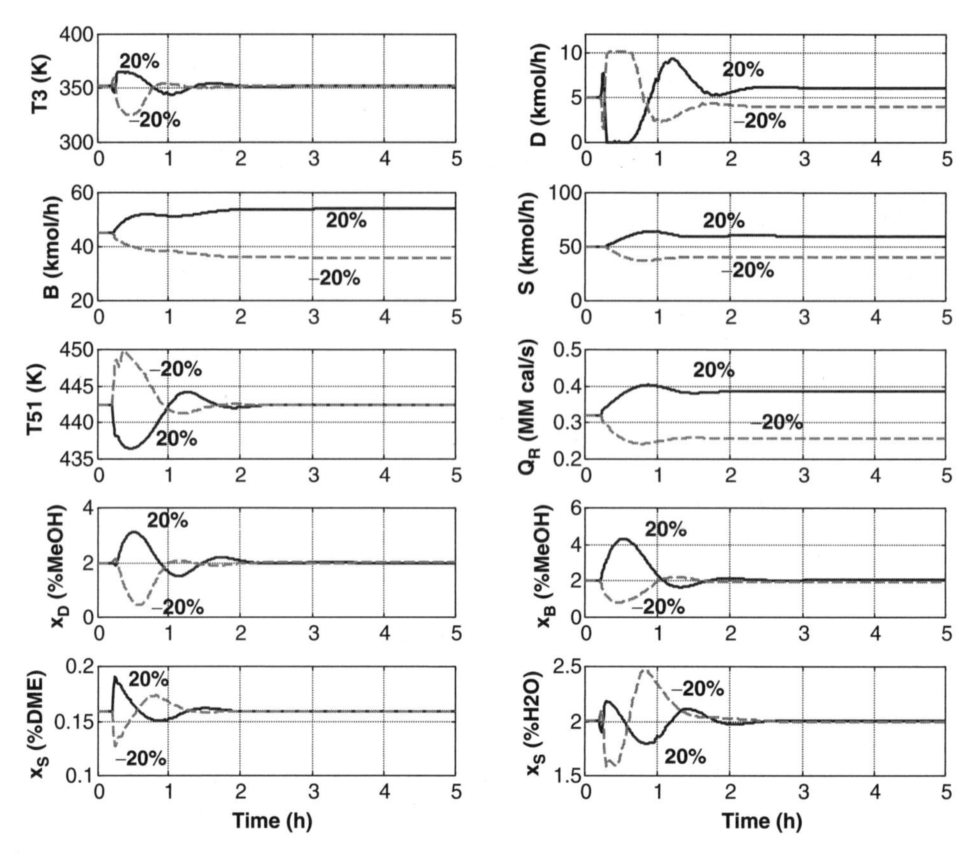

Figure 10.5 Liquid sidestream column; feed rate changes.

The effectiveness of this control structure is demonstrated in Figures 10.5 and 10.6. The responses of the system to 20% step increases and decreases in feed flowrate are shown in Figure 10.5. The compositions of all three products are returned to their specified values within ~1.5 h. The largest transient deviations occur in the bottoms composition. This could be improved by using a heat input to feed ratio.

Note that the increase in feed flowrate causes the distillate valve to go completely shut for about 20 min while the decrease in feed flowrate saturates the valve wide open. Remember that the distillate flowrate is very small. Note also in Figure 10.3 that the steady-state controller output signal to the distillate valve (V12) is only 13% instead of the normal 50%. The size of the control valve and the pump work exported from Aspen Plus were both increased to get more rangeability in the distillate flowrate. The valve size is changed by clicking the icon, then right-clicking and selecting *Forms* and *All Variables*. Then the value of *COmax* is increased (see Fig. 10.6).

To change the pump power, click the pump P12 icon, then right-click and select *Forms* and *Configure* (see Fig. 10.7). Use the dropdown arrow to change *UseCurves* from *True* to *False*, and increase the *EpowerR*.

Figure 10.8 gives responses for three feed composition disturbances. The solid lines show what happens when the DME in the feed is increased from 5 to 7 mol% DME (with MeOH reduced by a corresponding amount). The dashed lines show what

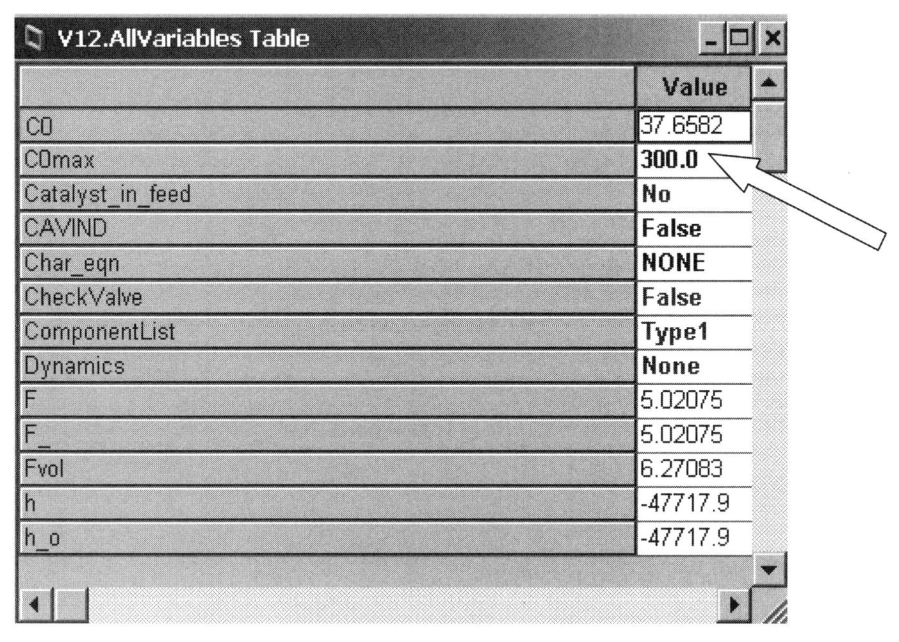

V12.AllVariables Table

	Value
C0	37.6582
C0max	300.0
Catalyst_in_feed	No
CAVIND	False
Char_eqn	NONE
CheckValve	False
ComponentList	Type1
Dynamics	None
F	5.02075
F_	5.02075
Fvol	6.27083
h	-47717.9
h_o	-47717.9

Figure 10.6 Increasing valve size.

P12.Configure Table

	Description	Value	Uni	Spec
ValidPhases	Valid phases	Liquid-Only		
ComponentList		Type1		
EffCalc	Efficiency is calculated	True		
ModelType	Model type	Pump		
UseCurves	Use performance curves	False		
PDriven	Model is pressure driven	True		
EpowerR	Specified electrical power	0.3	kW	Fixed
Eff	Pump efficiency	0.29565		Free
Meff	Driver efficiency	1.0		Fixed

Figure 10.7 Increasing pump work.

happens when the DME in the feed is decreased from 5 to 3 mol% DME. The dotted lines show an increase in the methanol concentration in the feed from 50 to 55 mol% (with water reduced by a corresponding amount). Product purities are well maintained.

10.2 VAPOR SIDESTREAM COLUMN

Vapor sidestream columns are frequently used when the feed stream is a ternary mixture and the concentration of the *heaviest* component is small. The other criterion is that the relative volatility between the intermediate component and the heaviest component must be fairly large.

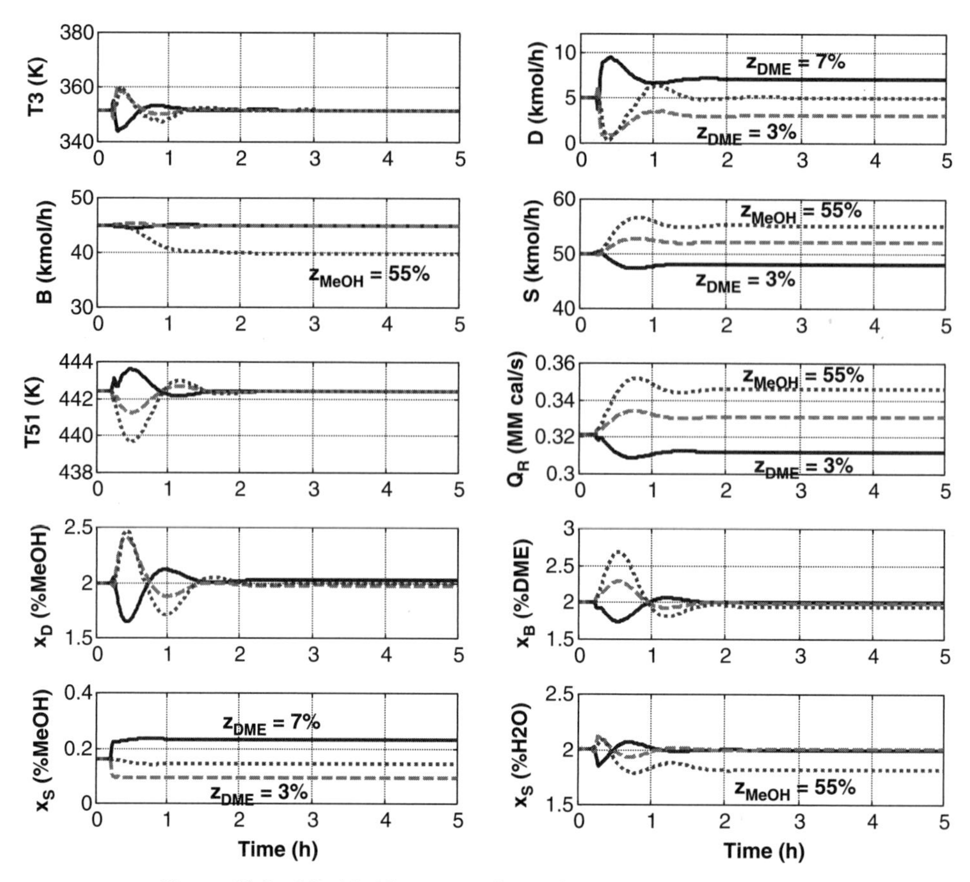

Figure 10.8 Liquid sidestream column; feed composition changes.

To illustrate the latter limitation, suppose that we take the same ternary mixture considered in the previous section. But now the feed composition is 45 mol% DME, 50 mol% MeOH, and 5 mol% water. Can a vapor sidestream column be used effectively? The normal boiling point of MeOH is 337.7 K. The normal boiling point of water is 373.2 K. This is a much smaller difference than is the case for DME and MeOH (248.4 vs. 337.7 K). The result is a relative volatility between MeOH and water of about 1.73.

All the water in the feed must flow down past the vapor sidestream drawoff tray. If the liquid composition on this tray is 5.4 mol% water, the vapor composition is 3.2 mol% water. Thus the purity of the sidestream is low. The only way to reduce the impurity of water in the sidestream is to reduce the water concentration in the liquid by drastically increasing the internal flowrates of the vapor and liquid in the column, specifically, increasing the reflux ratio. This renders the sidestream configuration uneconomical for this chemical separation.

10.2.1 Steady-State Design

In order to consider a reasonable system to illustrate a vapor sidestream column, we change the feedstream to contain *n*-butanol (BuOH) instead of water. The normal

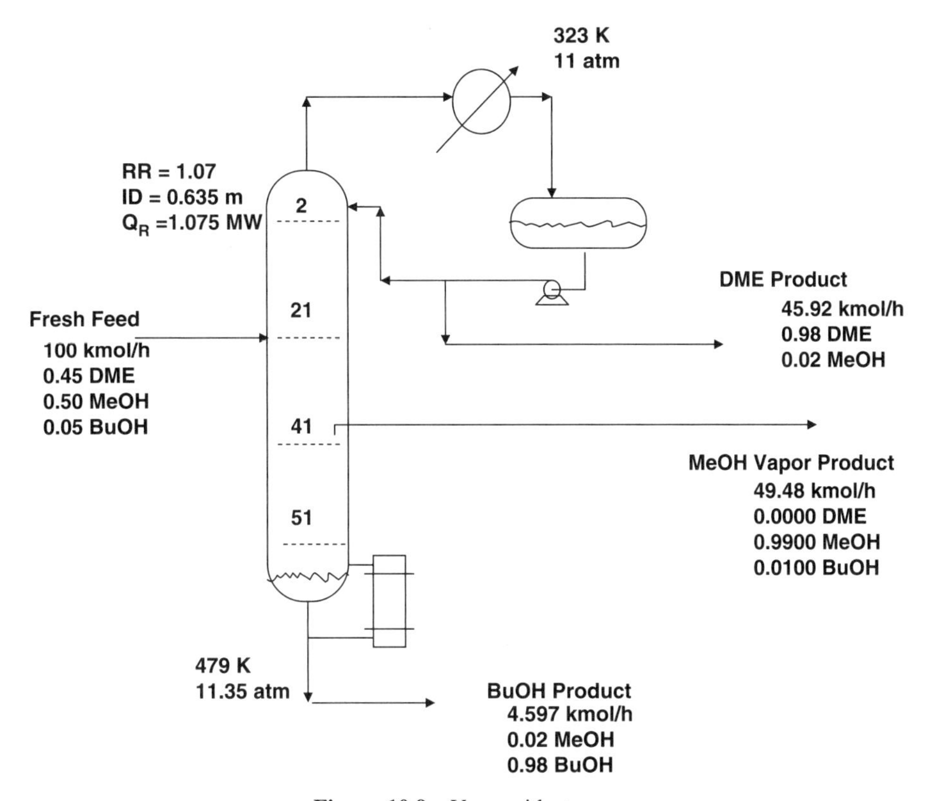

Figure 10.9 Vapor sidestream.

boiling point of *n*-butanol is 390.8 K, compared to 337.7 K for MeOH. This produces a relative volatility of ~4.4, so a *vapor* sidestream product with only 1 mol% BuOH can be produced with a reflux ratio of 1.07. The composition of the *liquid* on the sidestream drawoff tray is 4.3 mol% BuOH.

The column has 51 stages, and the feed is introduced on stage 21. The vapor sidestream is withdrawn from stage 41. The column operates at 11 atm. Figure 10.9 gives the flowsheet with stream conditions and design parameters for this vapor sidestream column with a ternary feedstream of composition 45 mol% DME, 50 mol% MeOH, and 5 mol% BuOH. The column diameter is 0.635 m. The reboiler heat input is 1.075 MW.

10.2.2 Dynamic Control

Since the reflux ratio is 1.07, the reflux drum level can be controlled by either distillate or reflux. We select distillate flow to control level to gain the advantage of proportional-only control smoothing out the disturbances to a downstream process. The reflux is manipulated to control the temperature on stage 3 at 339.5 K, which maintains the purity of the DME product. Figure 10.10 gives the temperature and composition profiles in the column. Note that the BuOH *vapor* composition y_{BuOH} is smaller than the BuOH *liquid* composition x_{BuOH} on all trays below the feed tray. This illustrates why the sidestream is removed as a vapor instead of a liquid.

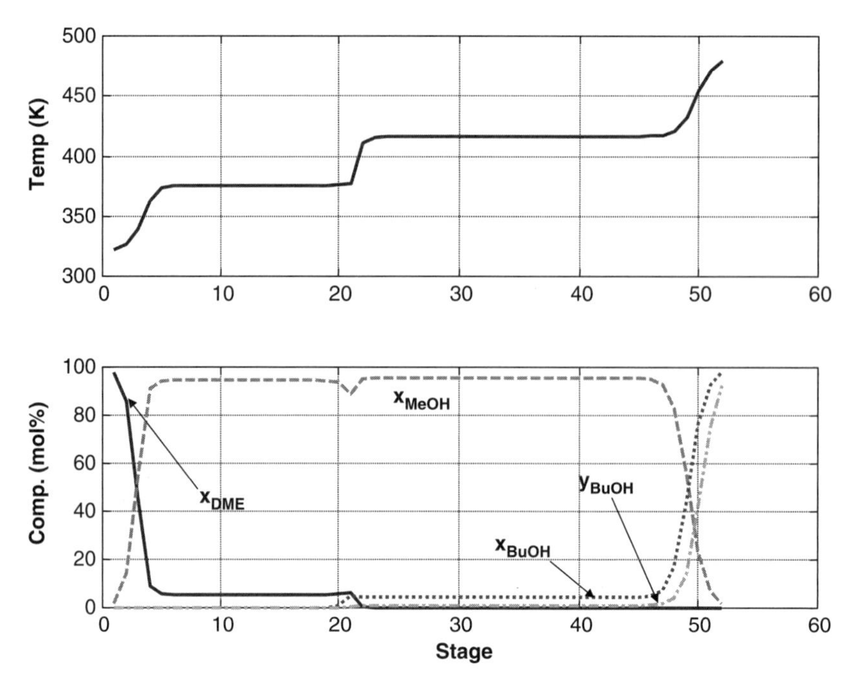

Figure 10.10 Profiles for vapor sidestream column.

The very small bottoms flowrate implies that the base level should be controlled by reboiler heat input and not by bottoms flowrate. The temperature on stage 50 is controlled by manipulating the bottoms flowrate. It is important to note that the base-level controller must be set on automatic for the temperature controller to work because changing the bottoms flowrate has no *direct* effect on stage 50 temperature. The level loop is "nested" inside the temperature loop.

The key control structure issue is how to set the vapor sidestream flowrate. It certainly must change as disturbances enter the column. One effective way to achieve this is to ratio the sidestream flowrate to the reboiler heat input. This control structure is shown in Figure 10.11. The two temperature controllers (TC3 and TC50) have 1-min deadtimes and 100 K temperature transmitter spans. They are tuned by running a relay–feedback test and using the Tyreus–Luyben settings. The TC3 controller manipulating reflux flowrate is tuned first with the TC50 controller on manual. Controller tuning constants are $K_C = 0.46$ and $\tau_I = 9.2$ min. Then the TC50 controller manipulating bottoms flowrate is tuned with the TC3 controller on automatic (sequential tuning). The tuning constants are $K_C = 3.8$ and $\tau_I = 30$ min. Note that this second loop is slower than the first. Controller faceplates are shown in Figure 10.12. The FCD flow controller is "on cascade" since its setpoint comes from the TC3 temperature controller. The FCS flow controller is also "on cascade" since its setpoint comes from the S/Q_R ratio. Note that the controller output signal for the bottoms flow controller FCB is only 14%. The control valve size and the bottoms pump power were increased to provide more rangeability of the very small bottoms flowrate.

The effectiveness of this control structure is demonstrated in Figures 10.13 and 10.14 for feed rate and feed composition disturbances. Responses to step increases and decreases

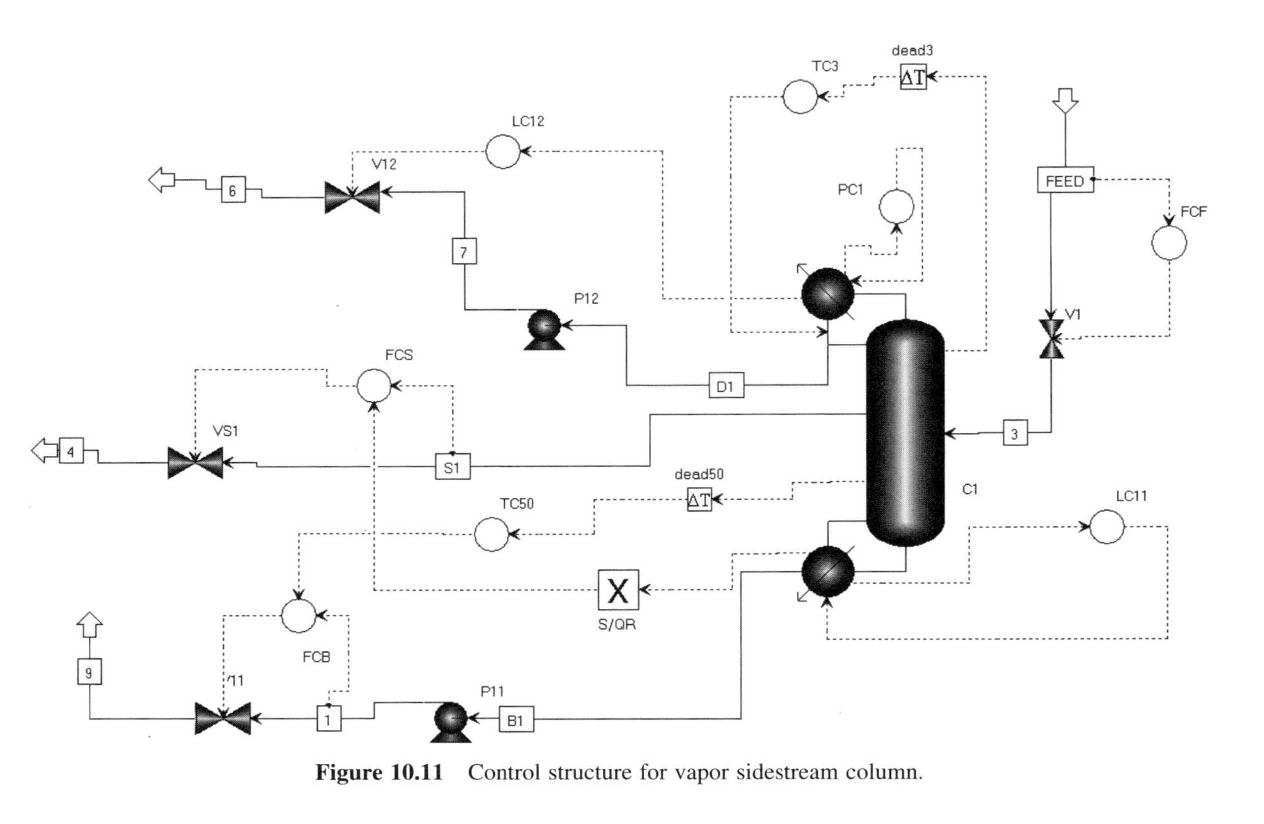

Figure 10.11 Control structure for vapor sidestream column.

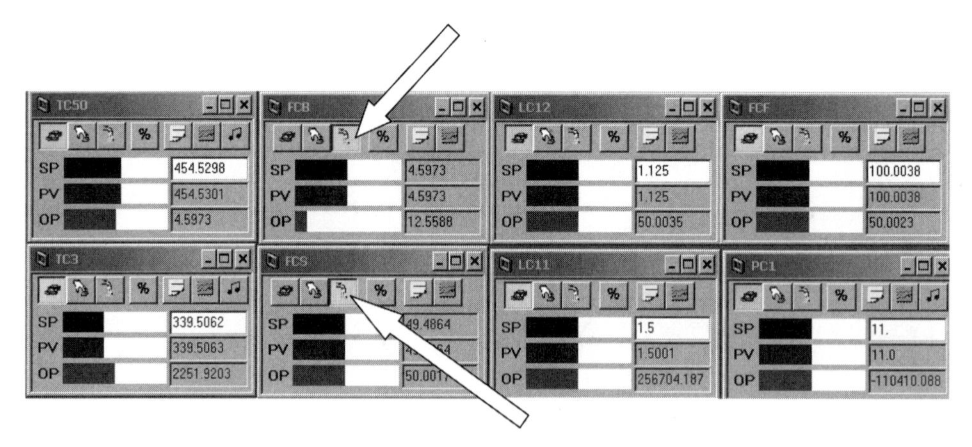

Figure 10.12 Controller faceplates.

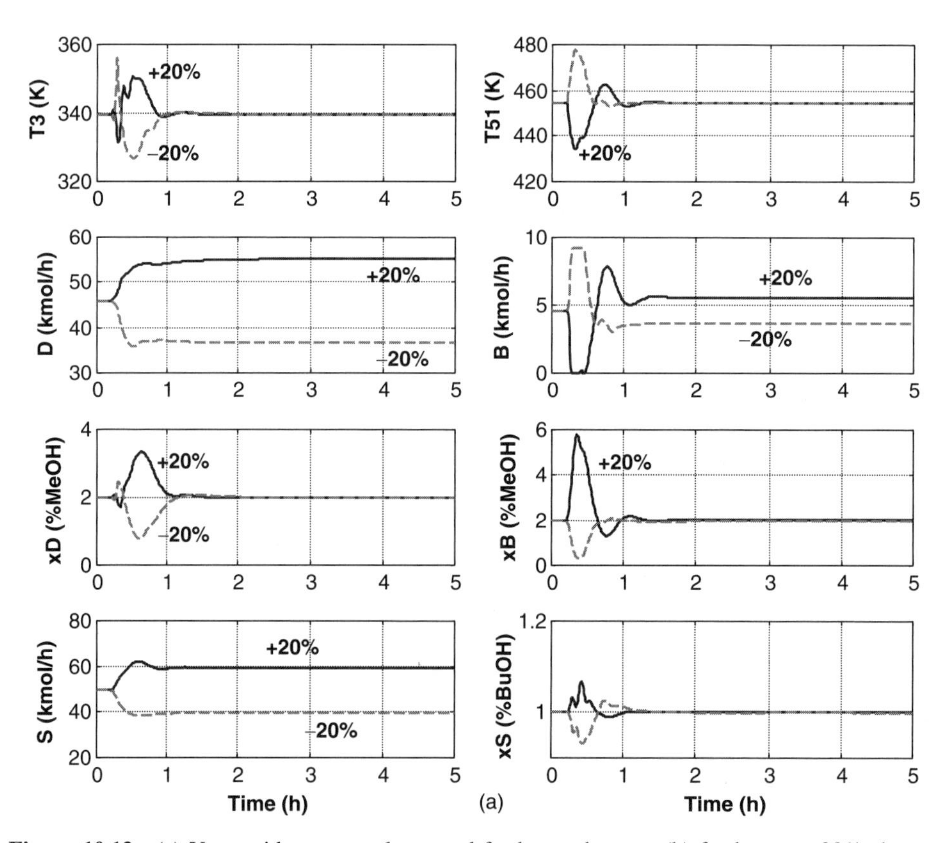

Figure 10.13 (a) Vapor sidestream column and feed rate changes; (b) feed rate + 20% change with and without Q_R/F ratio.

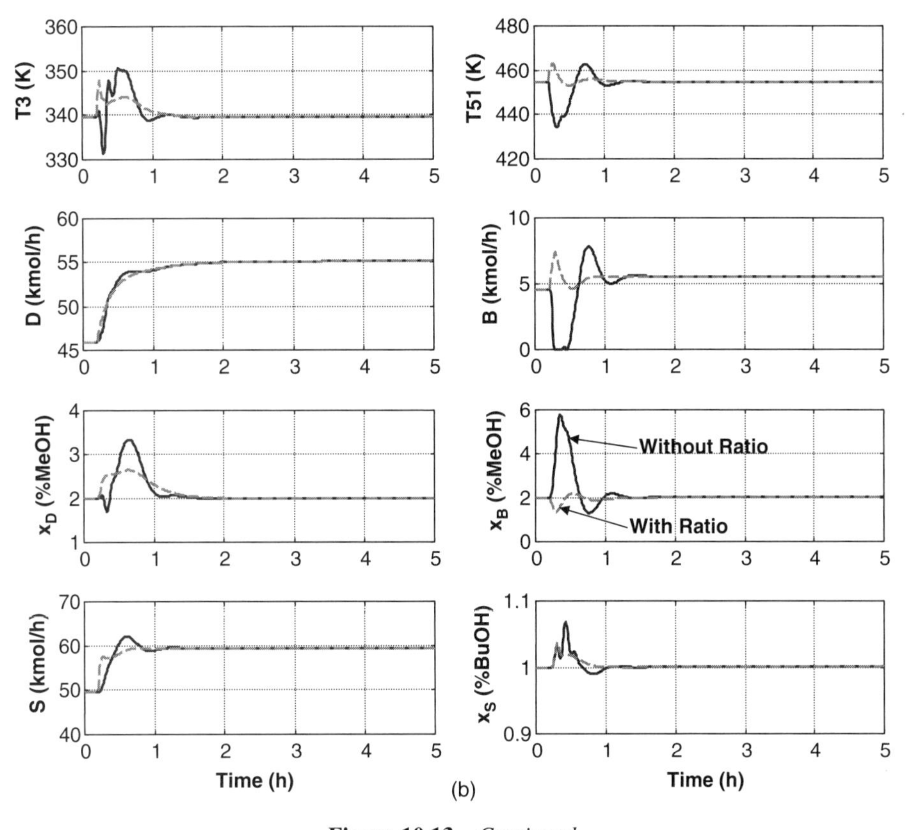

Figure 10.13 *Continued.*

in feed flowrate of 20% are shown in Figure 10.13a. Stable base-level regulatory control is achieved. Note that the bottoms flowrate saturates wide open or completely closed during these transients because of its small flowrate. Product purities return fairly close to their desired levels, but there is a large transient deviation in the bottoms purity for the +20% step increase.

The use of a steam to feed ratio greatly improves this response, as shown in Figure 10.13b. The base-level control output signal resets the reboiler heat input to feed ratio. The steady-state ratio is 0.038689 (in the required Aspen Dynamic units of "GJ/h" per "kmol/h"). The output range of the level controller is changed to 0–0.1, and the TC50 controller is retuned ($K_C = 3.2$ and $\tau_I = 29$ min). The deviation in bottoms purity is greatly reduced, and saturation of the bottoms valve is also avoided.

Responses to three feed composition disturbances are given in Figure 10.14. The solid lines indicate when the BuOH concentration in the feed is increased from 5 to 7 mol%, with the MeOH reduced accordingly. The dashed lines are when the BuOH concentration in the feed is decreased from 5 to 3 mol%, with the MeOH increased accordingly. The dotted lines indicate when the DME concentration in the feed is decreased from 45 to 40 mol%, with the MeOH increased accordingly. Product purities are held fairly close to their desired values, with the sidestream BuOH composition undergoing the largest changes.

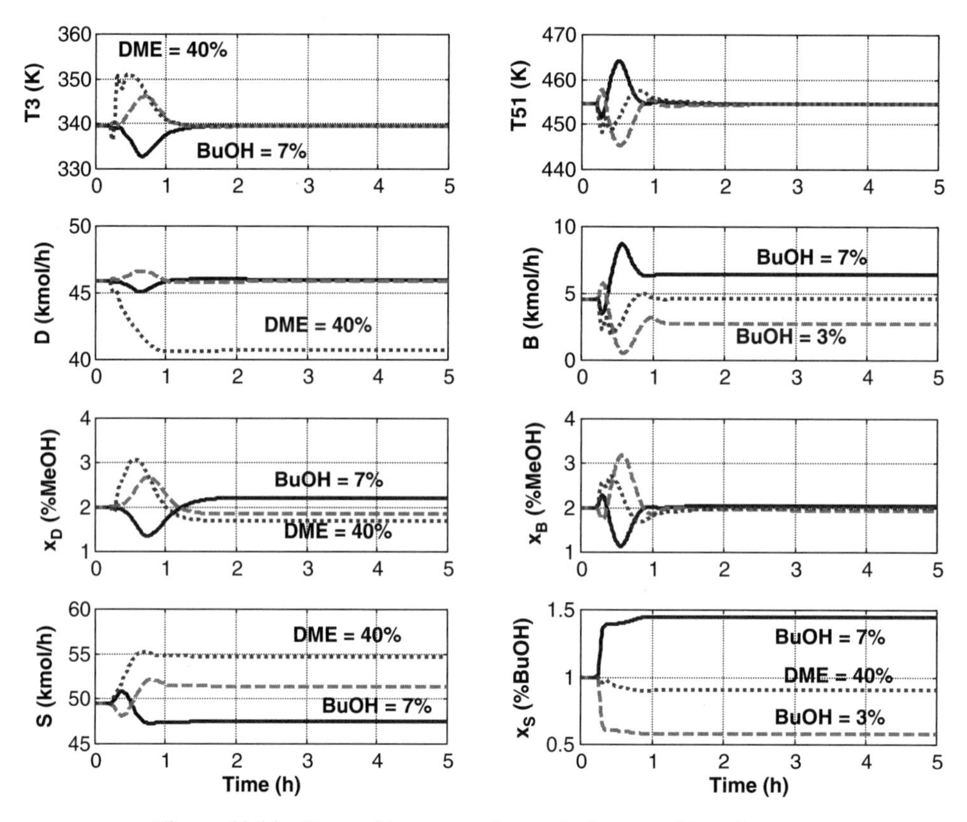

Figure 10.14 Vapor sidestream column; feed composition changes.

10.3 LIQUID SIDESTREAM COLUMN WITH STRIPPER

The flexibility of a sidestream column is greatly increased if additional separation columns are added. This can extend their economic applicability to feeds that have significant amounts of all components and to systems in which relative volatilities are not large. If the sidestream is a liquid, a stripper can be added that removes some of the light impurity in the liquid sidestream coming from the main tower.

To illustrate this configuration we take the same DME/MeOH/water system studied in Section 10.1, but increase the feed concentration of DME from a small 5 mol% up to a significant 35 mol%.

10.3.1 Steady-State Design

The feed flowrate is 100 kmol/h, and the feed is fed on stage 32 of a 52-stage column. A liquid sidestream is withdrawn from stage 12 and is fed into a small six-stage stripping column. The flowsheet is given in Figure 10.15. The bottoms from the stripper is the MeOH product. The vapor from the stripper is fed back to the main column.

The presence of a stripper provides additional degrees of freedom. In addition to being able to adjust the liquid sidestream drawoff rate from the main column, the heat input to the stripper reboiler can also be adjusted. Of course, the stripper can do nothing to change

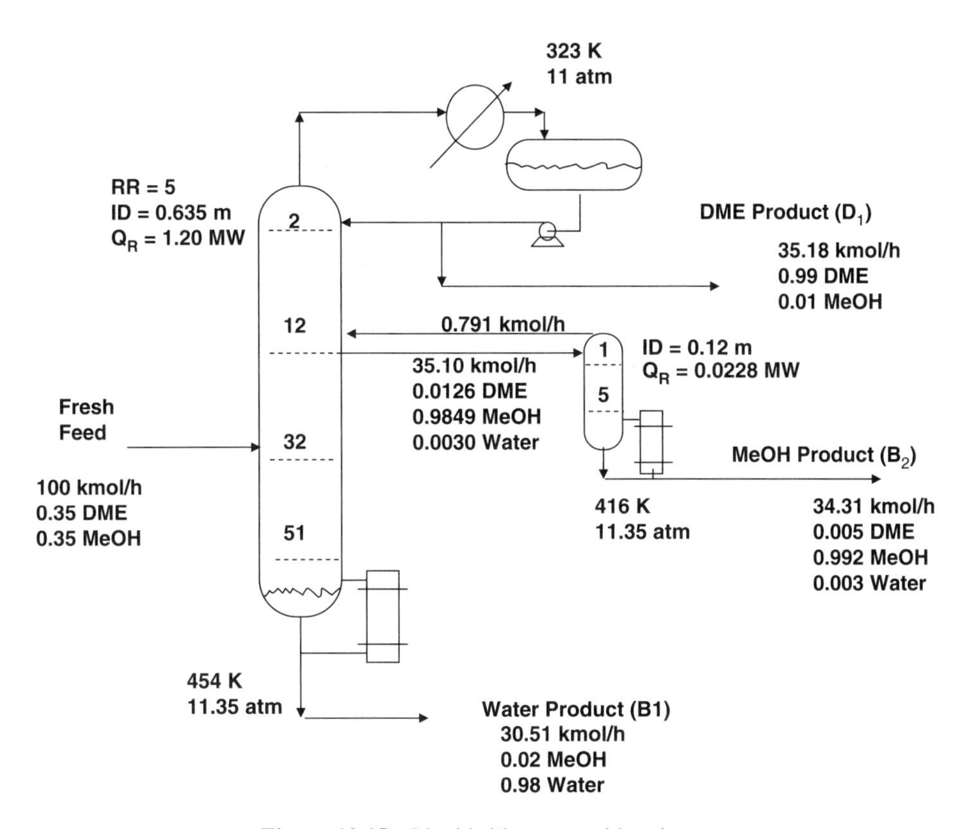

Figure 10.15 Liquid sidestream with stripper.

the amount of the heaviest component (water) in the stream that is fed to it. So the main column must be designed so that the water concentration in the liquid at the sidestream drawoff tray is small. In the design shown in Figure 10.15, the concentration of water is only 0.3 mol% at stage 12.

However, the stripper is able to adjust the amount of the lightest component that leaves as its bottoms product B_2. Therefore, larger amounts of DME cân be present in the liquid at the sidestream tray. This is why the sidestream/stripper configuration can handle larger concentrations of the lightest component in the feed. It is also why this configuration is often economical in systems where the relative volatility between the lightest and intermediate components is not large.

Achieving the steady-state design of this more complex system is not a trivial job. There is a recycle stream and 4 degrees of freedom. The design was successfully achieved by a sequential approach.

1. Initial guesses were made of the flowrates of the distillate, stripper bottoms and sidestream, and the reflux ratio of the main column.
2. A Design Spec/Vary function was set up to achieve a distillate impurity $x_{D1(\text{MeOH})}$ of 1 mol% MeOH by adjusting the distillate D_1 flowrate.
3. A second Design Spec/Vary was set up to achieve a main-column bottoms B_1 impurity $x_{B1(\text{MeOH})}$ of 0.5 mol% MeOH by adjusting the flowrate of the liquid sidestream S_1 withdrawn from the main column that is fed to the stripper.

4. A third Design Spec/Vary was set up to achieve a stripper bottoms B_2 impurity $x_{B2(\mathrm{DME})}$ of 2 mol% DME by adjusting the ratio of the stripper bottoms to the stripper feed (B_2/S_1).

The reflux ratio of 5 was held constant during these convergences. This additional degree of freedom could be used to adjust the water concentration in the sidestream fed to the stripper. A reflux ratio of 5 gives 0.3 mol% water in the sidestream, which permits a high-purity methanol product to be produced.

Figure 10.15 gives the steady-state conditions. The energy consumption in the main column is 1.20 MW, while in the stripper it is only 0.0228 MW. The diameter of the main column is 0.635 m, and the diameter of the stripper is 0.12 m.

10.3.2 Dynamic Control

The flowsheet shown in Figure 10.15 does not show the plumbing required to run a realistic pressure-driven dynamic simulation. The key feature is that the pressure in the stripper must be greater than that in the main column so that vapor can flow from the top of the stripper back to the main column. Therefore, in the simulation, a pump and a control valve are placed in the liquid sidestream. A control valve is also placed in the stripper overhead vapor line. All this plumbing is shown in Figure 10.16. In a real physical setup it is usually possible to use elevation differences to provide the necessary differential pressure driving force to get the liquid to flow from the main column into the stripper at a higher pressure and avoid the use of a pump.

The usual sizing calculations are performed for both the main column and the stripper (10-min liquid holdups in column bases and reflux drum). The flowsheet is pressure-checked, and the file is exported to Aspen Dynamics.

A control structure is developed for this more complex system. The various loops are described below. The key issues are how to manipulate the sidestream and how to maintain the compositions of the three product streams.

1. Feed is flow controlled.
2. With a reflux ratio of 5, the reflux drum level is controlled by manipulating reflux flowrate.
3. The temperature on stage 4 of the main column is controlled by manipulating distillate.
4. Base level is controlled by manipulating bottoms flowrate in both columns.
5. Pressure in the main column is controlled by manipulating condenser heat removal.
6. Pressure in the stripper is controlled by manipulating the valve in the vapor line V22.
7. The temperature on stage 51 in the main column is controlled by manipulating reboiler heat input to the main column.
8. The temperature on stage 5 in the stripper is controlled by manipulating stripper reboiler heat input.
9. The sidestream flowrate is ratioed to reflux flowrate.

This last loop is the most important feature of this control structure. It permits changes in the sidestream as disturbances enter the system.

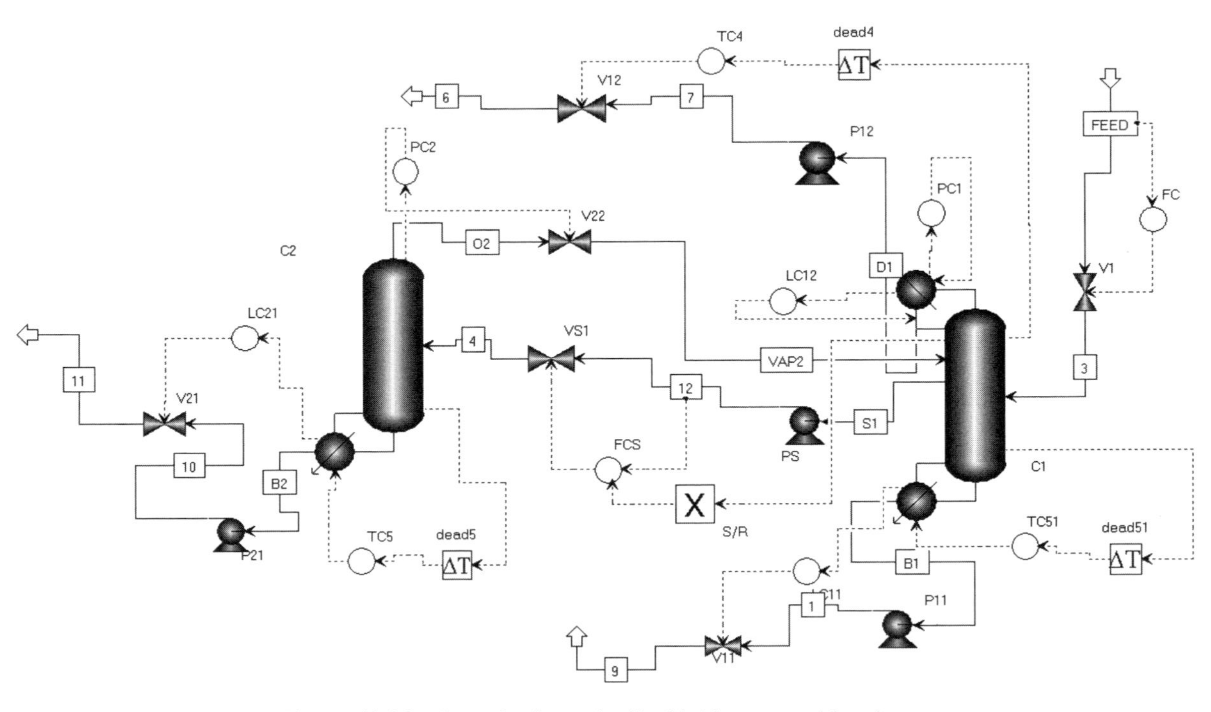

Figure 10.16 Control scheme for liquid sidestream with stripper.

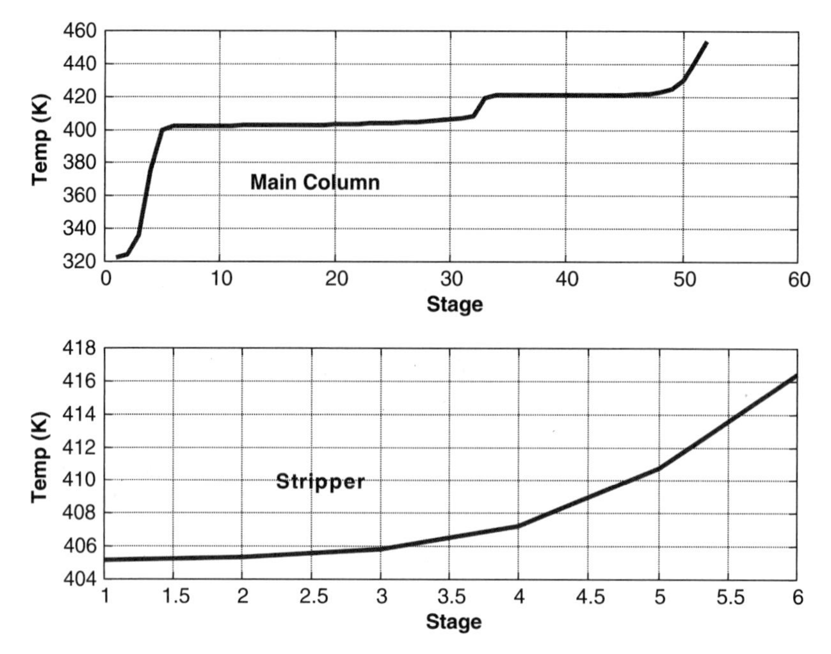

Figure 10.17 Temperature profiles.

Locations of the temperature control trays are selected by looking at the temperature profiles in the two columns, shown in Figure 10.17. The three temperature controllers (TC4 and TC51 in the main column and TC5 in the stripper) have 1-min deadtimes and 100 K temperature transmitter spans. They are tuned individually by running a relay–feedback test with the other two controllers on manual and using the Tyreus–Luyben settings. The controller tuning constants are given in Table 10.1.

Note that the TC4 temperature control loop is "nested" inside the reflux drum level controller since a change in distillate flowrate has no *direct* effect on stage 4 temperature. The tuning of the reflux drum level controller (LC12) affects the tuning of the TC4 temperature controller, as Table 10.1 shows. Performance is improved by tightening up on the level controller. This is illustrated in Figure 10.18 for a 20% increase in feed flowrate.

The responses for feed composition disturbances are shown in Figure 10.19. In Figure 10.19a the feed composition is changed from 35 mol% DME to either 40 or 30 mol% DME (with a corresponding change in MeOH). In Figure 10.19b the feed

TABLE 10.1 Stripper Temperature Controller Parameters

	TC4 with LC12 Gain = 2	TC4 with LC12 Gain = 5	TC51	TC5
K_U	3.67	2.65	2.98	18.0
P_U (min)	11.4	8.4	11.4	3.6
K_C	1.15	0.89	0.93	5.64
τ_I (min)	25.1	18.5	6.6	7.92

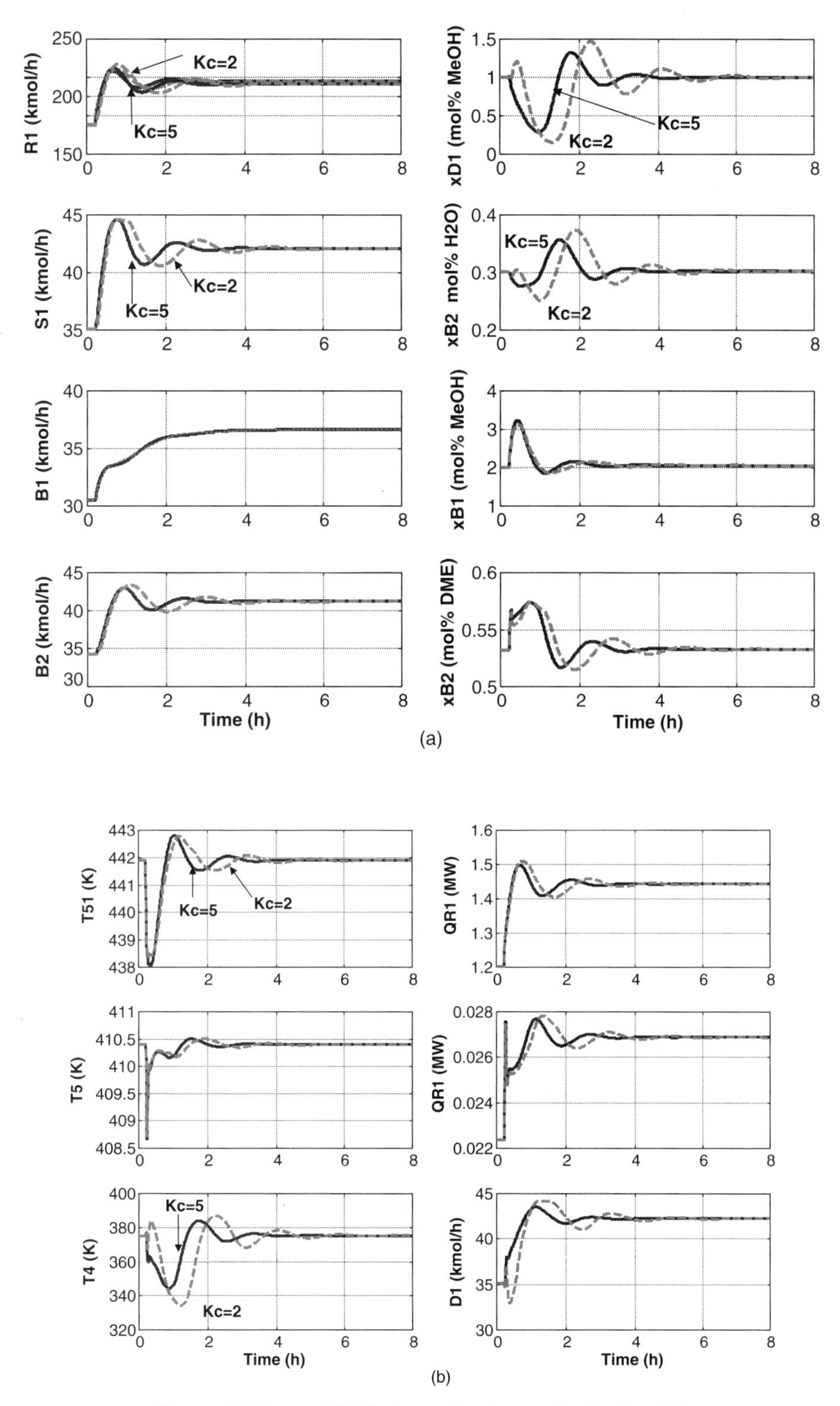

Figure 10.18 + 20% feed rate; level control gains 2 and 5.

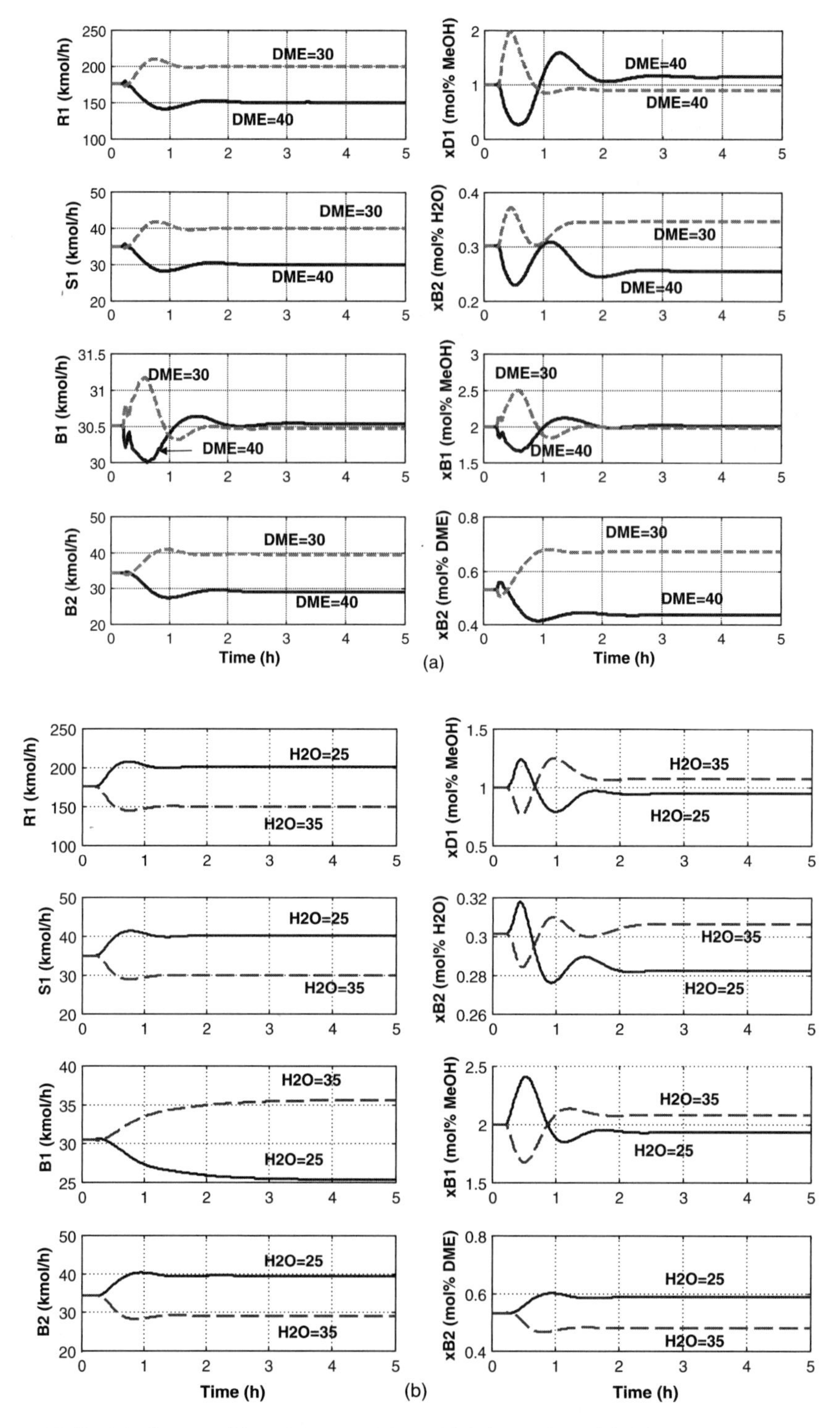

Figure 10.19 Feed composition changes: (a) DME 35 changed to 40/30; (b) water 30 changed to 35/25.

composition is changed from 30 mol% water to either 35 or 25 mol% water (with a corresponding change in MeOH).

The control structure provides stable base-level regulatory control for all these disturbances. Product purities are maintained fairly close to the desired values.

10.4 VAPOR SIDESTREAM COLUMN WITH RECTIFIER

If the sidestream is a vapor, a rectifier can be added that removes some of the heavy impurity in the vapor sidestream coming from the main tower. To illustrate this configuration, we take the same DME/MeOH/water system studied in the previous section but remove a vapor sidestream below the feed stage.

10.4.1 Steady-State Design

The feed flowrate is 100 kmol/h, and the feed is fed on stage 11 of a 52-stage column. A vapor sidestream is withdrawn from stage 31 and is fed into a 12-stage rectifier column. The flowsheet is given in Figure 10.20. The distillate D_2 from the rectifier is the MeOH product. The bottoms liquid stream from the rectifier is pumped back to stage 32 of the main column.

The rectifier must operate at a pressure lower than that of the main column so that the vapor sidestream can flow "downhill." The main column operates at 11 atm. The pressure

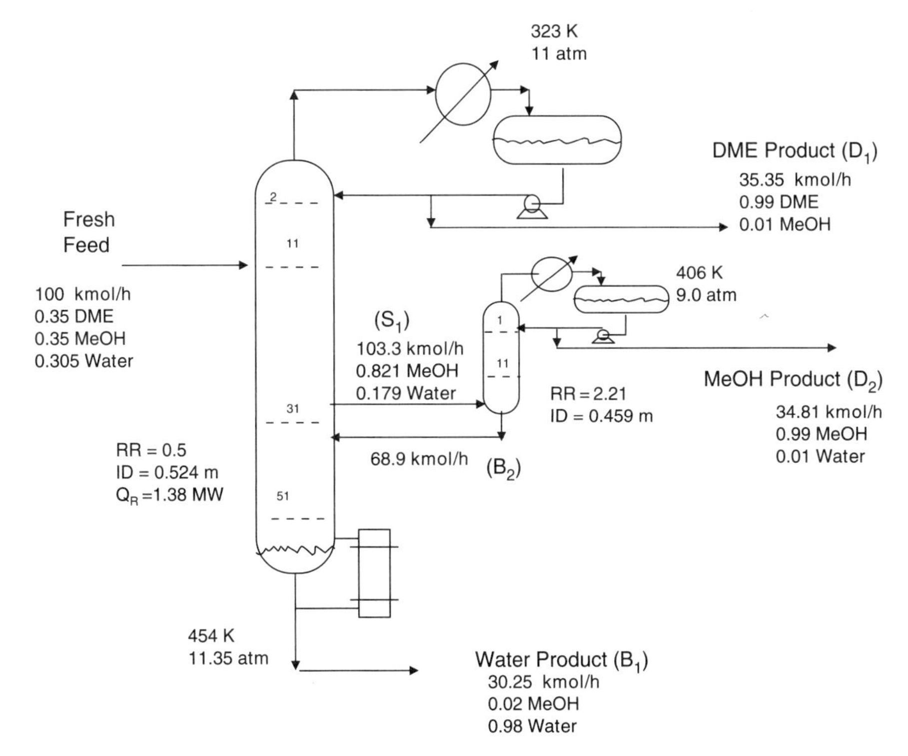

Figure 10.20 Vapor sidestream with rectifier.

in the rectifier is set at 9 atm to provide some pressure drop over the control valve in the vapor line.

The feed composition is 35 mol% DME, 35 mol% MeOH, and 30 mol% H_2O. The distillate from the main column D_1 is the DME product and has an impurity specification of 1 mol% MeOH. The bottoms from the main column B_1 is the water product and has an impurity specification of 2 mol% MeOH. The distillate from the rectifier D_2 is the methanol product and has an impurity specification of 1 mol% water. The DME concentration in this stream is negligible because there is very little DME in the main column below the feed tray. The feed is fed on stage 11 near the top. The vapor sidestream is withdrawn down at stage 31.

The steady-state design of this two-column system with recycle was achieved by a sequential approach. First, the flowrates of the distillates from the two columns were set equal to the molar flowrates of DME and MeOH, respectively, in the feed. The reflux ratio in the main column was set at 2. The flowrate of the vapor sidestream S_1 was set at 3 times the MeOH product rate, which gives a reflux ratio in the rectifier of 2. Note that there is only one degree of freedom in the rectifier, so setting the distillate flowrate completely specifies the column.

Then a guess was made of the composition of the recycle stream $x_{B2,j}$ flowing back to the main column (the bottoms of the rectifier B_2). The simulation was converged, and the difference between the guessed values of x_{B2} and the calculated values was observed. New guesses were made until there was little difference. Then the recycle loop was closed using a *Tear* specification.

The next step was to adjust the various degrees of freedom to attain the desired purities of the three products. Three Design Spec/Vary functions were used in a sequential manner:

1. The impurity of MeOH in the main-column distillate $x_{D1(\text{MeOH})}$ was fixed at 1 mol% MeOH by varying the distillate flowrate D_1.
2. The impurity of MeOH in the rectifier distillate $x_{D2(\text{MeOH})}$ was fixed at 1 mol% MeOH by varying the distillate flowrate D_2.
3. The impurity of MeOH in the main-column bottoms $x_{B1(\text{MeOH})}$ was fixed at 2 mol% MeOH by varying the flowrate of the vapor sidestream S_1.

Finally, the reflux ratio in the main column was reduced to see the effect on the DME impurity in the sidestream. Since DME is much more volatile than MeOH and is fed above the sidestream drawoff tray, the reflux ratio could be reduced to 0.5 without affecting sidestream composition. With a reflux ratio of 0.5, the liquid rate in the top section of the main column is quite small (17 kmol/h) compared to the liquid rates lower in the column (180 kmol/h). Therefore reflux ratios lower than 0.5 were not used.

Figure 10.20 gives design parameters and equipment sizes of this process. The reboiler heat input in the main column is 1.38 MW. It is interesting to compare this with the energy requirements of the sidestream/stripper flowsheet shown in Figure 10.15 ($1.20 + 0.022 = 1.22$ MW). The two processes produce essentially the same three product streams with the same purities. The energy consumptions of the two flowsheets are quite similar.

The diameter of the main column is 0.524 m, and the diameter of the rectifier is 0.459 m. Thus the size of the rectifier is significantly larger than the stripper (0.12 m) in the alternative flowsheet, indicating higher capital cost for the rectifier process. The inherent reason for this is that the MeOH/water separation (which takes place in the rectifier) is more difficult than the DME/MeOH separation (which takes place in the stripper).

10.4.2 Dynamic Control

The flowsheet shown in Figure 10.20 does not show the plumbing required to run a realistic dynamic simulation. The key feature is that the pressure in the rectifier must be less than that in the main column so that vapor can flow from the main column into the rectifier. Therefore, in the simulation, a control valve is placed on the vapor line and a pump and a control valve are placed in the liquid recycle linc from the bottom of the rectifier back to the main column. All this plumbing is shown in Figure 10.21. In a real physical setup it may be possible to use elevation differences to provide the necessary differential pressure driving force to get the liquid to flow from the rectifier into the main column at a higher pressure and avoid the use of a pump.

The usual sizing calculations are performed for both the main column and the rectifier (10-min liquid holdups in column bases and reflux drum). The flowsheet is pressure-checked, and the file is exported to Aspen Dynamics.

The development of a control structure for this complex system turned out to be more difficult that for the stripper flowsheet. The initial control scheme evaluated is shown in Figure 10.21. This is a logical extension of the control structure used for the vapor sidestream column in which the vapor sidestream is ratioed to the reboiler heat input. As we will demonstrate below, this structure worked well for some disturbances, but it could not handle decreases in the composition of methanol in the feed, resulting in a shutdown of the unit.

The various loops of the initial control structure are described below. The key issues are how to manipulate the sidestream and how to maintain the compositions of the three products:

1. Feed is flow controlled.
2. With a reflux ratio of 0.5, the reflux drum level is controlled by manipulating distillate flowrate.
3. The temperature on stage 4 of the main column is controlled by manipulating reflux flowrate. Stage 12 is tested as an alternative later.
4. Base levels are controlled by manipulating bottoms flowrates.
5. Pressure in the main column is controlled by manipulating condenser heat removal in the main-column condenser.
6. Pressure in the rectifier is controlled by manipulating condenser heat removal in the rectifier condenser.
7. The temperature on stage 51 in the main column is controlled by manipulating reboiler heat input.
8. The temperature on stage 9 in the rectifier is controlled by manipulating reflux flowrate in the rectifier.
9. The sidestream flowrate is ratioed to heat input to the reboiler.

The locations of the temperature control trays are selected by looking at the temperature profiles in the two columns, shown in Figure 10.22. The three temperature controllers (TC4 and TC51 in the main column and TC9 in the rectifier) have 1-min deadtimes and 100 K temperature transmitter spans. They are tuned individually by running a

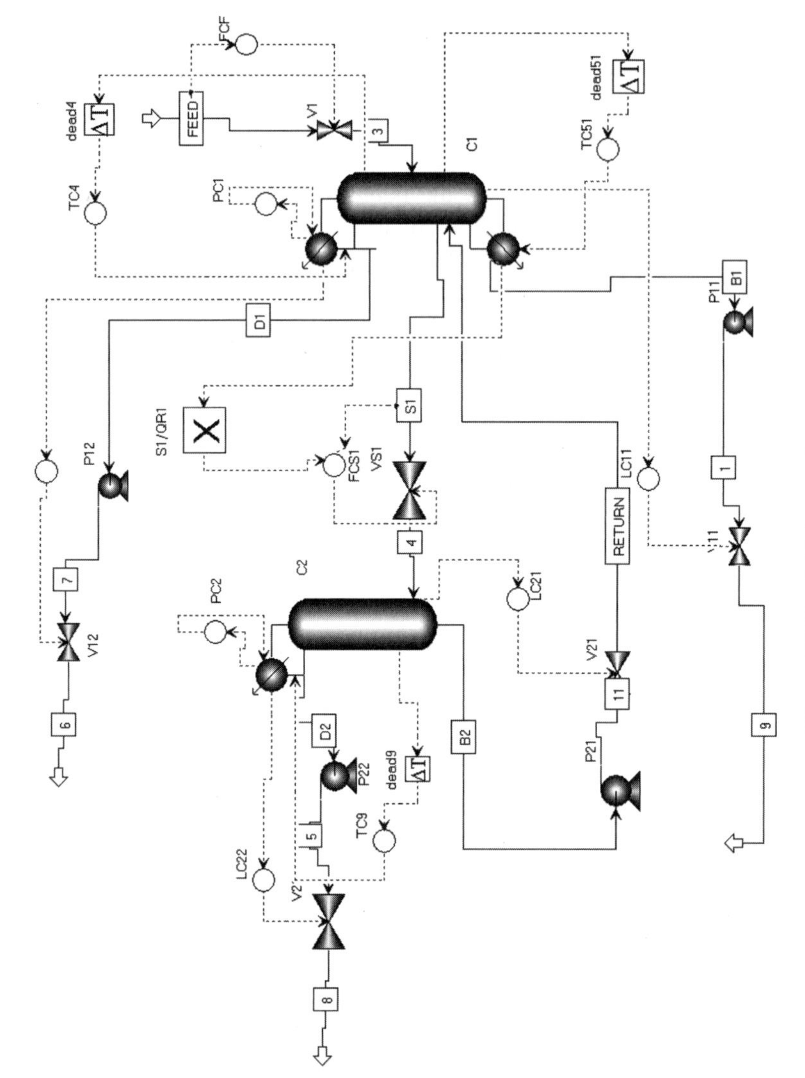

Figure 10.21 Initial control structure.

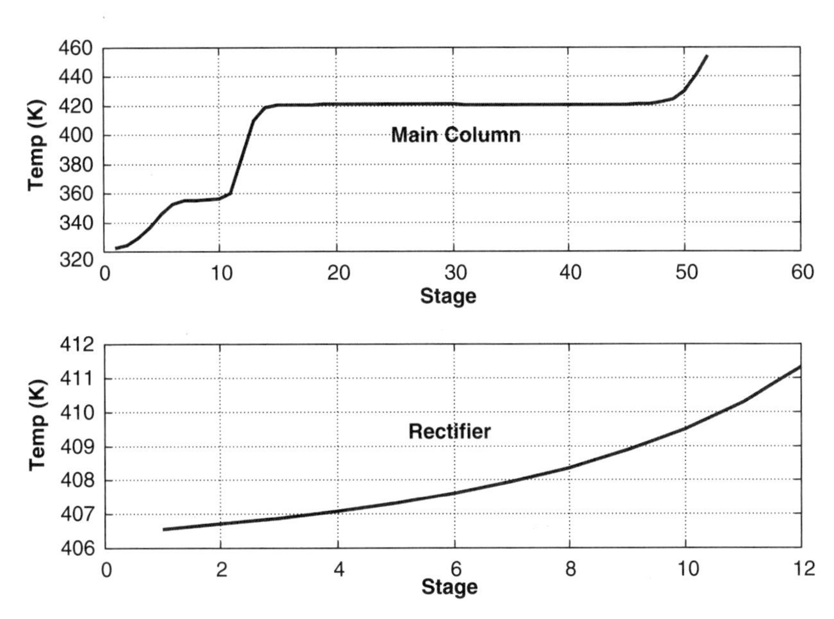

Figure 10.22 Temperature profiles.

relay–feedback test with the other two controllers on manual and using the Tyreus–Luyben settings. The controller tuning constants are given in Table 10.2. The only surprising result of this tuning is the very high gain of the TC9 controller.

The response of this control structure for changes in feed flowrate is quite acceptable, as shown in Figure 10.23a. However, when some types of feed composition disturbance are made, the unit shuts down.

For example, as shown in Figure 10.23b, the process shuts down after about 4 h, when the feed composition is changed from 35/35/30 DME/MeOH/H_2O to 40/30/30. The reflux flowrate R_2 in the rectifier goes to zero because the temperature controller responds to lowering temperatures in the rectifier. This occurs because too much DME has worked its way down the column to the vapor sidestream drawoff tray. Once this happens, the rectifier can do nothing about rejecting the light DME. Instead of using stage 3 for temperature control, stage 12 was tested, but the results were similar.

To overcome these problems, a new control structure was developed. The flowrate of the reflux in the main column is quite small, so instead of manipulating it to control a temperature, it is held constant for a given feed flowrate (a reflux to feed ratio is used). As shown in Figure 10.24, the major change in the control structure is to control the

TABLE 10.2 Rectifier Temperature Controller Parameters

	TC4	TC51	TC9
K_U	5.54	2.05	190
P_U (min)	4.2	3.6	8.4
K_C	1.73	0.64	59
τ_I (min)	9.24	7.92	18.5

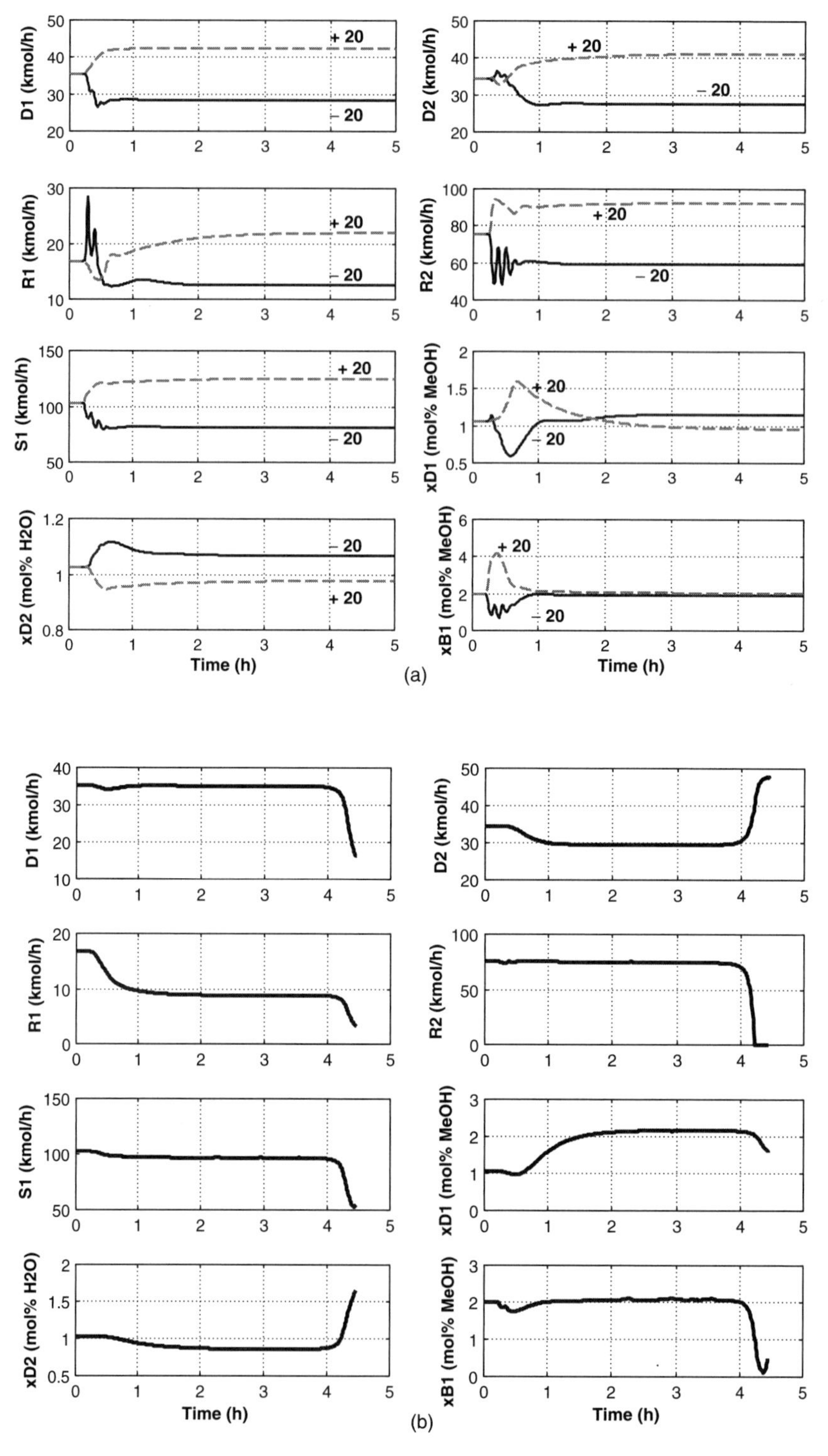

Figure 10.23 Initial control structure: (a) feed rate changes; (b) feed composition DME 35 changed to 40.

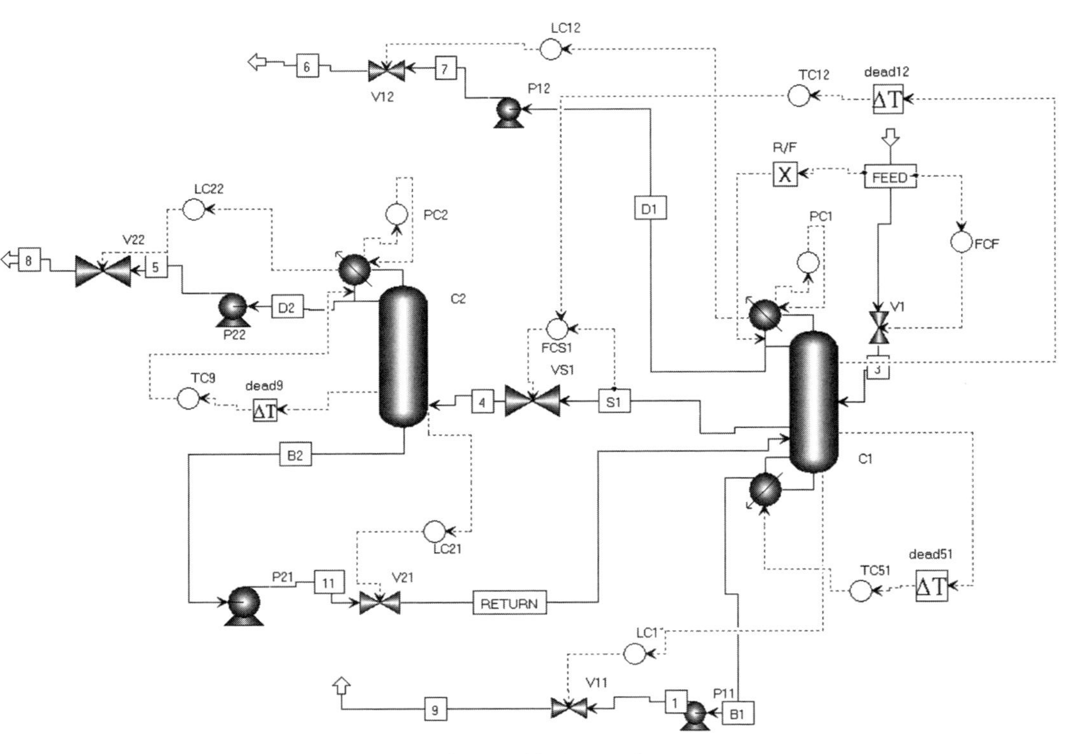

Figure 10.24 Modified control structure.

temperature on stage 12 by manipulating the vapor sidestream flowrate. This is a direct-acting controller; an increase in stage 12 temperature *increases* the vapor sidestream flowrate to the rectifier. The TC12 temperature controller is tuned by running a relay–feedback test with the other two controllers on manual and using the Tyreus–Luyben settings: $K_C = 0.22$ and $\tau_I = 12$ min.

The modified control structure provides stable regulatory control for both feed flowrate and feed composition. However, as shown in Figure 10.25, the 20% increase in feed flowrate produces a large transient disturbance in the purity of the distillate from the main column with a peak $x_{D1(\mathrm{MeOH})}$ of ~8 mol% MeOH. This can be improved by using feedforward control. Installing a sidestream to heat input ratio as shown in Figure 10.26, with the ratio reset by the TC12 temperature controller, improves the load response of the system. The temperature controller is retuned ($K_C = 0.10$ and $\tau_I = 12$ min). A maximum controller output of 40 is used since the normal S_1/Q_{R1} ratio is 21.

Figure 10.27 compares the responses to a 20% increase in feed flowrate for the two cases, with and without the use of the ratio. The improvement the purity of the DME product is striking. Figure 10.28a gives results for changes in the DME feed concentration from 35 mol% to either 40 or 30 mol% DME (with corresponding changes in MeOH). Figure 10.28b gives results for changes in the water feed concentration from 30 mol% to either 25 or 35 mol% H_2O (with corresponding changes in MeOH). The control structure handles these many disturbances quite effectively.

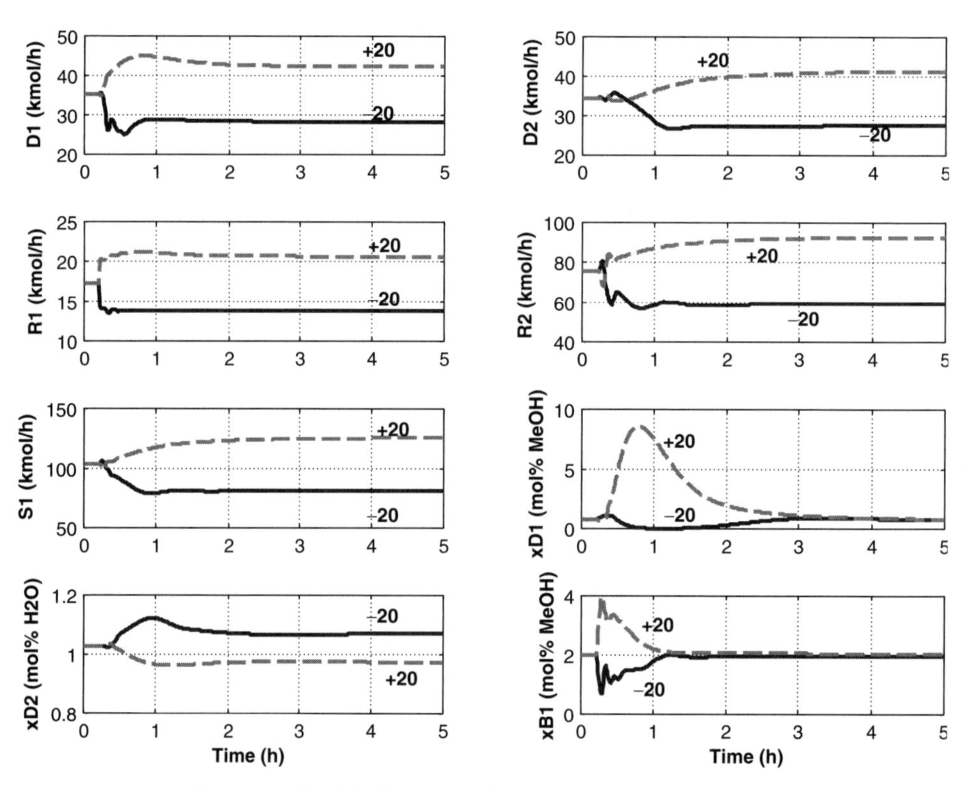

Figure 10.25 Modified control structure; feed rate changes.

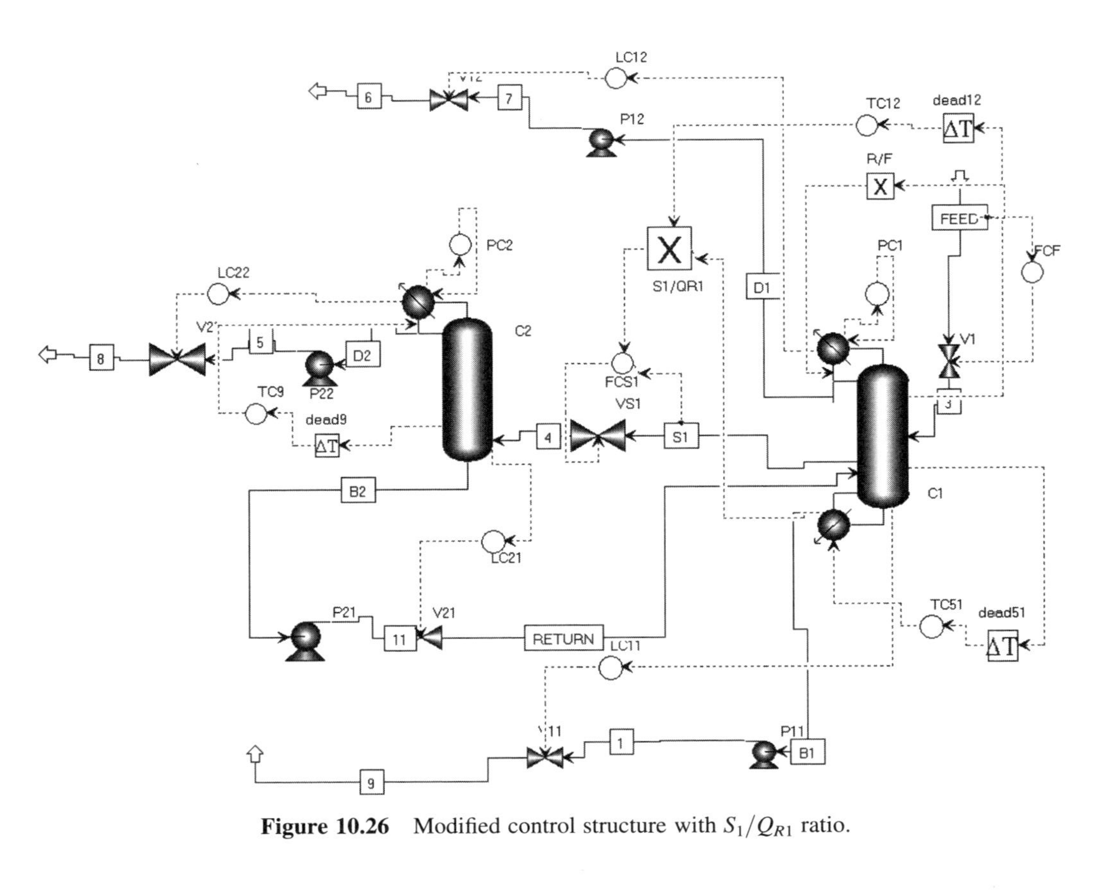

Figure 10.26 Modified control structure with S_1/Q_{R1} ratio.

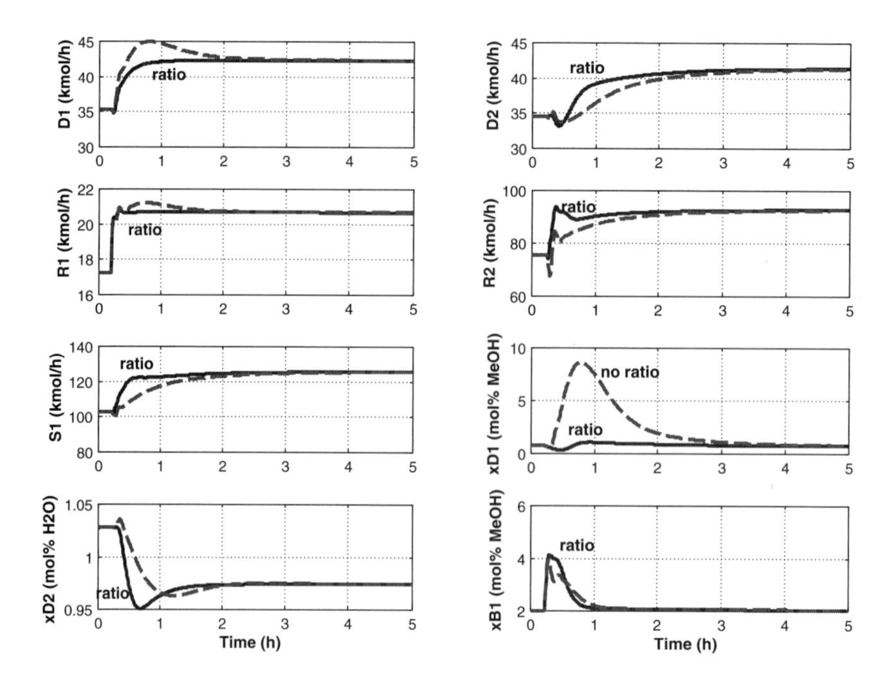

Figure 10.27 + 20% feed rate change with and without S_1/Q_{R1} ratio.

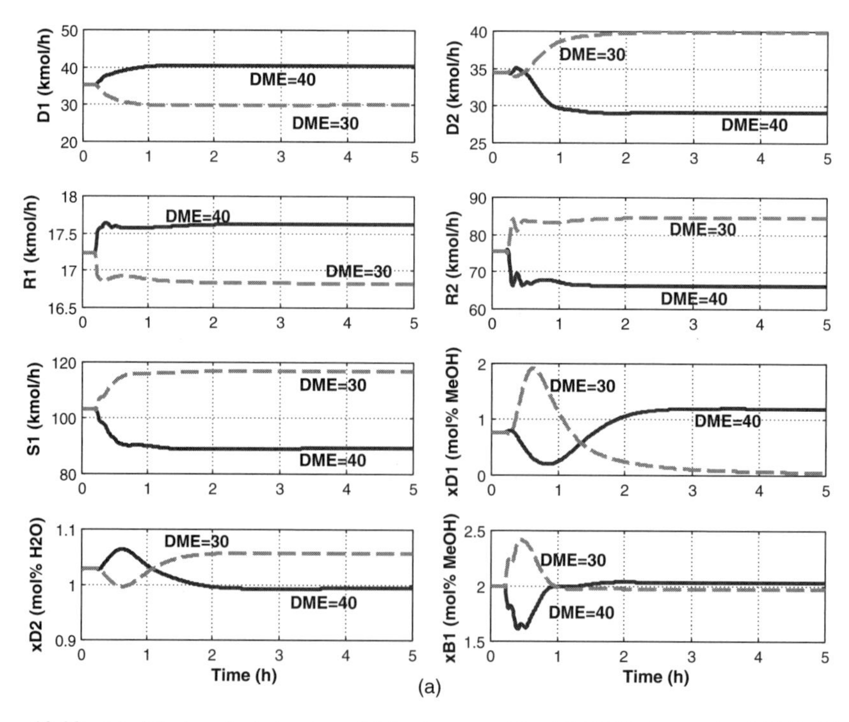

Figure 10.28 Modified control structure: (a) feed composition DME 35 changed to 30/40; (b) feed composition H_2O 30 changed to 25/35.

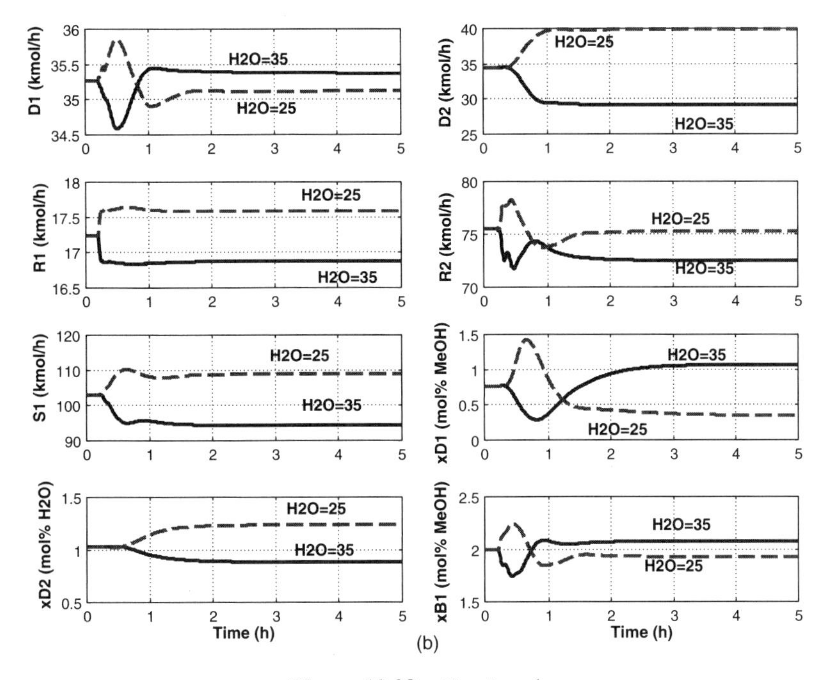

Figure 10.28 *Continued.*

Note that the reflux molar flowrate in the main column R_1 changes for the changes in feed composition. This occurs because a mass ratio of reflux-to-feed is used and the molar flowrate of the feed is held constant.

10.5 SIDESTREAM PURGE COLUMN

The final sidestream column configuration studied in this chapter is the case where there is a very small amount of the intermediate component in the feed. The lightest component is the distillate product. The heaviest component is the bottoms product. The intermediate component builds up in the middle of the column, forming a composition "bubble." The objective is to purge off a very small stream from the column at the location where the intermediate composition is fairly large so that there is only a small loss of valuable products. Thus a high-purity sidestream is not required.

10.5.1 Steady-State Design

The feed flowrate is 100 kmol/h and its composition is 49 mol% DME, 1 mol% MeOH, and 50 mol% water. The feed is fed on stage 16 of a 31-stage column. A small liquid sidestream is withdrawn from stage 12 at a flowrate of 1.21 kmol/h with a composition of 17.4 mol% DME, 81.7 mol% methanol, and 0.9 mol% water. Thus the purity of the sidestream is low, but this is not a problem because only small amounts of valuable products (the DME and the water) are lost in this small stream.

The purities of the distillate and bottoms products are very high (99.99 mol%). If the specifications for the two product purities were low (e.g., 1 mol% MeOH), the MeOH

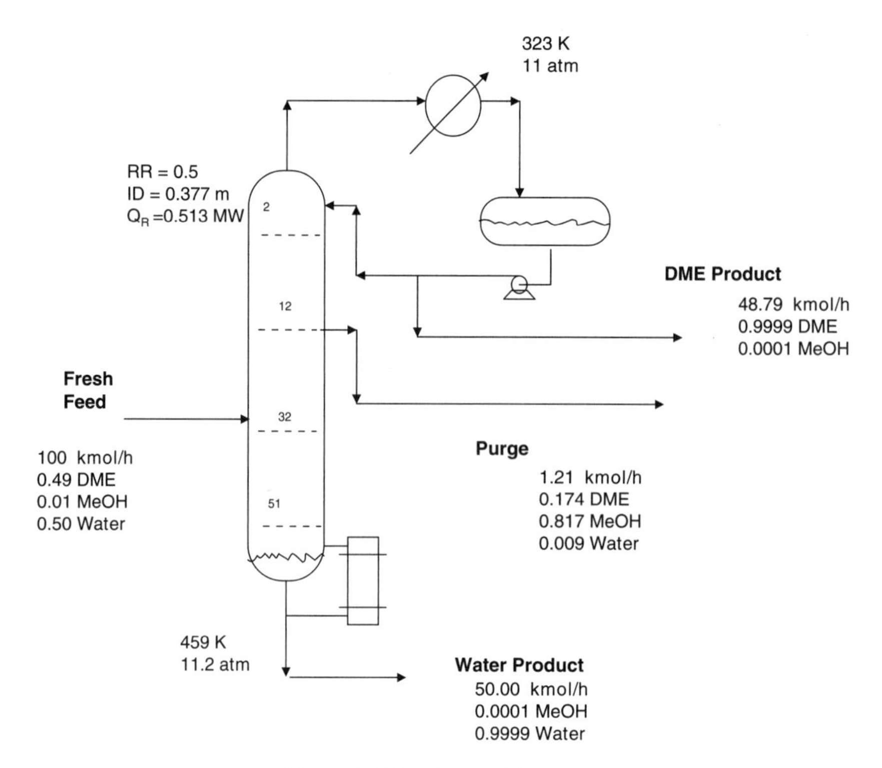

Figure 10.29 Purge sidestream.

in the feed could simply be removed in the distillate and bottoms. The column operates at 11 atm with a reflux ratio of 0.5 and a reboiler heat input of 0.513 MW. Figure 10.29 gives the flowsheet of the process with design parameters.

The steady-state simulation is converged by first setting the reflux ratio equal to 0.5. Then two Design Spec/Vary functions were used. The first function adjusted distillate flowrate to achieve 0.01 mol% MeOH in the distillate; the second one adjusted sidestream flowrate to achieve 0.01 mol% MeOH in the bottoms.

The purity of the sidestream and the shape of the composition profiles depend on the reflux ratio used. Figure 10.30 gives the composition profiles with a reflux ratio of 0.5. Note that there are two peaks in the methanol composition profile. The sidestream is withdrawn at stage 12, where the larger peak occurs. Figure 10.31 shows how the methanol composition profile changes with reflux ratio. The higher the reflux ratio, the higher the purity of the sidestream. The reflux ratio of 0.5 is selected because it gives a sidestream that is reasonable pure.

10.5.2 Dynamic Control

The control of this purge sidestream column is much more complex than one might expect. Since the sidestream is very small, we might assume that simply flow-controlling the sidestream at a rate high enough to remove all the intermediate in the worst case (highest feed concentration) might do the job.

For example, suppose that we set the sidestream flowrate at 3 kmol/h instead of the design 1.21 kmol/h. Unfortunately this reduces the concentration of the methanol in the sidestream from 81.7 mol% to 33.3 mol% under design conditions where the feed

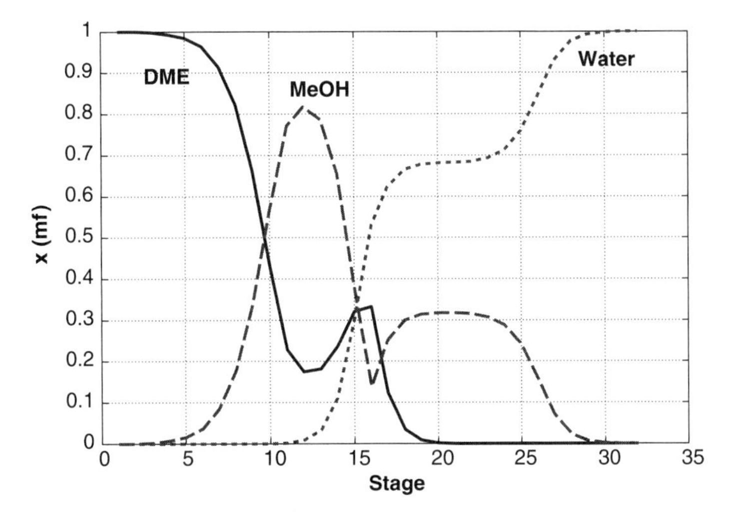

Figure 10.30 Purge sidestream column; $RR = 0.5$.

composition is 1 mol% MeOH. Let us consider a control structure in which the temperature on stage 17 is controlled by manipulating reboiler heat input and reflux flowrate is fixed. Now, if the feed composition is increased to 2 mol% MeOH, the sidestream composition changes to only 42 mol% MeOH. This is not enough to remove all the additional MeOH in the feed, so the distillate purity is severely affected (it increases to 1.55 mol% MeOH). So a simple control structure with a fixed sidestream flowrate does not provide effective product quality control. The control structure must be able to adjust the sidestream flowrate in some manner so that methanol cannot drop out the bottom or go overhead.

An apparently straightforward control structure is to measure the methanol composition of the sidestream and to control this composition by manipulating sidestream flowrate. As we demonstrate below, this scheme does not work well.

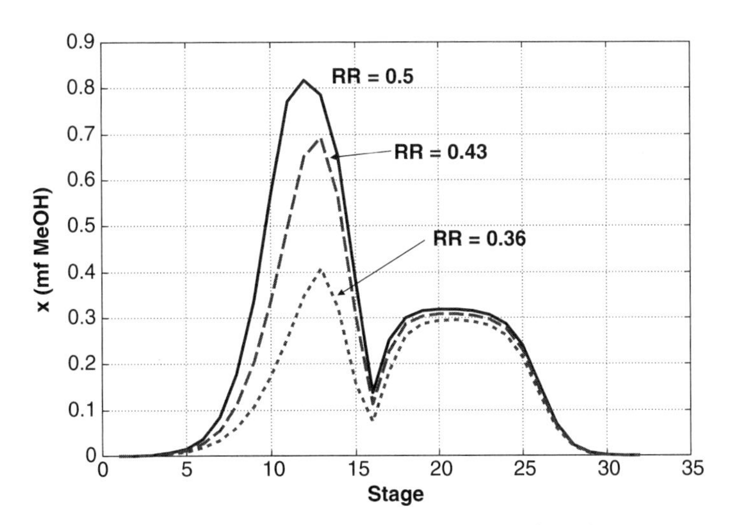

Figure 10.31 Purge sidestream column: effect of *RR*.

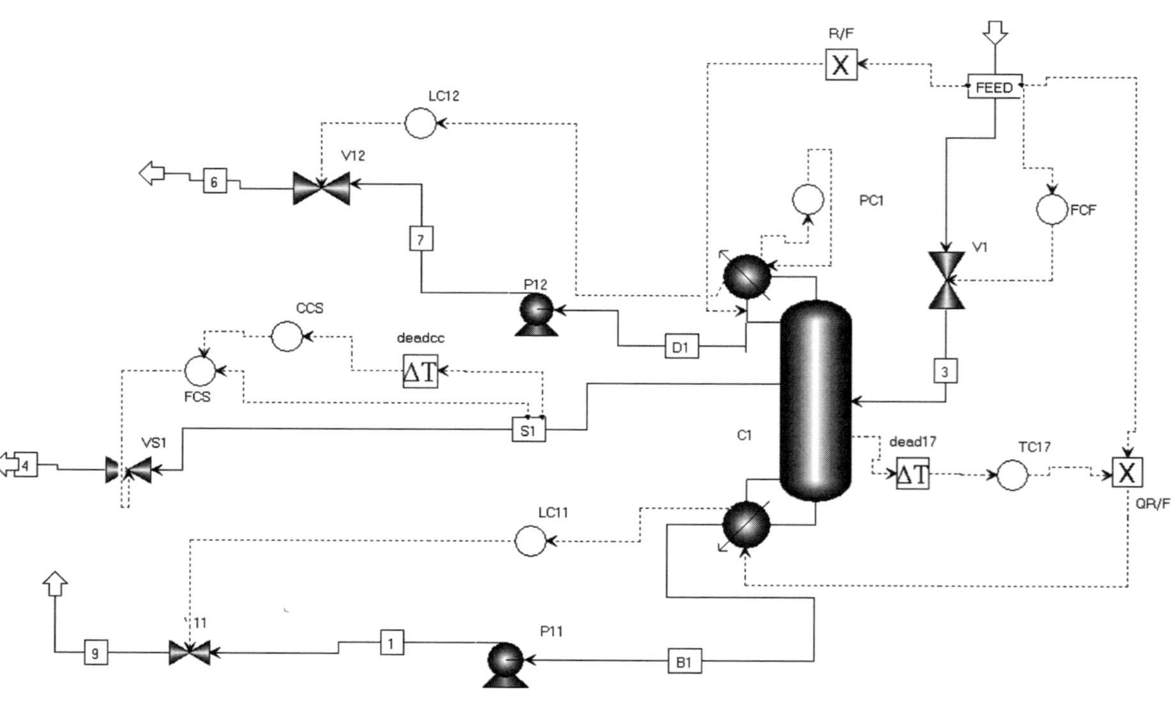

Figure 10.32 Control structure with $x_{S(MeOH)}$ control.

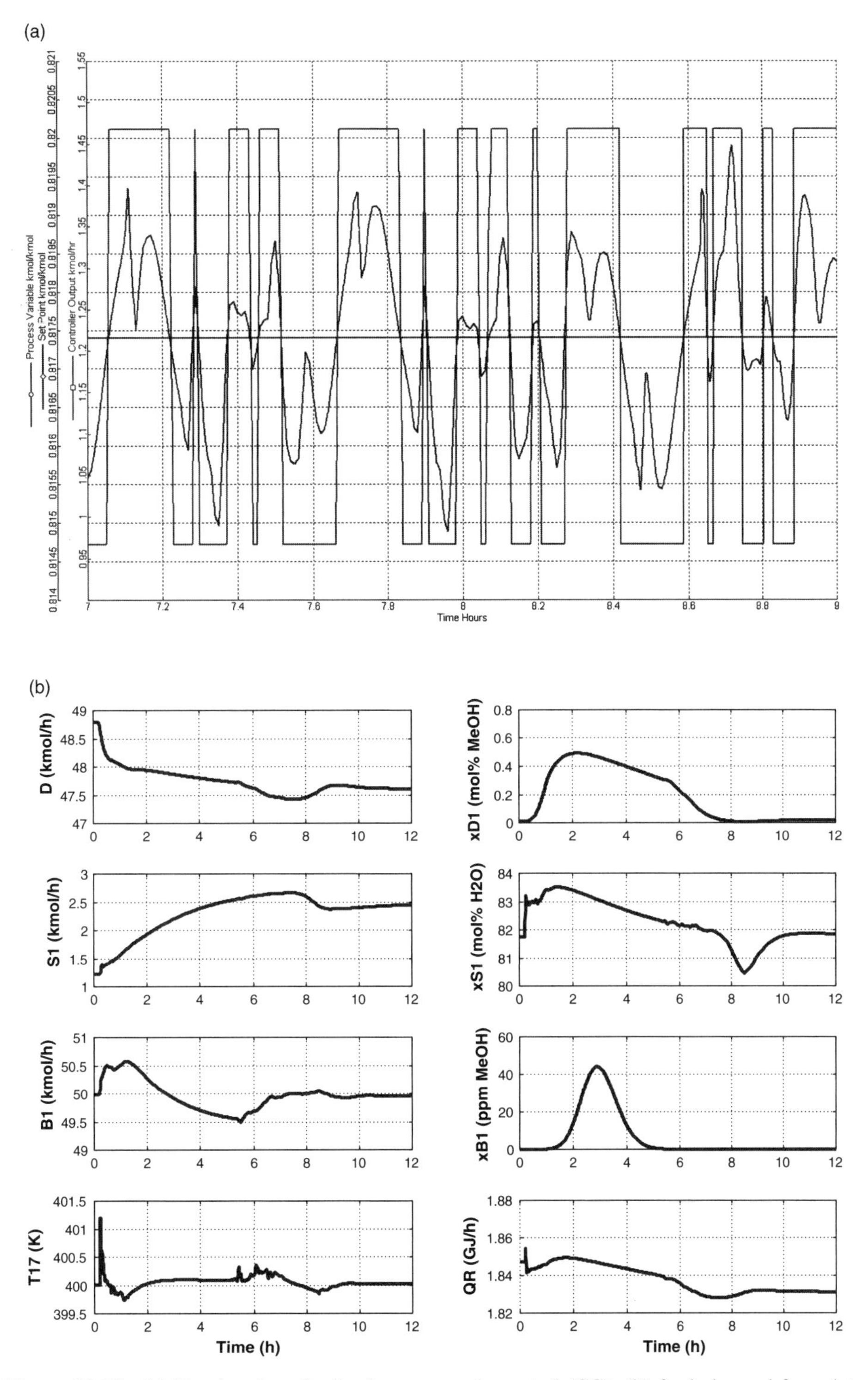

Figure 10.33 (a) Erratic relay–feedback test cascade control (CC); (b) feed changed from 1 to 2 mol% MeOH; CC sidestream with empirical tuning.

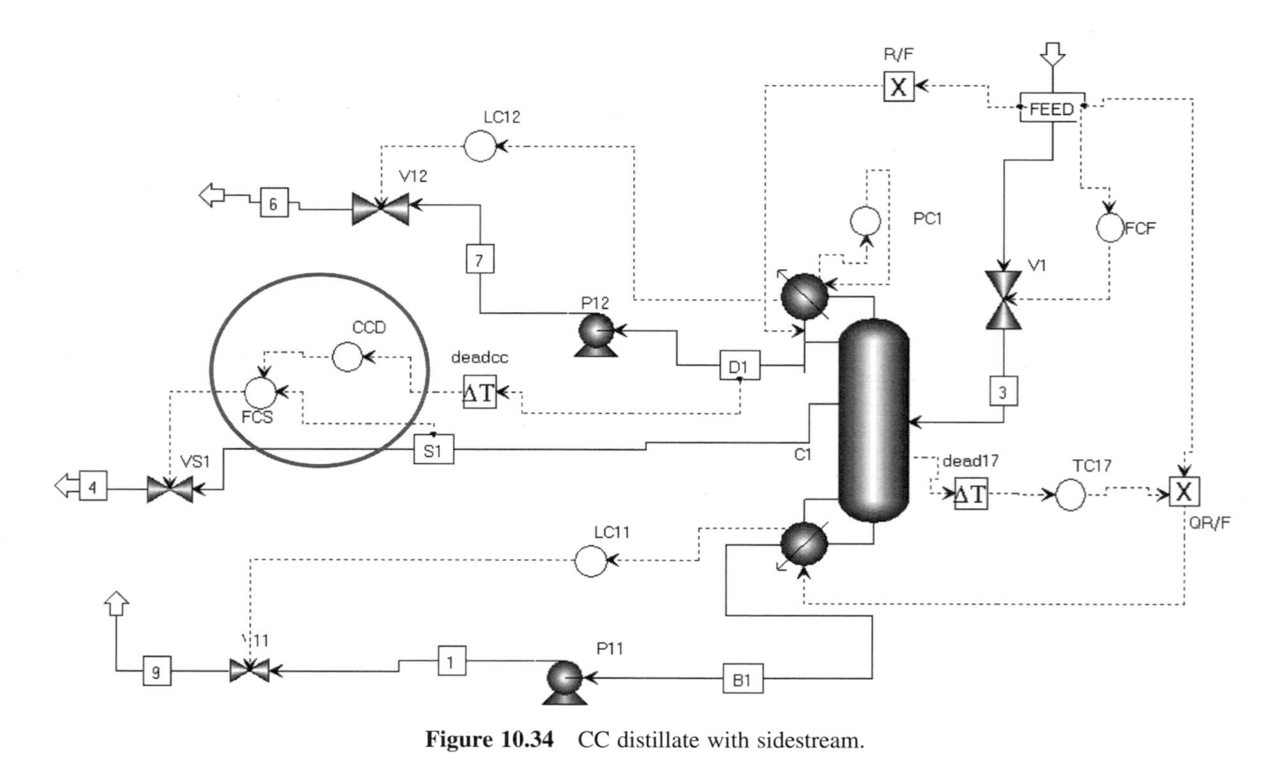

Figure 10.34 CC distillate with sidestream.

To gain some insight into the problem, singular value decomposition analysis was used. There are three manipulated variables: reflux, reboiler heat input, and sidestream flowrate. The steady-state gains between tray temperatures and the three inputs were obtained numerically. The resulting singular values are 119, 15.8, and 2.68. The large condition number indicates that the control of more than one temperature in the column will be difficult. This was found to be the case. Trying to use two temperature controllers resulted in severe interaction.

Therefore we assume that an analyzer is available to measure sidestream MeOH composition. A 3-min deadtime is used in this loop. Stage 17 temperature is controlled (the location of the steepest part of the temperature profile) by reboiler heat input through a heat input to feed ratio (see Fig. 10.32). The temperature control loop is tuned first with the composition loop on manual ($K_C = 0.159$ and $\tau_I = 7.9$ min). Then the composition loop is tuned with the temperature loop on automatic. Note that the flow controller on the sidestream flow is on "cascade" with its setpoint coming from the composition controller. A reflux to feed ratio is also used.

Unfortunately the relay–feedback test on the composition loop gives very erratic results, as shown in Figure 10.33a with no symmetric switches of the controller output and rapid oscillatory response of the process variable (sidestream methanol composition). The reason for this strange behavior remains a mystery. The resulting controller tuning constants ($K_C = 6.8$ and $\tau_I = 26$ min) gave unstable responses. Empirical tuning constants

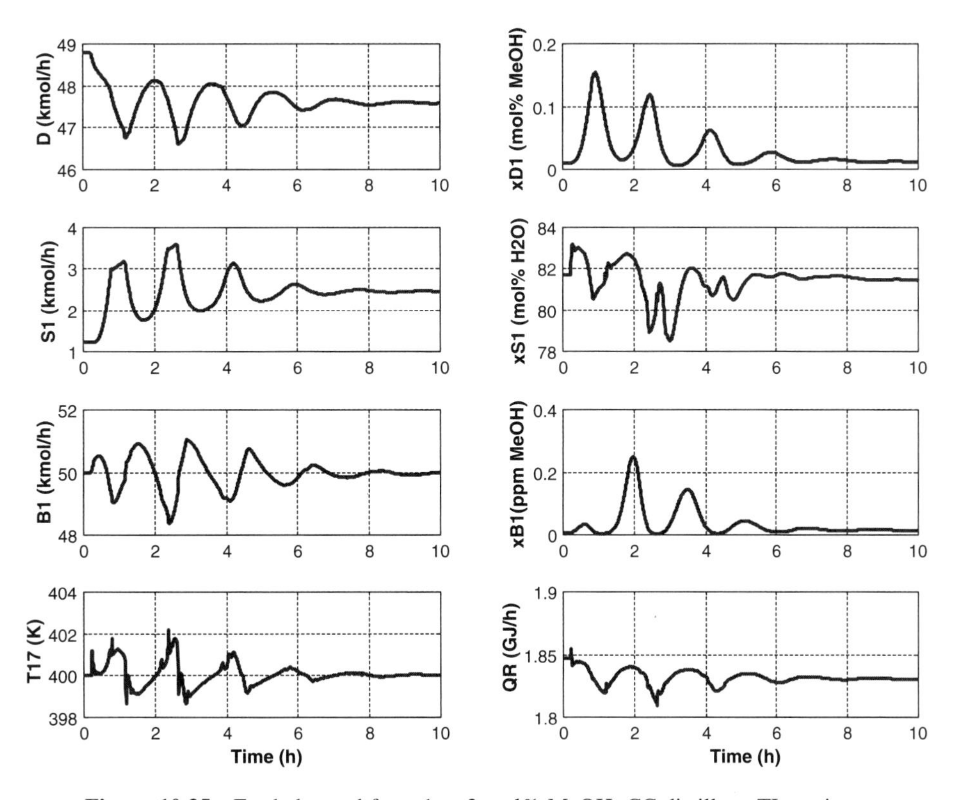

Figure 10.35 Feed changed from 1 to 2 mol% MeOH; CC distillate; TL settings.

of $K_C = 1$ and $\tau_I = 30$ min give stable response but poor transient control of product purities. For example, Figure 10.33b gives the response of the system when the methanol feed composition is increased from 1 to 2 mol%. Note that there is a huge transient increase in the methanol impurity in the distillate (upper right graph in Fig. 10.33b), which lasts for over 7 h. Clearly this performance is unacceptable.

A revised control structure was developed. Instead of controlling the methanol composition of the sidestream, the methanol composition of the distillate is controlled by manipulating the sidestream flowrate (see Fig. 10.34). The relay–feedback test of this loop gives reasonable responses ($K_U = 1.16$ and $P_U = 70$ min). If the Tyreus–Luyben settings are used ($K_C = 0.363$ and $\tau_I = 156$ min), the response is somewhat oscillatory when the methanol feed composition increases from 1 to 2 mol%, as shown in Figure 10.35. However, if the Ziegler–Nichols settings are used ($K_C = 0.527$ and $\tau_I = 59$ min), the response is quite good, as shown in Figure 10.36. The transient disturbance in the distillate purity is reduced because the sidestream is increased more quickly because of the higher gain and smaller reset time.

Figure 10.37 gives responses for 20% increase and decrease in feed flowrate, using Ziegler–Nichols settings in the distillate composition controller. Figure 10.38 gives responses when the DME in the feed is increased from 49 to 54 mol% and when it is decreased from 49 to 44 mol% (with a corresponding change in the water composition). The modified control structure handles all these disturbances quite well.

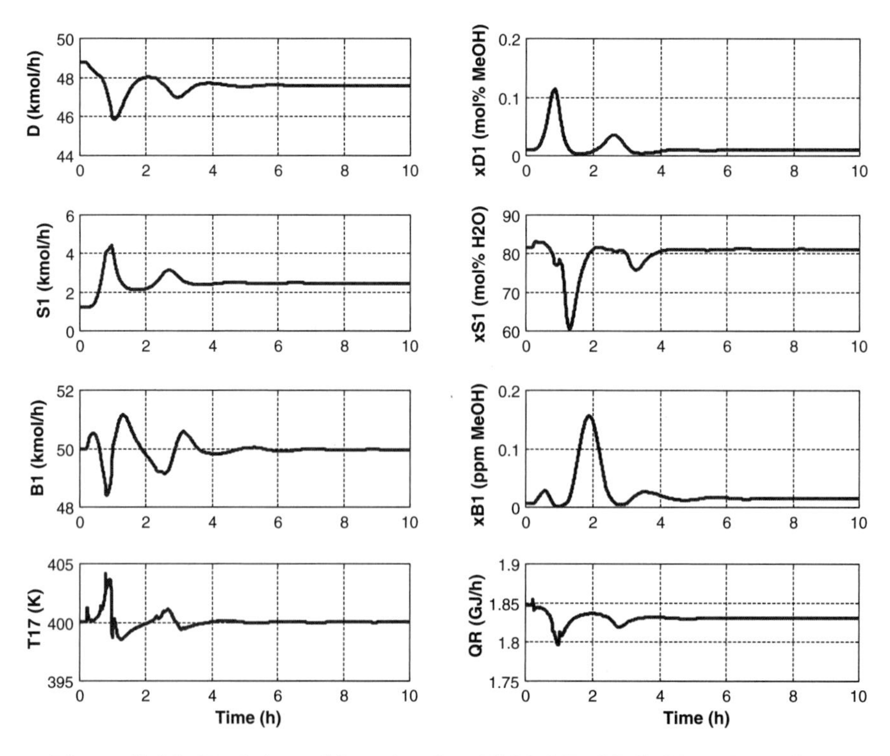

Figure 10.36 Feed changed from 1 to 2 mol% MeOH; CC distillate; ZN settings.

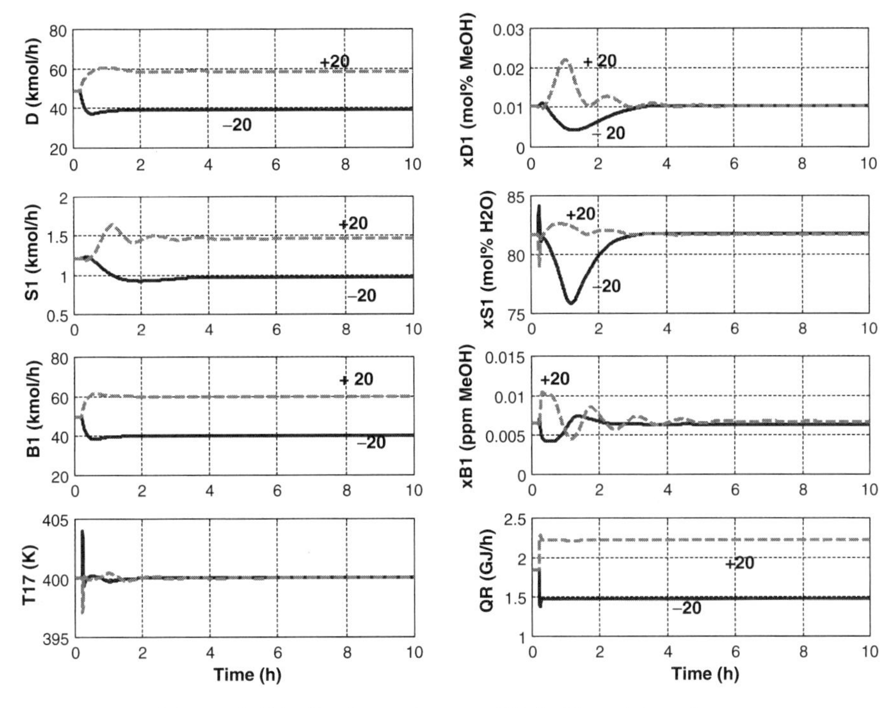

Figure 10.37 Feed rate changes; CC distillate; ZN settings.

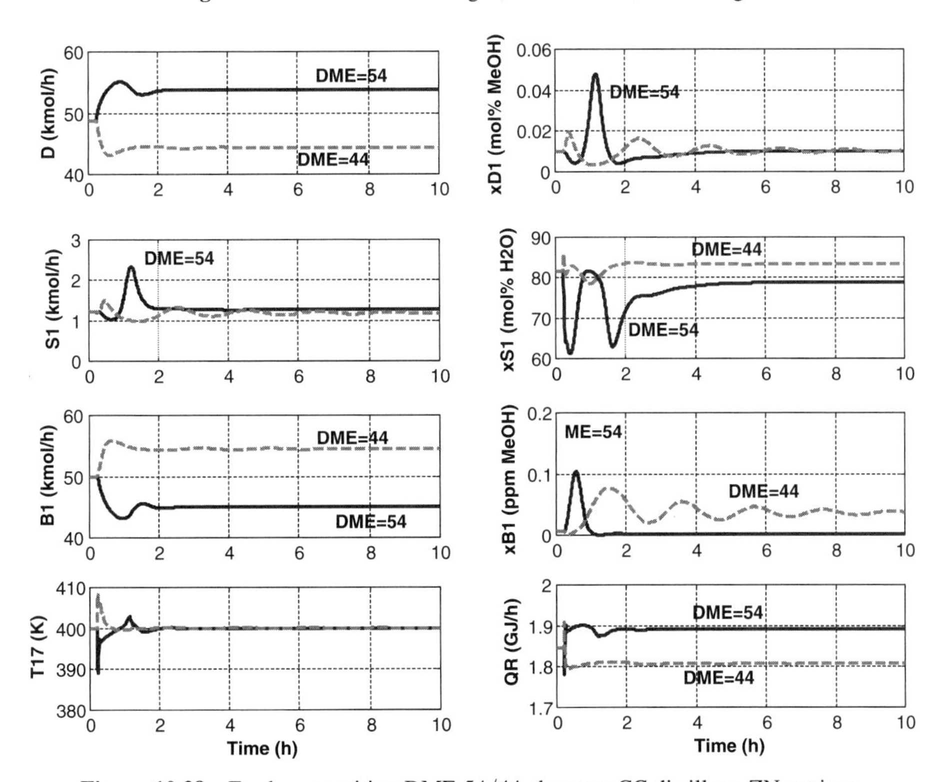

Figure 10.38 Feed composition DME 54/44 changes; CC distillate; ZN settings.

10.6 CONCLUSION

Several types of sidestream distillation columns have been considered in this chapter. The additional complexity of using sidestreams, with or without additional columns (strippers or rectifiers), makes the design and control of these processes more difficult.

CHAPTER 11

CONTROL OF PETROLEUM FRACTIONATORS

Up to this point in the book, we have looked at distillation columns that separate specific chemical components. In the refining of crude oil, mixtures of many thousands of components must be handled. These components vary from quite light hydrocarbons (methane, ethane, propane, etc.) to very high-molecular-weight components that boil at extremely high temperatures. Petroleum refineries have units that separate (by distillation) and transform (by a variety of reactions) these mixtures.

Crude oil is produced in hundreds of locations around the world. It is found underground, sometimes under high pressure and sometimes requiring pumping. A vast system of pipelines and huge supertankers transport the crude oil to refineries in which it is processed to make a large number of important products, such as gasoline, heating oil, jet fuel, asphalt, and wax. Most of the raw feedstocks for the chemical industry are produced in refineries, including ethylene, propylene, and benzene.

The initial separation of crude oil into several "cuts" is achieved in a very large distillation column called a "pipestill" or "atmospheric crude distillation" unit. These cuts have different boiling point ranges. Low-molecular-weight gas comes off the reflux drum as a vapor product from a partial condenser. The liquid product from the reflux drum is a light low-boiling naphtha. Products of jet fuel, heating oil, and a heavy high-boiling gasoil are removed as sidestreams from the column. The effluent from a catalytic cracking reactor is also a mixture of petroleum fractions, which are separated in a distillation column called a "cat fractionator." Many other units in a refinery must deal with these complex mixtures.

Crude oil as it comes from the ground is usually a mixture of saturated hydrocarbons: paraffins and naphthenes. The effluent of a cat cracker contains both saturated and unsaturated hydrocarbons: aromatics and olefins. These differences in the type of components affect the density and average molecular weight of the petroleum fraction for the same boiling point range.

As is true with all naturally occurring feedstocks, the composition or boiling range of crude oil varies greatly from production field to production field. This variability results in a very significant dynamic control problem in a refinery that feeds crude oil from a variety of sources, which is often the case.

Nelson[1] provides a thorough discussion of many aspects of the petroleum industry, including types and sources of crude, characterization of petroleum fraction, and types of refinery operations.

The following section is a brief introduction to how petroleum fractions are characterized and quantified so that petroleum fractionators can be designed and their control studied. In the next section we will discuss the details of setting up a steady-state simulation using the specified properties of the crude oil to a simple "preflash" column. Then we will look at the dynamic control of this column. In a later section we expand the process considered to include a pipestill (atmospheric crude unit). This large complex column has three sidestreams, which are withdrawn from sidestream strippers using open steam to remove the light hydrocarbons. It also has two "pumparounds" at intermediate locations up the column to remove heat. This complex system presents challenging design problems and challenging control issues.

11.1 PETROLEUM FRACTIONS

In the chemical industry, we deal with compositions (mole fractions). In petroleum refining, we deal with *boiling point ranges*. For example, suppose that we take a sample of heating oil and place it in a heated container at atmospheric pressure. The temperature at which the first vapor is formed is called the "initial boiling point." This corresponds to the bubblepoint of a mixture of specific chemical components. If we continue to heat the sample, more and more material is vaporized. The "5% point" is the temperature at which 5% of the original sample has vaporized. Liquid volume percents are traditionally used. The "95% point" is the temperature at which 95 liquid vol% of the original sample has vaporized. The "final boiling point" is the temperature at which all of the liquid disappears. This is somewhat similar to the dewpoint of a mixture of specific chemical components. Heating oil has a 5% point of about 460°F and a 95% point of ~620°F.

There are three types of boiling point analysis: ASTM D86 ("Engler"), ASTM D158 ("Saybolt") and true boiling point (TBP). The first and second are similar to the boiling of vapor as described in the previous paragraph. In the third, the vapor from the container passes into a packed distillation column and some specified amount is refluxed. Thus the third analysis exhibits some fractionation, while the first and second are just single-stage separations. The ASTM analysis is easier and faster to run. The TBP analysis gives more detailed information about the contents of the crude.

Figure 11.1 gives typical boiling curves for a light naphtha stream. The curve in Figure 11.1a is a TBP curve and that in Figure 11.1b, an ASTM D86 curve. The abscissa shows the volume percent distilled; the ordinate, temperature. Note that the initial and final parts of the curves are quite different because of the fractionation that occurs in the TBP distillation. The 50% boiling point is almost the same (249 and 243°F). Table 11.1 compares the results of these two methods.

[1]W. L. Nelson, *Petroleum Refinery Engineering*, 4th ed., McGraw-Hill, 1958.

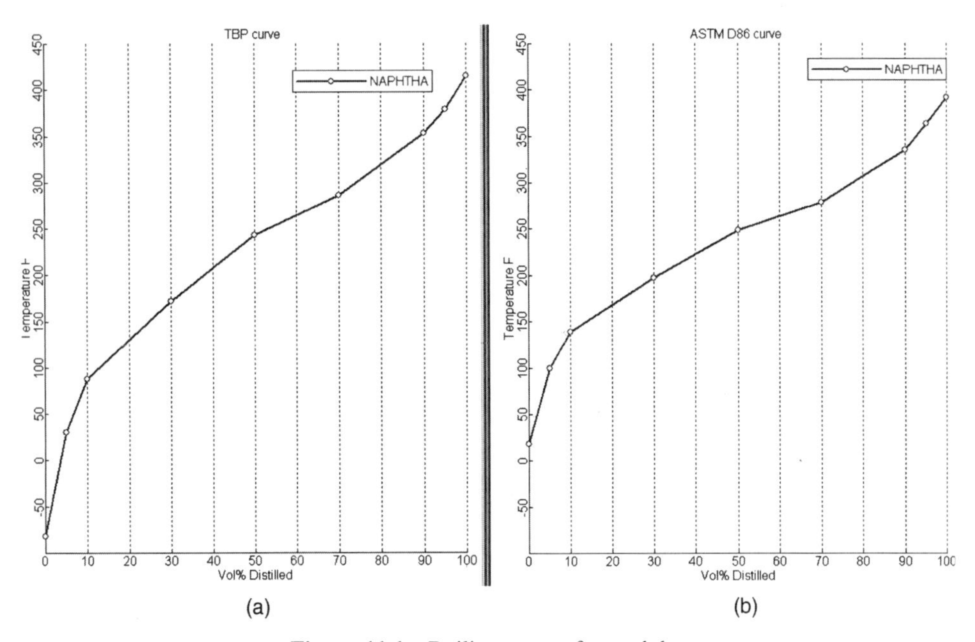

Figure 11.1 Boiling curves for naphtha.

Figure 11.2 gives typical curves for a crude oil. Note the very wide range of the boiling points from −100 to 1600°F. The TBP curve is wider than the ASTM curve. Table 11.2 gives numerical values of these curves. The 50% points are similar, but both ends of the curves are different.

These curves are obtained by using the Plot Wizard. As we will demonstrate in the next section, after the steady-state program has run, click on *Results Summary* in the *Data Browser*. Then click *Plot* on the top toolbar and select *Plot Wizard*. The window shown in Figure 11.3a opens. Clicking the *Dist Curve* and hitting *Next* open the window shown in Figure 11.3b. The stream of interest and the type of curve are selected. Clicking *Next* and then *Finish* produces a plot.

TABLE 11.1 Comparison of Boiling Point Methods for Naphtha

Vol% Distilled	ASTM D86	TBP[a]
IBP[b]	18	−81
5	100	31
10	139	88
30	197	172
50	249	243
70	279	286
90	336	353
95	369	379
FBP[c]	392	415

[a]True boiling point.
[b]Initial boiling point.
[c]Final boiling point.

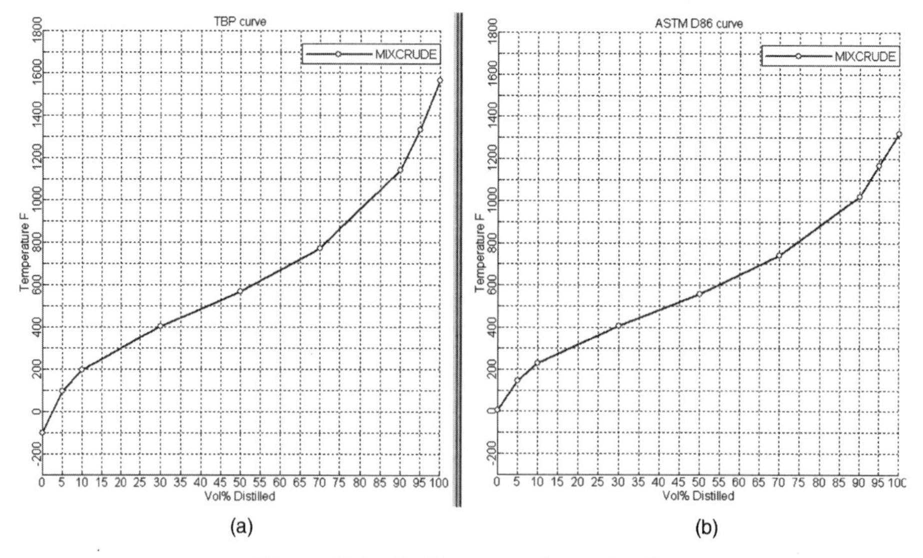

Figure 11.2 Boiling curves for crude oil.

The TBP plot of the crude gives an estimate of the amount of each petroleum cut in the crude. As shown in Figure 11.4, the boiling point range of each product from the pipestill is shown on the horizontal lines. For example, the boiling point range of the jet fuel in the crude is 280–460°F. By reading off the volume percentages at these temperatures (17 and 40 vol%, respectively), the yield of jet fuel can be estimated to be about 22% of the crude fed to the pipestill. So a 100,000-b/d (b/d = barrels per day) crude unit should produce about 22,000 b/d of jet fuel when fed this type of crude. Of course, the yields will be different for other crudes with different TBP curves.

The properties of a petroleum stream are not specified in terms of compositions. Instead, properties are used such as 5% point, final boiling point, Reid vapor pressure (RVP), flashpoint, and octane number.

The method for performing quantitative calculations with petroleum fractions is to break them into "pseudocomponents" with each having an average boiling point, specific gravity, and molecular weight. Aspen Plus generates these pseudocomponents given "assay" information about the crude oil.

TABLE 11.2 Comparison of Boiling Point Methods for Crude Oil

Vol% Distilled	ASTM D86	TBP
IBP	5	−99
5	146	97
10	227	196
30	408	403
50	554	569
70	742	772
90	1021	1143
95	1169	1331
FBP	1317	1563

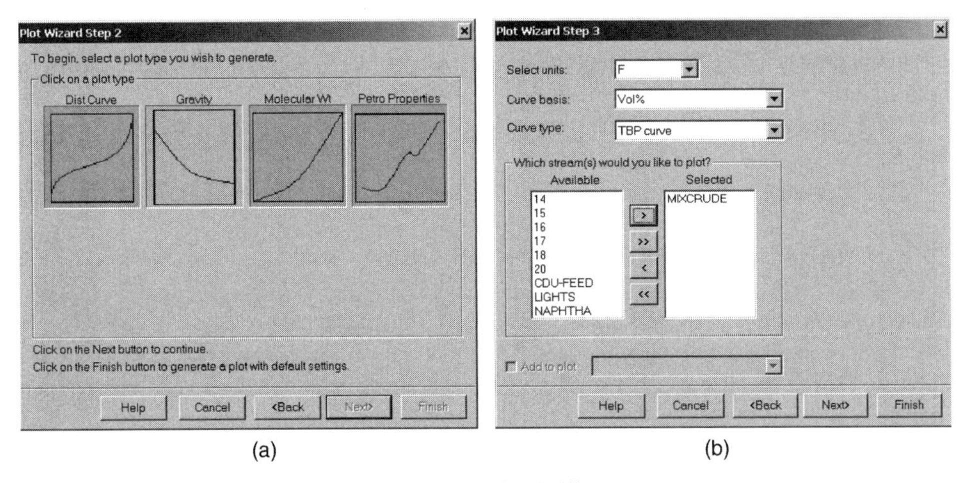

(a) (b)

Figure 11.3 Producing boiling curves.

Another difference in petroleum refining is how density is defined. In the chemical industry, density is defined as specific gravity or mass per unit volume. The density of a material in petroleum refining is traditionally reported in "degrees American Petroleum Institute" (°API). Specific gravity and °API are inversely related: the higher the specific gravity, the lower the °API:

$$\text{Degrees API} = \frac{141.5}{\text{specific gravity}} - 131.5$$

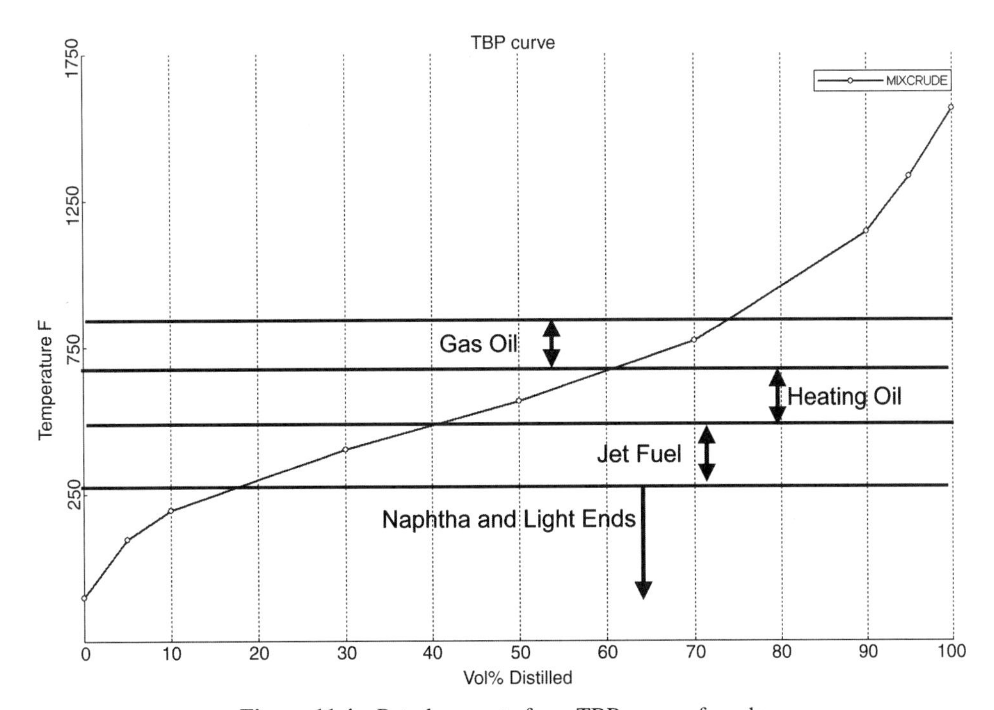

Figure 11.4 Petroleum cuts from TBP curve of crude.

For example, the crude considered above has a density of 34.1°API, which corresponds to a specific gravity of 0.854. The naphtha has a density of 61.7°API, which corresponds to a specific gravity of 0.732.

The traditional unit used for flowrate in petroleum refining in the United States is barrels per day (b/d). One barrel is 42 U.S. gallons.

11.2 CHARACTERIZATION OF CRUDE OIL

The information available for a crude oil typically consists of "assay" data: a TBP curve, a gravity curve, and a "light-ends analysis." These are illustrated below. The material in this section is taken from the very useful and detailed documentation available from Aspen Technology,[2] and the example used in this book is based on the example presented in that source. Petroleum English units are used in the steady-state design here because they are used in that report. When we move into dynamics, metric units will be used because the current version of Aspen Dynamics (version 12) does not offer Petroleum English units.

We will consider a crude oil called Oil-1. The assay data for Oil-1 are given in Table 11.3. We have already discussed TBP distillation information. In petroleum refining, the term "light ends" refers to specific light hydrocarbons, such as methane and ethane. You can see that there are small amounts of these light component dissolved in the crude oil, even though it is at atmospheric pressure and ambient temperature. The API gravity data give the density of the various cuts as they are produced in the TBP distillation.

A *Template* is set up in Aspen Plus that takes this data and generates pseudocomponents. These will then be used in simulating a simple petroleum distillation column called a *preflash unit* and a second, more complex petroleum pipestill with multiple sidestream products.

The program is started in the normal way by clicking on *Start, Programs, AspenTech, Aspen Engineering Suite, Aspen Plus 12.1*, and *Aspen Plus User Interface*. The window shown at the top of Figure 11.5 opens and the *Template* option is selected. Clicking *OK* opens the window shown at the bottom of Figure 11.5 with the *Simulation* page tab open. We select *Petroleum with English Units* from the list of options on the left. At the bottom right of the window, we select *Assay Data Analysis* and click *OK*.

A *Data Browser* window opens, and clicking *Setup* and *Specifications* gives the view shown in Figure 11.6. A title can be inserted in the appropriate box. Next, the *Components* item is clicked in the *Data Browser*, then *Specifications* is clicked, which opens the window shown at the top of Figure 11.7 with the *Selection* page tab selected. The individual components from water to *n*-pentane are typed in the first column (*Component ID*) and the third column (*Component name*). The *Type* and *Formula* columns are automatically filled in, as shown at the bottom of Figure 11.7.

Now a crude oil is specified. We will use the assay data for Oil-1 given in Table 11.3. On the next line of the *Selection* page tab we type in *OIL-1* in the first column. Clicking the second column opens a dropdown menu (shown in Fig. 11.8), and *Assay* is selected. Clicking the *Petroleum* page tab and sequentially clicking *Assay/Blend*, *OIL-1*, and *Basic Data* in the *Data Browser* opens the window shown at the top of Figure 11.9.

[2]Aspen Technology, Inc., *Getting Started Modeling Petroleum Processes.*

TABLE 11.3 Crude Oil Assay Data for Oil-1 (31.4°API)

TBP (Liquid Vol%)	Distillation Temperature (°F)	Light-Ends Component	Analysis (Liquid Volume Fraction)	API Curve (Midpoint Vol%)	Gravity (°API)
6.8	130	Methane	0.001	5	90.0
10	180	Ethane	0.0015	10	68.0
30	418	Propane	0.009	15	59.7
50	650	Isobutane	0.004	20	52.0
62	800	*n*-Butane	0.016	30	42.0
70	903	2-Methylbutane	0.012	40	35.0
76	1000	*n*-Pentane	0.017	45	32.0
90	1255	—	—	50	28.5
—	—	—	—	60	23.0
—	—	—	—	70	18.0
—	—	—	—	80	13.5

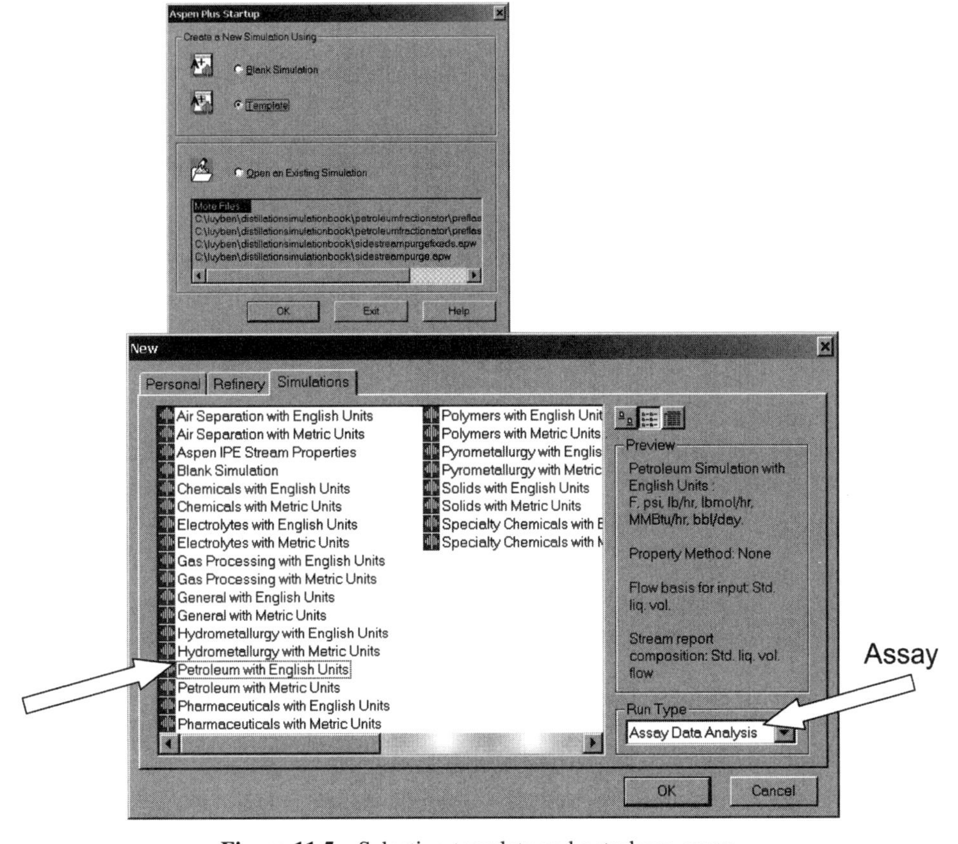

Figure 11.5 Selecting template and petroleum assay.

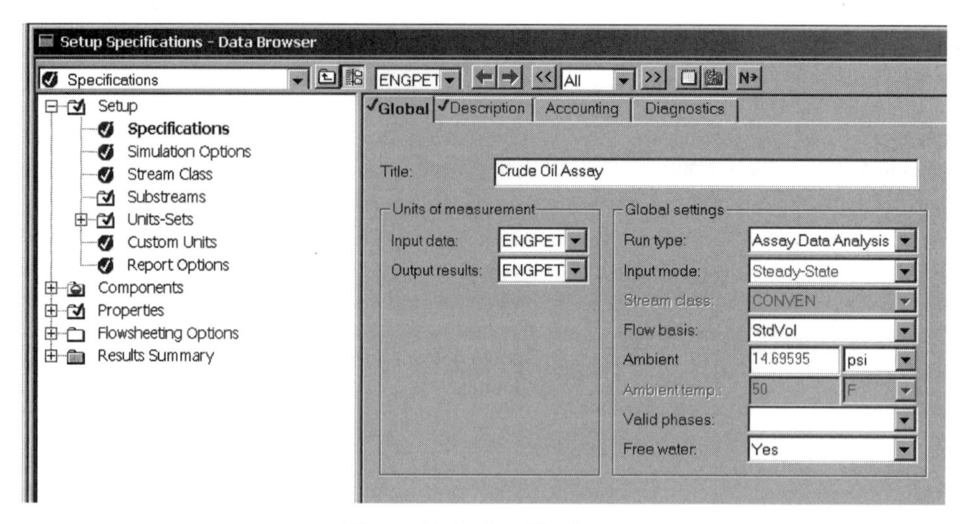

Figure 11.6 Specifications.

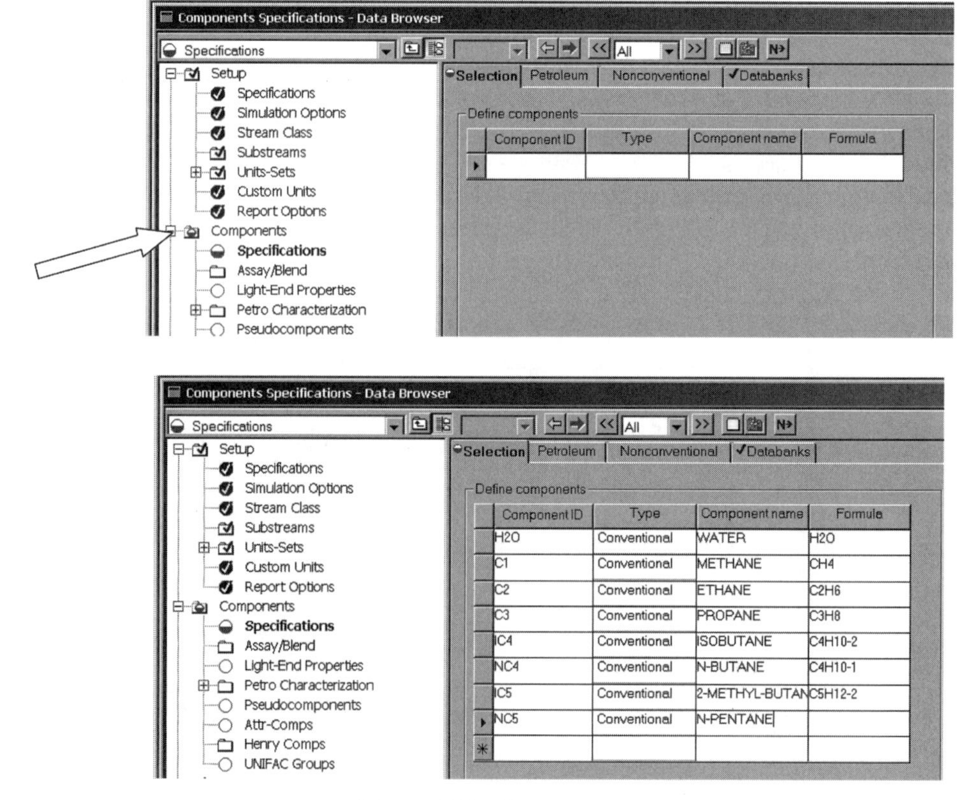

Figure 11.7 Components.

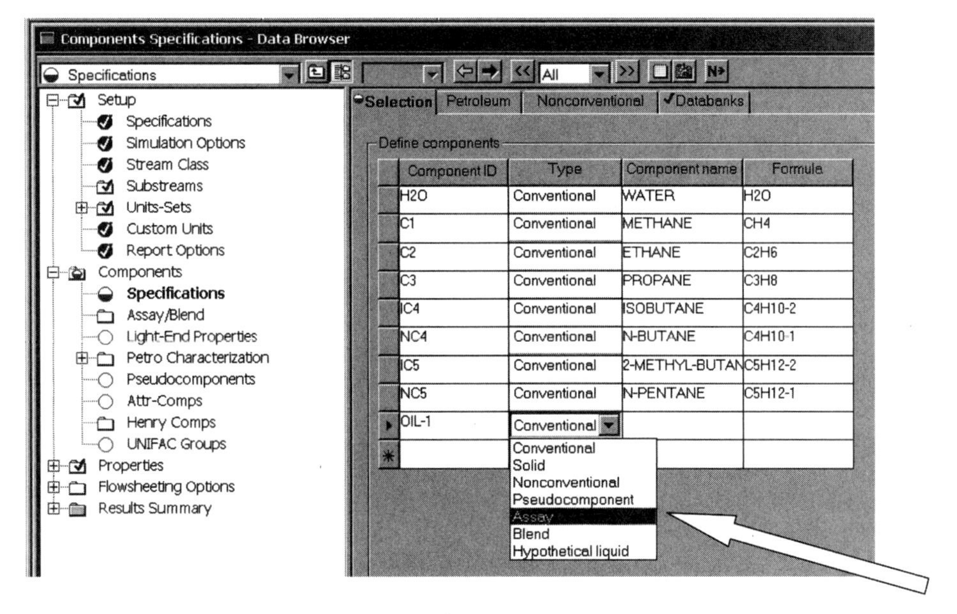

Figure 11.8 Crude oil assay.

In the *Distillation Curve Type* we use the dropdown menu to select *True boiling point (liquid volume basis)*. In the *Bulk gravity value* the number 31.4 is entered in the *API gravity* box.

The percent distilled and temperature data for Oil-1 from Table 11.3 are entered, as shown at the bottom of Figure 11.9. The *Light Ends* page tab is clicked and the data from Table 11.3 is entered (see Fig. 11.10). The *Gravity/UOPK* page tab is clicked, and the *API gravity* is selected as the type. The data from Table 11.3 are entered (see Fig. 11.11).

All the assay data have been entered, and we are ready to generate pseudocomponents. Clicking the blue *N* button opens the window shown in Figure 11.12a. We select *Specify options for generating pseudocomponents.* A new browser window opens called *Components Petro Characterization Generation* (see Fig. 11.12b). Clicking *New* opens the window shown at the top of Figure 11.13, on which we enter an identification name such as "Crude1." Clicking *OK* opens window shown at the bottom of Figure 11.13, where *OIL-1* is selected from the dropdown menu to be included. Finally the blue *N* button is clicked and *OK* is clicked on each of the windows that come up sequentially (shown in Fig. 11.14). This completes the generation of the pseudocomponents, called *Assay Data Analysis.*

The results can be seen by going to *Components*, *Petro Characterization*, and *Results* on the *Data Browser.* Figure 11.15 gives a partial list of the pseudocomponents. Note that each has a normal boiling point, density, molecular weight, and critical properties.

This same procedure is repeated for a second crude called Oil-2 whose assay data are given in Table 11.4. Note that this crude is somewhat lighter than Oil-1 (API gravity is 34.8 compared to 31.4 and 50% point is 450 vs. 650°F for Oil-1). We will feed both of these crude oils to a unit discussed in the following section.

We are now ready to proceed with our simulation. We will start by looking at a simple preflash column that is often used in refineries to remove some of the lightest material from the crude before sending it into the pipestill.

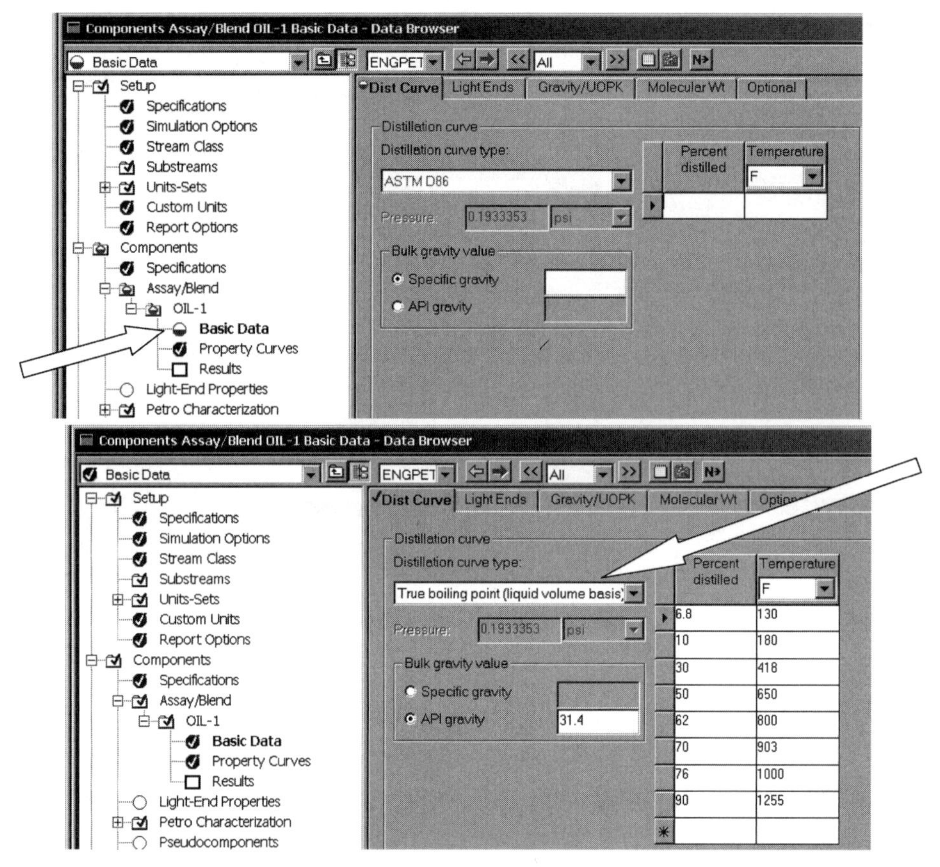

Figure 11.9 Crude oil basic data.

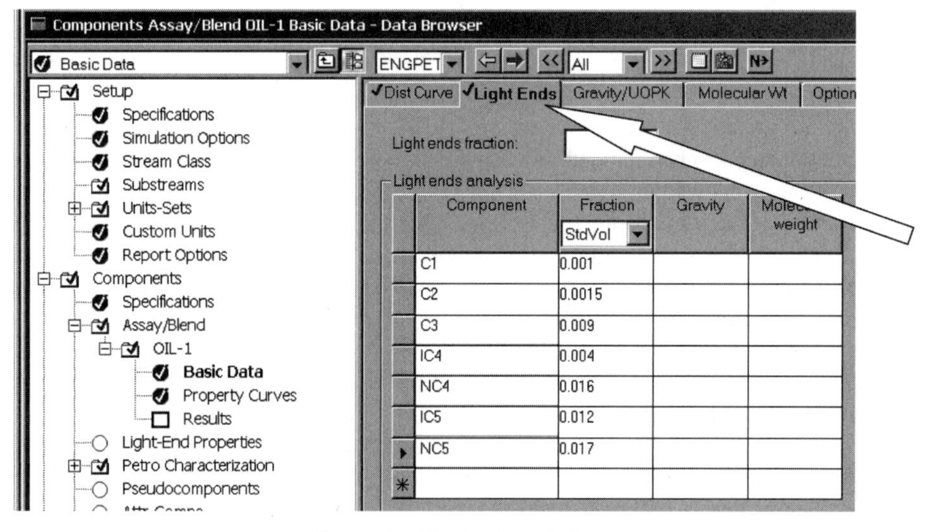

Figure 11.10 Light-ends data.

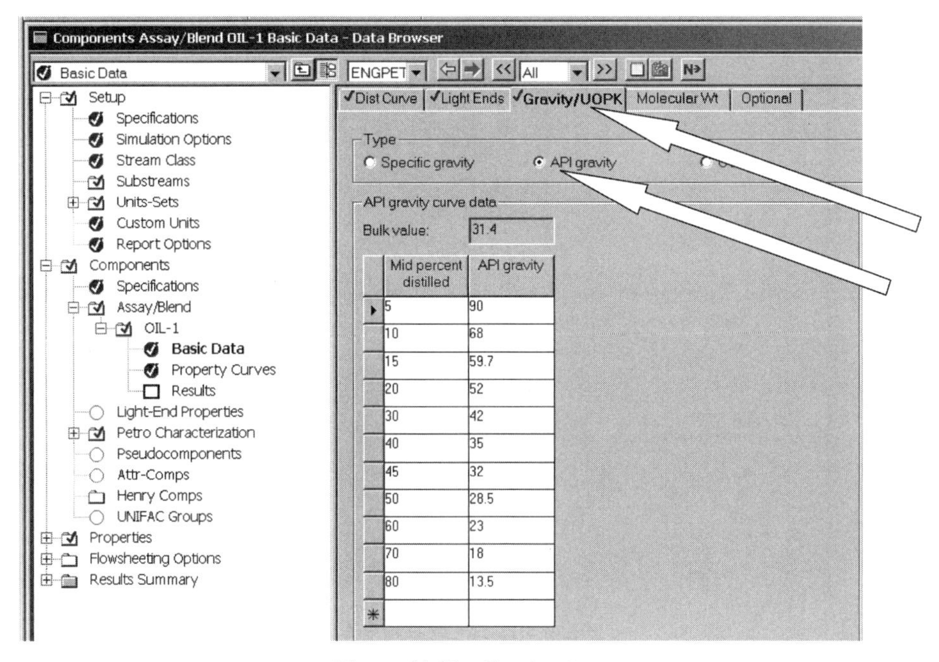

Figure 11.11 Gravity data.

(a)

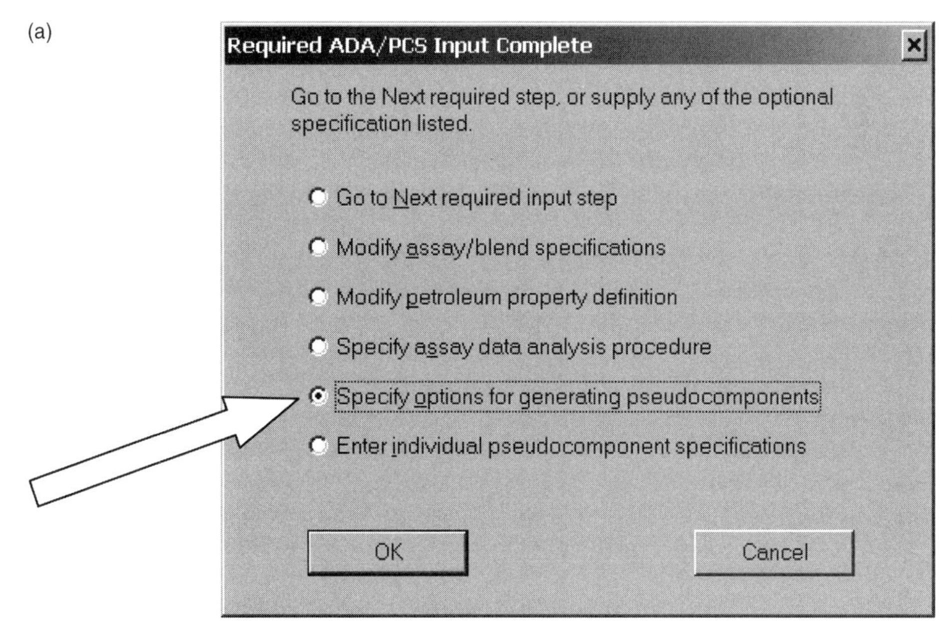

Figure 11.12 Generate pseudocomponents.

(b)

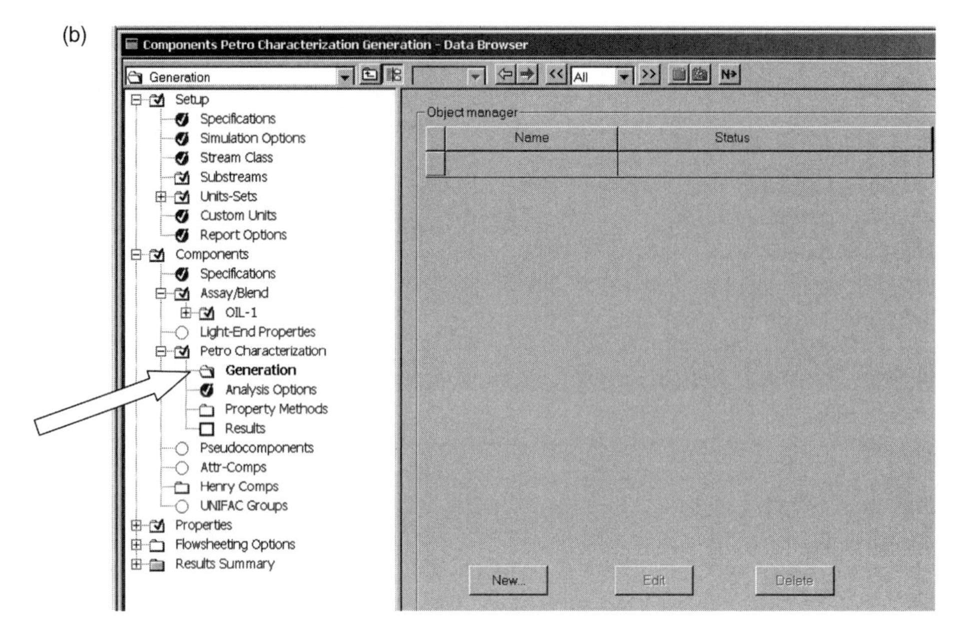

Figure 11.12 *Continued.*

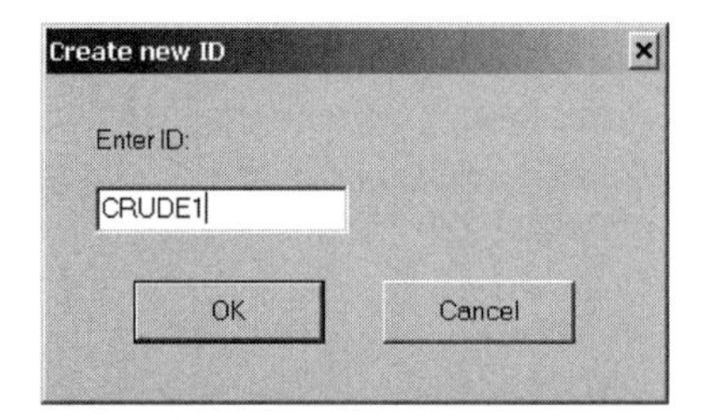

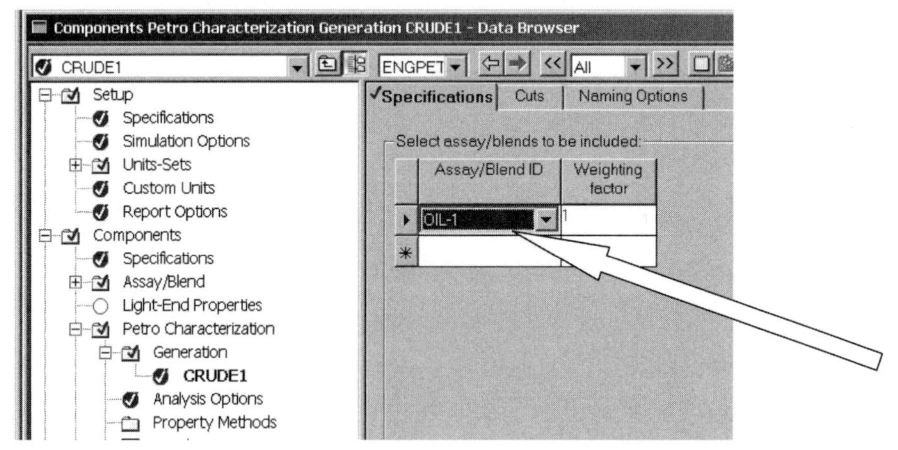

Figure 11.13 Petroleum characterization generation.

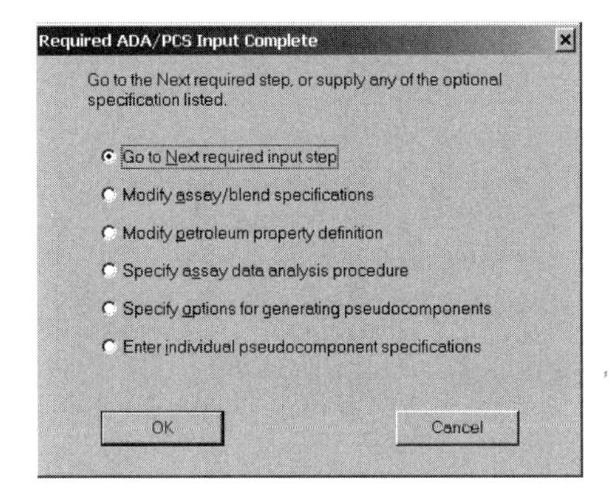

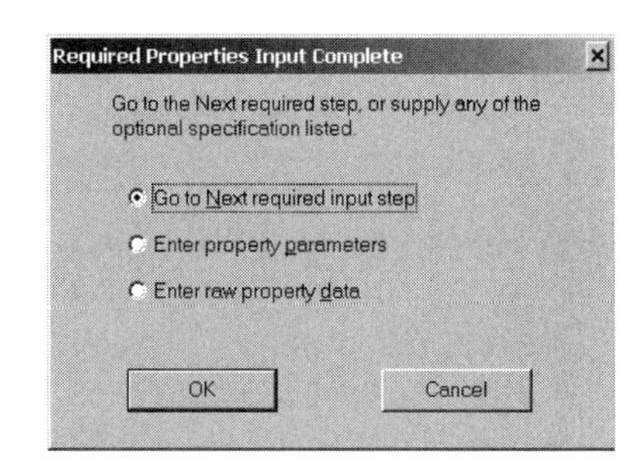

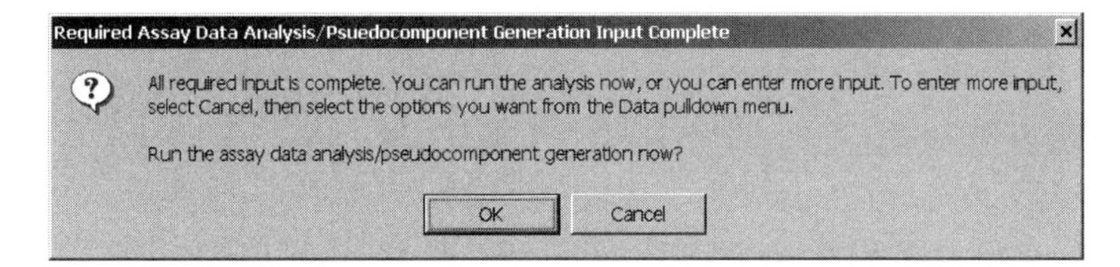

Figure 11.14 Completing petroleum characterization generation.

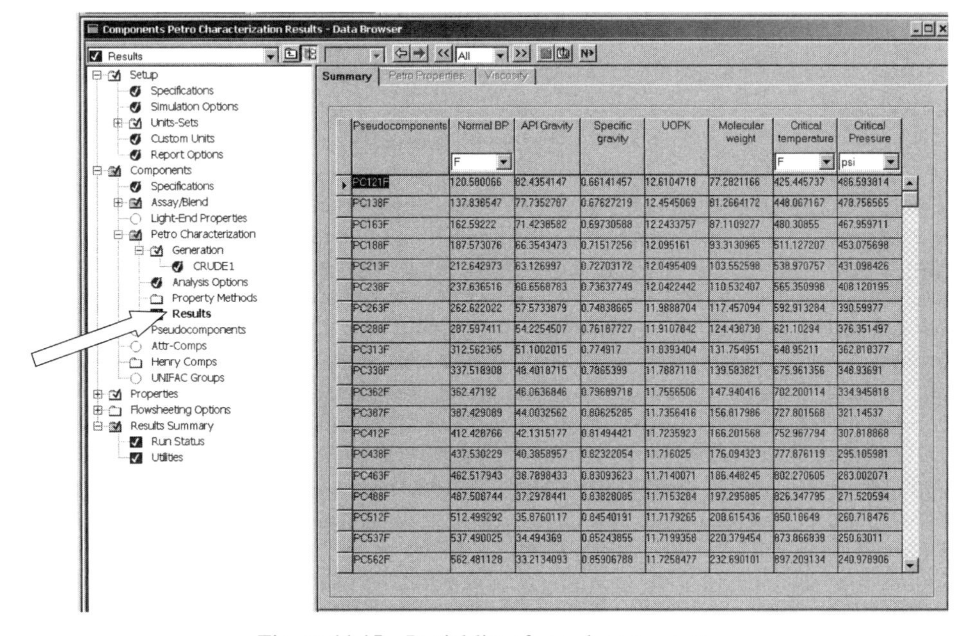

Pseudocomponents	Normal BP (F)	API Gravity	Specific gravity	UOPK	Molecular weight	Critical temperature (F)	Critical Pressure (psi)
PC121F	120.580066	82.4354147	0.66141457	12.6104718	77.2821166	425.445737	486.593814
PC138F	137.838547	77.7352787	0.67627219	12.4545069	81.2664172	448.067167	478.756565
PC163F	162.59222	71.4238582	0.69730588	12.2433757	87.1109277	480.30855	467.959711
PC188F	187.573076	66.3543473	0.71517256	12.095161	93.3130965	511.127207	453.075698
PC213F	212.642973	63.126997	0.72703172	12.0495409	103.552598	538.970757	431.098426
PC238F	237.636516	60.6568783	0.73637749	12.0422442	110.532407	565.350998	408.120195
PC263F	262.622022	57.5733879	0.74838665	11.9888704	117.457094	592.913284	390.59977
PC288F	287.597411	54.2254507	0.76187727	11.9107842	124.438738	621.10294	376.351497
PC313F	312.562365	51.1002015	0.774917	11.8393404	131.754951	648.95211	362.818377
PC338F	337.518908	48.4018715	0.7865399	11.7887118	139.583821	675.961356	348.93691
PC362F	362.47192	46.0636846	0.79689718	11.7556506	147.940416	702.200114	334.945818
PC387F	387.429089	44.0032562	0.80625285	11.7356416	156.817986	727.801568	321.14537
PC412F	412.428766	42.1315177	0.81494421	11.7235923	166.201568	752.967794	307.818868
PC438F	437.530229	40.3858957	0.82322054	11.716025	176.094323	777.876119	295.105981
PC463F	462.517943	38.7898433	0.83093623	11.7140071	186.448245	802.270605	283.002071
PC488F	487.508744	37.2978441	0.83828085	11.7153284	197.295885	826.347795	271.520594
PC512F	512.499292	35.8760117	0.84540191	11.7179265	208.615436	850.18649	260.718476
PC537F	537.490025	34.494369	0.85243855	11.7199358	220.379454	873.866839	250.63011
PC562F	562.481128	33.2134093	0.85906788	11.7258477	232.690101	897.209134	240.978906

Figure 11.15 Partial list of pseudocomponents.

TABLE 11.4 Crude Oil Assay Data for Oil-2 (34.8°API)

TBP (Liquid Vol%)	Distillation Temperature (°F)	Light-Ends Component	Analysis (Liquid Volume Fraction)	API Curve (Midpoint Vol%)	Gravity (°API)
6.5	120	Methane	0.001	2	150.0
10	200	Ethane	0.002	5	95.0
20	300	Propane	0.005	10	65.0
30	400	Isobutane	0.01	20	45.0
40	470	*n*-Butane	0.01	30	40.0
50	450	2-Methylbutane	0.005	40	38.0
60	650	*n*-Pentane	0.025	50	33.0
70	750			60	30.0
80	850			70	25.0
90	1100			80	20.0
95	1300			90	15.0
98	1475			95	10.0
100	1670			100	5.0

11.3 STEADY-STATE DESIGN OF PREFLASH COLUMN

The first petroleum fractionator simulated is a simple distillation column that removes some of the light material in the crude. Figure 11.16 gives the Aspen Plus flowsheet of this unit. There are two crude feedstreams that are combined and heated in a furnace in which the feed is partially vaporized before entering the bottom of the column. There is no reboiler. Live steam is introduced in the bottom of the column to strip out some of the light components in the bottoms stream, which is fed to a pipestill that is considered in the next section.

The valves and pumps are standard equipment, but the column is different from the typical *RadFrac* that is used when specific chemical components are employed. A petroleum fractionator is selected from the model library menu on the bottom of the Aspen Plus window by clicking *Columns* and then *PetroFrac*. Figure 11.17 shows the palette of possible configurations. We choose the fourth from the left on the top row, which is a rectifier with a furnace.

It is a little tricky to attach the stream from the tee where the two crudes are mixed to the furnace. When a material stream is selected at the bottom of the Aspen Plus window and the cursor is moved to the flowsheet, a red input arrow appears at the bottom of the column, as shown in Figure 11.18a. Place the cursor over the red arrow and drag it to the left until it points to the furnace. Then click this arrow, which attaches the stream (see Fig. 11.18b).

The total crude feed to the preflash column is 100,000 b/d, with each of the two crudes (Oil-1 and Oil-2) set at 50,000 b/d. Their temperatures are 200°F. The furnace outlet temperature is specified to be 450°F, which requires a heat input of 203×10^6 Btu/h. The crude is ~30 wt% vaporized in the furnace.

Selecting *Setup* under the column block (*PREFLASH*) opens the window shown in Figure 11.19, where we see page tabs of *Configuration*, *Streams*, *Pressure*, *Condenser*, and *Furnace*. The column is set up to have 10 stages, no reboiler, and a partial condenser.

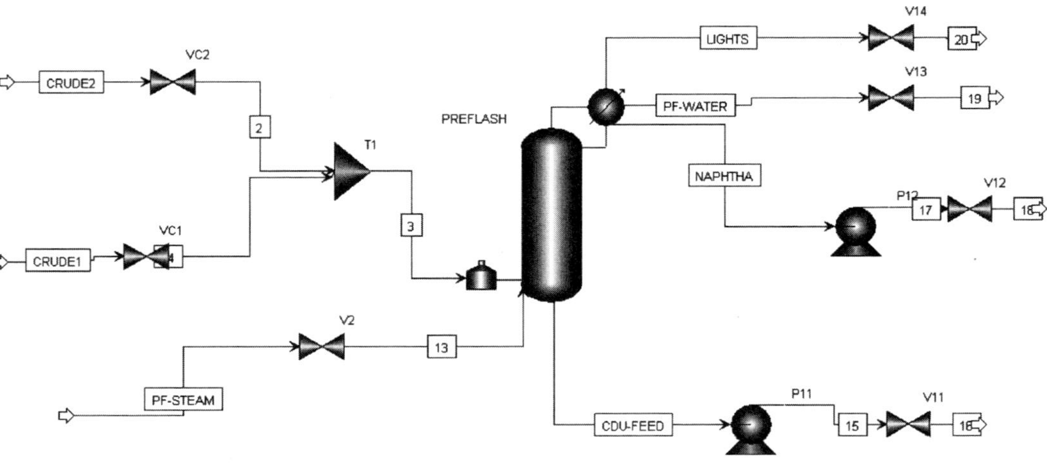

Figure 11.16 Preflash column.

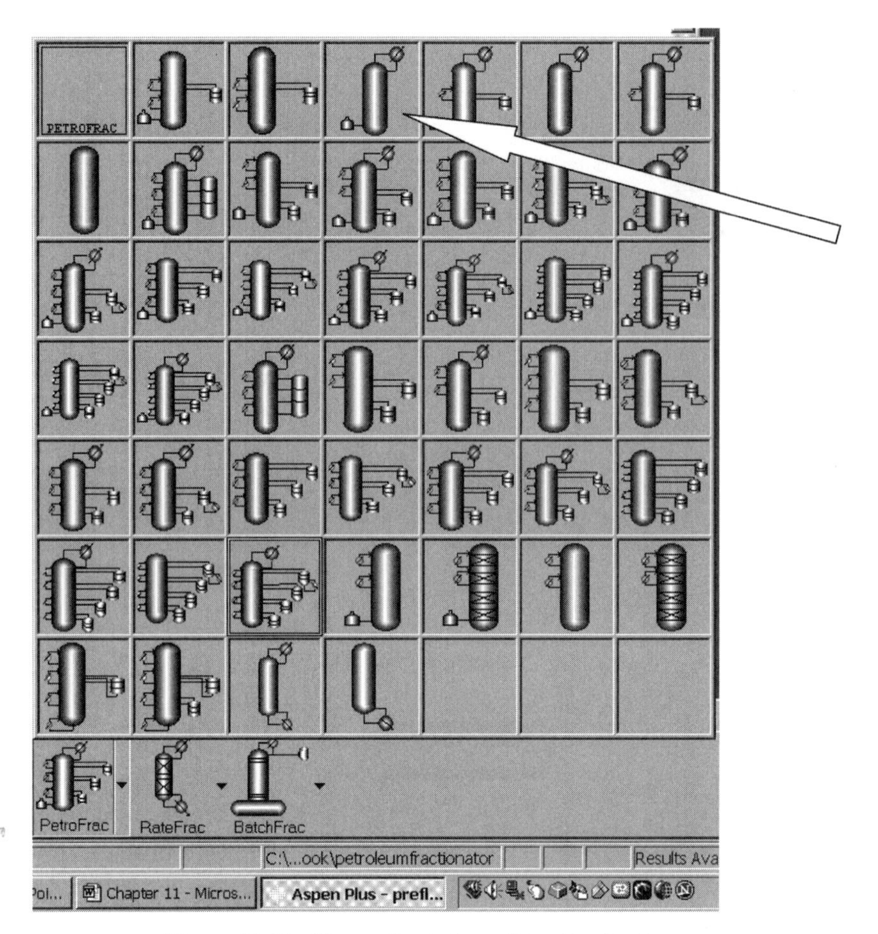

Figure 11.17 Types of petroleum (PetroFrac) columns.

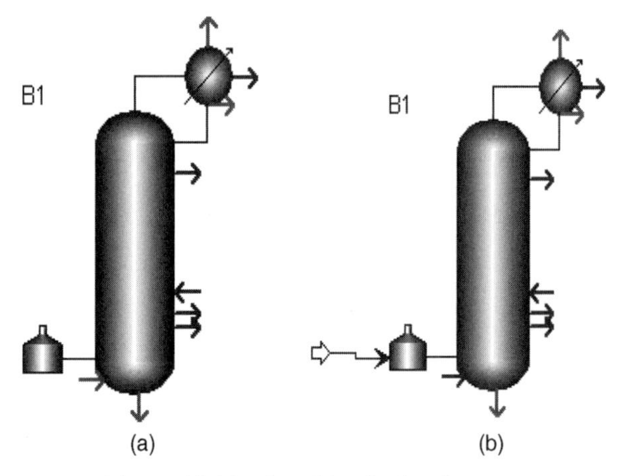

Figure 11.18 Attaching feed to furnace.

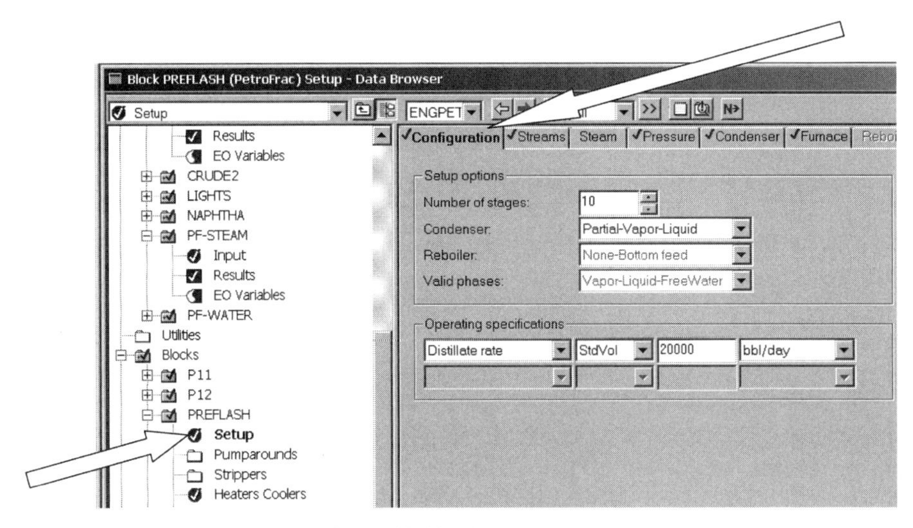

Figure 11.19 Column setup.

Both a gas stream and a liquid distillate are removed from the reflux drum. In addition, since live steam is fed into the bottom of the column and separates into an aqueous phase in the reflux drum, a water stream is removed from a small "boot" at the bottom of this drum that serves as a decanter. The stripping steam flowrate is 5000 lb/h, and its temperature is 400°F. Both the gas and the liquid products contain some water. The water decanted is 244 lb/h.

The distillate rate is set at 20,000 b/d. This will be adjusted later to obtain a desired ASTM 95% point of 375°F for the liquid distillate product, which is a light naphtha stream. Note that there is only one degree of freedom in this rectifying column since there is no reboiler. All the vapor coming up the column originates from the partially vaporized furnace effluent.

Clicking the *Stream* page tab opens the window shown in Figure 11.20, where the combined crude feedstream is specified to be fed to the furnace by using the dropdown menu. The stripping steam is fed on stage 10.

Opening the *Pressure* page tab (shown at the top of Fig. 11.21) permits setting pressures in the column. The pressure in the reflux drum is specified to be 39.7 psia. The condenser pressure drop is 2 psi, so the pressure on stage 2 is set at 41.7 psia. The pressure at the bottom of the column is specified to be 44.7 psia.

Clicking the *Condenser* page tab opens the window shown at the bottom of Figure 11.21, where the condenser temperature is specified to be 170°F. This is high enough to permit the use of air-cooled condensers, which conserves the use of cooling water.

Clicking the *Furnace* page tab opens the window shown in Figure 11.22, on which the temperature is set at 450°F and the pressure is set at 44.7 psia. This is the same pressure as that at the bottom of the column. If the pressure is set at a higher value, the steady-state simulation will run, but an error will occur when exporting a pressure-driven dynamic simulation file.

The final job in the steady-state design is to achieve the desired specification of an ASTM 95% point of 375°F (ASTM D86). An initial guess of 20,000 b/d for the liquid distillate flowrate gives an ASTM 95% point of 353°F. This is lower than the specification,

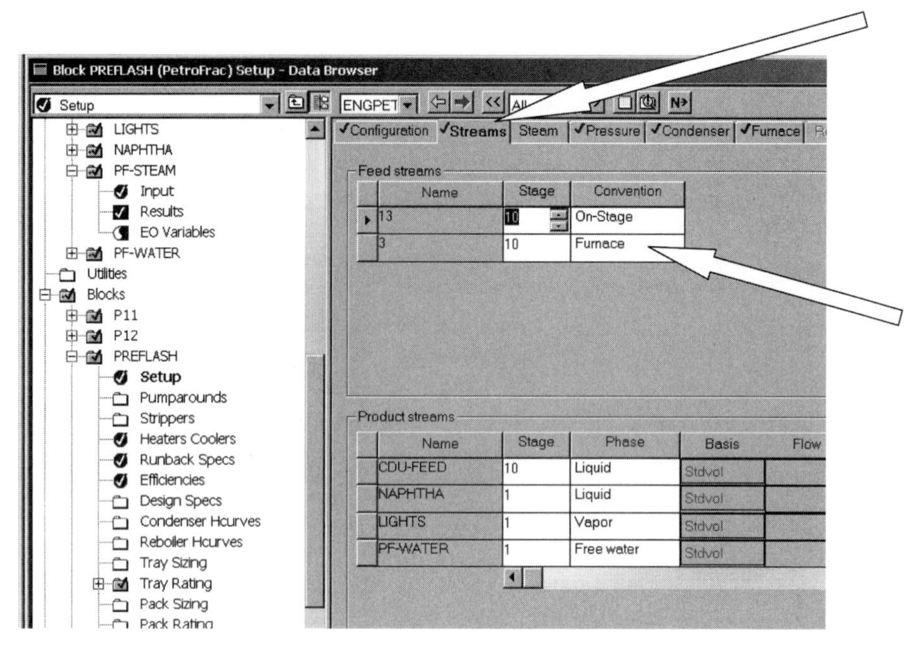

Figure 11.20 Stream locations.

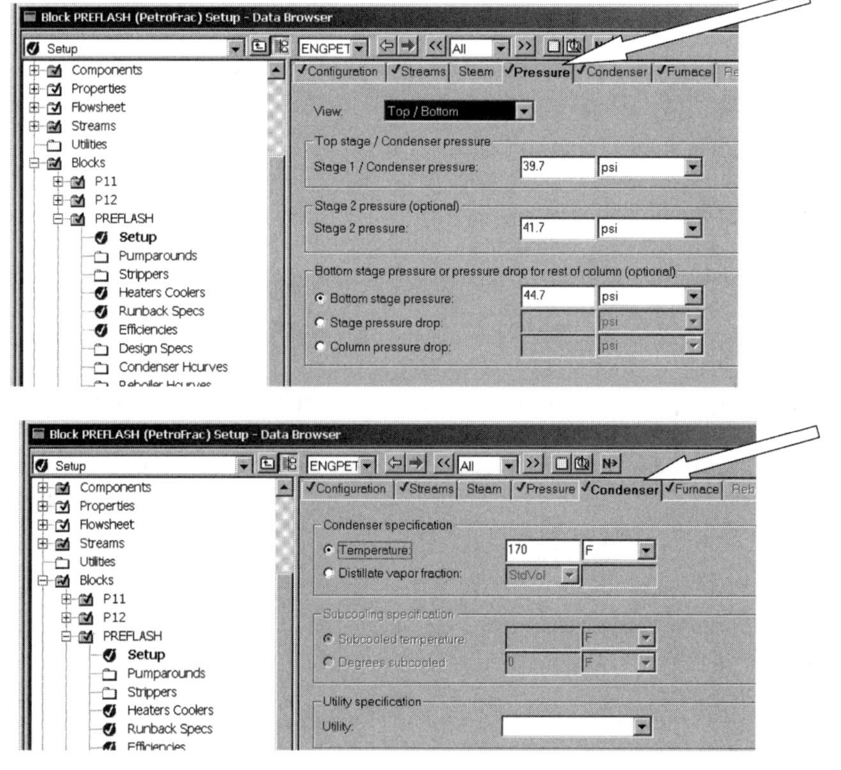

Figure 11.21 Setting pressures.

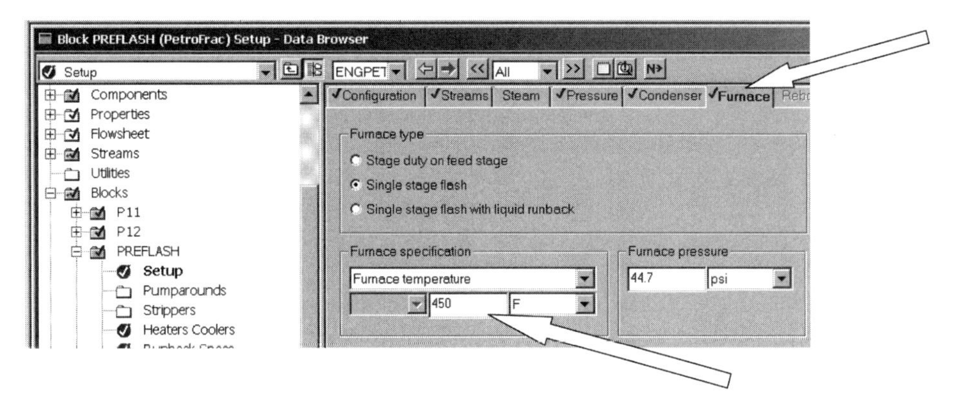

Figure 11.22 Setting furnace conditions.

which indicates that more material can be taken overhead. Increasing the flowrate carries more higher-boiling material into the naphtha product.

A "Design Specs" function can be used to achieve the specification. Clicking *Design Specs* under the *PREFLASH* column block, clicking *New*, and giving an identification label open the window shown at the top of Figure 11.23. The *Type* is specified at *ASTM D86 temperature* (*dry, liquid volume basis*). The *Target* is 375°F at a *Liquid %* of 95%. Click the *Feed/Product Streams* and select *NAPHTHA* as the *Selected Stream* (see the bottom of Fig. 11.23). Finally the *Vary* page tab is clicked and *Distillate flow rate* is

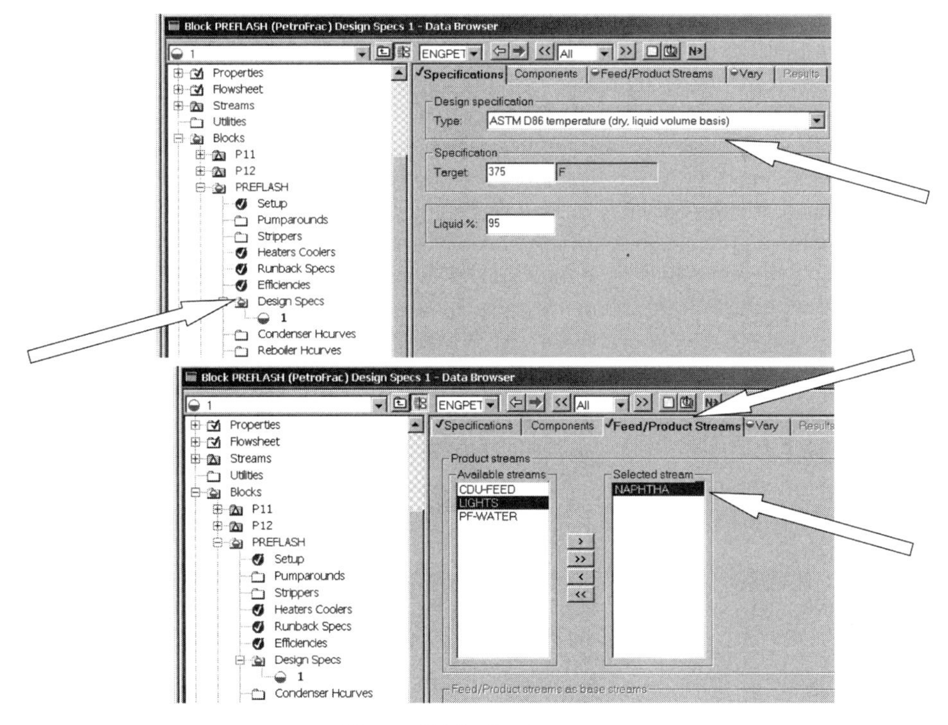

Figure 11.23 Design specs.

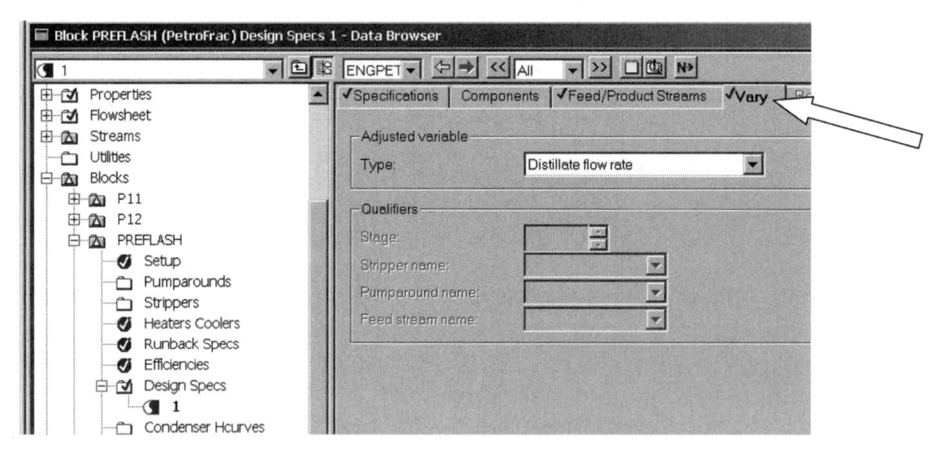

Figure 11.24 Set *Vary* in design specs.

selected as the *Type* (see Fig. 11.24). Running the simulation by clicking the blue *N* button produces a naphtha flowrate of 21,040 b/d to achieve the 375°F target.

One source of confusion is a difference between the b/d flowrates specified in the original *Setup* and the *Liq vol 60F bbl/day* reported in the *Stream Results*. The specified 20,000 b/d is at the actual flow conditions (temperature and pressure). The results given in the *Stream Results* are at standard conditions (60°F and 1 atm).

The diameter of the column is calculated in the normal way by using the *Tray Sizing* feature. The result is a diameter of 11.1 ft. The composition of the *LIGHTS* vapor stream from reflux drum is given in Table 11.5. Most of the light hydrocarbons that are in the crude oil feedstreams are removed in this vapor stream. They are sent to downstream units for separation into individual components.

The steady-state design is now complete. Figure 11.25a gives the flowsheet of the *PREFLASH* column with conditions and properties of the various streams. All the flowrates are given in barrels per day at standard conditions.

Since English petroleum units are not available in the current version of Aspen Dynamics, we will switch to metric units when we look at dynamic control. Figure 11.25b gives the flowsheet of the *PREFLASH* column in metric units (1T = 1000 kg).

TABLE 11.5 ***LIGHTS* Composition**

	Mole Fraction (%)	lb·mol/h	lb/h
H_2O	0.150	86.3	1554
C_1	0.0697	39.5	634
C_2	0.0854	49.1	1478
C_3	0.144	82.8	3652
iC_4	0.0918	52.8	3069
nC_4	0.154	88.3	5131
iC_5	0.0555	31.9	2302
nC_5	0.115	66.2	4777
Total	1.00	575	29,726

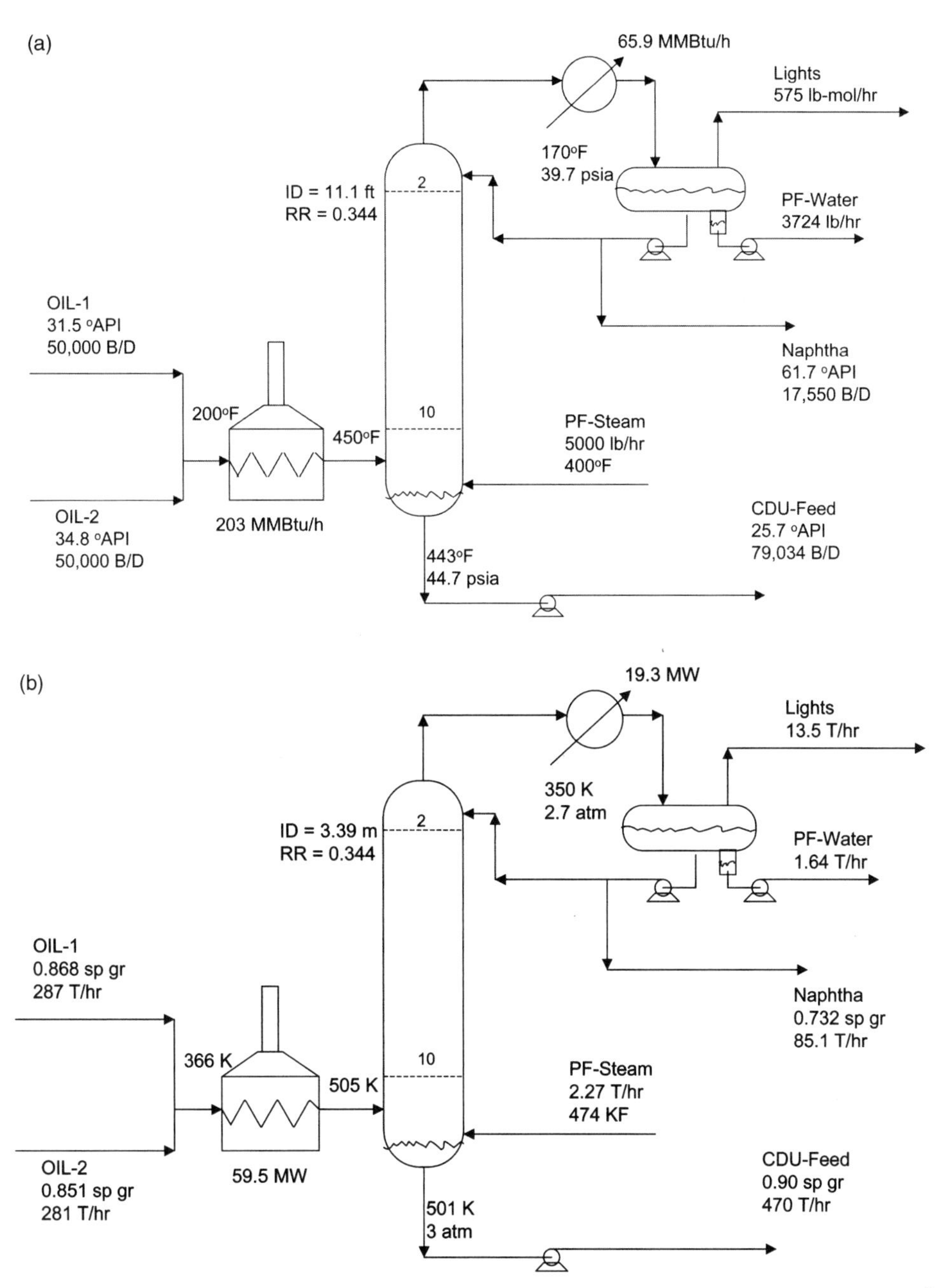

Figure 11.25 PREFLASH flowsheet: (a) engineering units (where 65.9 MMBtu/h = 65.9×10^6 Btu/h); (b) metric units.

11.4 CONTROL OF PREFLASH COLUMN

The reflux drum and column base are sized to provide 5 min of liquid holdup when half-full. The file is pressure checked and exported to Aspen Dynamics. The initial control scheme that opens is shown in Figure 11.26. Note that there is a level controller

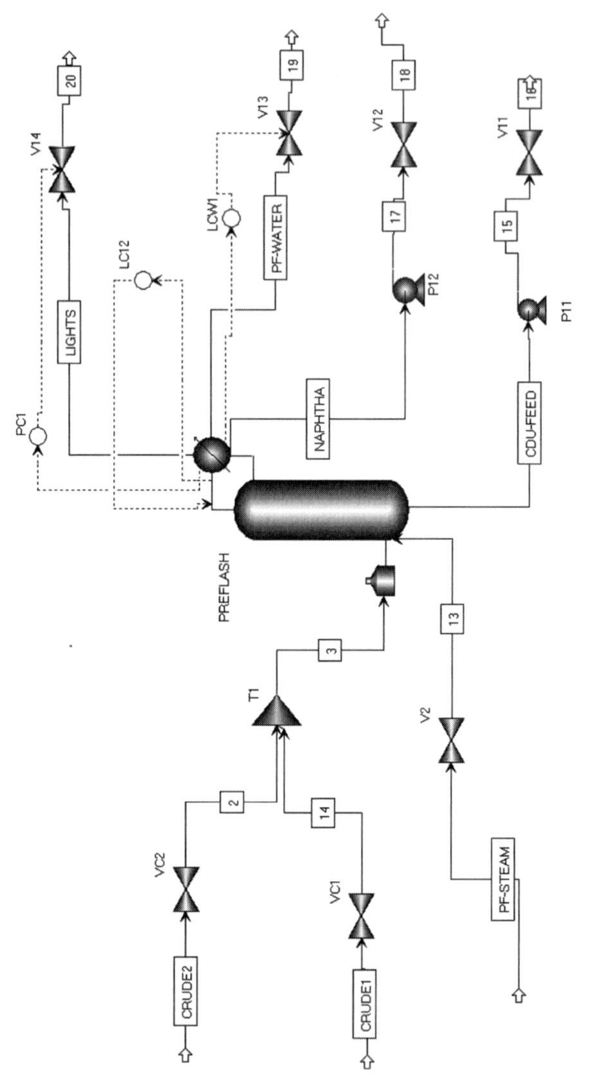

Figure 11.26 Initial control scheme.

(LCW1) that pulls off free water from the reflux drum. The other level controller (LC12) manipulates reflux flowrate to hold the liquid level of the organic phase in the reflux drum. Since the reflux ratio is only 0.344 in this column, we change the control structure to hold reflux drum level with the *NAPHTHA* flowrate (valve V12). Pressure controller PC1 manipulates the valve V14 in the vapor line.

Flow controllers are added to the two crude feeds and the stripping steam. A base-level controller is added that manipulates bottoms flowrate. A temperature controller is added that holds the furnace outlet temperature by manipulating furnace heat input.

The only remaining loop is a controller that manipulates the flowrate of the *NAPHTHA* to maintain a 95% point at 375°F (191°C in the metric units used in Aspen Dynamics). The default properties available in Aspen Dynamics for any stream do not include boiling point information. To obtain these data, we must turn on the "Stream Sensor." Select the *NAPHTHA* stream, right-click, select *Forms*, and select *Configure Sensor*. The view shown at the left in Figure 11.27 shows the window that opens on which we click the *Sensor On* and *Calculate Phase Properties* boxes, specify the *Valid Phases* to be *Liquid-Only*, and select the additional properties that we wish to have available. In our case the *ASTM D86 Temperature* is the item of interest, so it is moved to the right column under *Selected Properties*. The number "95" is entered in the *Liquid Volume % Distilled* box. Now this property is reported in the stream results for the *NAPHTHA* stream (see the right side of Fig. 11.27). This property can also be selected as an input to a controller ("95BPt"), which will manipulate reflux flowrate to maintain the desired ASTM 95% boiling point of the naphtha.

Figure 11.28 shows the control structure. All level controllers are proportional with gains of 2. The furnace temperature loop has a deadtime of 3 min instead of the normal 1-min deadtime used in temperature loops because the dynamics of a fired furnace are usually slower than those of a steam-heated reboiler. Relay–feedback testing and Tyreus–Luyben settings give controller tuning constants of $K_C = 0.465$ and $\tau_I = 13$ min (with a temperature transmitter range of 100–500°C and a maximum heat input of

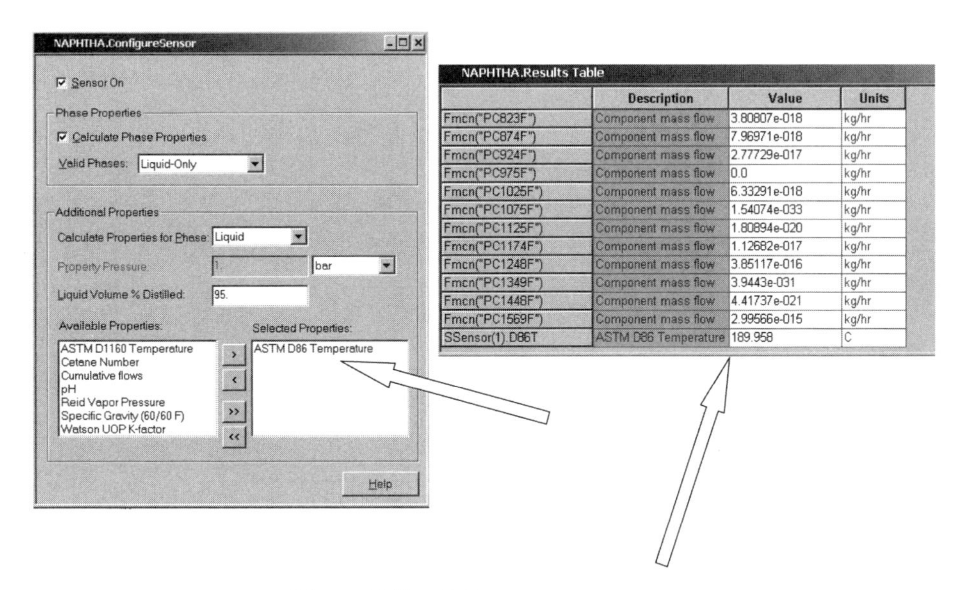

	Description	Value	Units
Fmcn("PC823F")	Component mass flow	3.80807e-018	kg/hr
Fmcn("PC874F")	Component mass flow	7.96971e-018	kg/hr
Fmcn("PC924F")	Component mass flow	2.77729e-017	kg/hr
Fmcn("PC975F")	Component mass flow	0.0	kg/hr
Fmcn("PC1025F")	Component mass flow	6.33291e-018	kg/hr
Fmcn("PC1075F")	Component mass flow	1.54074e-033	kg/hr
Fmcn("PC1125F")	Component mass flow	1.80894e-020	kg/hr
Fmcn("PC1174F")	Component mass flow	1.12682e-017	kg/hr
Fmcn("PC1248F")	Component mass flow	3.85117e-016	kg/hr
Fmcn("PC1349F")	Component mass flow	3.9443e-031	kg/hr
Fmcn("PC1448F")	Component mass flow	4.41737e-021	kg/hr
Fmcn("PC1569F")	Component mass flow	2.99566e-015	kg/hr
SSensor(1).D86T	ASTM D86 Temperature	189.958	C

Figure 11.27 Setting up stream sensor.

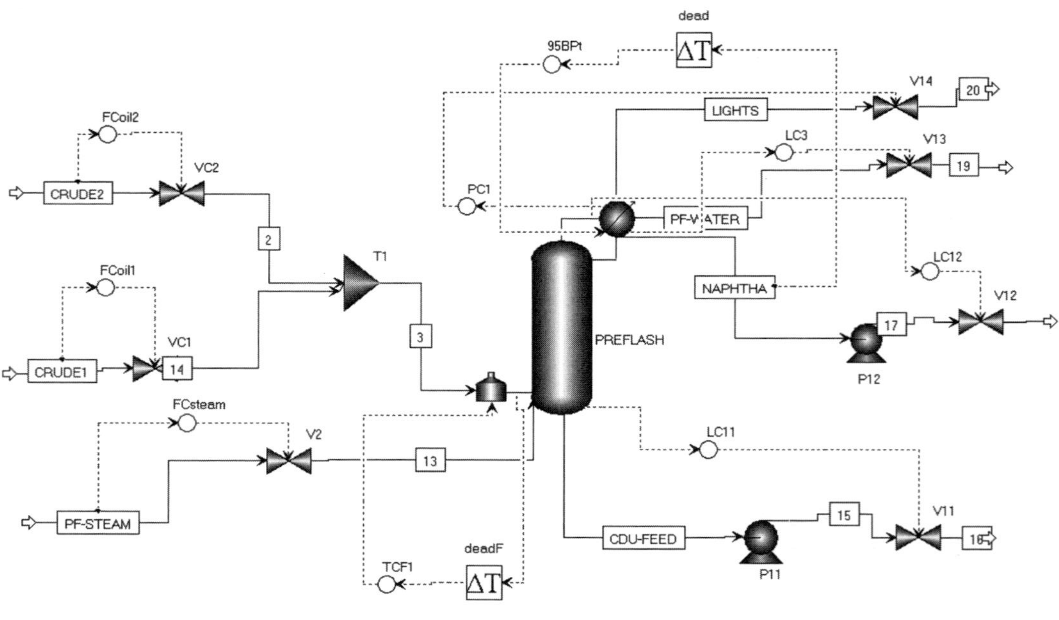

Figure 11.28 Control structure with 95% BP control.

428 GJ/h). The 95BPt controller has a deadtime of 3 min and is tuned in the same way as the temperature controller ($K_C = 0.821$ and $\tau_I = 26$ min) with a boiling point temperature transmitter range of 150–250°C and a maximum reflux flowrate of 70,000 kg/h.

The effectiveness of this control structure is demonstrated in Figure 11.29a for a 20% increase and a 20% decrease in the flowrates of both crude feeds and steam to the base of the column. The peak dynamic deviations in the 95% boiling point of the naphtha are about 12°C. The new steady state for an increase in feed has higher flowrates of reflux, lights, and bottoms (CDU Feed in Fig. 11.29a) but only a small increase in the flowrate of the naphtha. The new steady state for a decrease in feed has lower flowrates of reflux, naphtha, and bottoms (CDU Feed), but the reduction in lights is smaller than is the increase for the positive change in feed. Increasing the load on the column reduces the yield of naphtha at the expense of lights for the same naphtha 95% boiling point of 191°C (375°F).

Figure 11.29b gives results when there is a change in the ratio of the two crude oils. The solid lines are for a decrease in Crude1 from 280 to 224 T/h, while Crude2 is increased from 275 to 329 T/h. Since Crude2 is lighter than Crude1, there is an increase in the flowrate of the "lights" and a decrease in the flowrate of the bottoms. Unexpectedly, the naphtha decreases slightly. The dashed lines represent an increase in Crude1 from 280 to 336 T/h while Crude2 decreases from 275 to 221 T/h. Responses are almost the mirror image of the reverse disturbance.

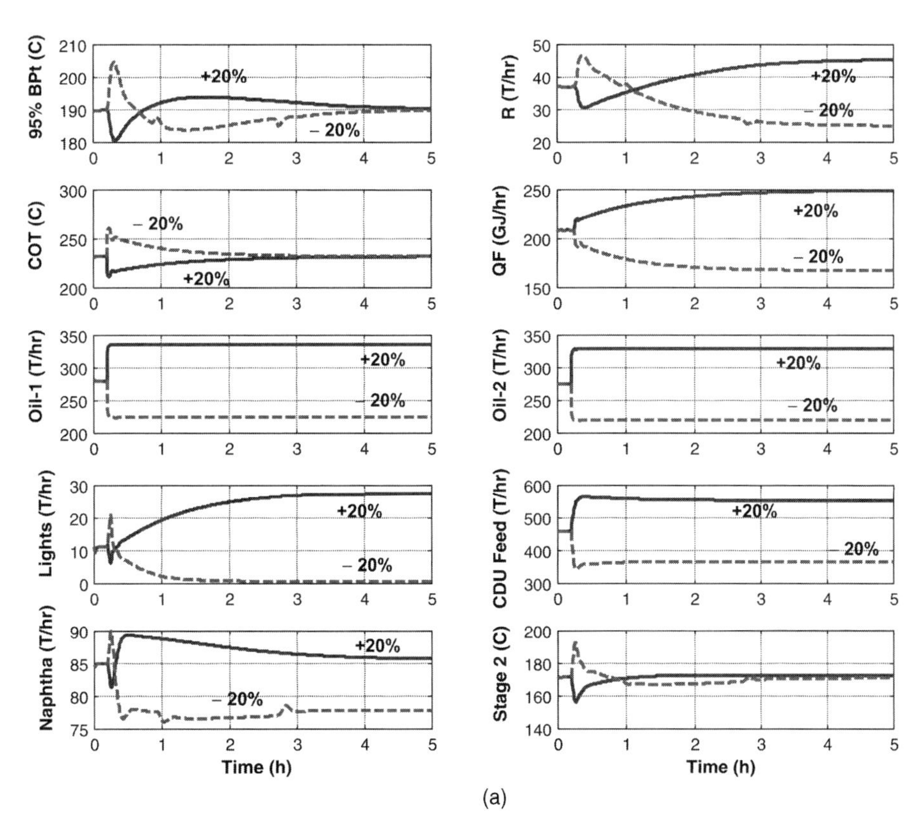

Figure 11.29 (a) Feed rate changes; (b) switching crudes.

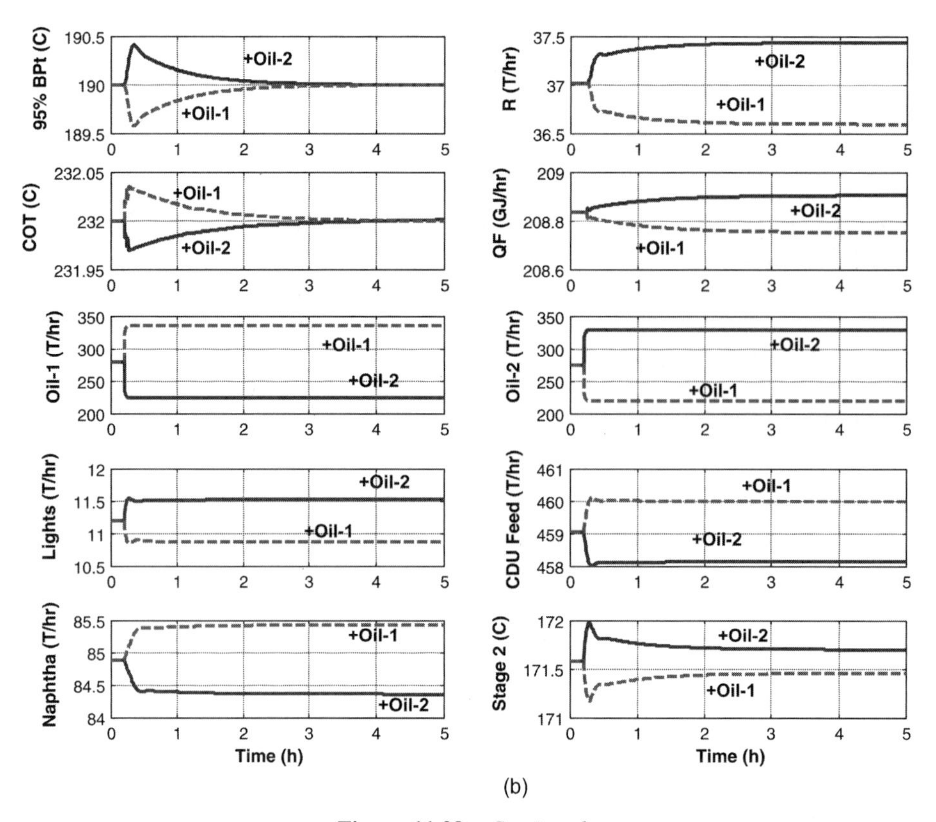

(b)

Figure 11.29 *Continued.*

Note that it takes about 3–4 h for the 95% boiling point loop to settle out. Note also that the temperature on stage 2, which is given in the bottom right graph in these figures, does not change much at the new steady-state conditions for these disturbances. This suggests that the use of temperature control instead of boiling point control might work pretty well.

To test this theory, the control structure is modified to manipulate reflux to hold stage 2 at 171.6°C, as shown in Figure 11.30. The TC2 temperature controller has a deadtime of 1 min and is tuned in the usual way, yielding tuning constants $K_C = 0.90$ and $\tau_I = 5.3$ min (with a temperature transmitter range of 150–250°C and a maximum reflux flowrate of 70,000 kg/h). Note that this reset time is much small than that of the 95% boiling point controller, so faster closedloop dynamics can be expected.

Figure 11.31 shows that this is indeed true. The responses with temperature control are the solid lines. The response with 95% boiling point control are the dashed lines. The disturbance is a swing in crude oils to less Oil-1 and more Oil-2. The process steadies out in about 90 min with temperature control. The naphtha boiling point is not held exactly at 191°C, but ends up at about 189.3°C. The naphtha yield is somewhat smaller, with more bottoms.

11.5 STEADY-STATE DESIGN OF PIPESTILL

Now that we have learned the fundamentals of working with petroleum fractionators by looking at a simple preflash column, we are ready to tackle the next downstream unit, a

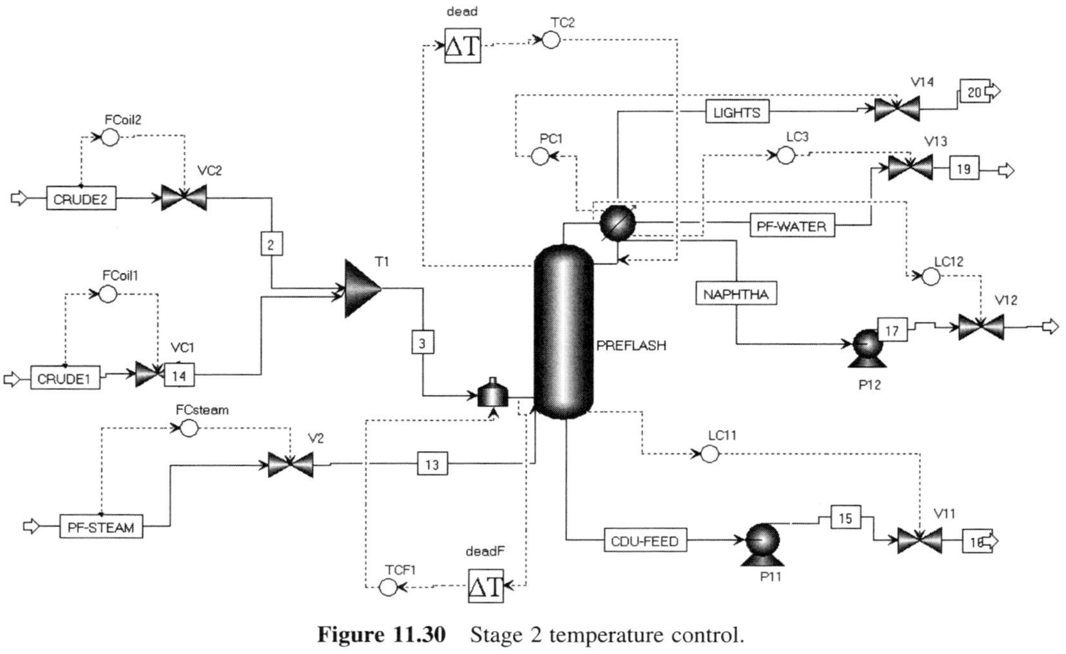

Figure 11.30 Stage 2 temperature control.

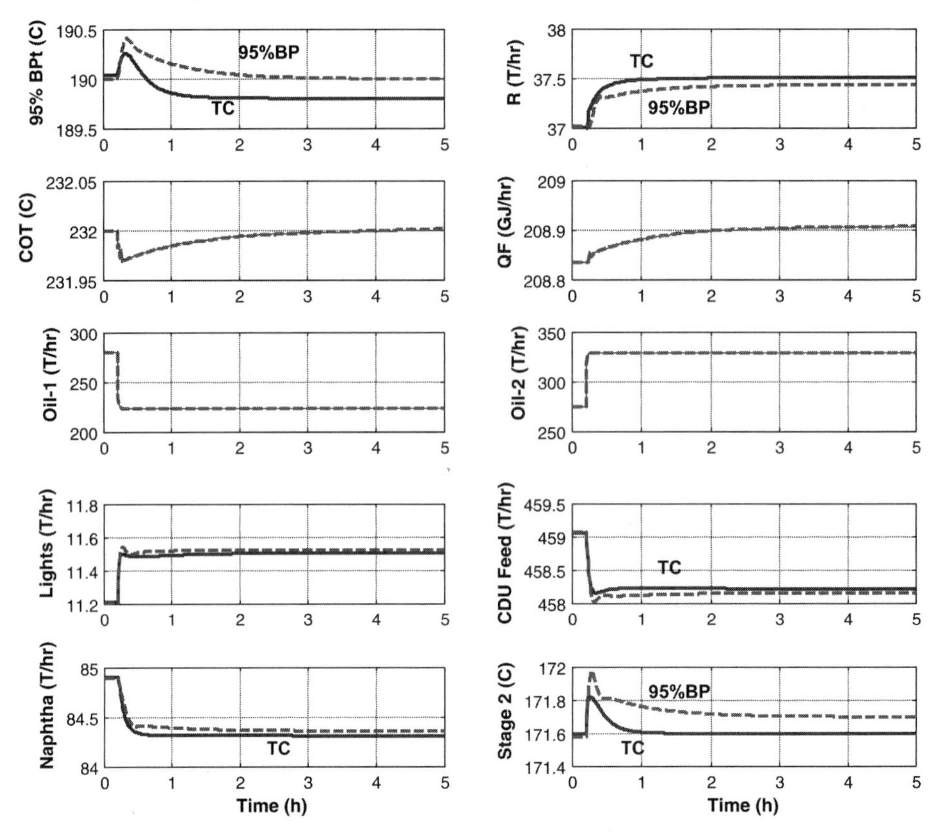

Figure 11.31 Swinging crudes with stage 2 temperature control.

pipestill. The preflash column has removed the light hydrocarbons from the crude feed as a vapor product and produced a light naphtha product as a distillate overhead product. The bottoms from the preflash column is pumped through a furnace in which about 70% of the material is vaporized (depending on the assay of the crude) and fed into a very large column. This column produces an overhead distillate product (heavy naphtha) and three sidestreams: kerosene, diesel, and atmospheric gasoil. The bottoms is called "reduced crude" and is fed to another downstream pipestill operating under vacuum so that more gasoil can be recovered.

Two "pumparounds" are used to recover some of high-temperature energy in the vapor stream flowing up the column. A pumparound takes hot liquid from a tray and pumps it through a heat exchanger that cools the liquid. The cooled liquid is returned back to a tray higher in the column. These pumparound trays are direct-contact heat exchangers. Typically the heat is used to preheat the crude feed to the unit with the objective of reducing furnace fuel consumption. The design of a pipestill incorporates an interesting trade-off between energy consumption and product purity and yield. If no pumparounds were used, all the vapor in the stream from the furnace would pass up the column and be condensed in the water-cooled or air-cooled condenser. No heat would be recovered, so furnace firing would be much larger. A large amount of reflux would be required, which would give higher L/V (liquid to vapor) ratios in the column and therefore better fractionation. This improves the separation between the sidestream products. So the tradeoff is between energy consumption and separation.

11.5.1 Overview of Steady-State Design

The flowsheet of the pipestill is shown in Figure 11.32. The column is very large in diameter (20.3 ft), operates with a top pressure of 15.7 psia, and has a total of 25 stages. The bottoms stream from the preflash column (see Fig. 11.25) is pumped to a furnace that heats the stream to 684°F. The higher the temperature of the furnace exit [coil outlet temperature (COT)], the more of the stream is vaporized and the more gasoil that can be recovered. However, there is a limit to the furnace temperature due to excessive thermal decomposition ("cracking") of the crude in the furnace. If the furnace tube wall temperatures are too high, coke will be formed. This interferes with heat transfer and eventually requires a shutdown of the unit to remove the coke. The heat duty in the furnace is 201×10^6 Btu/h.

The feed is partially vaporized: 2278 lb·mol/h of vapor with a feed of 3644 lb · mol/h. It is introduced into the flash zone on stage 22. There are three stages below the flash zone that are used to strip out any light material that is in the liquid leaving the flash zone. Open steam is fed to the bottom of the column at a rate of 12,000 lb/h. The bottoms stream from the pipestill ("reduced crude") goes to a downstream vacuum pipestill in which more gasoil is recovered. The low pressure in the vacuum furnace produces more vapor for the same furnace temperature.

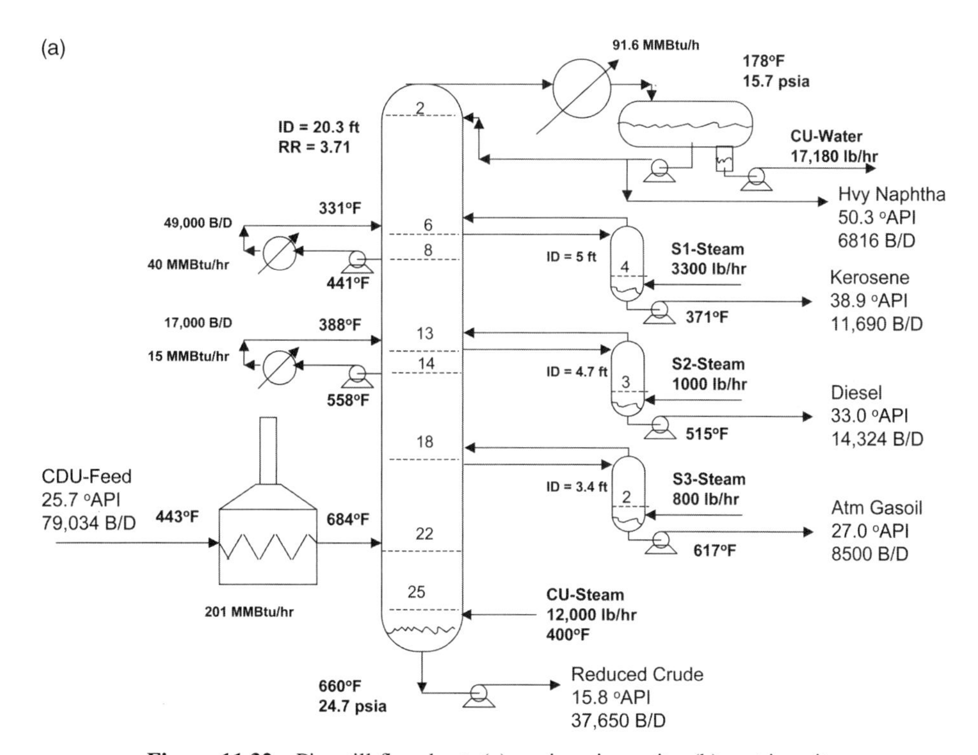

Figure 11.32 Pipestill flowsheet: (a) engineering units; (b) metric units.

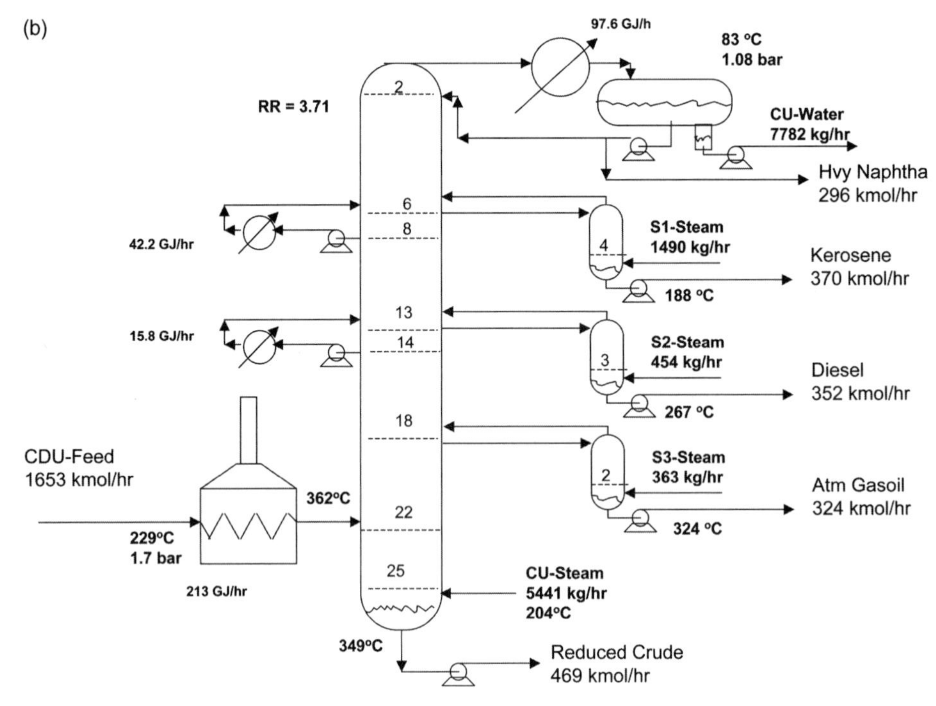

Figure 11.32 *Continued.*

The vapor from the pipestill flash zone flows up the column. At stage 14 a pumparound removes 15×10^6 Btu/h, which reduces the vapor flowing up the column and increases the liquid flowing down the column. This high-temperature heat (558°F) is used for feed preheating. At stage 8 a second pumparound removes 40×10^6 Btu/h, which further reduces the vapor flowrate and increases the liquid flowrate. This high-temperature heat (441°F) is also used for feed preheating.

The vapor leaving the top of the column is condensed in a water or air-cooled condenser. The liquid distillate is a heavy-naphtha stream, which is used for the production of gasoline. It has ASTM 5% and 95% boiling points of 195 and 375°F, respectively. In some refineries it is sent to a reforming unit to produce aromatics (benzene, toluene, and xylenes) and hydrogen. The condensed water is decanted off the reflux drum. Note that this water stream is quite large (17,180 lb/h) because of all the open stripping steam that is used in the column base and sidestream strippers.

The reflux ratio is 3.71. At stage 6 some liquid is withdrawn and fed to a four-stage stripper. Open steam (3300 lb/h) is used to strip light material from the liquid leaving the main column. A kerosene product is produced from the bottom of the stripper. It has ASTM 5% and 95% boiling points of 396 and 502°F, respectively.

In the distillation of distinct chemical components, we talk about separation in terms of the compositions of the impurities in the product streams. In the distillation of petroleum fraction, separation is expressed in terms of "gaps" and "overlaps." These terms refer to the difference between the 95% boiling point of a lighter product and the 5% boiling point of the adjacent next-heavier product. If there were perfect separation of the petroleum cuts, the final boiling point of a lighter product would be equal to the initial boiling point of the next-heavier product. But separation is not perfect. The 95%–5% difference is used as a measure of fractionation. It can be improved by using more trays or by increasing the liquid to vapor ratio in the section of the column in which the separation between the two cuts is occurring. For example, there is a *gap* of 21°F between the heavy naphtha (95% point of 375°F) and the kerosene (5% point of 396°F). This fairly good separation is achieved because of the 3.71 reflux ratio and the five trays between these two products. As we will see, the separations between the other products have overlaps instead of gaps because of the smaller liquid to vapor ratios in the lower sections of the column.

Liquid from stage 13 is withdrawn and fed to a three-stage stripper. Open steam (1000 lb/h) is fed to the bottom of the stripper. A diesel product is produced from the bottom of the stripper. It has ASTM 5% and 95% boiling points of 489 and 640°F, respectively. Note that there is a 13°F *overlap* between the 95% point of the kerosene (502°F) and the 5% point of the diesel (489°F).

A two-stage stripper at stage 18, using 800 lb/h of open steam, produces atmospheric gasoil (AGO) with ASTM 5% and 95% boiling points of 589 and 782°F, respectively. There is a 51°F *overlap* between the 95% point of the diesel (640°F) and the 5% point of the AGO (589°F). This sloppy separation between petroleum cuts is typical of petroleum separation. The values of the different products are usually not drastically different, and improved fractionation can seldom be justified.

11.5.2 Configuring the Pipestill in Aspen Plus

Installing all the equipment and setting up all the conditions discussed above is a fairly involved procedure. We will go through the details step by step. The pipestill is installed on the flowsheet by clicking *Columns* and *PetroFrac* on the *Model Library* toolbar at the bottom of the Aspen Plus window. As shown in Figure 11.33, the "CDU10F" icon is selected (third row, middle column) and pasted on the flowsheet. This configuration has a furnace, multiple sidestream strippers, and multiple pumparounds. Figure 11.34 shows the final flowsheet with all the equipment installed, including the pumps and valve required for dynamic control.

The first step is to connect the bottoms from the preflash column to the furnace of the pipestill. As we did with the preflash furnace, the red input arrow that appears at the bottom of the column (shown at the top of Fig. 11.35) must be dragged over to the furnace (shown at the bottom of Fig. 11.35).

A similar procedure of clicking an arrow and dragging it to the desired location must be performed on the strippers to connect the steam lines and the product withdrawal lines at the base of the stripper. Figure 11.36 gives an example in which a material stream is selected to be installed as steam to the third stripper. Figure 11.36a shows blue arrows appearing at the *top* stripper. We click on the input arrow pointing to the side of the stripper and drag it to the side of the *third* stripper. When the blue arrow is in the desired location (see Fig. 11.36b), the mouse button is released and the connection is made. This procedure is repeated for the steam inputs of the other strippers and for the product streams leaving the bottom of each stripper.

Material streams for the heavy-naphtha distillate (HNAPH), water from the reflux drum decanter (CU-WATER), steam to the column base (CU-STM), and the bottoms (RCRUDE) are made in the normal way.

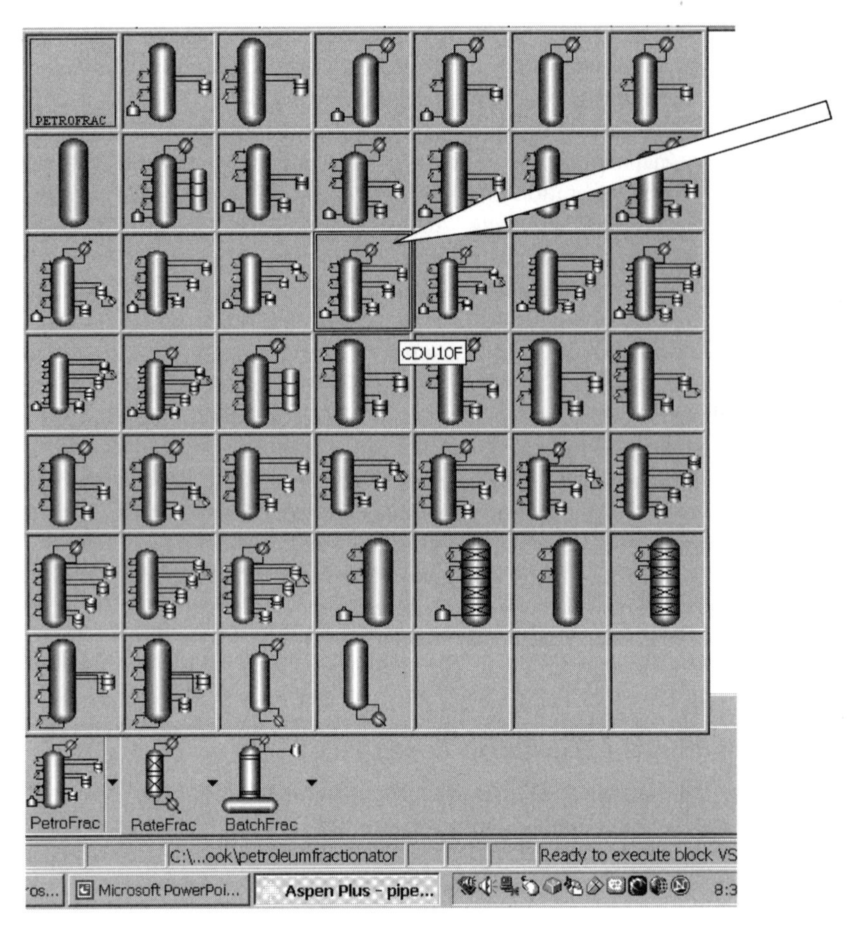

Figure 11.33 PetroFrac palette.

Now we must set up the pumparounds and the strippers. On the Aspen Plus *Data Browser* under the CDU block, there are items for pumparounds and strippers. Clicking the *Pumparounds* item opens a window on which *New* is clicked, an identification label is selected, and the window shown at the top of Figure 11.37a opens. On the *Specifications* page tab, a number of items are specified. For pumparound P-1, the stage from which the hot liquid is removed is stage 8. The stage to which the cool liquid is returned is higher in the column at stage 6. The flowrate is specified to be 49,000 b/d, and the heat removed is set at 40×10^6 Btu/h. Clicking the *Results* page tab (after the simulation has been run) gives useful information about pumparound P-1, as shown in the bottom window of Figure 11.37a. Note that the temperature of the return pumparound liquid is 331°F.

The setup and results for the second pumparound P-2 are shown in Figure 11.37b. The stage from which the hot liquid is removed is stage 14. The stage to which the cool liquid is returned is stage 13. The flowrate is specified to be 11,000 b/d, and the heat removed is set at 15×10^6 Btu/h.

The three strippers are installed in a similar manner. The *Stripper* item under the CDU block is clicked, the *New* button is clicked, the new stripper is identified, and the window

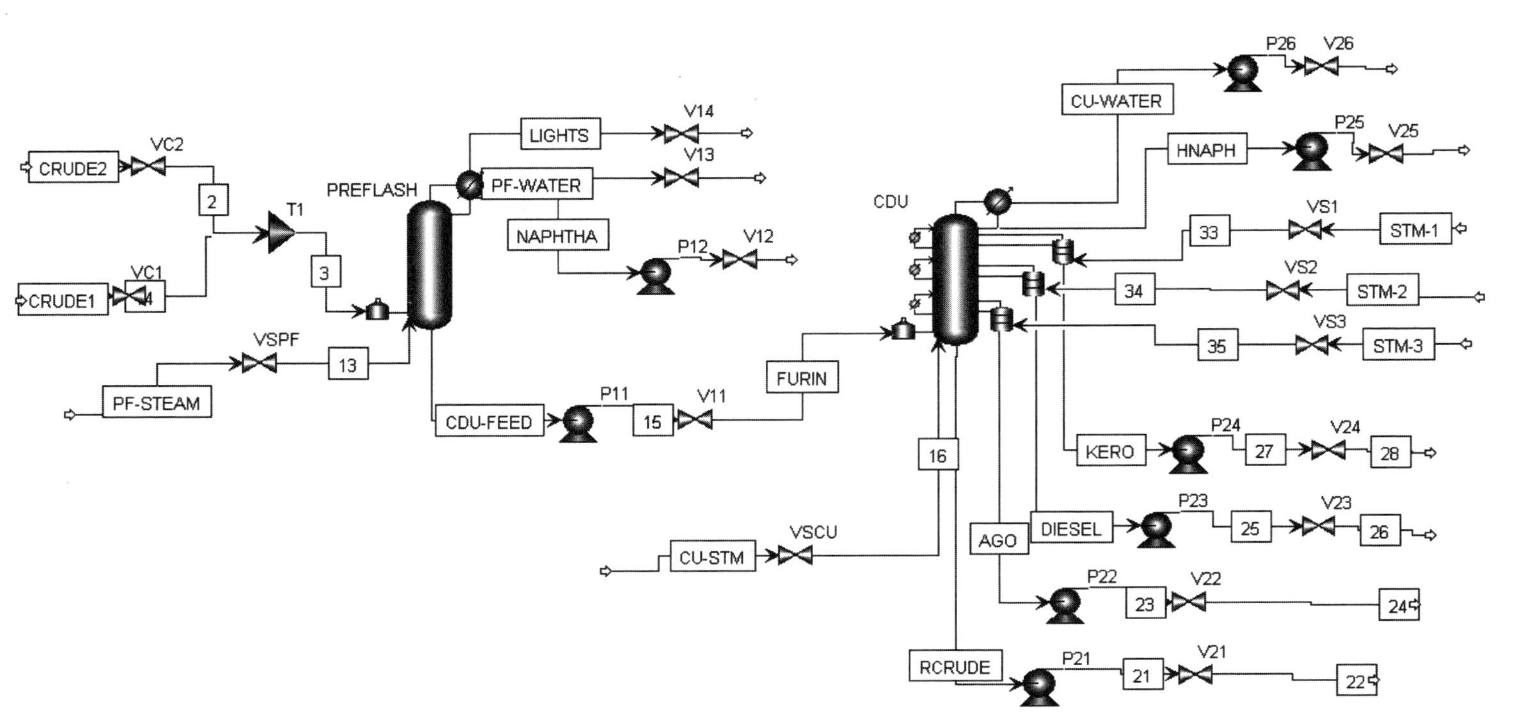

Figure 11.34 Preflash and pipestill.

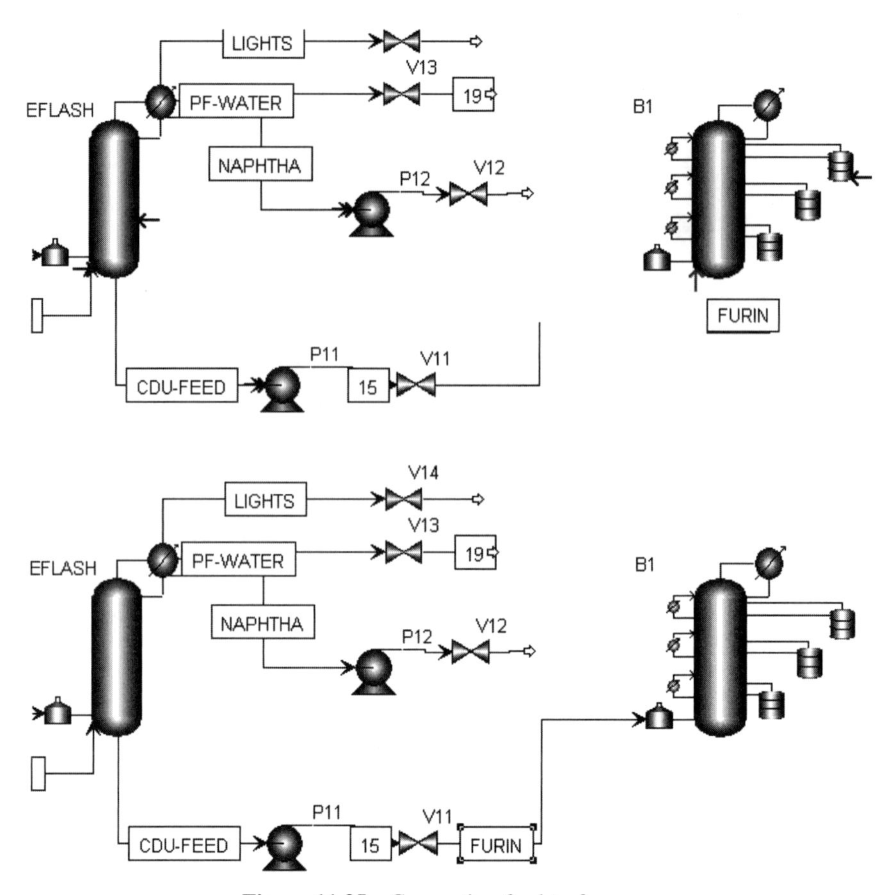

Figure 11.35 Connecting feed to furnace.

shown at the top of Figure 11.38 opens. With the *Configuration* page tab clicked, a number of parameters of the stripper are specified. For the top stripper S-1, the number of stages is 4, the product is *KERO*, the liquid drawoff stage from the main column is stage 6, and the vapor from the stripper is returned to stage 5. The steam stripping is done by the stream labeled "33." The flowrate of the bottoms product from the stripper is specified to be 11,700 b/d. Clicking the *Pressure* page tab opens the window shown at the bottom of Figure 11.38, on which the stage 1 pressure and the pressure drop per stage are set.

The pumps on the pipestill are specified in the normal way with 100 psi pressure rises. The valves are specified to have 50 psi pressure drops. The diameter of the main column and the diameters of each stripper are calculated using the *Tray Sizing* functions. The *Specifications* page tab is shown in Figure 11.39 for the top stripper (S-1).

Setting up the main column starts with clicking the *CDU* item under *Blocks* in the *Data Browser*. Then *Setup* is clicked and the *Configuration* page tab is pressed, which opens the window shown at the top of Figure 11.40a. The number of stages, a total condenser, and an estimate of the distillate flowrate are specified. Clicking the *Streams* page tab opens the window shown at the bottom of Figure 11.40a, on which the locations of the various input and output streams are specified. Note that the feed to the furnace is specified to

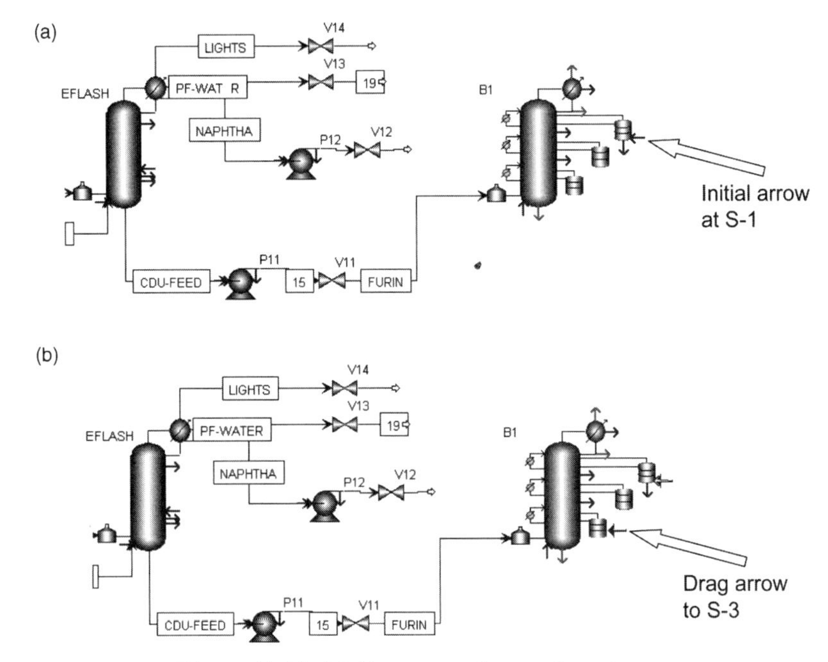

Figure 11.36 Making connections to the strippers.

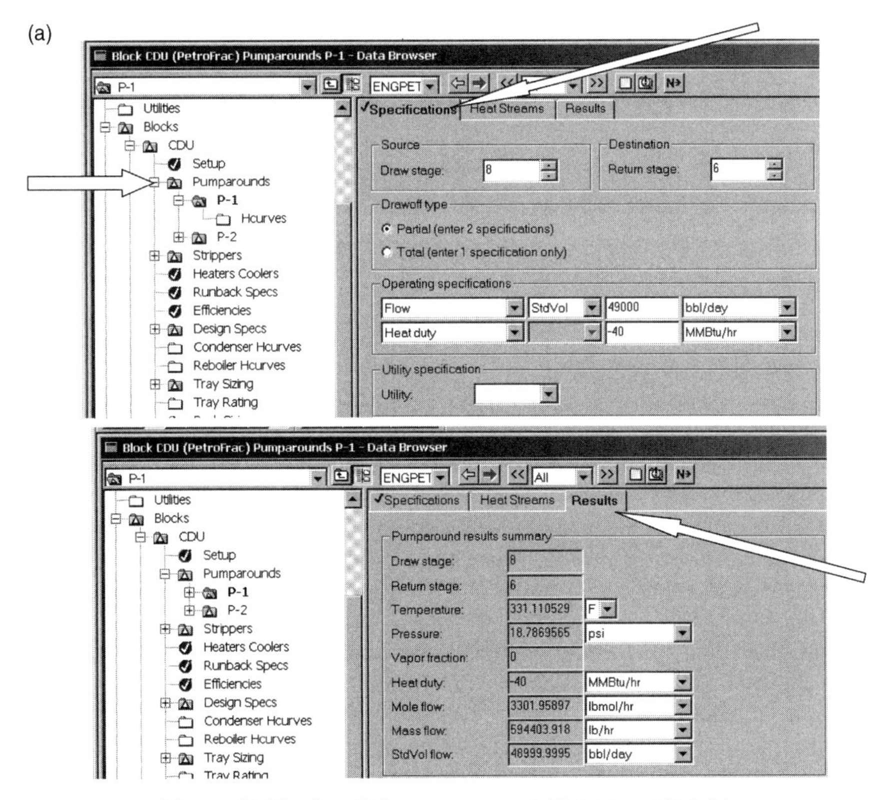

Figure 11.37 Specifying pumparounds P-1 (a) and P-2 (b).

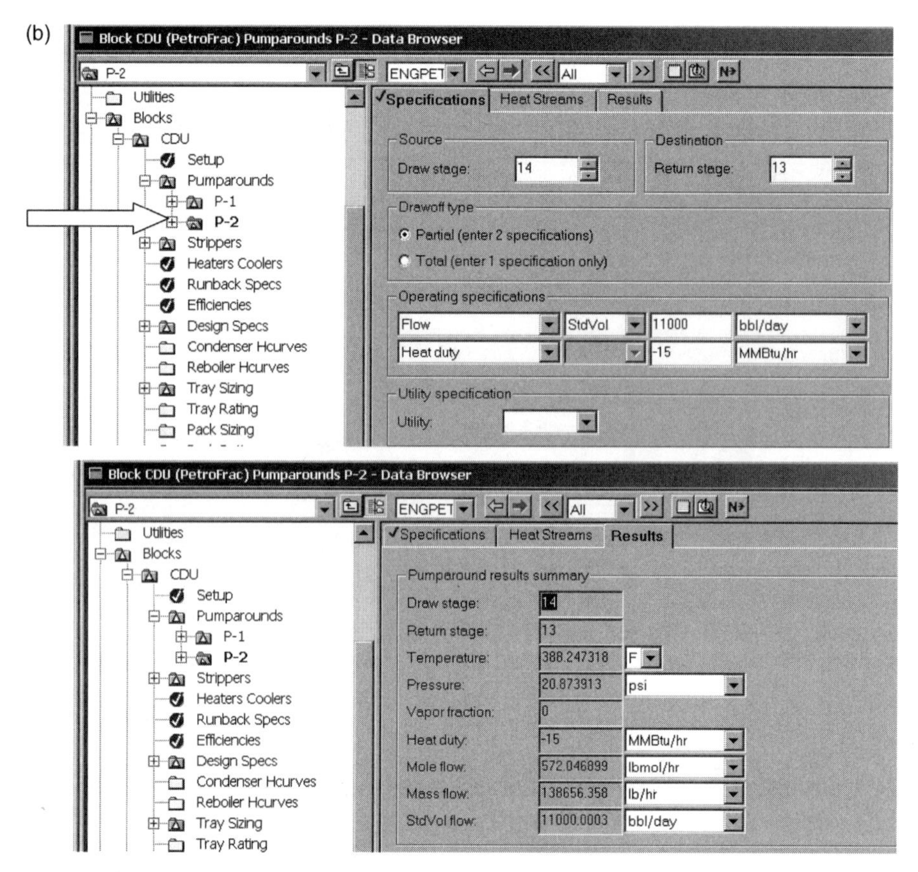

Figure 11.37 *Continued.*

enter on stage 22, while the stripping steam enters three trays lower (stage 25). Clicking the *Pressure* page tab gives the window shown at the top of Figure 11.40b, and the appropriate pressure information is entered.

Clicking the *Furnace* page tab opens the window shown at the bottom of Figure 11.40b. There are several options for setting up the furnace. For *Furnace type* we select the *Single stage flash with liquid runback.* For *Furnace Specification* we select *Fractional overflash* and specify it to be 0.03. The term "overflash" refers to the fraction of the vapor that is produced in the flash zone of the column that is returned as liquid to the flash zone from the tray above the flash zone. Most of the liquid coming down the column is withdrawn as sidestreams, but some liquid is needed on the trays below the lowest sidestream and above the flash zone to prevent entrainment of heavy liquid up the column with the vapor. Entrainment could drive the color of the gasoil off specification. It could also increase the concentration of metal contaminants in the gasoil because some crude oils contain small amounts of metals. The gasoil is usually fed to a catalytic cracking unit, and the catalyst in this unit is deactivated by metals. Therefore a small "wash" or "overflash" stream is needed.

The final part of the steady-state design is to set up two *Design Specs*. The first varies the flowrate of the heavy naphtha to achieve an ASTM 95% boiling point of 375°F. The second varies the flowrate of the diesel to achieve an ASTM 95% boiling point of 640°F.

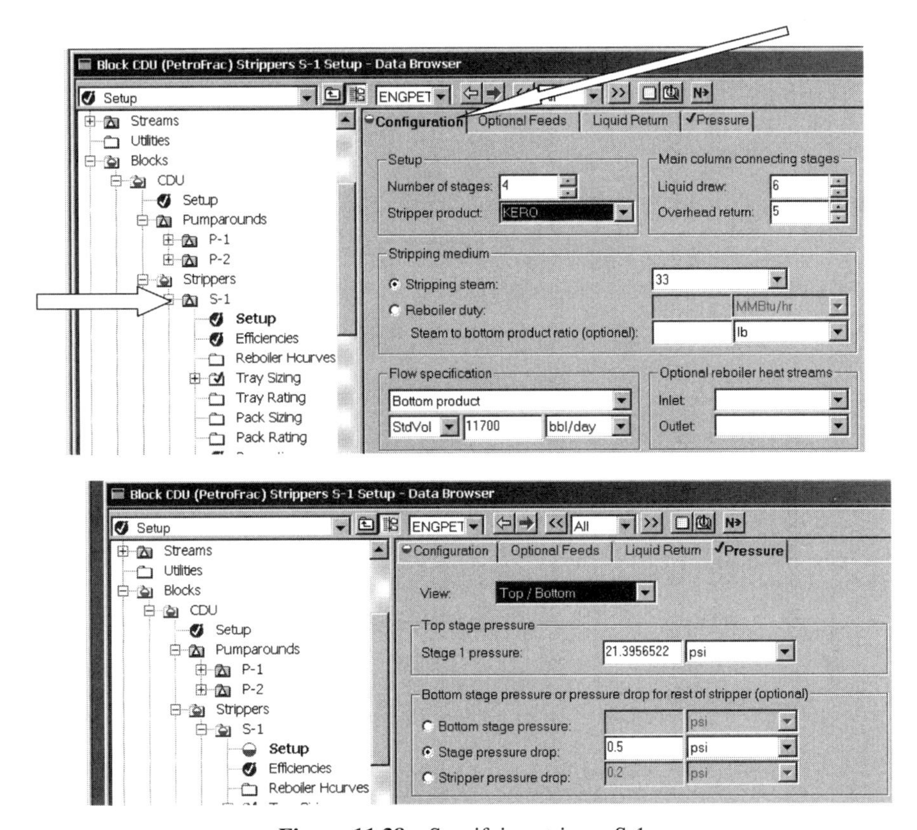

Figure 11.38 Specifying stripper S-1.

The setups for these are shown in Figure 11.41. The top panel in Figure 11.41a gives the *Specification* page tab for the heavy-naphtha design spec. The middle panel gives the *Feed/Product Streams* page tab. The bottom panel gives the *Vary* page tab. Note that *Distillate flow rate* is selected.

Figure 11.41b gives the same information for the diesel design specification. Note that the stream selected to vary is *Bottoms flow rate* from the S-2 stripper.

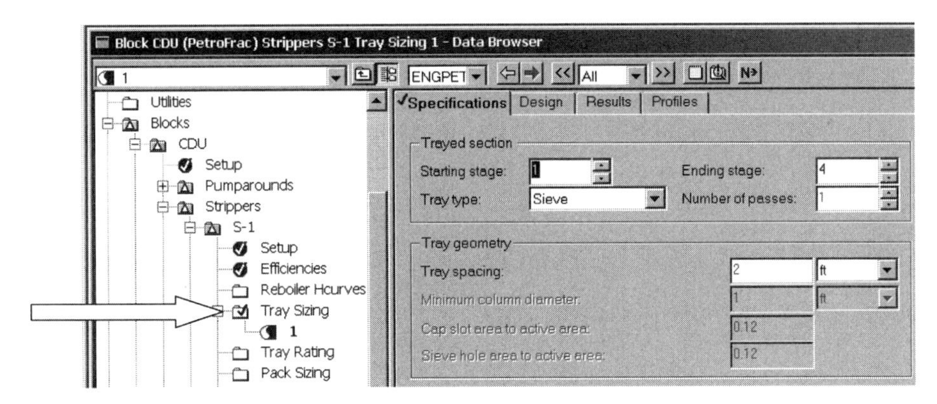

Figure 11.39 Tray sizing for stripper S-1.

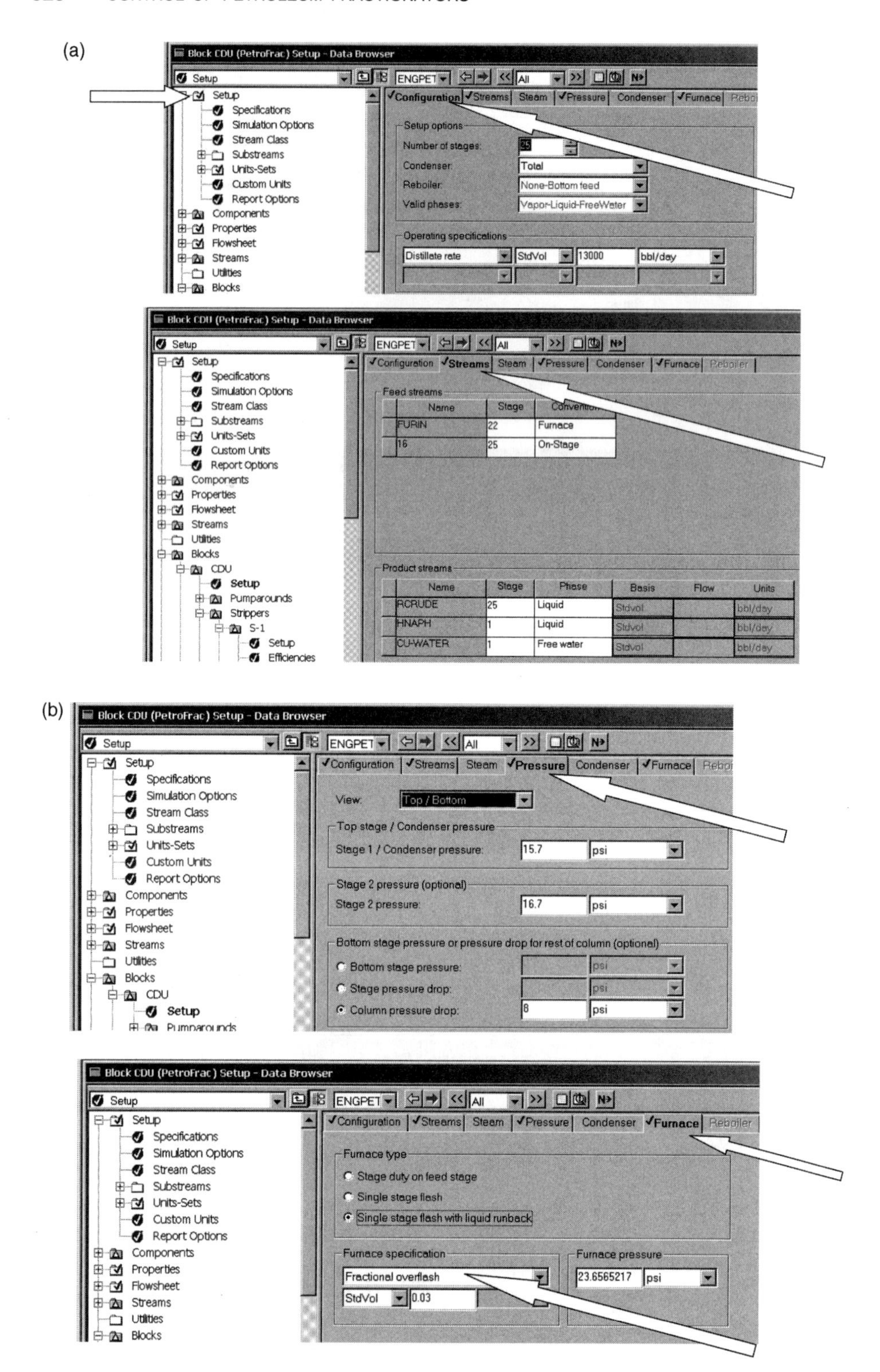

Figure 11.40 Setup for main column: (a) configuration and streams; (b) pressure and furnace.

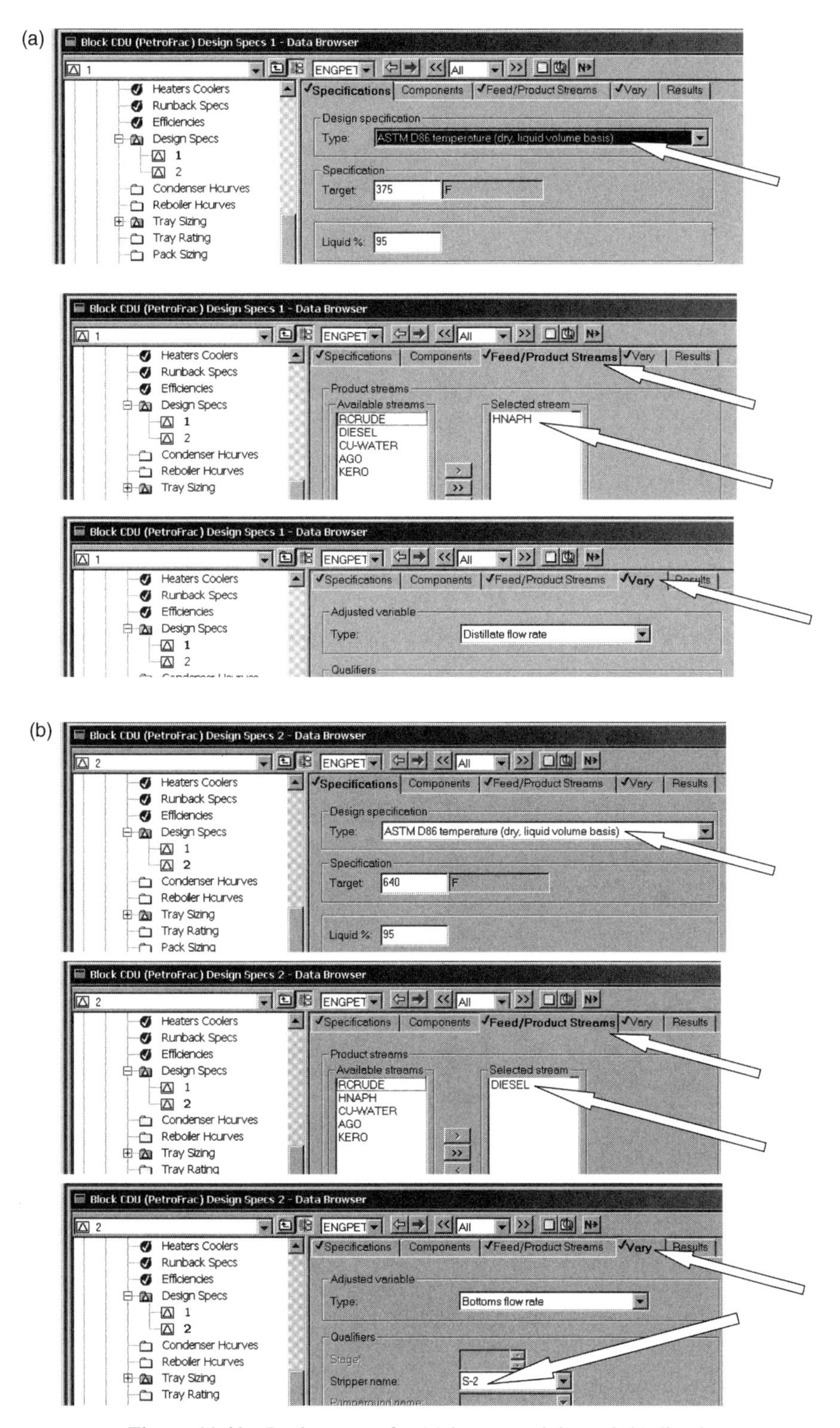

Figure 11.41 Design specs for (a) heavy naphtha and (b) diesel.

The ASTM curves for all the products from the pipestill are generated by going to *Results Summary* on the *Data Browser* window, clicking *Streams*, selecting the *Vol. % Curves* page tab, and selecting the *ASTM D86 curve* in the *Curve view* window, as shown in Figure 11.42.

Next, left-click the top of the *Vol%* column, which highlights this column. Go to the toolbar at the top of the Aspen Plus window, click *Plot* and click *X-Axis Variable*. Highlight each of the columns for the products that you want to plot by holding down the *Ctrl* key and clicking the top of each column. Then go to the toolbar, click *Plot*, and click *Y-Axis Variable*. Finally go to the toolbar, click *Plot*, and click *Display Plot*. Results are shown in Figure 11.43.

Two of the unique features of a petroleum fractionator are the large changes in temperature and flowrates from the bottom to the column. Figure 11.44 gives these profiles for the pipestill.

The temperature ranges from near 180°F at the top to over 650°F at the flash zone. Note that the temperature decreases slightly in the bottom three trays because of the stripping with 400°F steam.

The molar flowrates of the liquid and vapor increase as we move up the column from the flash zone. At the pumparounds the vapor rate decreases.

11.5.3. Effects of Design Parameters

Before we move on to dynamics and control, it may be useful to briefly explore the impact of some of the design parameters on the yields and the boiling points of the various products. This insight will be applied when a control structure is developed to maintain the several specifications in the face of disturbances.

In setting up the steady-state design, we have specified all the equipment parameters (the number of stages and locations of feeds and withdrawal points). In addition, we have specified 10 operating variables; that is, there are 10 operating degrees of freedom in this pipestill process:

1. The 95% boiling point specification on the heavy-naphtha stream is 350°F.
2. The diesel 95% boiling point specification is 640°F.
3. The flowrate of AGO is 8500 b/d.

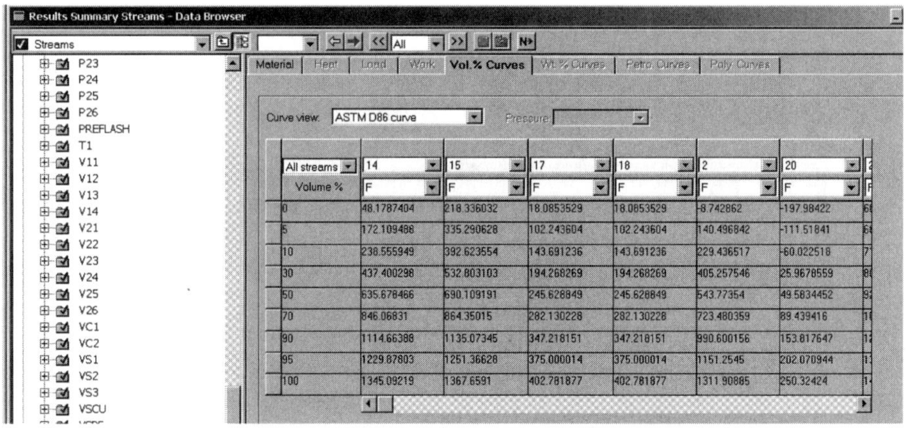

All streams / Volume %	14 (F)	15 (F)	17 (F)	18 (F)	2 (F)	20 (F)
0	48.1787404	218.336032	18.0853529	18.0853529	-8.742862	-197.98422
5	172.109488	335.290628	102.243604	102.243604	140.496842	-111.51841
10	238.555949	392.623554	143.691236	143.691236	229.436517	-60.022518
30	437.400298	532.803103	194.268269	194.268269	405.257546	25.9678559
50	635.678466	690.109191	245.628849	245.628849	543.77354	49.5834452
70	846.06831	864.35015	282.130228	282.130228	723.480359	89.439416
90	1114.66388	1135.07345	347.218151	347.218151	990.600156	153.817647
95	1229.87803	1251.36628	375.000014	375.000014	1151.2545	202.070944
100	1345.09219	1367.6591	402.781877	402.781877	1311.90885	250.32424

Figure 11.42 Generating ASTM curves for all products.

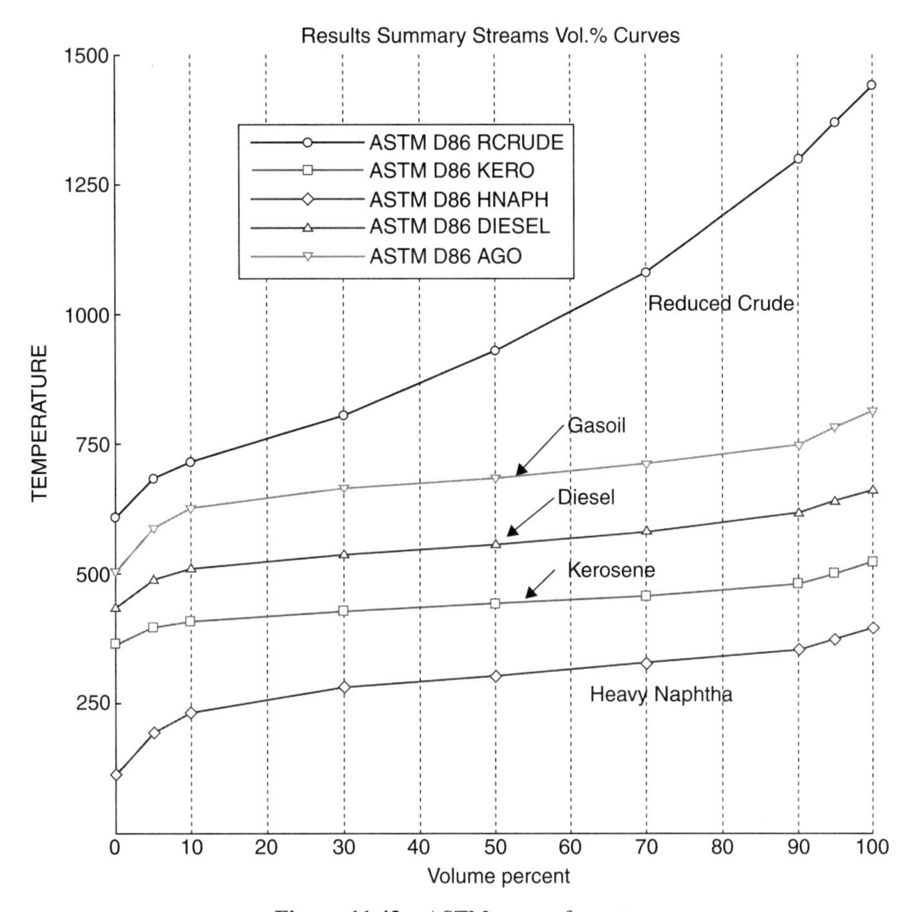

Figure 11.43 ASTM curves for cuts.

4. The fractional overflash specification is 0.03.
5. The four stripping steam flowrates are fixed (one to the base of the column and one to each of the three strippers).
6. The heat removals in the two pumparounds are fixed.

Let us explore the steady-state effects of changing some of these variables and establish some principles of operating a petroleum fractionator.

Effect of Changing a 95% Specification Suppose that we change the 95% boiling point specification on the heavy-naphtha stream from 375 to 350°F. The other degrees of freedom are unchanged. The result is a decrease in the heavy-naphtha flowrate from 6830 to 5425 b/d. The lower 95% boiling point means that a lower fraction of the crude is taken off as heavy naphtha. There is a corresponding reduction in the 5% boiling point of the kerosene, which changes from 395 to 380°F. This illustrates an important principle in the operation of sidestream petroleum fractionators.

Principle 11.1 The 95% and 5% boiling points of adjacent cuts cannot be **independently** set. For example, a decrease in the drawoff rate of a product results in a **decrease**

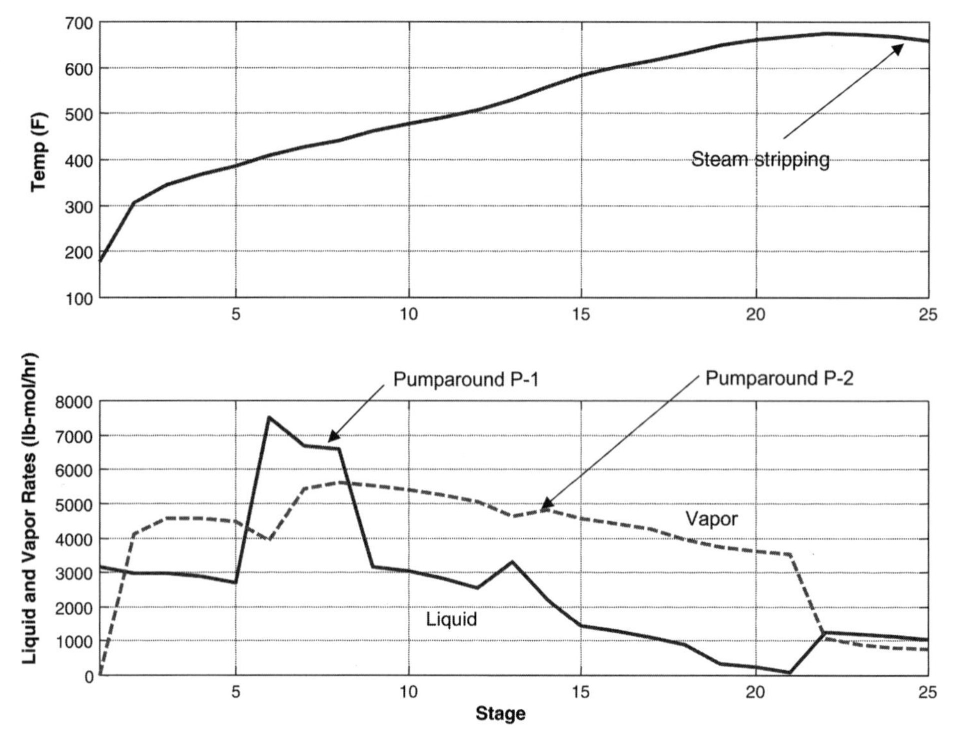

Figure 11.44 Pipestill profiles.

in its 95% boiling point and a **decrease** in the 5% boiling point of the next heavier product stream.

The reflux flowrate also increases (from 2517 to 2884 lb·mol/h). This and the lower distillate flowrate produce a higher reflux ratio (increasing from 3.875 to 5.41), which has the effect of providing more fractionation between the heavy naphtha and the kerosene. The "gap" between these cuts increases from $395 - 375 = 20°F$ to $380 - 350 = 30°F$.

In a similar way, suppose that the specification on the 95% boiling point of the diesel is reduced from 640 to 620°F. The flowrate of the diesel decreases from 14,363 to 12,112 b/d. This drops more light material into the lower AGO stream, so its 5% point drops from 589 to 500°F and its 95% point, from 782 to 765°F. The flowrate of the bottoms increases from 37,647 to 39,953 b/d, and its 5% point changes from 692 to 665°F.

Effect of Changing a Pumparound Changing a pumparound heat removal affects the vapor traffic in the column above the pumparound and the liquid traffic below the pumparound. Of course, it also affects the furnace firing rate because the temperature of the feed to the furnace changes. For example, suppose that we change the heat removal in the top pumparound (P-1) from 40 to 30×10^6 Btu/h. More vapor flows up through the top part of the column, which increases the reflux ratio from 3.875 to 4.466 and increases the condenser heat removal from 92 to 102×10^6 Btu/h. The higher liquid to vapor ratio provides better fractionation above the pumparound. The gap between the heavy naphtha and the kerosene increases from 395 − 375 = 20°F to 398 − 375 = 23°F. Thus there is a slight improvement in this separation.

The downside of this change is that more heat is rejected to cooling water in the condenser instead of being recovered by feed preheating. The effect on furnace firing depends on the configuration of the heat exchanger network used, which is not modeled in the simulation considered in this chapter.

Principle 11.2 Pumparounds affect separation between cuts and furnace firing in opposite ways. Reducing a pumparound heat removal improves separation between cuts above the pumparound, but increases furnace energy consumption.

Effect of Changing Stripping Steam Open steam is used in the strippers to remove the light material that is in the liquid withdrawn from the main column. Changing stripping steam flowrate affects the initial part of the boiling point curve, but has less of an impact on 5% point and essentially no impact on the 95% point and product flowrates. Of course, using more steam increases steam consumption and increases the load on water purification facilities required to handle the water decanted off the reflux drum.

For example, assume that the stripping steam to the top (kerosene) stripper is increased from 3300 to 5000 lb/h. All other degrees of freedom remain unchanged. The initial *TBP* boiling point of the kerosene changes from 311 to 321°F. The initial *ASTM* boiling point of the kerosene changes from 366 to 374°F. The ASTM 5% boiling point changes only from 395 to 399°F. The ASTM 95% point changes only from 502 to 503°F.

Principle 11.3 The flowrate of stripping steam affects the initial boiling point or the flashpoint of the cut.

The steady-state design is now complete. We are ready to investigate dynamics and control of this complex system.

11.6 CONTROL OF PIPESTILL

A petroleum fractionator, such as a pipestill or a cat fractionator, is almost overwhelmingly complex. In addition to the main column, there are strippers that have vapor and liquid streams going back to and coming from the main column. There are a very large number of control loops to set up. Let us enumerate the loops that we will set up, considering both the preflash column and the pipestill:

1. *Temperature loops*—temperatures of both furnaces (two loops)
2. *Flow loops*—two crude feeds, steam to two column bases, steam to three strippers, and stage 19 liquid in the pipestill (eight loops)
3. *Pressure loops*—condensers in both columns and three strippers (five loops)
4. *Level loops*—base level of two columns, water level in two reflux drums, organic level in two reflux drums, and base levels in three stripper bases (nine loops)
5. *ASTM boiling points*—95% boiling point of light naphtha, 95% boiling point of heavy naphtha, 5% boiling point of diesel, 95% boiling point of diesel (four loops)

There are 28 controllers to set up on these two columns!

The equipment associated with the preflash column has already been sized. The diameters of the pipestill column and the three strippers are sized using the *Tray Sizing* feature of Aspen Plus for each vessel. The results are

Vessel	Diameter (ft)
Pipestill	20.3
Stripper S-1	5.0
Stripper S-2	4.7
Stripper S-3	3.4

The reflux drum, the column base, and the bases of the three strippers are sized to give 5 min of holdup at a 50% level.

The file is pressure-checked and exported into Aspen Dynamics. When it opens, using version 12 of Aspen Dynamics, there is an error message that the system is overspecified. This problem was corrected by Aspen Support using a "hot fix."[3]

The initial control structure set up by Aspen Dynamics is shown in Figure 11.45. The flowsheet is quite congested with many process lines, control lines, and equipment. It takes a fair amount of artwork to rearrange the drawing to make it readable.

The two condenser pressure controllers, two organic-phase level controllers, and the water-phase level controller in the preflash column have been installed. In addition, pressure controllers on the three strippers are set up. Note that the vapor flows from the strippers back to the main column are manipulated. No control valve is shown in the vapor line, indicating that a "flow-driven" assumption is made in this flow. Also note that the three stripper pressure signals all appear to come from the first stripper. These will be moved in the final flowsheet to start from the appropriate stripper for each pressure controller. This is done by clicking the control signal line, clicking the blue arrow at the point of origin, and dragging it to the correct location when the arrow turns red. The original flowsheet showed the three pressure controller output signals all going to the vapor line of the top stripper. These lines have been relocated to show them going to the correct vapor line of each stripper.

Installing the level controller to hold the levels in the base of each column and the two levels in each reflux drum is straightforward. Note that the organic level is controlled by manipulating the reflux flowrate in both columns because the reflux ratios are large. When selecting the *PV* signal for the levels in the reflux drums, the organic phase is *Level 1* and the water phase is *Level 2*. Proportional controllers with gains of 2 are used on all levels. The two furnace temperature controllers are installed in the conventional way. Deadtimes of 1 min are used in these loops, and temperature transmitter ranges are 100–500°C. Relay–feedback testing and Tyreus–Luyben tuning give controller gains of 0.6 and integral times of 4 min in both controllers.

Flow controllers are installed on the steam to the base of the two columns. These flowrates are ratioed to the feedflows to the respective column by using multipliers. The molar steam to feed ratio in the preflash column is $125.9/2722 = 0.04625$. The total crude feed is

[3]The extensive help of Arabella Geser (Global Customer Support & Training, Aspen Tech Europe) in getting the pipestill dynamic simulation to run is gratefully acknowledged. The new version of Aspen 2004 has incorporated this correction.

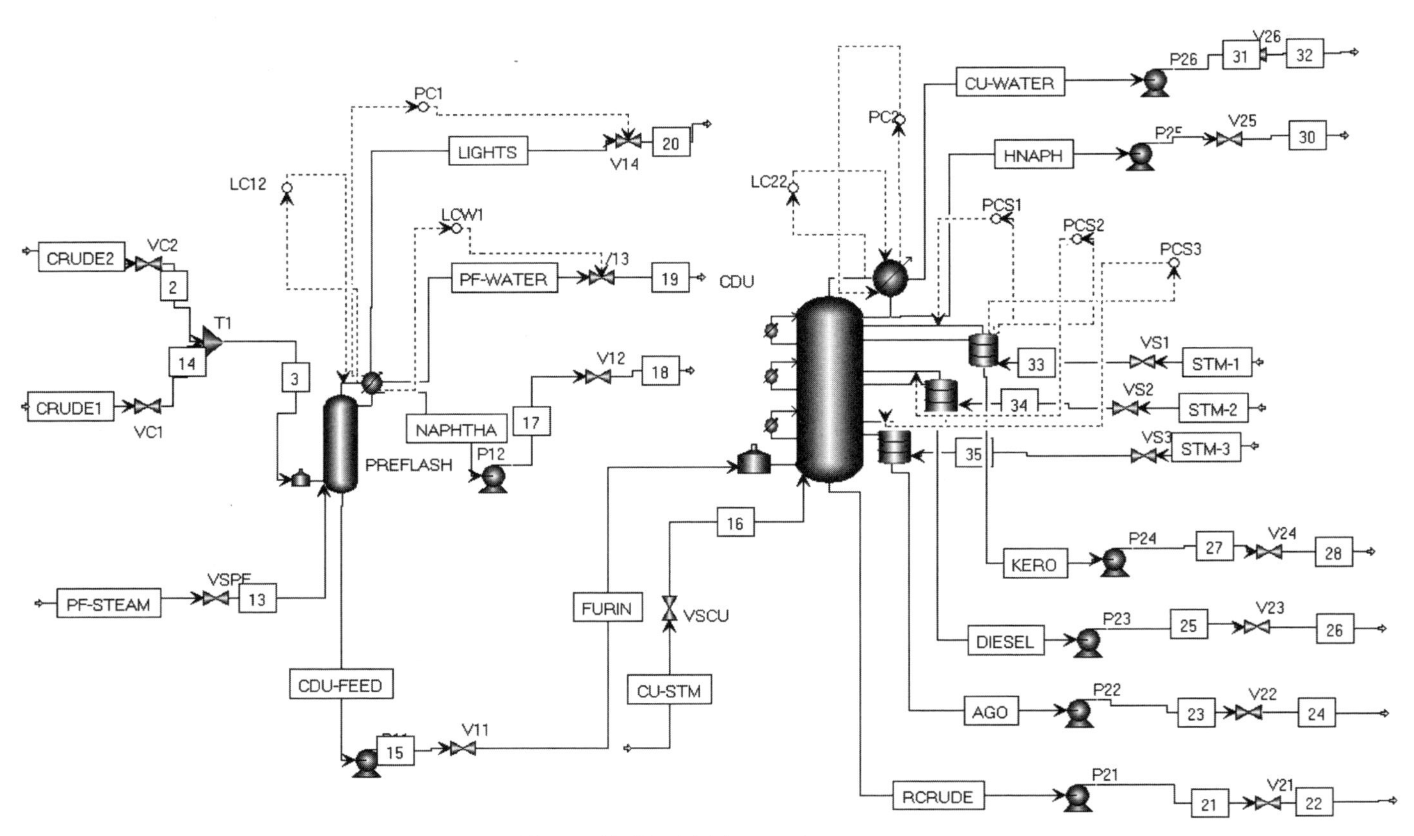

Figure 11.45 Initial control structure.

used (after the summer). The molar steam to feed ratio in the pipestill is 302.1/1654 = 0.1827. The two steam flow controllers are "on cascade" with their setpoints coming from multipliers set up with the appropriate constant and with the appropriate flow signal input.

Setting up the stripper base-level controllers requires a little graphical skill. All the input and output arrows appear on the top stripper and must be moved to the correct location. For example, Figure 11.46 shows an arrow pointing to the liquid line between the main column and the top stripper. This is the correct location for the *Stripper Draw* (*S-1*) when the output signal for the stripper S-1 level controller is being set up. For the other strippers, the arrow must be moved to the correct location. Figure 11.47 shows the selection of the manipulated variable (level controller *OP* signal). In Aspen Dynamics, the *OP* signal is called the *Control Variable* instead of the less confusing terminology of calling it the manipulated variable. In this book the "control**led**" variable is the *PV* signal.

Specifying the *PV* signal to the stripper S-1 level controller is shown in Figure 11.48. The stripper has four stages, so the level on stage 4 is selected. Each of the three stripper level controllers is set up in the same way. The diagram is quite congested. Figure 11.49 gives an enlarged view of the stripper section of the flowsheet showing the three pressure and three level controllers with the *PV* and *OP* signals coming from the correct locations on the appropriate stripper vessels. Note that the level controllers are "reverse"-acting since they control level by changing the flows of material *into* the strippers.

A flow controller is installed to control the "overflash" flow of liquid below the AGO drawoff tray. This is achieved by manipulating the control valve V22 in the AGO line. Remember that the liquid drawn from the column to the stripper controls the liquid level in the base of the stripper. Manipulating the AGO flow changes the liquid drawoff rate and therefore the amount of liquid that is *not* drawn off and, as a result, flows down the column. The FCwash flow controller must be "direct"-acting; specifically, if there is too much stage 19 liquid, the AGO flow should be increased.

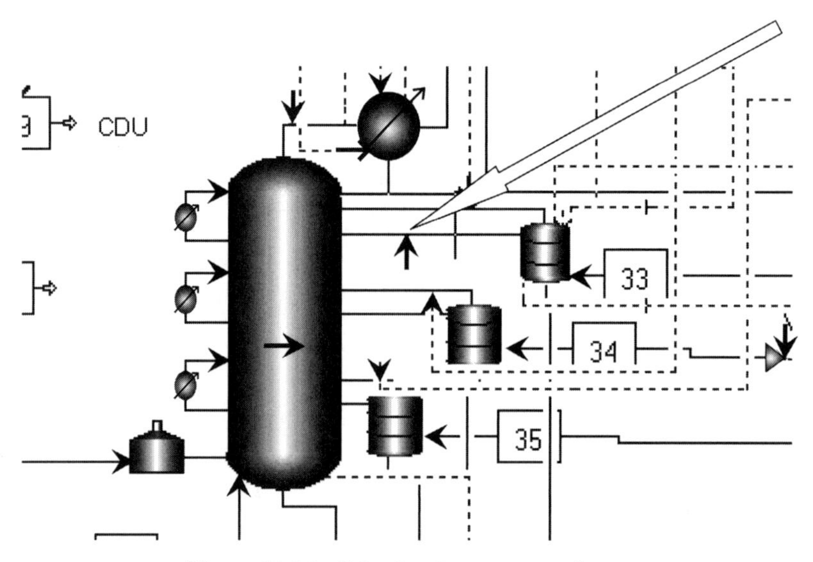

Figure 11.46 Selecting draw rate to stripper.

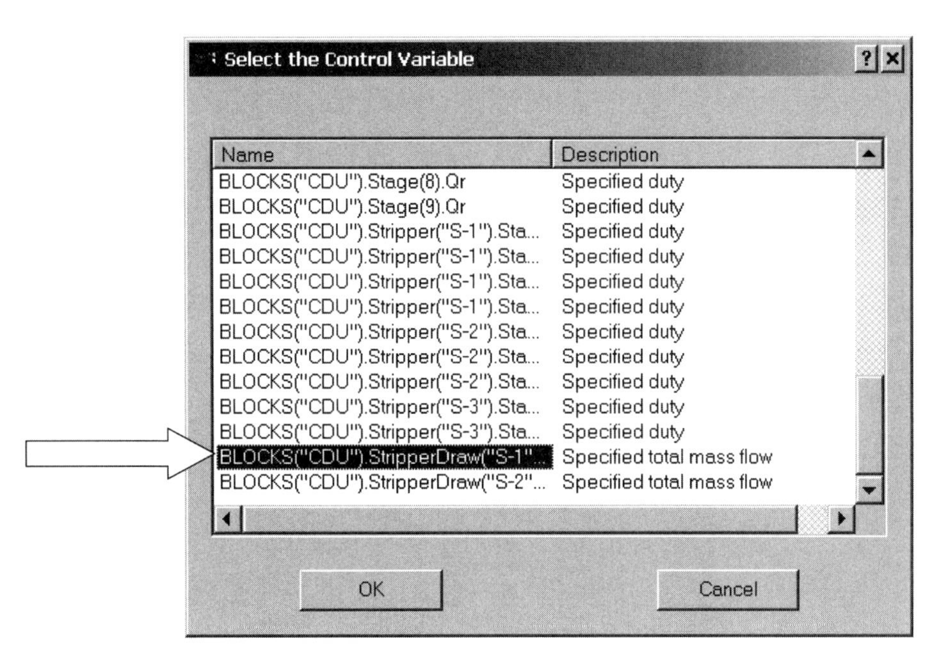

Figure 11.47 Selecting draw rate to stripper S-1.

The internal liquid and vapor flowrates are not listed in the possible output variables to be controlled, so a "flowsheet equation" is used. Figure 11.50 shows the equation used. The mass flowrate of liquid leaving Stage 19 ("Fml_out") of the main column ("CDU") is defined as the *PV* signal to a flow controller ("FCwash"). When this equation is compiled, the red light at the bottom of the window indicates that the system is overspecified. This is corrected by changing the *PV* variable in the FCwash controller from "fixed" to "free." The appropriate ranges of the variables are inserted in the controller. The steady-state flowrate of stage 19 liquid is 48,800 kg/h.

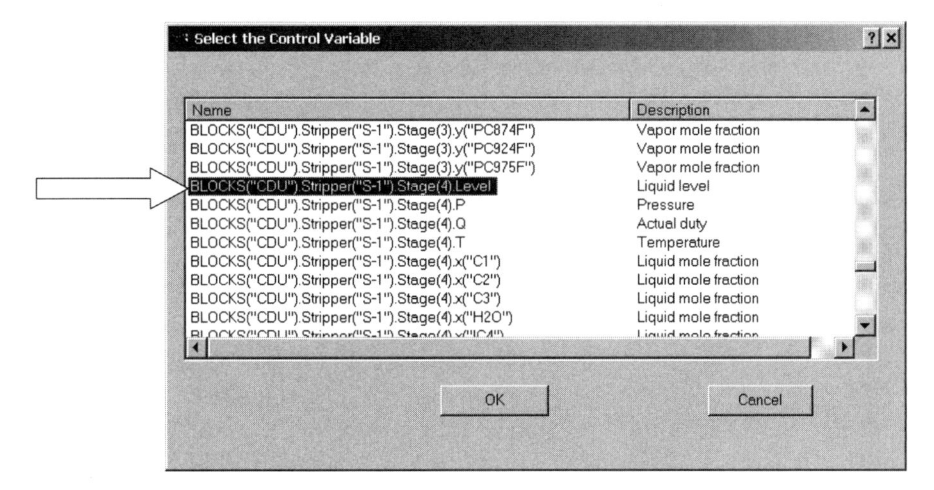

Figure 11.48 Selecting stage 4 level in stripper S-1.

CDU
PCS3
PCS2
PCS1
33
LCS1
34
LCS2
35
LCS3

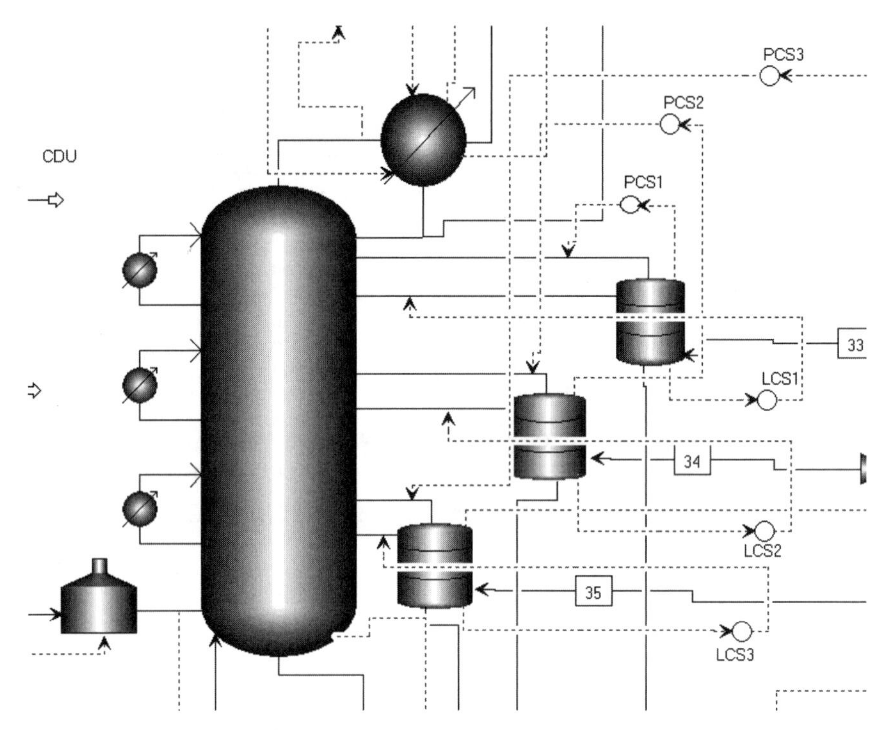

Figure 11.49 Pressure and level control setup of strippers.

The final controllers to install are the ASTM boiling point controllers. The appropriate boiling points are selected as discussed earlier in this chapter by using the *Configure Sensor* feature for each stream. The light-naphtha flow in the preflash column is manipulated to control its 95% boiling point at 190.6°C. The heavy-naphtha flow in the pipestill is manipulated to control its 95% boiling point at 190.6°C. The diesel flow is manipulated to control its 95% boiling point at 338°C. All of these controllers are reverse-acting.

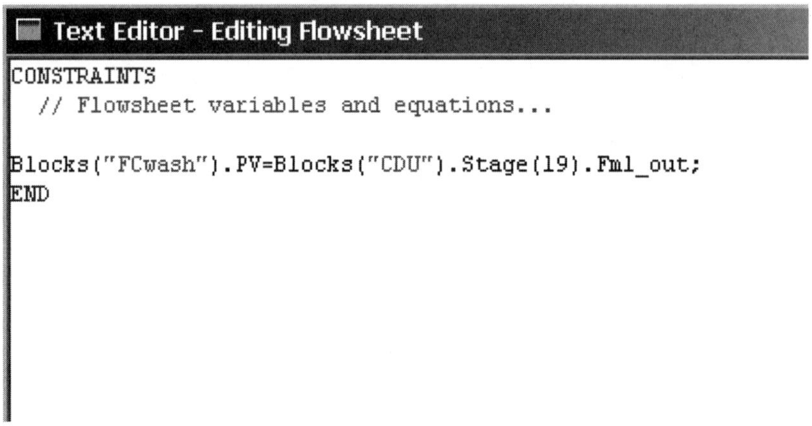

Text Editor - Editing Flowsheet
CONSTRAINTS
// Flowsheet variables and equations...
Blocks("FCwash").PV=Blocks("CDU").Stage(19).Fml_out;
END

Figure 11.50 Flowsheet equation for controlling wash flowrate.

TABLE 11.6 Boiling Point Controller Tuning

	SP (°C)	TT (°C)	D (min)	K_u	P_u (min)	K_C	τ_I (min)
Naphtha 95%	190.6	150–250	3	1.49	6	0.46	13
H Naphtha[a] 95%	190.6	150–250	3	6.63	21	2.1	46
Diesel 5%	254	200–300	3	7.03	23	2.0	51
Diesel 95%	338	300–400	3	4.06	23	1.2	51

[a]Heavy naphtha.

We also want to control the 5% boiling point of the diesel. This is achieved by manipulating flowrate of the kerosene. To get this 5% boiling point, another *Configure Sensor* is used, looking at stream 25 of the flowsheet, which has the same composition as does the diesel. Note that this controller is reverse-acting. If the 5% boiling point of the diesel is too high, more light material needs to be dropped down into this sidestream. This means that the kerosene flow should be decreased.

Relay–feedback tests are performed on each loop individually with the other boiling point controllers on manual. Tuning results are given in Table 11.6.

All the controller faceplates are shown in Figure 11.51. There are 25 controllers. There should be three more flow controllers, one on each of the steam to the strippers. The final flowsheet is given in Figure 11.52. Not installed or shown are the steam flow controllers on the strippers because they add more congestion to an already cluttered picture.

The effectiveness of this control scheme is demonstrated in Figure 11.53. The disturbances are step changes in the setpoints of the two crude oil flow controllers at time equal to 0.2 h. The responses to both positive and negative 20% changes are shown. The maximum

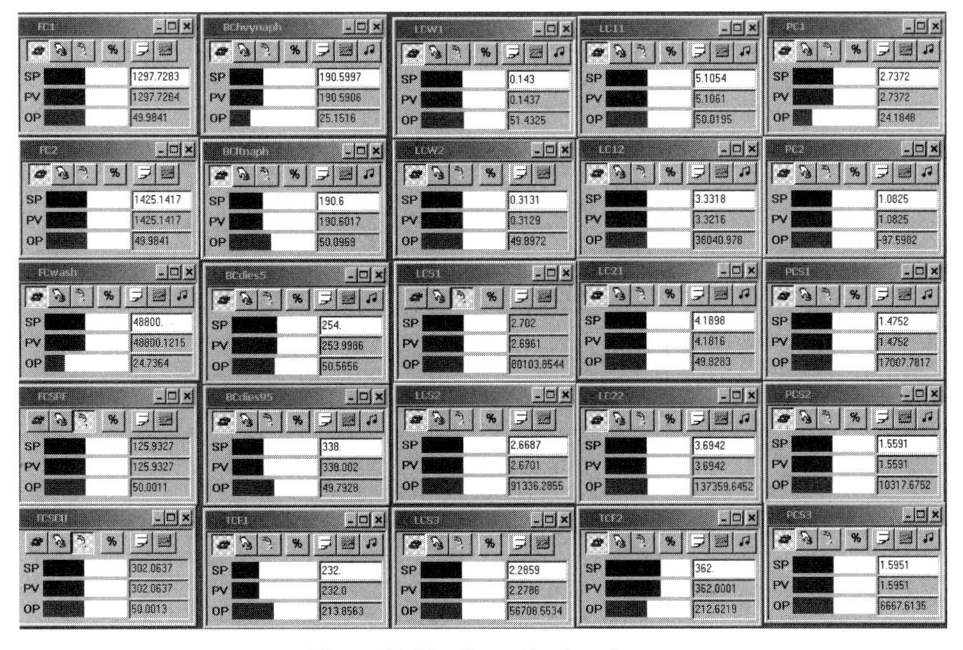

Figure 11.51 Controller faceplates.

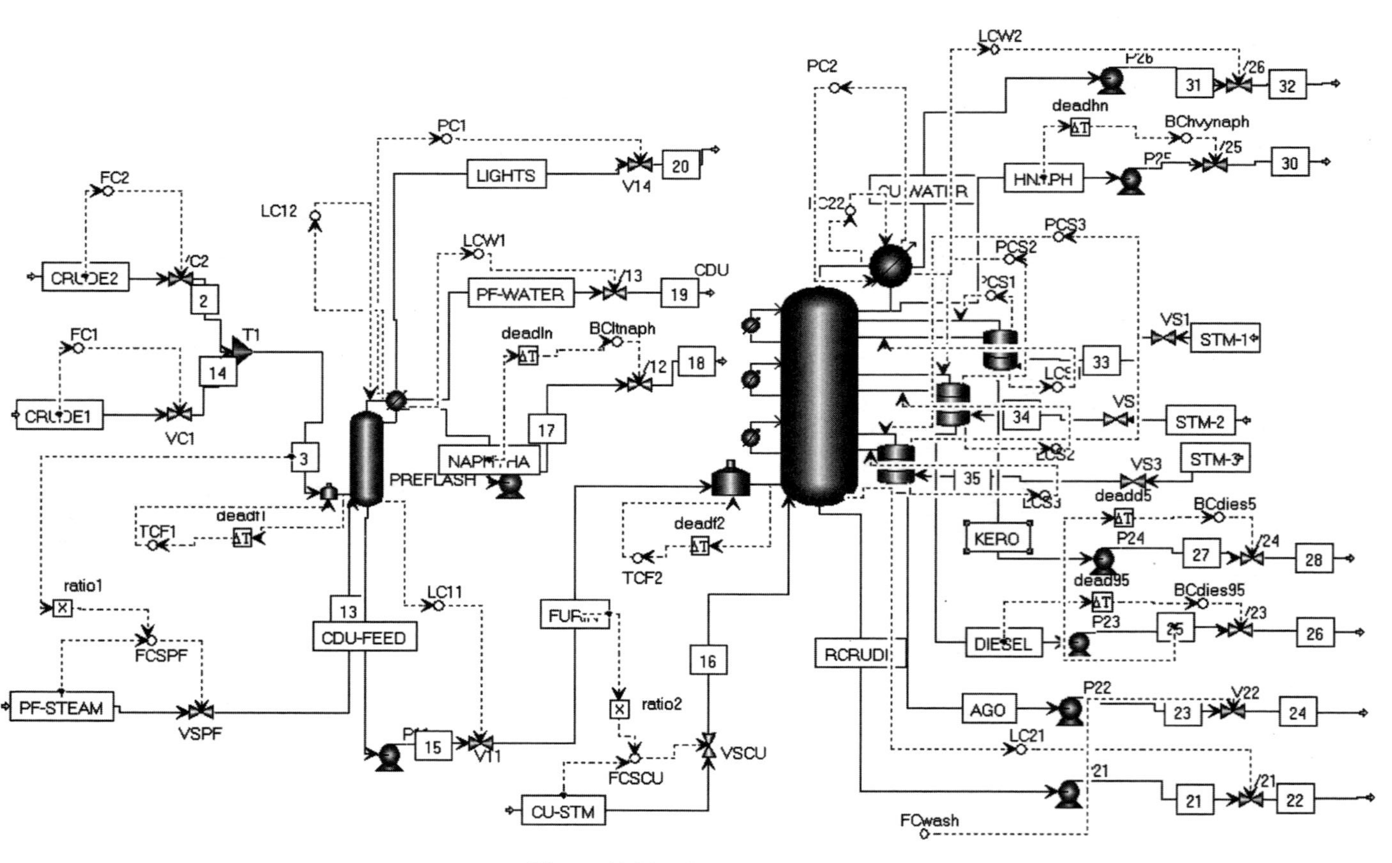

Figure 11.52 Control structure.

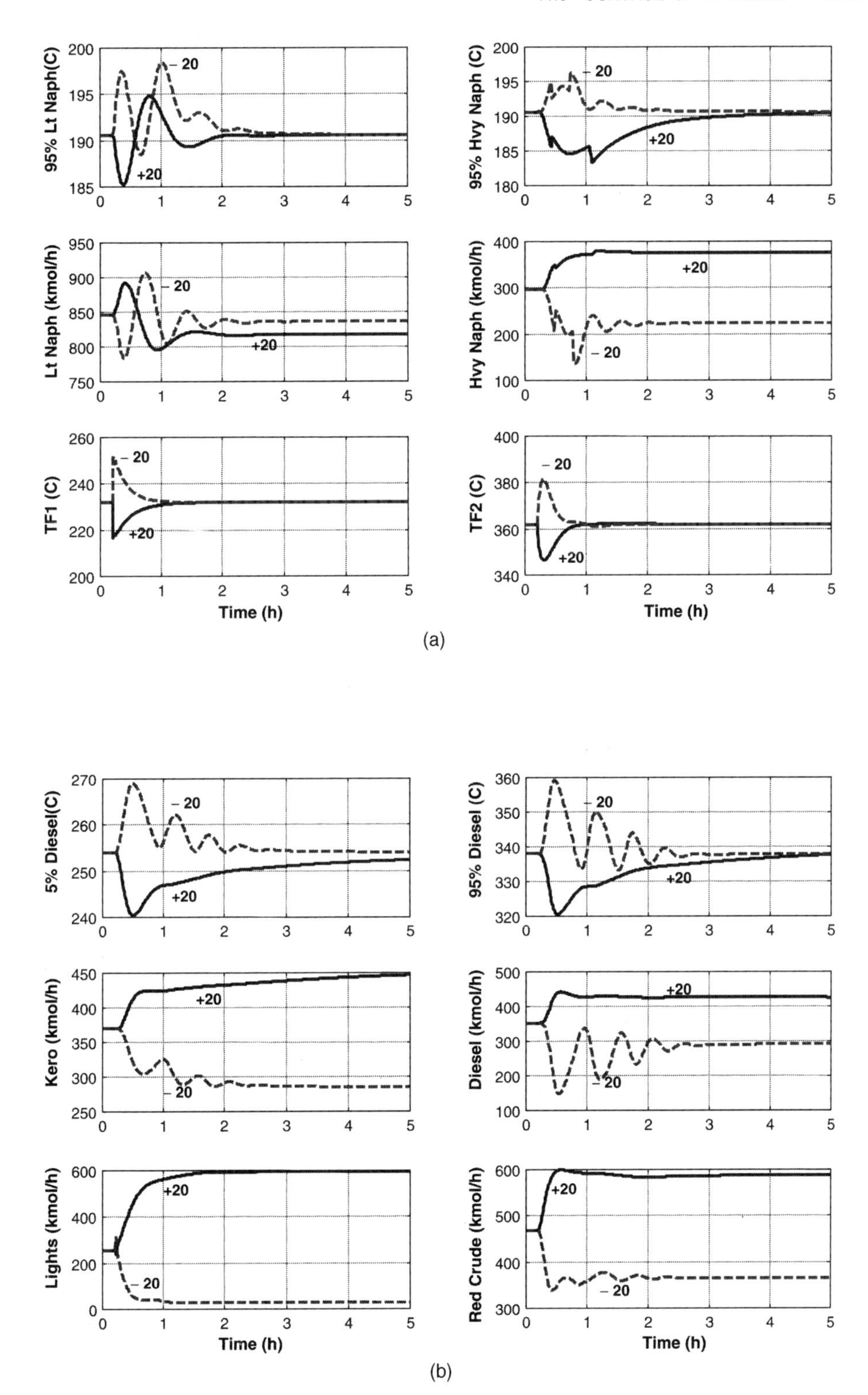

Figure 11.53 Feed flowrate disturbances.

deviations in the 95% boiling points of the light- and heavy-naphtha products are about 6°C. The deviations in the 5% and 95% boiling points of the diesel product are about 20°C for these quite large disturbances.

The 20% increase in feedflow resulted in the saturation of several control valves when the original design size was used (valves 50% open). The valve sizes were doubled to remove these limitations. The valves modified were V14 (*LIGHTS* from the preflash column), V25 (*HNAPHTHA*), and V22 (*AGO*). The steady-state positions of these three valves are now about 25% open, as can be seen on the faceplates of the "BChvynaph" 95% boiling point controller, the "PC1" pressure controller in the *PREFLASH* column, and the "FCwash" flow controller, which manipulates the AGO to hold constant the stage 19 liquid flowrate (see Fig. 11.51).

11.7 CONCLUSION

We have concluded this book with an example of a petroleum fractionator. The handling of petroleum cuts by studying boiling points has been reviewed. The control problem is to maintain the desired boiling point specifications.

The pipestill is a very complex column with multiple sidestreams, which come from stripping columns attached to the main column. In addition to the normal base and reflux drum levels and column pressures, the levels and pressures in these strippers must also be controlled.

I hope that the examples presented in this book have demonstrated the usefulness of simulation in the steady-state design and dynamic control of distillation columns. Working with distillation columns is challenging, important, and fun. Enjoy!

INDEX

Distillation Design and Control Using Aspen™ Simulation, By William L. Luyben